Metallurgy and Metallurgical Engineering Series

ROBERT F. MEHL, *Consulting Editor*

PHYSICAL CHEMISTRY OF METALS

PHYSICAL CHEMISTRY OF METALS

Lawrence S. Darken, Ph.D.
Research Laboratory
United States Steel Corporation

Robert W. Gurry, Ph.D.
Research Laboratory
United States Steel Corporation

With a Collection of Problems by

Michael B. Bever, Sc.D.
Department of Metallurgy
Massachusetts Institute of Technology

CBS Publishers & Distributors Pvt. Ltd.
New Delhi • Bengaluru • Chennai • Kochi • Kolkata • Mumbai
Hyderabad • Nagpur • Patna • Pune • Vijayawada

ISBN: 81-239-1479-2

First Indian Edition: 1987
Reprint: 2002

This edition has been reprinted in India by arrangement with McGraw-Hill Book Company, Inc., New York, USA

Published by **Satish Kumar Jain** and produced by **Varun Jain** for
CBS Publishers & Distributors Pvt. Ltd.,
4819/XI Prahlad Street, 24 Ansari Road, Daryaganj, New Delhi - 110002
delhi@cbspd.com, cbspubs@airtelmail.in • www.cbspd.com
Ph.: 23289259, 23266861, 23266867 • Fax: 011-23243014

Corporate Office: 204 FIE, Industrial Area, Patparganj, Delhi - 110 092
Ph: 49344934 • Fax: 011-49344935
E-mail: publishing@cbspd.com • publicity@cbspd.com

Branches:

- ***Bengaluru:*** 2975, 17th Cross, K.R. Road, Bansankari 2nd Stage, Bengaluru - 70 • Ph: +91-80-26771678/79 • Fax: +91-80-26771680 E-mail: cbsbng@gmail.com, bangalore@cbspd.com
- ***Chennai:*** No. 7, Subbaraya Street, Shenoy Nagar, Chennai - 600030 Ph: +91-44-26681266, 26680620 • Fax: +91-44-42032115 E-mail: chennai@cbspd.com
- ***Kochi:*** Ashana House, 39/1904, A.M. Thomas Road, Valanjambalam, Ernakulum, Kochi • Ph: +91-484-4059061-65 Fax: +91-484-4059065 • E-mail: cochin@cbspd.com
- ***Kolkata:*** 6-B, Ground Floor, Rameshwar Shaw Road, Kolkata - 700014 Ph: +91-33-22891126/7/8 • E-mail: kolkata@cbspd.com
- ***Mumbai:*** 83-C, Dr. E. Moses Road, Worli, Mumbai - 400018 Ph: +91-9833017933, 022-24902340/41 • E-mail: mumbai@cbspd.com

Representatives:

- Hyderabad: 0-9885175004
- Patna: 0-9334159340
- Vijayawada: 0-9000660880
- Nagpur: 0-9021734563
- Pune: 0-9623451994

Printed at:
Neekunj Print Process, Delhi (India)

PREFACE

This book is designed to suit the needs of the senior and graduate metallurgist in acquainting him with the fundamentals of the physical chemistry and the thermodynamics of metals and metallurgical processes. The thermodynamic method is particularly stressed, as it is felt that thermodynamics has more to offer metallurgy than has been commonly realized. In spite of the considerable number of texts on chemical thermodynamics now available, none seems particularly well suited to the needs of the student metallurgist. This statement may appear rather odd if one takes the viewpoint that thermodynamics is thermodynamics. However, rather pronounced differences, not only subjectively but experimentally, set off the substances of metallurgical interest from those (particularly aqueous solutions) which are traditionally of interest to the physical chemist. Only recently has a physical chemistry of the metallic state begun to develop. In this book the attempt is made to include this physical chemistry in simple terms and to present the fundamentals of chemical thermodynamics so that the student may be able to understand the thermodynamic method and apply it to metallurgical problems.

It is also hoped that this book may serve as a useful reference. The first portion summarizes modern views on the chemistry of the metallic state and briefly sketches the chemistry of other states with which the metallurgist must perforce deal. The later portions deal with thermodynamics, the treatment being mostly in the classical manner but with metallurgical applications and limitations always in mind. The sections on heterogeneous equilibrium discuss in some detail phases of variable composition, since these are commonplace at elevated temperature, particularly in metallurgical systems. This treatment, though following substantially that of Gibbs, is more detailed than that in most elementary texts. The chapters on the thermodynamics of the iron-carbon and iron-nitrogen systems are believed more comprehensive than can be found in the present literature. Although a large share of this book is devoted to the treatment of equilibrium, an introduction to reaction-rate phenomena in the final chapters serves at least as token acknowledgment of the great importance of this rapidly growing field.

Although an effort is made to develop each topic from fundamentals, the student cannot expect to appreciate the subject matter fully without some background in both physical chemistry and metallurgy. However, it is hoped that this book will be useful both to the metallurgist with meager physicochemical experience and to the physical chemist with little training in metals. Some knowledge of the elements of calculus on the part of the reader will be found useful. No quantitative quantum mechanics or statistical mechanics appear, although results therefrom are occasionally employed.

It was originally intended to incorporate, as appendixes, tabular thermodynamic data for numerous particular substances. However, the compilations of K. K. Kelley and of the Bureau of Standards, listed with others at the end of the introduction, are so exhaustive that a selection therefrom seems somewhat pointless when the originals are so readily available. These are highly recommended to the student and research worker as indispensable sources of numerical data.

The authors gratefully acknowledge the physical assistance and moral support of Margaret F. Darken in the preparation of the manuscript. This opportunity is also taken to express thanks for the cooperation of the stenographic and photographic departments of the Research Laboratory of the United States Steel Corporation.

The authors share Professor Bever's gratitude to those who have generously contributed ideas for the problems and who have given advice and criticism in elaborating them.

Lawrence S. Darken
Robert W. Gurry

Kearny, N.J.
January, 1953

CONTENTS

CHAPTER 1

INTRODUCTION

Metallurgical processes have been carried out for many centuries; yet the science of metals and metallurgy is relatively new. Although the new techniques and advances are many, it is still true that many metallurgical processes are carried out in much the same manner that they were in the days when metallurgy was essentially an art. A metallurgist of the older school may justifiably remind us that steel is still made by processes invented about a century ago and that the blast furnace has an even more venerable history. Although this metallurgist of the older school doubtless wishes to imply that metallurgical procedure will continue in the future as in the past, it is already apparent that he is whistling in the dark and that such an attitude of complacency reflects a total unawareness of the fact that during the last few decades a vast change in the method of attacking metallurgical problems has transformed metallurgical art into a science.

It seems quite general that significant advances in any particular field of learning frequently result from the application of knowledge from a different field. Thus in biology and medicine we find that many advances are introduced by the chemist, and in chemistry, by the physicist. It is not surprising to discover that both chemistry and physics have invaded the field of metallurgy. There can be no doubt that a large portion of the knowledge contributed to the metallurgical field from these other fields is sound and must be incorporated in the general background of metallurgical knowledge. There can be little doubt that the theoretical metallurgist who is to advance his science must have a thorough background in physics and physical chemistry.

The Concept of State. It is particularly noteworthy that the experience of the physicist and chemist has differed from that of the metallurgist in one important respect. This may be illustrated by observing that a chemist would consider the state of a sample of natural water to be adequately described by a statement of its temperature, pressure, and composition. On the other hand no metallurgist would consider that he had adequately described a piece of steel by stating these same properties. Even the metallography of the specimen may not be regarded as completing the description. Traditionally all this information is supplemented with a history of the specimen from the time it was last melted or perhaps even before. The historical mode of thinking is viewed askance by the chemist or physicist who is accustomed to thinking in terms of state properties which are measurable at the time of interest, no matter what the past history may have been. The historical approach is used in part for convenience but in part from ignorance or inability to describe a steel adequately. Herein lies one of the major unsolved metallurgical problems; present evidence indicates that the number of variables sufficient to describe an ordinary gas or liquid is not adequate to describe a common metal in its solid state.

There is a rather important gap, therefore, between traditional physicochemical experience, according to which a substance is described in terms of its present properties, or state, and metallurgical experience, which so far has led in part to a historical description. The bridging of this gap is one aim of the new metallurgy in which the concept of state must play a dominant role. The state variables to be considered are in many cases quite foreign to the physicist or chemist engaged in other fields, and in some cases a beginning has hardly been made on the understanding of the nature of the variables involved. Thus the new metallurgy represents somewhat of a break from the older but hardly can be described with fairness in terms of the application of a well-developed physics and chemistry. Rather it is a new science founded on metallurgical experience and on physical and physicochemical experience and theory.

Atomistics and Thermodynamics. In general, there are two types of scientific approach: one, typified by atomistics, is a detailed analysis leading step by step from a starting point to a finish; the other, typified by chemical thermodynamics, tells us that under certain conditions a particular initial state leads inexorably to a certain definite final state independent of the intervening processes. These two approaches constitute major subdivisions of the new science of metallurgy, and it is the purpose of this book to present the fundamentals thereof in terms of their bearing on metallurgy. To compare the nature of the two we may consider as an atomic analogy several moving balls on a tilted billiard table. One method of finding the ultimate condition would be to trace in detail the path of each ball until it ultimately came to rest; the other is to conclude immediately that all the balls will ultimately come to rest at the lowest corner of the table. These two methods represent the atomistic and the thermodynamic approaches, respectively.

The atomistic approach has in recent times been popularized, in particular by "that ultimate arbiter of matters crystallographic"—X-ray diffraction. In general it seems to have great intellectual appeal, and there is no doubt that it has been fruitful in attacking problems not amenable to thermodynamics. However, the powerful methods of thermodynamics are all too frequently ignored. A considerable part of this book is devoted to classical thermodynamics because it is believed that this science has much more to offer to the development of metallurgy than is commonly realized. As implied above, thermodynamics provides not only a set of equations, which unfortunately have sometimes been used blindly, but also a simple and effective manner of reasoning which is difficult to appreciate except by experience therewith. Perhaps no better recommendation could be given for the thermodynamic method than the following quotation from Einstein:[1]

> A theory is the more impressive the greater the simplicity of its premises is, the more different kinds of things it relates, and the more extended is its area of applicability. Therefore the deep impression which classical thermodynamics made upon me. It is the only physical theory of universal content concerning which I am convinced that, within the framework of the applicability of its basic concepts, it will never be overthrown. . . .

It is our hope that the student will gain not only an operational ability, important as this is, but an insight into the fundamental method of thermodynamics.

Although the great strength of thermodynamics lies in its non-atomic approach, this is also a weakness from the viewpoint of an integrated understanding of the subject matter. Certainly, in view of the strong presupposition of the unity of the sciences, nothing is to be gained by ignoring any branch. Although statistical mechanics and a general treatment of rate phenomena are beyond the scope of this book, it should be realized that the nature of the atomistic structure is capable of serving as an excellent guide in the fruitful utilization of thermodynamics. Thus the thermodynamics of gases leans heavily on the ideal-gas law, and our understanding thereof is greatly enhanced by the kinetic theory. Similarly, Raoult's law and modern solution theory are intimately related to the thermodynamics of solutions. Ordering effects in liquid and solid solutions have their counterpart in otherwise unexpected behavior of thermodynamic functions. The concept of the third law of thermodynamics and its limitations is greatly enhanced by an understanding of atomistics. The vacancy mechanism for metallic diffusion immediately raises a question concerning the circumstances under which we may expect the equilibrium number of vacancies (empty lattice sites) and the manner in which we are to describe the state of a metal which is out of internal equilibrium in this respect. Dislocation theory sheds some light on the cold-worked state and signals caution in the use of thermodynamics here. Modern theory of reaction rate, as developed in terms of the activated complex, leans heavily upon thermodynamics.

[1] Albert Einstein, "Autobiographical Notes," appearing in "Albert Einstein: Philosopher—Scientist," Tudor Publishing Company, New York, 1949.

Thus an understanding of the atomistics of metallic phases, and others of metallurgical interest, if not indispensable, is at least very helpful in the application of thermodynamics to metallurgy. The atomistic nature of phases in internal equilibrium is described in the following four chapters; a large part of the remainder of this book is devoted to the thermodynamic treatment of equilibrium. A correlation with elementary atomistics is included where this seems enlightening. The future clearly gives promise of the advancement of metallurgical science through the fruitful combination of atomistics and thermodynamics.

The following will be found useful as sources of thermodynamic data:

F. D. Rossini, D. D. Wagman, W. H. Evans, S. Levine, and I. Jaffe, Selected Values of Chemical Thermodynamic Properties, *Natl. Bur. Standards Circ.* 500, U.S. Government Printing Office, Washington, D.C., 1950.

"Selected Values of Chemical Thermodynamic Properties," issued currently in looseleaf form by the National Bureau of Standards, Washington, D.C.

F. D. Rossini, K. S. Pitzer, W. J. Taylor, J. P. Ebert, J. E. Kilpatrick, C. W. Beckett, M. G. Williams, and H. G. Werner, Selected Values of Properties of Hydrocarbons. *Natl. Bur. Standards Circ.* 461, U.S. Government Printing Office, Washington, D.C., 1947.

F. R. Bichowsky and F. D. Rossini, "Thermochemistry of the Chemical Substances," Reinhold Publishing Corporation, New York, 1936.

K. K. Kelley, Contributions to the Data on Theoretical Metallurgy, I–XI, U.S. Bur. Mines Bulls. 350, 371, 383, 384, 393, 394, 406, 407, 434, 476, 477, 1932–1950.

"International Critical Tables," McGraw-Hill Book Company, Inc., New York, 1926–1933.

"Landolt-Börnstein-Roth-Scheele Physikalisch-Chemische Tabellen," 5th ed., Springer-Verlag, Berlin, 1923–1936.

"Annual Tables of Constants and Numerical Data," Hermann & Cie, Paris, 1910–1948.

E. Justi, "Spezifische Wärme, Enthalpie, Entropie, und Dissoziation technischer Gase," Springer-Verlag, Berlin, 1938.

O. Kubaschewski and E. L. Evans, "Metallurgical Thermodynamics," Academic Press, Inc., New York, 1951.

F. Weibke and O. Kubaschewski, "Thermochemie der Legierungen," Springer-Verlag, Berlin, 1943.

John Lumsden, "Thermodynamics of Alloys," Institute of Metals, London, 1952.

Carl Wagner, "Thermodynamics of Alloys," Addison-Wesley Press, Inc., Cambridge, 1952.

G. Masing, "Handbuch der Metallphysik," Akademische Verlagsgesellschaft, Becker u. Erler Kom.-Ges., Leipzig, 1935–1941.

Committee on Physical Chemistry of Steelmaking, "Basic Open Hearth Steelmaking," American Institute of Mining and Metallurgical Engineers, New York, 1951.

C. D. Hodgman, "Handbook of Chemistry and Physics," Chemical Rubber Publishing Co., Cleveland, Ohio, 1950.

N. A. Lange, "Handbook of Chemistry," Handbook Publishers, Sandusky, Ohio, 1949.

"Chemical Engineers' Handbook," edited by J. H. Perry (ed.), McGraw-Hill Book Company, Inc., New York, 1950.

CHAPTER 2

GASES

EXPERIMENTAL APPROACH— AVOGADRO'S, BOYLE'S, AND CHARLES' LAWS

An equation of state is a relation between temperature, pressure, and composition. The first step toward the determination of an equation of state for gases was made by Boyle, who found experimentally (1660) that at constant temperature the volume of a fixed mass of gas is inversely proportional to the pressure. Later (1787) it was found by Charles that at constant volume the pressure exerted upon a fixed mass of gas is a linear function of the temperature. This law is now commonly expressed by the statement that at constant volume the pressure is proportional to the absolute temperature. The concept of absolute temperature will be discussed in detail in a later chapter. These two laws may be written:

At constant T:

$$P = \frac{A}{V} \qquad \text{Boyle's law} \tag{2-1}$$

At constant V:

$$P = BT \qquad \text{Charles' law} \tag{2-2}$$

A and B being constants. From these the following two partial derivatives are obtained:

$$\left(\frac{\partial P}{\partial V}\right)_T = -\frac{P}{V}$$

$$\left(\frac{\partial P}{\partial T}\right)_V = \frac{P}{T}$$

Using the fundamental equation for partial derivatives,

$$dP = \left(\frac{\partial P}{\partial V}\right)_T dV + \left(\frac{\partial P}{\partial T}\right)_V dT$$

and substituting the above values,

$$dP = -\frac{P}{V}dV + \frac{P}{T}dT$$

or

$$\frac{dP}{P} + \frac{dV}{V} = \frac{dT}{T}$$

Integrating,

$$\ln P + \ln V = \ln T + I$$

where I is an integration constant, or

$$PV = Te^I$$

If the mass of gas considered is 1 gram mole, the constant e^I is universally designated R, the *gas constant*, and the foregoing is the *ideal-gas law*.

$$PV = RT \tag{2-3}$$

The fact that the constant R has the same value independent of the gas under consideration follows from Avogadro's observation (1811) that under fixed conditions of temperature and pressure equal volumes of all gases contain the same number of molecules. Since at constant T and P the volume is proportional to the quantity of gas under consideration, the equation assumes the general form

$$PV = nRT$$

where n is the number of moles.

From the fact that the volume of 1 mole of ideal gas at 0°C (273.16°K) and 1 atm pressure is 22.4141 liters, it follows that

$$R = \frac{PV}{nT} = \frac{1(22.4141)}{1(273.16)} = 0.082055$$

liter atmospheres per degree per mole or 82.057 cubic centimeter atmospheres per degree per mole. Since the PV product, expressed as cubic centimeter atmospheres in this case, is clearly a unit of energy,[1] this may be converted to any other energy unit by use of the appropriate conversion factor. For example, R is also equal to 1.9872 cal/(mole)(deg) and 8.3144 (absolute) joules/(mole)(deg).

KINETIC THEORY OF GASES

The foregoing observed behavior of gases as well as other considerations led to the formulation of the atomic hypothesis (Dalton, 1803–1808). Although the use of the word "hypothesis" may have a slightly

[1] Pressure has the units of (force)/(area) or (force)/(length)^2, and volume the units of (length)^3; the pressure-volume product has the units of force times length, which is work or energy.

humorous ring to the present-day student, it must be clearly borne in mind that the hypothesis of atoms and molecules is not subject to experimental verification in the same sense as are the laws of Boyle, Charles, or Avogadro. It is subject to verification only indirectly through its effects (or deductions therefrom). This is the general nature of scientific advance: Experimental observations and generalizations such as the foregoing lead to postulates, hypotheses, or theories which are not directly verifiable by experiment. In their development these postulates, hypotheses, or theories suggest or predict further experiments or generalizations which themselves are subject to experimental verification. Extended experimental verification of the consequences of a theory lead to the general acceptance of the theory. Although the consequences of the atomic hypothesis have been amply verified in many ways, the word "hypothesis" is still judicially used. The term "construct" has been used by P. W. Bridgman to designate such postulates as products of the human mind—as distinguished from entities observable by the senses.

Having thus postulated the existence of atoms and molecules, let us now consider a cubic box of edge l containing, at temperature T, n molecules of mass m and velocity u, the same for all molecules. Let us further consider that one-third of the molecules travel in the direction of each of the three principal coordinate axes. The time required for a molecule to hit one end of the box, travel to the opposite end, and return to the original position and direction is $2l/u$; the number of round trips per second is the reciprocal, or $u/2l$. Each reflection at the wall of the cube is assumed to be perfectly elastic, and hence the momentum change (from mu to $-mu$) is $2mu$. The total momentum change per second at any one side of the cube is then the product of the number of trips per molecule per second, the momentum change per reflection, and the number of molecules traveling in that direction, which is

$$\frac{u}{2l}(2mu)\frac{n}{3} = \frac{1}{3}\frac{nmu^2}{l}$$

From the fundamentals of classical dynamics the force exerted upon a wall is the total change of momentum per second. Since by definition the pressure is the force per unit area, the pressure on the wall is given by the expression

$$P = \frac{1}{3}\frac{nmu^2}{l^3}$$

or, since $l^3 = V$,

$$PV = \tfrac{1}{3}nmu^2$$

As the kinetic energy of one molecule is $\frac{1}{2}mu^2$ and of n molecules is $\frac{1}{2}nmu^2$, the kinetic energy E_k of 1 gram mole (in which case $n = 6.02 \times 10^{23}$, designated N and known as *Avogadro's number*) is

$$E_k = \tfrac{1}{2}Nmu^2 = \tfrac{1}{2}Mu^2$$

M being the weight of 1 mole (the molecular weight). Substituting in the above relation it follows that

$$PV = \tfrac{2}{3}E_k$$

By comparison with the ideal-gas law it is seen that

$$E_k = \tfrac{3}{2}RT$$

and

$$u = \sqrt{\frac{2E_k}{M}} = \sqrt{\frac{3RT}{M}}$$

PROPERTIES OF GASES DERIVED FROM THE KINETIC THEORY

Heat Capacity of a Monatomic Gas. For a monatomic gas, one whose atoms may be regarded as point masses, the kinetic energy of translation considered above constitutes the total energy. (For a polyatomic gas this is generally not true, since the rotational and vibrational energies are usually significant.) Since, as will be shown in Chap. 6, the heat capacity at constant volume is equal to $(\partial E/\partial T)_V$ it follows from the above relation that for a mole of a monatomic gas

$$C_V = \tfrac{3}{2}R$$

This deduction from the kinetic theory has been experimentally verified for the inert gases, mercury, and other metallic vapors.

Effusion and Graham's Law. Let us consider the same box described above in the development of the kinetic theory and the same simplifying assumptions. Let us consider in one of the thin walls a small hole leading to a vacuum. The escape of molecules from such a hole is called *effusion*. The escape of molecules through a porous diaphragm in some cases approximates this type of flow rather closely. If the hole is sufficiently small that the average molecular velocity in its vicinity is the same as elsewhere in the container, *i.e.*, there is no fluid flow, then at fixed temperature and pressure the rate of escape, defined as the number of molecules escaping per second, is clearly proportional to n and u. If this same experiment is repeated with another gas at the same temperature and pressure (hence at the same n, in view of Avogadro's law), then the ratio of rates of escape is equal to the ratio of the molecular velocities. Since, as has been shown, the translational kinetic energy of a mole,

$\frac{1}{2}Mu^2$, is equal to $\frac{3}{2}RT$ for all gases, the kinetic energies of the above gases can be equated.

$$\tfrac{1}{2}M_1u_1^2 = \tfrac{1}{2}M_2u_2^2$$

where the subscripts 1 and 2 refer to the first and second gases. Whence

$$\frac{u_1}{u_2} = \sqrt{\frac{M_2}{M_1}}$$

and the rates of escape v_1 and v_2 are inversely proportional to the square root of the molecular weights.

$$\frac{v_1}{v_2} = \sqrt{\frac{M_2}{M_1}}$$

This is known as *Graham's law* (1829).

This same equation may be shown to be valid even if the gas does not escape into a vacuum but to some other pressure lower than that in the box, for in this case the number of molecules leaving the box per second is identical with that in the previous case, under the postulate that the hole be sufficiently small. Let us designate the efflux rates for the two gases as v_1' and v_2'. Then

$$\frac{v_1'}{v_2'} = \sqrt{\frac{M_2}{M_1}}$$

A similar relation holds for the influx of molecules.

$$\frac{v_1''}{v_2''} = \sqrt{\frac{M_2}{M_1}}$$

Combination of these two gives

$$\frac{v_1' - v_1''}{v_2' - v_2''} = \sqrt{\frac{M_2}{M_1}}$$

The net rates of escape of molecules, $v_1' - v_1'' = v_1$ and $v_2' - v_2'' = v_2$, are then related by the same expression as for escape into vacuum:

$$\frac{v_1}{v_2} = \sqrt{\frac{M_2}{M_1}}$$

Suppose now that different gases initially at the same pressure are present on opposite sides of the small hole. By the foregoing reasoning it is clear that the gas of lower molecular weight will pass through more rapidly than the one of higher molecular weight, thus tending to give rise to a pressure difference on the two sides. This phenomenon is in marked contrast to the ordinary flow through large holes, which always tends to equalize the pressure. Thus when one gas is separated from another by a porous wall such as the brickwork or sand seal of an industrial furnace or the wall of a refractory furnace tube, we may expect passage of each of the gases through the pores, even though the pressure and temperature are the same on both sides. Even raising the pressure slightly on one side will not prevent this. If the pores differ appreciably in size, and if some are large enough to tend to equalize the pressure by ordinary fluid flow, then the porous wall acts as a mixer—the two gases moving in opposite directions by effusion (the lighter gas having the faster rate) and the gas mixture moving by ordinary fluid flow in the direction to equalize the total flow.

Thermal Transpiration. Let us consider a container with a dividing wall containing a small hole similar to that in the preceding section. The same gas is put into both sides of the container, but one side is maintained at a temperature T_1, while the other side is at T_2. We wish to determine the relation between the pressures on the two sides after the system has been left long enough for the pressures to become constant, *i.e.*, for steady state to be reached. When this state has been attained, the number of molecules passing per second through the hole in one direction equals the number passing in the reverse direction. The number passing from either side is proportional to the number of molecules per unit volume on that side and to the velocity. From the ideal-gas law the number per unit volume, n/V, is equal to P/RT; also from the foregoing expression ($\frac{1}{2}Mu^2 = \frac{3}{2}RT$) the velocity u is $\sqrt{3RT/M}$; hence the number passing through the hole from either side is proportional to the product, which, dropping constants, is $P/\sqrt{T}$. Thus, at steady state,

$$\frac{P_1}{P_2} = \sqrt{\frac{T_1}{T_2}}$$

This same effect is achieved by the use of a porous diaphragm in a thermal gradient. Under such conditions the diaphragm, as we have seen, may be regarded as a "thermal pump," which pumps gas from the side at lower temperature to that at higher temperature.

Usually the steady-state condition discussed above is not attained even approximately. For example, both sides of the diaphragm may be at substantially atmospheric pressure, as in the case of the refractory wall or sand seal of a furnace; in such a case, of course, the pressure on the hot side cannot build up as required by the equation above. However, gas still passes through in the attempt, as it were, to establish this pressure differential. The rate at which various gases will pass through a given diaphragm under this condition is proportional to the molecular velocity, as discussed under effusion; *i.e.*, the rate of passage is inversely proportional to the square root of the molecular weight.

If the holes of the diaphragm are very small, of the order of 10^{-5} cm for gases at 1 atm, pressures may be obtained comparable to those calculated from the equation. For large holes ordinary flow occurs and the pressure is less. Although maximum pressures are obtainable with very small holes, clearly high transport of gas cannot be obtained in this case because of the high resistance. With large holes the rate of transport is low because of small pressure. At some intermediate size of pores the rate of transpiration will be a maximum. The usual porous diaphragm will contain openings of a variety of sizes so that thermal transpiration will occur in the smaller ones and ordinary fluid flow in the opposite direction through the larger ones, thus internally "short-circuiting" the diaphragm to a large extent. Therefore, the total transport of gas through the diaphragm usually is considerably less than might otherwise be anticipated. For example, when 2 or 3 in. of diatomaceous earth was rammed into a 1-in. porcelain tube, one end of which was left open to the atmosphere and heated to 1000°C and the other end closed and connected to a manometer, the pressure fell in a few minutes to 1 cm Hg less than atmospheric. When both ends were left open to the atmosphere, the rate of flow of air was about 20 cc/min. Most commercial bricks showed the effects to only about one-tenth the magnitude observed for diatomaceous earth. This effect may become important in industrial furnaces when it is necessary to maintain a controlled atmosphere, as in annealing or heat-treating.

Distribution of Velocities. Thus far, our simplified picture of gas molecules as mass points (since collisions have not been considered), one-third of which move in each of the three directions with a single constant velocity (at a given temperature), has led to several fruitful results. A more complete picture considers the molecules to have finite diameter and to move in all possible directions with all velocities but still does not consider them to occupy an appreciable fraction of the space or to have any forces acting between them. Some of the results of this extended kinetic theory will be given without derivation.

At constant temperature the probability that a given molecule has a velocity u at any instant is proportional to $e^{-mu^2/2kT}$, where k is Boltzmann's constant, that is, R, the gas constant, divided by N, Avogadro's number. If we let n_u be the number of molecules with velocity between zero and u, then dn_u is the number having velocity between u and $u + du$, which is clearly proportional to the above probability. A plot of dn_u/du vs. u is shown in Fig. 2-1. This ordinate has been chosen, since it is proportional to the probability of a given u and hence to the number of molecules with this u. The area under the curve from the origin to any velocity u is

$$\int_0^u \frac{dn_u}{du}\,du = \int_0^{n_u} dn_u = n_u$$

The total area under the curve is equal to n, the total number of molecules. Shown in the figure are the most probable velocity (the maximum of the curve) u_p, the arithmetic mean (average) velocity $\bar{u}$, and the root-mean-square velocity u_{rms}. The latter, which is the square root of the mean value of the squares of the velocities, is the most important in considerations of energy, since the mean kinetic energy of a molecule is $\frac{1}{2}mu_{\text{rms}}^2$.

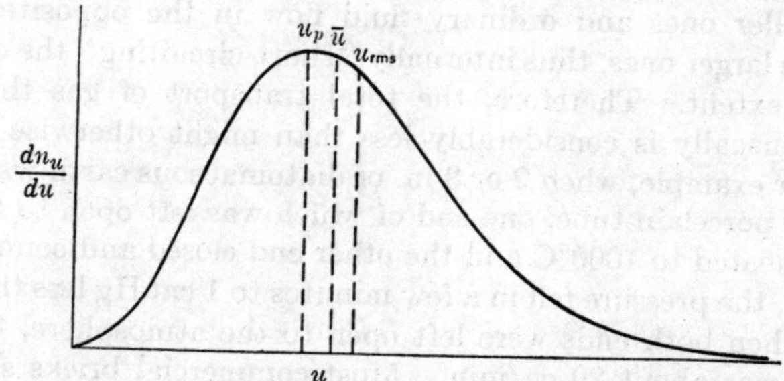

FIG. 2-1. The Boltzmann distribution of velocities of gas molecules. The ordinate is proportional to the number of molecules with a given velocity u. u_p, $\bar{u}$, and u_{rms} are the most probable velocity, the arithmetic-mean velocity, and the root-mean-square velocity, respectively.

This extended kinetic theory gives the rms velocity u_{rms} equal to $\sqrt{3kT/m} = \sqrt{3RT/M}$; this is seen to be identical with the expression derived previously under simplifying assumptions. The arithmetic mean velocity $\bar{u}$ is equal to $\sqrt{8kT/\pi m}$ or $\sqrt{8RT/\pi M}$, and the most probable velocity u_p is $\sqrt{2kT/m} = \sqrt{2RT/M}$. These quantities u_p, $\bar{u}$, and u_{rms} for a given gas are seen to bear the following relation to one another: $\sqrt{\frac{2}{3}}:\sqrt{8/3\pi}:1$. The three velocities are compared at 0°C in Table 2-1 for hydrogen, nitrogen, oxygen, and carbon dioxide.

TABLE 2-1. MOLECULAR VELOCITIES OF A FEW GASES AT 0°C
(In centimeters per second)

Gas	Most probable velocity, u_p	Average velocity, $\bar{u}$	Root-mean-square velocity, u_{rms}
H_2	15.02×10^4	16.94×10^4	18.39×10^4
N_2	4.03	4.54	4.93
O_2	3.77	4.25	4.61
CO_2	3.21	3.62	3.93

Velocity of Sound. Since the propagation of sound waves in gaseous media involves the motion of molecules, it would be reasonable to guess that the speed of such sound waves would be near the thermal agitation velocity we have been discussing. Since the molecules which carry the sound move in random directions, this reasoning points to the conclusion that the (linear) velocity of sound is less than the mean thermal agitational velocity of the molecules. It can be shown that the velocity of sound u_s is related to the average velocity $\bar{u}$ by the relation

$$\frac{\bar{u}}{u_s} = \sqrt{\frac{8}{\pi\gamma}}$$

where γ is C_P/C_V, the ratio of the heat capacity at constant pressure to that at constant volume. For monatomic gases $\gamma = 1.667$, and for diatomic gases $\gamma = 1.40$ (approximately), so that

$$\frac{\bar{u}}{u_s} = 1.236 \text{ for monatomic gases}$$

and

$$\frac{\bar{u}}{u_s} = 1.349 \text{ for diatomic gases}$$

Thus it is seen that the velocity of sound in a medium is indeed close to (but slightly less than) the molecular velocity and, moreover, is independent of pressure and proportional to $\sqrt{T}$.

Collision Frequency. The collision frequency, designated ω, is the number of collisions experienced by one molecule in one second. Calculation on the basis of the above assumptions gives this as $16nr^2\sqrt{\pi RT/M}$, where r is the atomic radius and n is the number of molecules per cubic centimeter. The mean free path l is the average distance traveled by a molecule between collisions, which is $\bar{u}$ divided by ω; hence $l = 1/(\sqrt{8}\,\pi nr^2)$. To the approximation that the radius is independent of temperature and pressure, it is seen that l is directly proportional to the absolute temperature at constant pressure and inversely proportional to pressure at constant temperature (since n is proportional to P and inversely proportional to T).

Coefficient of Viscosity. Consider a fluid streaming in the x direction, different layers of which are traveling at different velocities u_x; the rate of change of velocity in the y direction (normal to the x direction) is du_x/dy. For example, the fluid may be contained between two parallel plates. Clearly such a fluid is subject to a shearing stress in the x direction designated X_y. The coefficient of viscosity η is defined as $X_y/(du_x/dy)$; that is, it is the shearing force required to produce unit gradient of velocity in the fluid. The expression derived is $\eta = (1/6r^2)\sqrt{mkT/\pi}$. It follows from this that the viscosity of a gas is independent of pressure, a fact verified by experiment, absurd as it may seem. The foregoing expression also predicts that the viscosity varies as $\sqrt{T}$; this prediction is not borne out by experiment, for it is found that the viscosity increases more rapidly with temperature than this. The apparent failure of the theory is due to the erroneous assumption that r (one-half the closest distance of approach on impact) is independent of T; because of the increase of u with T, there is at higher T a lessened effect of the attractive forces on the paths of the molecules as well as a closer approach on impact. For this very reason the equation above is useful in evaluating the change of r with T. Various properties[1] of several gases at 25°C and 1 atm pressure are given in Table 2-2.

Diffusivity (Coefficient of Diffusion). Consider a mixture of two gases which is not uniform and is such that the composition changes in one direction only; for example, we may imagine a gas contained in a tube, the composition varying from one end to the other but remaining constant at any cross-sectional plane. The diffusivity D at any one cross section is defined as the amount of either substance which crosses the plane per square centimeter per unit time per unit concentration gradient. Since concentration is the amount per cubic centimeter, D is independent of the units used to express the amount; the units of D in the cgs system are centimeters squared per second. From the kinetic theory of gases the following expression is derived for D:

$$D = \frac{1}{16}\sqrt{\frac{kT}{\pi}}\left[\frac{(n_2/n_1r_1^2\sqrt{m_1}) + (n_1/n_2r_2^2\sqrt{m_2})}{n_1 + n_2}\right]$$

This can be written in more simple form, though not in fundamental quantities:

$$D = \frac{\pi}{8}\frac{n_2\bar{u}_1l_1 + n_1\bar{u}_2l_2}{n_1 + n_2}$$

The experimental value of D for various pairs of gases is given in Table 2-3.[1] It will be observed that for *self diffusion*, *i.e.*, isotopic diffusion, this reduces to $D = (\pi/8)\bar{u}l$; this formula may be regarded as approximately true for mixtures of real gases provided that the differences in atomic radius and molecular weight are not great. If n_2 is small, then $D = (\pi/8)\bar{u}_2l_2$. Experimental values of D for hydrogen, carbon monoxide, oxygen, and carbon dioxide at 0°C and 1 atm are given in Table 2-4.

[1] From S. Dushman, "Vacuum Technique," John Wiley & Sons, Inc., New York, 1949.

TABLE 2-2. MOLECULAR PROPERTIES OF GASES AT 25°C

Gas	Molecular weight, M	Average velocity, $\bar{u}$, cm/sec	Mean free path, l, cm		Molecular diameter, $2r$, cm*	Collision frequency at 1 atm, ω, sec^{-1}	Coefficient of viscosity, η, dyne-sec/(cm²), (poise)	C_P, cal/(mole)(deg)
			At 1 mm	At 1 atm				
H_2	2.016	17.70×10^4	9.31×10^{-3}	12.26×10^{-6}	2.75×10^{-8}	14.45×10^9	8.92×10^{-5}	6.89
He	4.003	12.56	14.72	19.36	2.18	7.16	19.86	4.97
Ne	20.18	5.594	10.45	13.75	2.60	1.68	31.66	4.97
Air	28.98	4.668	5.09	6.69	3.74	6.98	18.45	6.83
O_2	32.00	4.442	5.40	7.10	3.64	6.26	20.59	6.41
A	39.94	3.976	5.31	6.67	3.67	5.70	22.61	4.97
CO_2	44.01	3.787	3.34	4.40	4.65	8.61	14.96	8.89
Kr	83.7	2.747	4.06	5.34	4.15	6.48	25.02	4.97
Xe	131.3	2.193	2.98	3.93	4.91	5.71	23.08	4.97

* Calculated from viscosity.

TABLE 2-3. DIFFUSIVITY OF GAS MIXTURES AT 0°C AND 1 ATM*

Mixture	*Observed diffusivity D, cm²/sec*
H_2-air	0.661
H_2-O_2	0.679
O_2-air	0.177_5
O_2-N_2	0.174
CO-H_2	0.642
CO-O_2	0.183
CO_2-H_2	0.538
CO_2-air	0.138
CO_2-CO	0.136
N_2O-H_2	0.535
N_2O-CO_2	0.098_3

* From S. Dushman, "Vacuum Technique," John Wiley & Sons, Inc., New York, 1949.

TABLE 2-4. THERMAL CONDUCTIVITY AND SELF-DIFFUSIVITY OF GASES AT 0°C AND 1 ATM

Gas	Thermal conductivity κ, cal/(cm)(sec)(deg)*	Self-diffusivity D, cm²/sec†
H_2	41.3×10^{-5}	1.31
He	34.3	
Ne	11.12	
CO	5.37	0.174
Air	5.76	
O_2	5.83	0.189
A	3.82	
CO_2	3.43	0.109
N_2O	3.61	

* From S. Dushman, "Vacuum Technique," John Wiley & Sons, Inc., New York, 1949.
† From J. H. Jeans, "An Introduction to the Kinetic Theory of Gases," The Macmillan Company, New York, 1940.

Thermal Conductivity. This is the amount of heat transferred per second across a unit plane (1 cm² in area and at right angles to the direction of heat flow) when the thermal gradient is 1°/cm. The thermal conductivity κ for monatomic gases may be expressed as follows:

$$\kappa = \frac{\pi}{8} \frac{\bar{u} l C_V}{V} = 2.5 \frac{\eta C_V}{M}$$

where C_V and V are the molar heat capacity and molar volume, respectively. Experimental values of the coefficient in this expression (obtained by measurement of κ, η, and C_V) range from 1.5 to 2.5 for a variety of common gases. For diatomic and triatomic gases the value varies from 1.5 to 2.0 instead of being 2.5 as in the above expression. In fact the expression $\kappa = [(9\gamma - 5)/4](\eta C_V/M)$, where $\gamma = C_P/C_V$, is found empirically to hold very well for the gases which have been investigated. The thermal conductivity of several gases at 0°C and 1 atm pressure are included in Table 2-4.

Thermal Diffusivity. This is the amount of heat transferred across a similar unit plane per second when there is unit gradient of the heat content per unit volume, *i.e.*, concentration of heat. Its units are centimeters squared per second—much simpler than those of κ. Thus the thermal diffusivity is equal to the coefficient of thermal conductivity divided by C_V/V and is therefore $(\pi/8)\bar{u}l$, which, it will be noted is identical with the isotopic diffusivity. This identity of the diffusivity of heat and the diffusivity of matter, although surprising, is not entirely unexpected, since both are due to the same motion of the molecules. One might think that the heat would travel slightly faster because the random bouncing back of the molecules imparts thermal agitation without transmitting the molecule itself. However, the results of the extended kinetic theory are in good agreement with experiment as to the identity of these two diffusivities. This fact is of some practical importance in that one gas can heat another, essentially, only as it mixes with it by convection and diffusion, providing there is no diaphragm or appreciable transfer by radiation.

Applications. In the bessemer process for making steel by blowing air through the metal, the air toward the end of the blow reacts principally with carbon to form CO and CO_2. It might be thought that too rapid blowing would result in extraction of heat by part of the air passing through without reaction. However, in view of the above principle it is clear that unreacted air does not carry off appreciable heat (N_2 and O_2 being transparent to radiation), a fact verified by thermal analysis of the bessemer process. This phenomenon undoubtedly also occurs in many combustion processes, *e.g.*, the blowing of air on hot coals; the air which does not combine owing to lack of convection also does not abstract heat from the embers, although it may dilute and cool the flue gases. A process was once proposed for drying air by passing moist air over refrigerated tubes, the intent being to condense out the water without cooling the air. It is readily seen from the foregoing that the nature of the process is such that to remove half of the moisture, in effect, all the air must be cooled halfway to the temperature of the tubes. Any radiation effects would make this dehumidifier even less efficient.

It will be noted that the value of D at room temperature is 0.1 to 1 cm^2/sec for common gases. It can be shown that the time t required for 93 per cent of the gas to diffuse out of a cylinder of height x is given by the formula $t = x^2/D$. Thus if a beaker 10 cm high is filled with CO_2 and placed uncovered in still air (D is about $\frac{1}{7}$) the time required for the 93 per cent to diffuse out will be 700 sec or about 12 min. By the same token, if the initial temperature of the CO_2 is 50°C above room temperature, the average temperature after the same time interval will be 3.5°C (7 per cent of 50°C) above room temperature. In view of the fact that the time of a diffusion process is proportional to the square of the distance over which the diffusion is to occur, it is clear that for distances of the order of a centimeter diffusion of gases is relatively rapid, requiring about 7 sec for 93 per cent diffusion or about $\frac{1}{2}$ min for essentially complete diffusion. Thus, if it is required to remove oxygen from a flue gas by reaction with hot copper, a copper tube 1 cm in radius would be adequate provided the rate of flow were adjusted so that the gas were contained in the hot portion of the tube for about this same time. On the other hand, for large-scale meteorological phenomena, as clouds, it is seen that the effects of diffusion are negligible; for diffusion to extend to the extent of 93 per cent over a distance of 100 m would take about 22 years.

Going back to the formula for κ, it will be noticed that thermal conductivity is independent of pressure, since η and C_V are independent of pressure. The question which naturally arises at this point is how then do we account for the insulating property of the Dewar flask or Thermos bottle if the thermal conductivity of the gas between the walls is independent of the pressure. The resolution of this paradox lies in the fact that the formulas presented here apply only when the number of collisions of the molecules with one another is great compared with the number of collisions with the walls of the container, *i.e.*, when the mean free path is short compared with the dimensions of the container. The value of l for air at normal temperature and pressure (NTP) is 6×10^{-6} cm. Hence, for a Dewar vessel designed with walls 1 cm apart, since l is inversely proportional to pressure at constant temperature, it is clear that the pressure must be reduced to six-millionths of an atmosphere to produce an insulating effect appreciably better than that at 1 atm pressure. The vacuum must be considerably better than this to produce a good insulating effect. In fact, below this pressure the thermal conductivity is roughly proportional to the number of molecules or to the pressure, and to reduce the conductivity of air to one-thousandth of that which it has at atmospheric pressure, the pressure must be reduced to the order of 10^{-8} atm or 10^{-5} mm Hg.

DEPARTURES FROM THE IDEAL-GAS LAW

So far we have considered properties of ideal gases only. Experimentally, an ideal gas is one which obeys the relation $PV = RT$ for 1 mole. In the theoretical approach there were hypothesized molecules between which no forces acted and which filled no appreciable portion of the total volume. The development of this theory produced many fruitful results, some of which have been already discussed, and suggested many new lines of experimental investigation, thus amply filling the only role of any theory.

It has, of course, been realized from both angles of approach that the concept of an ideal gas is an oversimplification which may be approached more or less closely by an actual gas under certain conditions. Experimentally, the behavior of a real gas departs more or less seriously from that of an ideal gas at high pressure or low temperature—in short, when the density becomes great. Theoretically, the two limiting conditions mentioned above are not strictly fulfilled by any actual gas, since the size of the molecules is finite and the same forces which hold a liquid or solid together also operate between the molecules of a gas, although to an appreciable extent only near the instant of collision. The general problem of the behavior of gases at high pressure and at low temperature and of their transition to liquid still awaits adequate interpretation in terms of molecular behavior. In spite of the great amount of work done to date on the behavior of gases, no simple equation of state has been found which will represent the behavior of all gases over the range of temperature and pressure which has been investigated experimentally.

Van der Waals' Equation. The famous equation proposed by van der Waals (1880), $[P + (a/V^2)](V - b) = RT$ for 1 mole, is one of the simplest equations of state which represent the general behavior of gases at high density and the transition to the liquid state. It is only semiquantitative. In this expression the constant b is regarded as taking account of the volume occupied by the gas molecules and a as characterizing the attractive forces between the molecules. The constant b may be expressed $\frac{2}{3}\pi d^3 N$, which is four times the volume occupied by the molecules. Here d is the molecular diameter and N is Avogadro's number.

The general form of the graph of this equation is shown in Fig. 2-2. It will be noted that the isotherms at higher temperature approximate the rectangular hyperbola of the ideal-gas law whereas the isotherms at low temperature depart markedly therefrom. At low temperature, V is seen to have three values for each value of P; this corresponds to the fact that the equation is cubic in V. Over the dotted portion of the curve the system is unstable; in fact the part with positive slope can have no physical counterpart. There is an intermediate temperature where the curve exhibits a horizontal inflection; this temperature is known as the *critical temperature*, and the point of inflection as the *critical point*

Relationship between Gas and Liquid. It will be observed that at and above the critical temperature continuous compression results in continuous transition from the gas to the liquid without condensation; in other words, the distinction between gas and liquid has disappeared above the critical temperature. Below the critical temperature two curves are obtained experimentally, as shown by the solid lines, one for the liquid, the other for the gas. "However," to quote Lewis and Randall,[1] "it would be difficult to avoid the conclusion that these two branches are really parts of a single curve, and that a truly satisfactory mathematical equation for one would also fit the other." This point was first made by Thomson (1872) and was so well received that practically all elementary texts for over half a century have made the approach to the liquid state in this manner, through its relation to the gas.

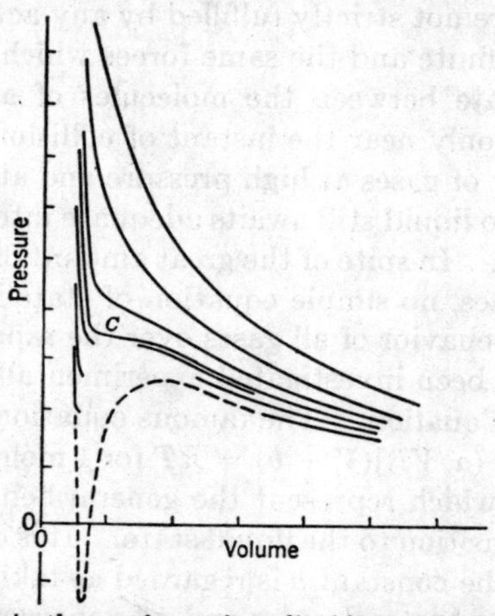

FIG. 2-2. Pressure-volume isotherms for carbon dioxide as calculated from van der Waals' equation. At a temperature below that of the critical point C, the system is unstable over the dotted portion of the curve.

Although the fruitfulness of this concept of continuity between liquid and gaseous states should not be belittled, particularly for those interested in the properties in the vicinity of the critical temperature, it should be emphasized that in most metallurgical processes the liquids involved usually exist under conditions which are removed very far indeed from the critical point. For example, iron melts at 1539°C and steelmaking temperatures are in the vicinity of 1600°C; to suggest that at perhaps 10,000°C the behavior of liquid iron would bear a vague resemblance to that of a compressed gas is not only useless but outright misleading. At 1600°C the chemical behavior of liquid iron and its atomic constitution resemble those of the solid much more closely than those of the gas. This point of view will be developed in detail in Chap. 3.

Lest there be misunderstanding, our viewpoint may be expressed in this

[1] Lewis and Randall, "Thermodynamics and the Free Energy of Chemical Substances," McGraw-Hill Book Company, Inc., New York, 1923.

way: To a steam engineer interested in the properties of water substance in the vicinity of its critical point, van der Waals' and Thomson's concept sheds considerable light, but to the metallurgist, interested in metals slightly above their melting point, the concept of the continuity between gaseous and liquid states contributes little but the incorrect inference that the atoms of the molten metal behave in the random manner characteristic of gases. For the present it will suffice to mention that the chemical forces between atoms or molecules are of very short range and play a predominant role in condensed phases (liquids or solids) but a very minor role in gases at a pressure of 1 atm or less.

In most metallurgical operations, involving gases at a pressure in the vicinity of 1 atm and particularly at elevated temperature, the ideal-gas law is adequate to describe the pressure-volume relations. As a general rule it may be expected that at low temperature the departures from this law are more pronounced for gases with higher boiling point. Thus, the following series is arranged in order of increasing boiling point: hydrogen, carbon monoxide, nitrogen or oxygen, carbon dioxide, water; and this is also the order of increasing deviation from ideality at low temperature. In the vicinity of room temperature and pressure the departure of carbon dioxide from ideality is 0.6 per cent (*i.e.*, the volume occupied by a mole is 0.6 per cent less than that calculated from the ideal-gas law), and that of carbon monoxide is 0.1 per cent. If 100 cc of pure carbon monoxide is burned to carbon dioxide, as is done in gas analysis, only 99.5 cc of carbon dioxide is obtained, thus resulting in $\frac{1}{2}$ per cent error in the analysis if no correction is made. Such a departure from ideality at low temperature is attributable principally to the intermolecular forces represented by the a constant of van der Waals' equation. Since these are the same forces which hold a liquid together, it is readily seen why these departures correlate with the boiling point and why they are negative in sign in the sense that the observed volume is less than the ideal-gas volume. At high temperature (steelmaking temperature) the departure from ideality becomes smaller but now is attributable principally to the molecular size, represented by van der Waals' b constant, and is therefore positive. As only the simpler (smaller) molecules are stable at such temperatures, the departures from ideality cannot be expected to exceed 0.01 to 0.02 per cent. At some intermediate temperature the positive and negative effects are equal and the gas behaves as a perfect gas up to a pressure of many atmospheres.

GAS MIXTURES

When more than one component is present in a phase under consideration, the state is obviously not completely specified unless the relative amounts of the various components are designated. Of the several ways of expressing composition, such as weight per cent or grams per cubic centimeter, the *mole fraction*, designated N with a subscript to designate the species under consideration, is found to be one of the most useful, particularly for gaseous mixtures. The mole fraction of a certain molecular species is defined as the number of moles of that species divided by the total number of moles of all species present. The sum of the mole fractions of all the species present is seen to be unity.

Partial Pressure. The partial pressure of any molecular species is defined as the product of the mole fraction of that species and the total pressure. Thus the partial pressure of component 1 is written

$$p_1 = N_1P$$

or in general

$$p_i = N_iP$$

The sum of all the partial pressures equals the total pressure, for

$$p_1 + p_2 + \cdots = (N_1 + N_2 + \cdots)P = P$$

In a mixture which obeys the ideal-gas law (*i.e.*, each component does), each component gas exerts the same partial pressure as though the others were absent. This statement is known as *Dalton's law*.

Dalton's law is experimentally verified with the aid of the so-called semipermeable membrane, which is composed of a substance that transmits certain elements or molecules more readily than others. For example, palladium is well known to be permeable to hydrogen and relatively impermeable to other substances. Iron is also more permeable to hydrogen than to other substances. At elevated temperature iron is quite permeable to nitrogen and impermeable, or nearly so, to helium and the other inert gases. Silver at elevated temperature is permeable to oxygen. Metals are generally permeable to gases which have a high solubility in them. Cellophane is particularly permeable to water. Rubber is very permeable to carbon dioxide and is more permeable to oxygen than it is to nitrogen; thus a rubber tire will be found to contain after a time a much lower percentage of oxygen than the air pumped into it. A piece of laboratory rubber tubing when filled with carbon dioxide and left overnight with ends closed will be found collapsed in the morning. If an evacuated silver tube is heated in air, it is found that oxygen passes through the wall until the pressure of oxygen on the inside is equal to the partial pressure of oxygen in the air on the outside, thus verifying Dalton's law. Another method of verification consists in introducing two gases into a vessel and measuring the volume at constant temperature and pressure before and after mixing. The percentage change in volume on mixing has been found to range from a few hundredths to a few tenths of a per cent, this change being completely explained by departures from the ideal-gas law.

In terms of the kinetic theory it is easy to see that the partial pressure of one component of a gas mixture is the same as though that component alone occupied the space, for, in this theoretic approach, pressure is taken as the rate of change of momentum per unit area and the velocity and collision frequency with the walls of any particular species are independent of the presence or absence of other species. It is to be noted, however, that it is only in the case of an ideal gas that the partial pressure as estimated by diaphragm methods is identical with that obtained from the definitional relation $p_i = N_iP$. If we had defined, by convention, the partial pressure to be that obtained by the semipermeable-diaphragm method, then in the case of a nonideal gas we would be confronted with the awkward situation that the total pressure would not equal the sum of the partial pressures. Combination of the definitional relation for p_i with the ideal-gas law for the mixture yields the relation $p_iV = n_iRT$, the ideal-gas law for a single component.

Departure from the Gas Laws Due to Chemical Reaction. In all the preceding part of this chapter, it has been tacitly assumed that the gas under consideration does not undergo chemical reaction during the course of our observation or consideration of it. Since gas reactions are of importance, we shall now consider the modification required if chemical reaction occurs. Let us first consider this from the experimental viewpoint. Suppose that 1 mole of hydrogen is heated to about 3000°K at about 1 atm pressure in a suitable vessel. Then it is found that the pressure is not proportional to the absolute temperature as it would be if Charles' law were obeyed but rises much more rapidly. If the temperature is held constant and the gas is compressed, the volume decreases more rapidly than would be anticipated from Boyle's law. Moreover, if the ideal-gas law is assumed, it is found that at all temperatures and pressures the number of moles n is between 1 and 2. This type of departure from ideal behavior is different in nature from the van der Waals' type, since hydrogen behaves very nearly ideally at room temperature and the van der Waals' departure decreases with rising temperature.

The obvious resolution of the difficulty is to postulate that the diatomic molecules partially dissociate into monatomic molecules. In fact, if it is assumed that the ideal-gas law is obeyed and that n varies, then the mole fraction of each of the molecular species H_2 and H can be calculated. Denoting the total number of moles as n, those of H as n_H, and those of H_2 as n_{H_2}, we have

$$n = n_H + n_{H_2}$$

Also, the total number of gram atoms of hydrogen, 2, does not change during any chemical reaction, so that

$$2 = n_{\mathrm{H}} + 2n_{\mathrm{H}_2}$$

Thus, using n obtained by applying the ideal-gas law, n_H and n_{H_2} result from simultaneous solution of these two equations. It is then found that $n_{\mathrm{H}}^2/n_{\mathrm{H}_2}$ is constant at any fixed temperature over a large range of the total pressure. As this is in agreement with the well-known mass law, it would seem reasonable to conclude that our hypothesis of dissociation of H_2 into H is justified. However, the reasoning is somewhat cyclic, since the experimental verification of the mass law is by means of a number of experiments of this type.

Of course, auxiliary evidence in favor of this conclusion could be obtained from the heat of reaction. In our hypothetical experiment, it would be observed that the heat required to raise the temperature of the system is much greater than would normally be expected. Similarly the heat liberated upon compression would also be much greater than normal. Such large heat effects are traditionally associated with chemical reaction. Cyclic reasoning is not completely absent here either, for this traditional association is founded on a multitude of similar cases. Also, the color change (spectroscopic data) is capable of giving quantitative evidence of the dissociation.

Thus, whenever it seems unreasonable to suppose that the gas under consideration departs appreciably from the ideal-gas law by reason of proximity to the critical point (van der Waals' type of departure), it is common to suppose that a chemical reaction is occurring and that the ideal-gas law holds for each molecular species. The only modification required then is that due account be taken of the variation of the number of moles of each constituent n_i in the equation $p_iV = n_iRT$.

The student may think that the point made here is so elementary and obvious that it hardly merits his attention. However, it cannot be overemphasized that such reasoning about experiments with gases is the primary basis of our present-day concept of molecular constitution. It should be quite clear that in the case of liquids a similar establishment of molecular constitution is impossible, since in this case there exists no general equation of state analogous to the ideal-gas law. Traditionally it has been assumed that the molecular structure established for a gas persists also in the corresponding liquid. This assumption is unwarranted in many cases, particularly in the case of molten metals. The constitution of liquids will be discussed at length in Chap. 5.

CHAPTER 3

SOLIDS

The Periodic Table. Mendeleev (1869) was the first to design a periodic table. He arranged the atoms in order of atomic weight and was the first to observe a periodicity in chemical properties in such an arrangement. Subsequent investigation showed that the correct order was that of atomic number, *i.e.*, order of number of electrons outside the nucleus. A variety of interesting arrangements of the periodic table have been devised;[1] one of the more common is shown in Table 3-1. The most obvious feature of such a table is that chemically similar elements form *groups* numbered from zero to eight (Newland's "law of octaves," 1864). In general a group divides into *A* and *B* *subgroups* with different properties; these are indicated in Table 3-1 by a slight lateral shift, *A* on the left and *B* on the right. For example, the alkali metals lithium, sodium, potassium, rubidium, and cesium, which have very pronounced chemical similarities, constitute the *A* subgroup of the first group, and the halogens fluorine, chlorine, bromine, iodine, and astatine occur in the *B* subgroup of the seventh group.

It will be observed that repetition of chemical properties does not occur at equal intervals of atomic number; the interval is 8 elements for the two *periods* from lithium through chlorine, 18 for the second two periods from potassium through iodine, and 32 for the period including the rare earths. This gives rise to the subgroups; *e.g.*, copper, silver, and gold occur in the same group as lithium, sodium, potassium, rubidium, and cesium but in a different subgroup in accord with their chemical properties.

Elementary Particles and the Atom. The present viewpoint of the structure of matter is that there are three normal constituents of matter: the *electron*, the *proton*, and the *neutron*. The mass of the proton and neutron is very nearly the same (1.67×10^{-24} g), being about 1850 times that of the electron. The electron bears a negative charge of 4.8022×10^{-10} esu, the proton bears a positive charge of identical magnitude, and the neutron is electrically neutral. Various other particles, the positron the mesons and *V* particles, and the neutrino, are evidenced in nuclear phenomena; however, these may be regarded in part as the debris result

[1] C. A. Zapffe, *Trans. ASM*, **38**, 239 (1947).

TABLE 3-1. PERIODIC ARRANGEMENT OF THE ELEMENTS

Period \ Group	0	I	II	III	IV	V	VI	VII	VIII		
1		H 1 1.0080									
2	He 2 4.003	Li 3 6.940	Be 4 9.02	B 5 10.82	C 6 12.010	N 7 14.008	O 8 16.0000	F 9 19.00			
3	Ne 10 20.183	Na 11 22.997	Mg 12 24.32	Al 13 26.97	Si 14 28.06	P 15 30.98	S 16 32.066	Cl 17 35.457			
4	A 18 39.944	K 19 39.096	Ca 20 40.08	Sc 21 45.10	Ti 22 47.90	V 23 50.95	Cr 24 52.01	Mn 25 54.93	Fe 26 55.85	Co 27 58.94	Ni 28 58.69
		Cu 29 63.54	Zn 30 65.38	Ga 31 69.72	Ge 32 72.60	As 33 74.91	Se 34 78.96	Br 35 79.916			
5	Kr 36 83.7	Rb 37 85.48	Sr 38 87.63	Y 39 88.92	Zr 40 91.22	Nb 41 92.91	Mo 42 95.95	Tc 43 99.	Ru 44 101.7	Rh 45 102.91	Pd 46 106.7
		Ag 47 107.880	Cd 48 112.41	In 49 114.76	Sn 50 118.70	Sb 51 121.76	Te 52 127.61	I 53 126.92			
6	Xe 54 131.3	Cs 55 132.91	Ba 56 137.36	La 57* 138.92	Hf 72 178.6	Ta 73 180.88	W 74 183.92	Re 75 186.31	Os 76 190.2	Ir 77 193.1	Pt 78 195.23
		Au 79 197.2	Hg 80 200.61	Tl 81 204.39	Pb 82 207.21	Bi 83 209.00	Po 84 210.	At 85 211.			
7	Rn 86 222.	Fr 87 223.	Ra 88 226.05	Ac 89† 227.05							

* Rare Earths
Group III-*A*, Period 6

Ce 58	140.13	Pm 61	147.	Gd 64	156.9	Ho 67	164.94	Yb 70	173.04
Pr 59	140.92	Sm 62	150.43	Tb 65	159.2	Er 68	167.2	Lu 71	174.99
Nd 60	144.27	Eu 63	152.0	Dy 66	162.46	Tm 69	169.4		

† Actinides
Group III-*A*, Period 7

Th 90	232.12	Np 93	237.	Cm 96	242.
Pa 91	231.	Pu 94	239.	Bk 97	(243.)
U 92	238.07	Am 95	241.	Cf 98	(244.)

ing from atomic distintegration rather than as a normal constituent[1] of intact atoms. It is well to bear in mind that all of these particles are constructs, as are the atom and the molecule, invented to interpret observed results; however, these constructs and the theories associated with them have been so fruitful and of such high predictive value that they are commonly regarded as being on a par with directly observable objects.

An atom is composed of a positively charged *nucleus* of protons and neutrons and *orbital electrons,* equal in number to the charge on the nucleus, whose distance from the nucleus is large compared with the size of the nucleus. The total number of protons and neutrons is approximately equal to the *atomic weight.* The number of orbital electrons (or the number of protons in the nucleus) is known as the *atomic number.* Chemical properties are determined almost exclusively by the number of orbital electrons (the atomic number). Atoms with the same atomic number but different atomic weight are known as *isotopes.* Slight differences in chemical properties of isotopes are observable, notably in the case of hydrogen and deuterium. The orbital electrons may be regarded as arranged in "shells" of which the outermost, containing the *valence electrons,* is most important in determining the chemical properties. Elements with similar chemical characteristics, *e.g.,* fluorine, chlorine, bromine, and iodine, contain the same number of valence electrons. It is believed that all the forces acting between the electrons and the nuclei in all aggregations of atoms, and hence all the chemical properties, are ultimately interpretable in terms of Coulomb's law of electrostatic attraction or repulsion, stating that the electrical force between two charges is equal to the product of the charges divided by the square of the distance between them. Even with the aid of wave mechanics the general problem of interpretation of chemical behavior presents prohibitive difficulties, though great strides have been made in this direction.

ATOMIC STRUCTURE

The concept of an electron moving in its orbit about an atomic nucleus like a planet around the sun is somewhat of an aid to the imagination but must be regarded as a very crude analogy. In fact, modern theorists sometimes find it convenient to regard the electron sometimes as a particle and sometimes as a wave. To inquire as to its true nature is meaningless,

[1] The term "elementary particle" has not been mentioned, as it is not clear what meaning is to be attached to this. An isolated neutron is now reported to transform spontaneously to a proton and an electron. Thus this "building block" by itself decomposes but when incorporated in the structure does not.

since these constructs are invented only to represent certain phenomena and to predict new ones. The electron is frequently pictured as a rather "fuzzy" entity to which it is impossible to assign at any given instant a location and a velocity, but only a probability of being in a certain region and having a certain velocity.

Some of these difficulties we may circumnavigate by observing that the electron (if regarded as a mass point) has three *degrees of freedom;*[1] *i.e.*, three coordinates are required to define its position. Also, any particle in motion has an energy. If the coordinates are properly chosen, a portion of the total energy may be associated with each coordinate. Each of these energies is quantized; *i.e.*, it is regarded that each such energy cannot assume any value but is restricted to certain integral multiples, the integers being known as *quantum numbers.* A large quantum number thus corresponds to a high energy, and a small one to a low energy. Since each portion of the energy is quantized, it is apparent that three quantum numbers are required. For the orbital electrons these are designated n, the principal quantum number; l, the secondary quantum number; and m, the magnetic quantum number. Although these three would be adequate for a mass point, it is found that the electron does not quite behave as such a mass point and that a fourth quantum number s, corresponding to a "spin," is required; s assumes only the values $+\frac{1}{2}$ and $-\frac{1}{2}$. The integral value of n, the principal quantum number, corresponds to the shell occupied by the electron in the simpler picture. It is found, in the case of an orbital electron, that l may have any value from zero to $n - 1$ and m any value from $-l$ to $+l$. Values of l equal to 0, 1, 2, and 3 are usually denoted by the letters s, p, d, and f, respectively; thus the valence electron of potassium for which $n = 4$ and $l = 0$ is designated $4s$.

The *Pauli exclusion principle* states that no two electrons of an atom may have the same set of quantum numbers. Thus, the maximum total number of orbital electrons with principal quantum number 1, corresponding to the first shell, is two, since the only permissible value for l and m is zero, and for s is $+\frac{1}{2}$ and $-\frac{1}{2}$. The two possible sets of values are as follows:

[1] The number of degrees of freedom of any material object is the number of coordinates necessary to describe its position completely at any instant. It will be noticed that this number will be independent of the choice of coordinates. For instance a point on a plane may be located from its Cartesian coordinates x and y, from its polar coordinates r and θ, or by means of any of a number of coordinate systems. In every case only two coordinates are required. Similarly in three-dimensional space three coordinates are required. A rigid body in three-dimensional space requires five coordinates—three to fix the position of its center of gravity and two to establish the direction of its major axis.

n	l	m	s
1	0	0	$\frac{1}{2}$
1	0	0	$-\frac{1}{2}$

For $n = 2$, *i.e.*, the second shell, there are eight possible sets of quantum numbers:

n	l	m	s
2	1	1	$\frac{1}{2}$
2	1	1	$-\frac{1}{2}$
2	1	0	$\frac{1}{2}$
2	1	0	$-\frac{1}{2}$
2	1	-1	$\frac{1}{2}$
2	1	-1	$-\frac{1}{2}$
2	0	0	$\frac{1}{2}$
2	0	0	$-\frac{1}{2}$

Table 3-2 shows the number of electrons in the shells and subshells, *i.e.*, those with various values of n and l, for each of the chemical elements. It will be noted that the maximum total number of electrons in any shell is $2(n)^2$, or 2, 8, 18, 32, ... Strictly this table refers to an isolated atom, *i.e.*, one with no other atom in the vicinity. The arrangement given in each case corresponds to the lowest possible energy state, and hence the most stable state, for a given number of orbital electrons. This is seen from the fact that in the first three periods no electron with higher value of n (corresponding to higher energy) appears until all possible lower quantum states are occupied. Although n is called the principal quantum number in the sense that it is predominant in defining the energy state, the other quantum numbers also define a contribution to the energy. Thus, although the lowest energy state is always the most stable, it does not always happen that lower shells are completely filled before a higher one may be occupied; electrons enter the fourth and higher shells before the preceding ones are completely filled. This phenomenon gives rise to the so-called *transition elements* of the long periods (scandium through nickel and yttrium through palladium) and also to the group of elements known as the *rare earths*. The arrangement of Table 3-2 is quite consistent with the periodic chart of Table 3-1, it being noted that elements with electrons only with principal quantum number 1 fall in the first period, those with electrons of principal quantum number 2 (but no greater) fall in the second period, etc. Also the group number corre-

TABLE 3-2. ARRANGEMENT OF ORBITAL ELECTRONS FOR THE ELEMENTS*

	Shell	K	L		M			N				O				P				Q			
	n	1	2		3			4				5				6				7			
	Orbit	s	s	p	s	p	d	s	p	d	f	s	p	d	f	s	p	d	f	s	p	d	f
Atomic No.	Element / l	0	0	1	0	1	2	0	1	2	3	0	1	2	3	0	1	2	3	0	1	2	3
1	H	1																					
2	He	2																					
3	Li	2	1																				
4	Be	2	2																				
5	B	2	2	1																			
6	C	2	2	2																			
7	N	2	2	3																			
8	O	2	2	4																			
9	F	2	2	5																			
10	Ne	2	2	6																			
11	Na	2	2	6	1																		
12	Mg	2	2	6	2																		
13	Al	2	2	6	2	1																	
14	Si	2	2	6	2	2																	
15	P	2	2	6	2	3																	
16	S	2	2	6	2	4																	
17	Cl	2	2	6	2	5																	
18	A	2	2	6	2	6																	
19	K	2	2	6	2	6	—	1															
20	Ca	2	2	6	2	6	—	2															
21	Sc	2	2	6	2	6	1	2															
22	Ti	2	2	6	2	6	2	2															
23	V	2	2	6	2	6	3	2															
24	Cr	2	2	6	2	6	5	1															
25	Mn	2	2	6	2	6	5	2															
26	Fe	2	2	6	2	6	6	2															
27	Co	2	2	6	2	6	7	2															
28	Ni	2	2	6	2	6	8	2															
29	Cu	2	2	6	2	6	10	1															
30	Zn	2	2	6	2	6	10	2															
31	Ga	2	2	6	2	6	10	2	1														
32	Ge	2	2	6	2	6	10	2	2														
33	As	2	2	6	2	6	10	2	3														
34	Se	2	2	6	2	6	10	2	4														
35	Br	2	2	6	2	6	10	2	5														
36	Kr	2	2	6	2	6	10	2	6														
37	Rb	2	2	6	2	6	10	2	6	—	—	1											
38	Sr	2	2	6	2	6	10	2	6	—	—	2											
39	Y	2	2	6	2	6	10	2	6	1	—	2											
40	Zr	2	2	6	2	6	10	2	6	2	—	2											
41	Nb	2	2	6	2	6	10	2	6	4	—	1											
42	Mo	2	2	6	2	6	10	2	6	5	—	1											
43	Tc	2	2	6	2	6	10	2	6	6	—	1											
44	Ru	2	2	6	2	6	10	2	6	7	—	1											
45	Rh	2	2	6	2	6	10	2	6	8	—	1											
46	Pd	2	2	6	2	6	10	2	6	10	—	—											
47	Ag	2	2	6	2	6	10	2	6	10	—	1											
48	Cd	2	2	6	2	6	10	2	6	10	—	2											
49	In	2	2	6	2	6	10	2	6	10	—	2	1										
50	Sn	2	2	6	2	6	10	2	6	10	—	2	2										

TABLE 3-2. ARRANGEMENT OF ORBITAL ELECTRONS FOR THE ELEMENTS.*—(*Continued*)

	Shell	*K*	*L*		*M*			*N*				*O*				*P*				*Q*			
	n	1	2		3			4				5				6				7			
Atomic	Orbit	*s*	*s*	*p*	*s*	*p*	*d*	*s*	*p*	*d*	*f*	*s*	*p*	*d*	*f*	*s*	*p*	*d*	*f*	*s*	*p*	*d*	*f*
No.	Element / *l*	0	0	1	0	1	2	0	1	2	3	0	1	2	3	0	1	2	3	0	1	2	3
51	Sb	2	2	6	2	6	10	2	6	10	—	2	3										
52	Te	2	2	6	2	6	10	2	6	10	—	2	4										
53	I	2	2	6	2	6	10	2	6	10	—	2	5										
54	Xe	2	2	6	2	6	10	2	6	10	—	2	6										
55	Cs	2	2	6	2	6	10	2	6	10	—	2	6	—	—	1							
56	Ba	2	2	6	2	6	10	2	6	10	—	2	6	—	—	2							
57	La	2	2	6	2	6	10	2	6	10	—	2	6	1	—	2							
58	Ce	2	2	6	2	6	10	2	6	10	2	2	6	—	—	2							
59	Pr	2	2	6	2	6	10	2	6	10	3	2	6	—	—	2							
60	Nd	2	2	6	2	6	10	2	6	10	4	2	6	—	—	2							
61	Pm	2	2	6	2	6	10	2	6	10	5	2	6	—	—	2							
62	Sm	2	2	6	2	6	10	2	6	10	6	2	6	—	—	2							
63	Eu	2	2	6	2	6	10	2	6	10	7	2	6	—	—	2							
64	Gd	2	2	6	2	6	10	2	6	10	7	2	6	1	—	2							
65	Tb	2	2	6	2	6	10	2	6	10	9	2	6	—	—	2							
66	Dy	2	2	6	2	6	10	2	6	10	10	2	6	—	—	2							
67	Ho	2	2	6	2	6	10	2	6	10	11	2	6	—	—	2							
68	Er	2	2	6	2	6	10	2	6	10	12	2	6	—	—	2							
69	Tm	2	2	6	2	6	10	2	6	10	13	2	6	—	—	2							
70	Yb	2	2	6	2	6	10	2	6	10	14	2	6	—	—	2							
71	Lu	2	2	6	2	6	10	2	6	10	14	2	6	1	—	2							
72	Hf	2	2	6	2	6	10	2	6	10	14	2	6	2	—	2							
73	Ta	2	2	6	2	6	10	2	6	10	14	2	6	3	—	2							
74	W	2	2	6	2	6	10	2	6	10	14	2	6	4	—	2							
75	Re	2	2	6	2	6	10	2	6	10	14	2	6	5	—	2							
76	Os	2	2	6	2	6	10	2	6	10	14	2	6	6	—	2							
77	Ir	2	2	6	2	6	10	2	6	10	14	2	6	7	—	2							
78	Pt	2	2	6	2	6	10	2	6	10	14	2	6	9	—	1							
79	Au	2	2	6	2	6	10	2	6	10	14	2	6	10	—	1							
80	Hg	2	2	6	2	6	10	2	6	10	14	2	6	10	—	2							
81	Tl	2	2	6	2	6	10	2	6	10	14	2	6	10	—	2	1						
82	Pb	2	2	6	2	6	10	2	6	10	14	2	6	10	—	2	2						
83	Bi	2	2	6	2	6	10	2	6	10	14	2	6	10	—	2	3						
84	Po	2	2	6	2	6	10	2	6	10	14	2	6	10	—	2	4						
85	At	2	2	6	2	6	10	2	6	10	14	2	6	10	—	2	5						
86	Rn	2	2	6	2	6	10	2	6	10	14	2	6	10	—	2	6						
87	Fr	2	2	6	2	6	10	2	6	10	14	2	6	10	—	2	6	—	—	1			
88	Ra	2	2	6	2	6	10	2	6	10	14	2	6	10	—	2	6	—	—	2			
89	Ac	2	2	6	2	6	10	2	6	10	14	2	6	10	—	2	6	1	—	2			
90	Th	2	2	6	2	6	10	2	6	10	14	2	6	10	—	2	6	2	—	2			
91	Pa	2	2	6	2	6	10	2	6	10	14	2	6	10	2	2	6	1	—	2			
92	U	2	2	6	2	6	10	2	6	10	14	2	6	10	3	2	6	1	—	2			
93	Np	2	2	6	2	6	10	2	6	10	14	2	6	10	4	2	6	1	—	2			
94	Pu	2	2	6	2	6	10	2	6	10	14	2	6	10	5	2	6	1	—	2			
95	Am	2	2	6	2	6	10	2	6	10	14	2	6	10	6	2	6	1	—	2			
96	Cm	2	2	6	2	6	10	2	6	10	14	2	6	10	7	2	6	1	—	2			
97	Bk																						
98	Cf																						

* These electronic configurations are for unexcited *isolated atoms*. As discussed in the text, the electronic configuration for atoms in a crystal may differ greatly from that tabulated.

sponds to the number of electrons with the largest principal quantum number, *i.e.*, the number in the outer shell. The postulates stated, including the Pauli exclusion principle, were necessitated by the requirement (among others) that there be such correspondence between this theoretically derived table and Mendeleev's observations.

THE STRUCTURE OF THE MOLECULE

Since the concept of the molecule was originally derived from consideration of the gaseous state, it is simpler to restrict our initial discussion of the molecule to this state. As already noted, the forces which bind atoms together to form any stable configuration, such as a molecule, crystal, or even a liquid, presumably derive fundamentally from Coulomb's law of attraction between positively and negatively charged particles (the protons and electrons). This is so broad a generalization that relatively little specific information can be gained from it. However, in certain cases, such as sodium chloride vapor, the interpretation of the chemical combination of sodium with chlorine is that the sodium atom gives up its valence electron to the chlorine atom so that the outer shell of each attains a stable configuration of eight electrons, the electronic configurations of the atoms then being similar to those of the rare gases neon and argon, respectively. However, the sodium atom now bears a unit positive charge and the chlorine atom a unit negative charge; as a result of their charge they are called *ions*, an ion being any atom or group of atoms which bears a net electrical charge. The force holding the atoms together, known in general as the *chemical bond*, is in this and similar cases regarded as being mainly electrostatic. It derives almost exclusively from simple coulombic attraction and is sometimes called the *coulombic force*. Such compounds are said to be *polar* or *ionic*.

In other cases, *e.g.*, the chlorine molecule, the bearing of Coulomb's law on the situation is much less obvious. In this case an interpretation of the interaction of the atoms was suggested by G. N. Lewis. Each atom having initially seven electrons in the outer shell may form the stable configuration of eight electrons by sharing an electron with the other atom. Thus two electrons are shared in common by the two chlorine atoms. Such atoms are said to be held together by a *covalent* or *homopolar* linkage (or bond). This type of bond is most common in organic compounds as well as O_2, H_2, N_2, etc. More than one pair of covalent bonds may be involved; *e.g.*, the oxygen molecule has two such pairs. The two electrons associated with a covalent bond have identical values of quantum numbers l, m, and n but have opposite spins, that is, s is $+\frac{1}{2}$ for one and $-\frac{1}{2}$ for the other, in accord with the Pauli exclusion principle that the two do not have identical sets of quantum numbers.

Resonance and Mixed Bond Types. The structure of the chlorine molecule may be represented by $:\underset{\cdot\cdot}{\overset{\cdot\cdot}{Cl}}:\underset{\cdot\cdot}{\overset{\cdot\cdot}{Cl}}:$ and of oxygen by $:\overset{\cdot\cdot}{O}::\overset{\cdot\cdot}{O}:$, the dots around each symbol representing electrons and the dots between symbols shared electrons. The total number associated with each atom is seen to be eight. In some cases, such as nitrous oxide, a variety of ways of writing the bonding arrangement is possible:

$$:\overset{\cdot\cdot}{N}::N::\overset{\cdot\cdot}{O}: \qquad (A)$$

$$:N:::N:\underset{\cdot\cdot}{\overset{\cdot\cdot}{O}}: \qquad (B)$$

$$:\underset{\cdot\cdot}{\overset{\cdot\cdot}{N}}:N:::O: \qquad (C)$$

It so happens that the configuration (*C*) and others not shown are unstable (correspond to higher energy states) compared with (*A*) and (*B*). However, (*A*) and (*B*) are almost equally stable, so that any given molecule of nitrous oxide sometimes has the configuration (*A*) and sometimes (*B*). In fact it passes very rapidly back and forth between the two. It is conventional to speak of such a molecule as *resonating* between the structures (*A*) and (*B*). The benzene molecule offers perhaps the best known example of resonance.

Resonance is a phenomenon which is not at all understandable in terms of classical mechanics but which is interpretable in terms of wave mechanics. A molecule which resonates between two such structures is more stable than it would be if only one possible structure existed; its energy is less by an amount known as the *resonance energy.* The phenomenon of resonance is very important and offers an interpretation of the otherwise inexplicable stability of many molecules, *e.g.*, benzene, and is of great importance in the interpretation of the stability of metallic crystals, as will be discussed later.

Thus far we have discussed only bonds which are purely ionic or purely covalent, and resonance in the latter type. This classification is somewhat arbitrary, since mixed bond types are more common. For example, according to Pauling[1] the energy of the bond in the H_2 molecule may be divided in the following way:

Resonance of the two electrons between the nuclei (covalent bond)	80%
Ionic bond and resonance thereof (H^+H^- and H^-H^+)	5
Complex interactions included under the term *deformation*	15

[1] L. Pauling, "The Nature of the Chemical Bond," Cornell University Press, Ithaca, New York, 1939.

Pauling also gives the percentage of ionic character of the bond in HI, HBr, HCl, and HF as 5, 11, 17, and 60, respectively. In H_2O the bond has 39 per cent ionic character. From the general point of view, all types of chemical combination may be interpreted on the basis of finding the most stable electronic configuration (lowest energy state). However, as a first approximation, it is found convenient to divide bonds into the two classes described above, since in the majority of cases the distinction between ionic and homopolar linkages is significant.

BONDS IN CRYSTALS

Crystals with Ionic or Covalent Bonds. Bonds which hold atoms together to form gas molecules were discussed in the preceding section. Molecules are much more clearly visualized in the gaseous state, where they are sufficiently far apart that there is little interaction between them. Indeed, as mentioned, the concept of the molecule originated from the chemical reactions and equilibria of gases. In proceeding now to consider the forces which hold crystalline substances together, it is only natural to expect that the constructs invented to interpret the nature of gases may be extended to interpret the nature of solids. Such is found to be the case, and the chemical bonds in terms of which the strength of a solid is interpreted may readily be classified in the same way as the bonds between the atoms in a molecule. Thus, the sodium chloride crystal is composed of Na^+ and Cl^- ions, and the forces contributing to its strength are very similar to those which hold the gaseous molecule together. It will be noted that in the gas molecule each sodium ion is associated with only one chlorine ion whereas in the crystalline solid each sodium ion is surrounded by six chlorine ions and each chlorine ion by six sodium ions. The idealized configuration is represented in Fig. 3-1. The forces holding such a crystal together are readily visualized as being essentially coulombic in nature, and the crystal is said to be *ionic* or *polar*.[1]

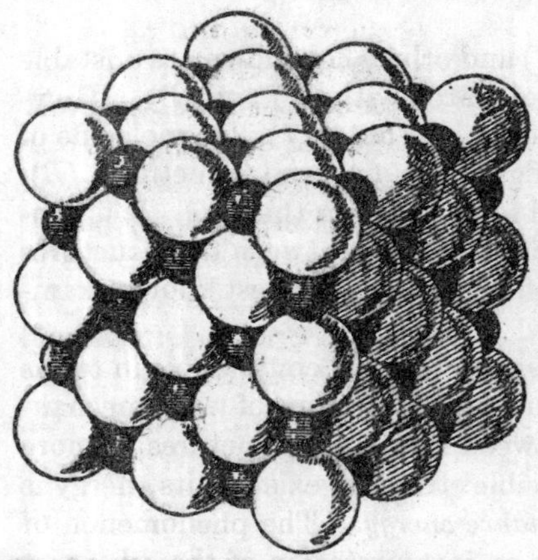

FIG. 3-1. The idealized arrangement of ions in the sodium chloride crystal. The larger spheres represent chlorine ions. [*From W. Barlow, Z. Krist.*, **29**, 433 (1898).]

[1] The above description of the sodium chloride crystal as consisting of Na^+ and Cl

The forces which hold the carbon atoms together in diamond are essentially homopolar or covalent in nature, just as are the forces in a large variety of organic molecules. In diamond each carbon atom is surrounded by four other equally spaced atoms; the type of crystal arrangement is known as *tetrahedral,* since these four atoms are situated at the corners of a regular tetrahedron. Each of the four valence electrons of a carbon atom is shared with one of the four adjacent carbon atoms, so that an electron-pair bond is formed between every pair of adjacent carbon atoms. Such crystals embodying paired electrons are called *homopolar* or *covalent.*

As is the case of bonds in molecules, the bonds in crystals are usually of mixed type, being neither purely covalent nor purely ionic. The Si—O bond in silica is about half ionic and half covalent.

Metallic Crystals. The nature of the bond in metallic crystals seems on first consideration to be quite different from the two types already discussed. In order to account for one of the most obvious properties of metals, the electrical conductivity, the *free-electron-gas* theory was proposed by Drude (1900) and extended by Lorentz (1916). This rather crude theory, which ignored the interaction between the charged electrons (according to Coulomb's law), achieved some success in interpreting Ohm's law as well as the law of Wiedemann-Franz, the relation between electrical and thermal conductivity. The thermal emission of electrons was also partially explained thereby. However, the deficiencies were many. Particularly, a grave discrepancy arose as regards the heat capacity; the free electrons do not contribute to this as required by the classic law of equipartition of energy. Magnetic properties and the dependence of electrical conductivity on temperature were inadequately explained.

Sommerfeld (1928) translated the above classical theory of the free-electron gas into the language of the quantum physics. This significant achievement removed the old anomaly of the heat capacity (the contribution of the electrons to the heat capacity was now found to be proportional to the absolute temperature, except at very high temperature), and yielded an interpretation of the thermoelectric potential, the Thomson and Peltier effects, and the Hall effect. The temperature dependence of the electrical and thermal conductivities was still unexplained as were the

---

ions is perhaps somewhat oversimplified. A complete description would specify the (average) electron density of the bonding electrons. If this density does not drop to zero in the vicinity of the sodium atom, it is misleading to speak of an Na^+ ion, for it has, in a sense, drawn back part of the charge from the Cl^- ions. The extent of this effect is not fully agreed upon, but it is usually thought to be small for the alkali halides.

optical properties; no light was shed on the distinction between electrical conductors and insulators.

A major advance in this field was achieved by Bloch and Brillouin, who developed the model of the electron gas in a periodic potential (of the atoms on the lattice sites). The wave nature of the electron plays an important role in the theory based on this model. It is found that, for a given direction of motion of an electron, the energy values are concentrated in certain allowed energy intervals, separated by forbidden intervals. This leads, in three-dimensional space, to zones bounded by surfaces at which the energy is discontinuous. These zones, known as *Brillouin zones*, are associated with the selective reflection of electron waves by the crystal lattice—a phenomenon similar to the Bragg reflection of X rays. This new method, which has been further developed by many physicists[1] and which has been generally adopted, has made it possible to handle many problems left unsolved by the free-electron-gas theory.

This new electron-gas theory thus provides a good model and an interpretation of many aspects of metallic behavior. It offers a particularly appealing interpretation of electrical conductivity: In view of the wave nature of the electron it is easily understood that any departure from perfect periodicity in the lattice is accompanied by a decrease in the conductivity.[2] The following factors which lead to departures from perfect periodicity also result in lower conductivity: (1) elevated temperature (increased thermal agitation), (2) addition of alloying element, (3) plastic deformation or cold work. Despite the many virtues of this theory it has so far not brought forth any satisfactory interpretation of the nature of the chemical bond in the metallic state. In simple, if inexact, language it has contributed much to physics but little to chemistry. Since our interest focuses on the latter, the electron-gas theory will not be discussed at any further length; let us rather turn our attention to a theory more directly concerned with the chemistry of the metallic state.

[1] A brief résumé is given by R. Kronig [*Physica*, **15**, 1 (1949)]. Cottrell ("Theoretical Structural Metallurgy," Longmans, Green & Co., Inc., New York, 1948) and Raynor ("An Introduction to the Electron Theory of Metals," The Institute of Metals, London, 1947) give a more detailed exposition.

[2] This effect may be interpreted in terms of an analogy. Imagine a rope with one end free which is rigidly held at evenly spaced points, corresponding to lattice points in a perfect metallic crystal. A standing wave (the electron) with the fixed points as nodes may be generated throughout the length of the rope by judicious manipulation of the free end. If, however, the fixed points are irregularly spaced, the generation of a standing wave is much more difficult, if not impossible, and the resulting waves are of much smaller amplitude. In the first case of evenly spaced nodes, energy is readily transmitted along the rope. In the second case of unevenly spaced nodes, energy is rapidly dissipated, corresponding to ohmic loss in a metal.

Pauling's[1] Theory of the Metallic Bond. This concept should not be regarded as entirely contradictory to the foregoing. Pauling amasses considerable evidence supporting the fact that the chemical bond in the metallic state consists of a multiplicity of *resonating covalent bonds.* According to this viewpoint, which is here adopted, an atom in a metallic lattice at one instant shares its electron with one of the neighboring atoms and at a later instant shares it with another neighboring atom. As mentioned previously, the energy associated with such resonance is negative and contributes a considerable degree of stability to the configuration. In fact the energy of a resonating bond in a metallic crystal is about twice that of a nonresonating bond.[2] This would lead us to expect that, of the various conceivable crystalline forms, that one would be most stable which afforded the highest degree of resonance, *i.e.*, that one in which any arbitrarily selected atom has the largest number of nearest neighbors. That such is actually the case will be demonstrated later.

This concept of Pauling's is also fruitful in interpreting the strength and ductility of metals. The resonance energy is the additional energy which must be supplied to break the bond and hence contributes to the cohesive strength. Also the malleability and ductility, characteristic of pure metals, which are associated with the ability to yield under shear stress without failure, are quite consistent with the ability of a bond thus to shift from one atom to the adjacent atom or from one pair of atoms to another pair.

According to this viewpoint the metallic bond is very closely related to the covalent shared electron-pair bond; each atom in a metallic lattice forms covalent bonds which resonate between the various possible positions. It will be noted that this theory is very similar to the electron-gas theory in interpreting the high electrical conductivity of metals. An electron which may be shared first by one pair of atoms, then another and another, etc., clearly has a high mobility as would an electron in the electron gas. This high mobility would not be possessed by an electron which is shared by two particular atoms or by an electron, such as those involved in the resonating bonds of benzene, which is shared by the particular group of atoms constituting a molecule.

[1] L. Pauling, The Nature of the Chemical Bond, *Phys. Rev.*, **54,** 899 (1938); *J. Am. Chem. Soc.*, **69,** 542 (1947); *J. chim. phys.*, **46,** 276 (1949); *Nature*, **161,** 1019 (1948). L. Pauling and A. M. Soldate, *Acta Crystal.*, **1,** 212 (1948). L. Pauling and F. J. Ewing, *Rev. Modern Phys.*, **20,** 112 (1948). L. Pauling, *Proc. Roy. Soc.* (*London*), **A196,** 343 (1949).

[2] The enthalpy change accompanying the reaction K(bcc) = K(g) is about twice that accompanying the reaction $\frac{1}{2}K_2(g) = K(g)$, thus illustrating that the breaking of a resonating bond requires about twice the energy necessary to break a nonresonating bond.

Typical nonmetallic covalent bonds of a given atom tend to be rather strongly directed, *i.e.*, to form certain angles one with the other. This will be discussed in more detail in the next section. No such tendency is noted in the case of metallic bonds. Although the pronounced similarity between covalent and metallic bonds may be regarded as reasonably well established by Pauling, it is obviously still desirable for purposes of classification to consider the metallic bond as a distinct type.

The Nature of the Metallic Bond. As just discussed, the most satisfactory picture of the metallic bond seems to be the resonating covalent bond as described by Pauling. The resonance contributes a great stabilizing influence, and in view of this, the question naturally arises as to the interpretation of the fact that such resonance occurs only in certain substances (particularly the metals) and not in others. Pauling also answers this question. He points out that, to have free resonance of the valence bonds among the interatomic positions, as in a metal, it is necessary for each of the atoms that receive a bond to have an unoccupied *orbital*[1] for the electron that jumps to it from a neighboring atom. Thus, metallic resonance requires an extra orbital, in general unused, for each atom.[2] A detailed analysis of the conditions for resonance by means of the wave function indicates that the energy is a minimum, not when every atom has an extra orbital, but when approximately three-quarters of the atoms have an extra orbital, the remaining atoms making use of all their orbitals for bond formation or for occupancy by unshared electron pairs. To quote Pauling, "The reason that the diatomic molecules of alkali metals condense to a metallic crystal whereas the diatomic molecules of hydrogen do not may be simply stated as resulting from the presence of extra orbitals in the valence shell of the alkali metal atoms and the absence of an extra orbital for the hydrogen atom."

It seems informative to follow Pauling by representing schematically a few examples. Possible formulations of the electronic configuration of a tin atom (in solid tin) are represented below.

	Valence	4*d*	5*s*	5*p*	Gray tin	White tin
Sn *A*	4	⇅ ⇅ ⇅ ⇅ ⇅	•	• • •	100%	25%
Sn *B*	2	⇅ ⇅ ⇅ ⇅ ⇅	⇅	• • ○	—	75%
Sn *C*	0	⇅ ⇅ ⇅ ⇅ ⇅	⇅	⇅ ○ ○	—	—
		Kr-like core				

[1] By orbital is meant a set of three quantum numbers.

[2] Other types of resonance are also possible but are of less significance.

The symbol ⇅ indicates a pair of atomic or nonbonding electrons, • a bonding electron, and ○ a metallic orbital, normally open and accommodating the resonance. The second formulation, Sn *B*, may be imagined as derived from the first, Sn *A*, by the dropping of one bonding electron from 5*p* to 5*s*, forming there an electron pair; this leaves a vacant orbital in Sn *B*. By a similar step Sn *C* may be imagined as derived from Sn *B*; it will be noted that the configuration Sn *C* corresponds to that given for an isolated Sn atom in Table 3-2. Gray tin with its diamond-type structure and absence of metallic properties is assumed to consist exclusively of tetravalent Sn *A*. White tin might at first be thought to consist exclusively of bivalent Sn *B* atoms, as Sn *B* possesses the two requirements for the metallic bond—bonding electrons and an extra orbital. In accord with the principle mentioned in the preceding paragraph, however, only three-quarters of the atoms need have an extra orbital—the energy of two resonating bonds being roughly the same as that of four nonresonating bonds. Hence it seems likely that white tin is composed of 75 per cent Sn *B* atoms and 25 per cent Sn *A* atoms, leading to a valence of $(\frac{3}{4})(2) + (\frac{1}{4})(4) = 2.5$. This valence seems in accord with the properties of white tin, in particular with the valence (2.44, given later in Table 3-4) derived from the interatomic distances with the aid of Eq. (3-1), as will be discussed later in this chapter. It seems worth noting at this point that Pauling in his later work considers lead in the pure state to have a valence of 2.0 rather than 2.5, all atoms being analogous to Sn *B* with none of the *A* type; this is in accord with the softer and more metallic nature of lead.

One more example will be given. As in the case of tin, possible electronic configurations for zinc are shown below.

	Valence	3*d*	4*s*	4*p*
Zn *A*	6	⇅ ⇅ ⇅ • •	•	• • •
Zn *B*	4	⇅ ⇅ ⇅ ⇅ •	•	• • ○
Zn *C*	2	⇅ ⇅ ⇅ ⇅ ⇅	•	• ○ ○
Zn *D*	0	⇅ ⇅ ⇅ ⇅ ⇅	⇅	○ ○ ○

Clearly, metallic zinc cannot consist exclusively of Zn *A* atoms, for these have no extra orbitals to accommodate resonance. Similarly Zn *D* is excluded, as it has no bonding electrons at all. It might be expected that metallic Zn consists principally of Zn *B* atoms, as this type has the largest number of bonding electrons consistent with an extra orbital to accommodate resonance. In fact Pauling now considers that metallic zinc consists nearly entirely of these *B* type atoms, corresponding to a valence of 4.

PLASTIC DEFORMATION OF METALS

The foregoing concept of the metallic bond affords a simple interpretation of the high ductility usually associated with pure metals. If a particular bond (electron pair) is stressed to a high-energy state, it can readily move to another position of lower energy; the availability of this alternative position is guaranteed by the open metallic orbitals which accommodate the resonance. In nonmetallic substances like sulfur or quartz a strained bond has no alternative position to which it can move; sufficient strain results in rupture without appreciable deformation, *i.e.*, brittle failure.

An older interpretation of the ductility of metals lies in the undirected nature of the metallic bond as contrasted to the directed nature of the common covalent bond. By a *directed bond* is meant one which has a very strong tendency to form a particular angle with the other bonds radiating from the same atom. The bonds joining the carbon atoms in diamond are directed at an angle of 109°28′; the two bonds of oxygen have a very strong tendency to form an angle of 105°. Crystals with directed bonds may be expected to possess an angular rigidity not possessed by metals in which the bonds do not have preferred angles.

In order to form a clearer picture it is first necessary to realize that ductility, malleability, softness, etc., are all properties very closely related to the ability to yield plastically[1] under a shear force. It has been observed that the shear force required to produce plastic flow in single crystals of very pure metals is low indeed, so low that it is inconceivable that the mechanism involved can be represented by the temptingly simple picture of one plane of atoms sliding as a whole over the adjacent plane. The force required to produce this latter phenomenon is calculable from elastic constants and is found to exceed by a large factor the observed force necessary to produce plastic deformation. In spite of this disparity there is overwhelming experimental evidence that the most common type of plastic deformation, known as ***slip***, produces the same over-all result as though one plane of atoms slipped over the adjacent plane.[2] Figure 3-2 shows schematically the appearance of a crystal before and after the application of a tensile stress. The change in direction of the slip planes is apparent.

The discrepancy between the observed and calculated force, mentioned

[1] If sufficiently large (nonhydrostatic) forces are applied to a substance, it remains partly deformed after the forces are released; this deformation is known as *plastic deformation.*

[2] In addition to the type of plastic flow known as slip and interpreted here in terms of dislocations, there are also two other known types of plastic deformation; one of these is known as twinning and the other involves a sliding at the grain boundaries.

above, has led to the hypothesis of *dislocations*, a subject upon which there is now a considerable literature and which is discussed in detail in other books of this series.[1] A dislocation is an extended lattice imperfection; one type is shown in Fig. 3-3, which illustrates in section the deformation of a body under a shearing stress by the movement of a dislocation. It is seen that such a process involves a wavelike motion somewhat similar to that of a caterpillar in motion or that of wave induced in a rope lying on the ground by snapping one end. The similarity of interest in these cases lies in the fact that a displacement of the whole is produced by the over-all process and yet only a small fraction of the whole is in motion at any instant. The analogy between the motion of the dislocation and the wave in the rope is still closer, in that appreciable work is required for the initial displacement but much less (per step) is required to propagate the dislocation once it is started. Thus the theory offers an interpretation of the previously mentioned small shear force required for plastic deformation of a single pure crystal.

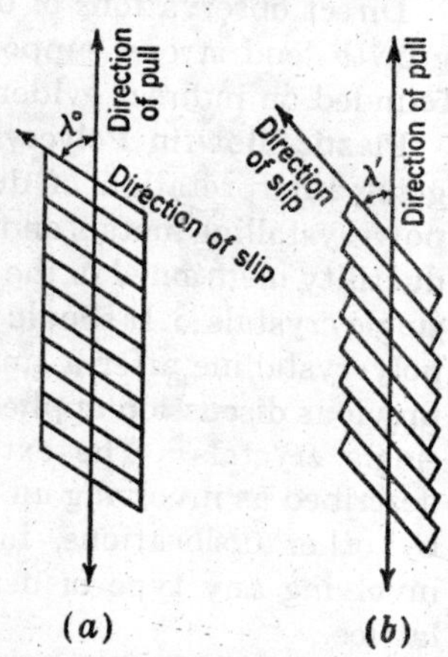

Fig. 3-2. Schematic diagram illustrating the orientation of slip planes in a crystal (*a*) before and (*b*) after straining in tension. (*From F. Seitz, "Physics of Metals," McGraw-Hill Book Company, Inc., New York, 1943.*)

It should be noted from the figure, however, that the production of the dislocation requires, in addition to the shortening and lengthening of

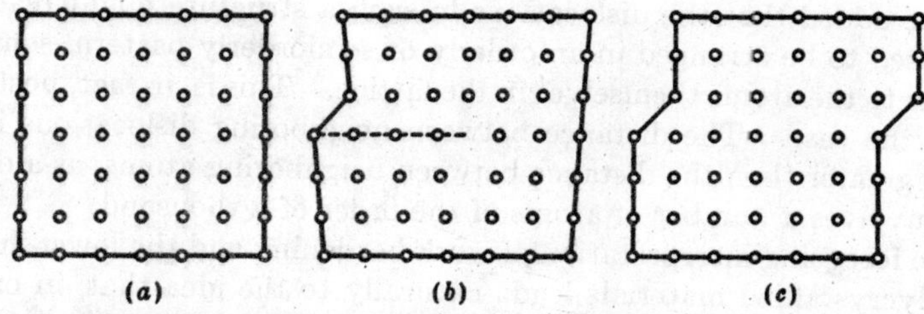

Fig. 3-3. Drawings illustrating the deformation of a crystal by the movement of a dislocation. (*From Seitz.*)

bonds, the bending of bond angles. This is in accord with the conclusion arrived at intuitively at the beginning of this section, namely, that such

[1] F. Seitz, "The Physics of Metals," McGraw-Hill Book Company, Inc., New York, 1943. C. S. Barrett, "Structure of Metals," McGraw-Hill Book Company, Inc., New York, 1943.

dislocations may be formed more readily in a metallic lattice where bonds are undirected, thus accounting for the ductility of a metal.

Direct observations of dislocations and their important role in crystal growth lend strong support to dislocation theory, which was originally founded on indirect evidence.

Plastic Flow in Polycrystalline Metals; Work Hardening. The foregoing interpretation of ductility needs some extension to apply to the polycrystalline metals and alloys usually encountered, since the very high ductility mentioned in the previous section is found only in highly purified single crystals. It should be noted also that even a single crystal becomes polycrystalline after a small amount of plastic deformation, so that the previous discussion applies only to the very first stages of deformation of single crystals. The extension of the theory of dislocations may be described as involving an interference with the movement of dislocations by other dislocations, inclusions, grain boundaries or by any means involving any type of departure from the geometrically perfect crystal lattice.

A dislocation cannot move so readily across grain boundaries. Thus we may picture that, when a polycrystalline material is strained, the dislocations initially present, or first formed, become stuck at grain boundaries and that subsequent dislocations are stuck as they approach these first immobilized ones. With continued cold work, the number of stuck dislocations increases until eventually the whole material is filled with them. This construct naturally involves a structure with a maximum amount of cold work; *i.e.*, as far as the structure of the metal is concerned, the process of cold working cannot be indefinitely continued. It might also be added that the dislocations in such a structure might readily be supposed to be arranged in an orderly or semiorderly pattern, somewhat similar to the atoms themselves in the lattice. This is, in fact, postulated to be the case. The distance between neighboring dislocations is very much greater than the distance between neighboring atoms, as a dislocation involves a number of atoms of the order of a thousand.

The foregoing interpretation of work hardening and the lower ductility of polycrystalline materials leads naturally to the idea that, in order to deform a material which is already filled with dislocations, a very much larger force is required, since a large number of interfering dislocations must be moved at the same time. The greatly enhanced resistance to shear thus acquired may result in ultimate failure of the metal by a mechanism other than shear, in case the metal is strained to the point of failure. Failure by a pure shear mechanism may be illustrated by that of a very viscous liquid (which has a very small resistance to shear as compared to a solid); *e.g.*, when taffy is pulled, it becomes attenuated to an ever thinner filament which eventually breaks, but only when very thin. Another type of failure is the so-called brittle type, where relatively little plastic flow is involved. This may be pictured approximately as involving a direct breaking of chemical bonds and is called failure by cleavage. It should be noted, however, that, in most actual failures which are so classified, the energy required (work done) to produce the failure is much larger than the energy of the chemical bonds broken. It seems reasonable to suppose that in such cases some energy is first supplied to produce the dislocations associated with work hardening; the material thus becomes sufficiently resistant to plastic deformation that it fails by cleavage.

Among the other factors which determine the nature of the failure is the nature of the force applied to the material, which in turn depends, in the case of a test specimen, upon the test employed. If the test involves a large shear force and small tensile force the specimen may fail in shear; if, on the other hand, the test involves a large tensile force and small shear force, a specimen of the same material may fail by cleavage. Of the types of test commonly employed, let us consider the tensile test, the torsion test, and the impact test. The nature of the forces involved in these tests is shown schematically in Fig. 3-4. An analysis of the ratio of the greatest normal (cleavage) stress to the greatest shear stress in these three cases discloses that this ratio is 2:1 for the tensile test, 1:1 for the torsion test, and considerably greater than 2:1 (but somewhat indefinite) for the impact test.[1] Temperature is also an important factor in determining the nature of the failure. The force required to produce cleavage is much less sensitive to temperature than that required to produce plastic deformation, the latter being greater the lower the temperature. Hence the tendency to fail by cleavage is greater the lower the temperature.

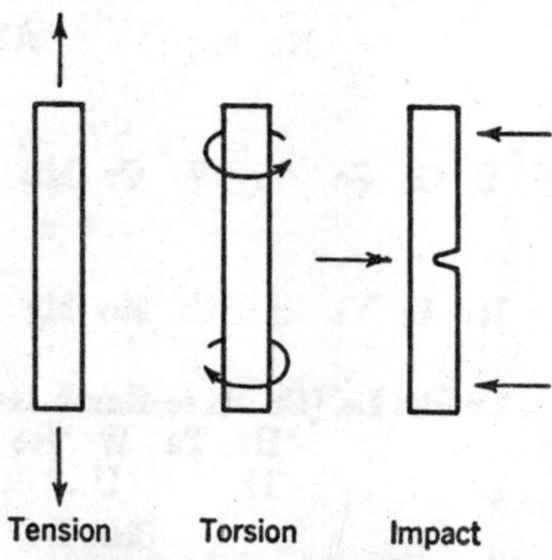

Fig. 3-4. Schematic diagram of forces in the tension, torsion, and impact tests.

HUME-ROTHERY'S CLASSIFICATION OF THE ELEMENTS

Hume-Rothery has divided the elements, except the rare gases, into three classes as shown in Fig. 3-5. Each class corresponds roughly to one of the traditional classifications as follows: class I, metals; class II, intermediate elements; class III, nonmetals.

[1] K. Heindlhofer, *Trans. AIME*, **116**, 232 (1935).

Class I Elements. The elements of this class constitute about two-thirds of the known elements. With the exception of one form of manganese and one form of wolfram, every class I element crystallizes in one (or more) of the three following lattice structures: (1) face-centered cubic (fcc), (2) hexagonal close-packed (hcp), (3) body-centered cubic (bcc). These three types are shown in Fig. 3-6. A *unit cell* of a crystal is defined

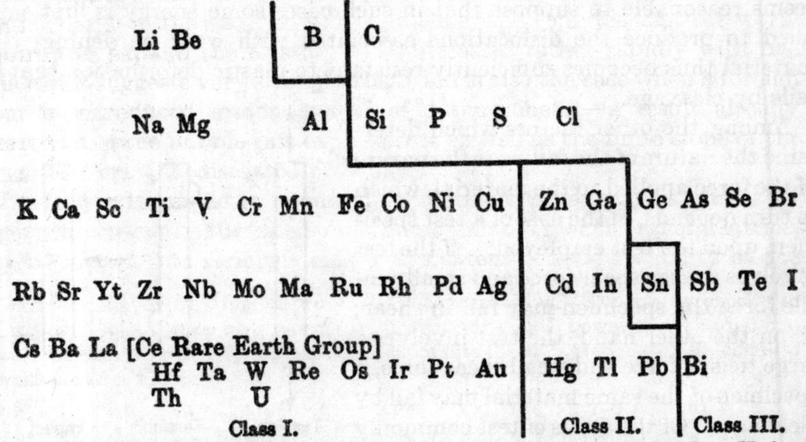

FIG. 3-5. Hume-Rothery's classification of the elements: class I, metals; class II, intermediate elements; class III, nonmetals. (*From W. Hume-Rothery, "Structure of Metals and Alloys," Institute of Metals, London,* 1944.)

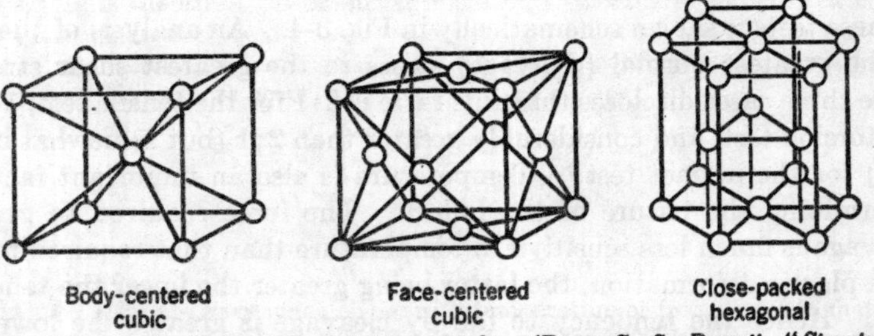

FIG. 3-6. The principal structures of the metals. (*From C. S. Barrett, "Structure of Metals," McGraw-Hill Book Company, Inc., New York,* 1952.)

as the smallest unit by reproduction of which an entire lattice may be conceived as generated. It is particularly worthy of note that the first two types of crystal lattice correspond to close-packed arrangements; the face-centered cubic structure is sometimes called close-packed cubic.

It is a well-known geometrical fact that the close-packed arrangement of circles in a plane (or of marbles in a tray) is hexagonal, *i.e.*, the arrangement in which each circle is tangent to 6 others. The corresponding close-packed arrangement of spheres in three dimensions is that in which each sphere is tangent to 12 others. Of the various ways of visualizing this the easiest is to consider 6 of the 12 coplanar with the central sphere, and two sets of three, one in a plane above and the other below. There are two different arrangements which satisfy this condition. In one, each of the spheres in the upper set of three is directly above one in the lower set of three; the other arrangement may be obtained from the first by rotating one set of three through an angle of 60°, it being noted that there are six possible positions for the three spheres. These two possible arrangements correspond to the *hexagonal close-packed* and *face-centered cubic* structures, respectively.

The reader who is not thoroughly familiar with these facts will find it well worth while to convince himself of the foregoing geometry by securing 13 marbles or balls and actually arranging them in the manner described.

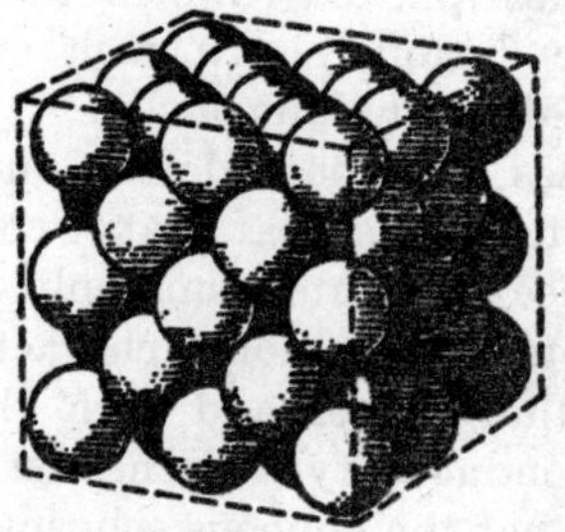

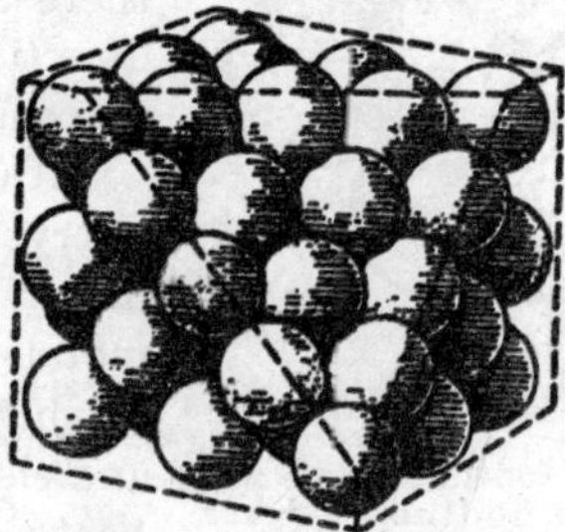

FIG. 3-7. Illustration of the close-packed arrangement of spheres in fcc array. (*After Barlow.*)

The close-packed arrangement in fcc crystals is illustrated in the exposed plane resulting from the removal of a corner of the cube as in Fig. 3-7. In such cases where each atom has 12 nearest neighbors, the *coordination number* is said to be 12 (designated CN12). In general the coordination number of an atom is the number of nearest neighbor atoms.

In the hcp arrangement of spheres the ratio of lattice parameters c/a may be shown geometrically to be equal to $\sqrt{\frac{8}{3}} = 1.633$. X-ray measurements of several crystalline elements in this system give a value which is in excess of this. For example, for zinc c/a is 1.856. In such a case the atoms might be conceived as ellipsoids rather than spheres; an interpretation of this ratio is given later. In the cubic systems this deformation is rare. The concept of atoms as hard spheres, implied in this whole discussion, is, of course, a gross simplification, but nevertheless the picture has been fruitful; *e.g.*, it is compatible with the observed low compressibility of solids. This construct of an atom with a rather definite radius equal to one-half the interatomic distance will serve as a basis for the understanding of the nature of condensed phases (solids and liquids).

In both the hcp and cubic lattices, close-packed layers of atoms occur in four planes. For example, in the fcc structure these are disclosed, as mentioned previously, by removal of corner atoms as shown in Fig. 3-7. This process may be repeated on each of the eight corners; however, the planes formed on diagonally opposite corners are parallel and thus duplicate each other. The resulting solid is an octahedron, and the close-packed planes are known as *octahedral planes*. The Miller indices[1] of these planes are (111). Each such plane contains three sets of close-packed rows of atoms, the total number of directions of these (for the four planes) being six. In plastic deformation slip occurs parallel to the close-packed planes and in the direction of the close-packed rows. In the more open structure of the bcc lattice there are no close-packed rows in the (111) direction (cube diagonal). In such crystals there is no well-defined plane of slip.

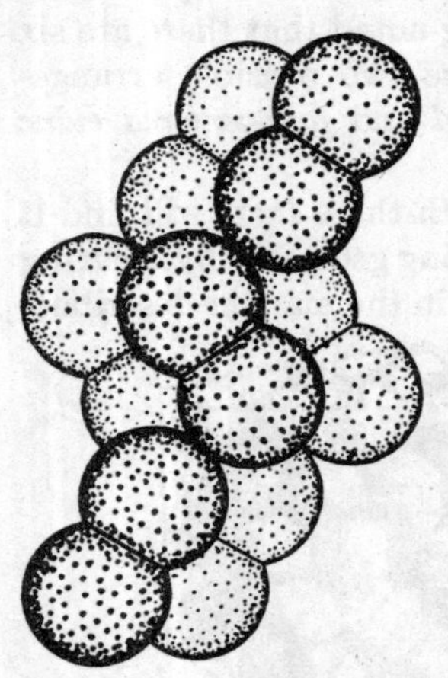

FIG. 3-8. Model of the crystal lattice of iodine. (*After Wyckoff, from Barrett.*)

Although class I has been defined in terms of crystal structure, it is clear that it contains in general those elements commonly called metals. A metal is commonly characterized by high (metallic) luster and high thermal and electrical conductivity; the chemist would hasten to add that the aqueous solution of its oxide is alkaline.

Class III Elements. The elements of the third class (Fig. 3-5) have been shown by Hume-Rothery to crystallize in such a fashion that the coordination number is equal to $(8 - N)$, N being the group number in the periodic table; they are therefore said to obey the "$(8 - N)$ rule." For example, the structure of the crystalline halogens (group VII*B*) may perhaps best be understood by imagining that the atoms are arranged in a line as in Fig. 3-8 for iodine, where it will be noted that each atom has $(8 - 7)$ or 1 nearest neighbor. This string is then twisted and intertwined into a lattice model. The elements of the VI*B* group (selenium and tellurium) crystallize in a somewhat similar manner except that the atoms are equally spaced along a helix so that the coordination number is $(8 - 6)$ or 2. The arrangement is shown in Fig. 3-9. In group V*B* the crystalline arrangement is such that each atom has three $(8 - 5)$ close neighbors and three more at a somewhat greater distance. The well-

[1] The *Miller indices* of a crystallographic plane are the reciprocals of the intercepts of this plane on the coordinate axes. The unit in which the intercepts are expressed is the lattice parameter.

known diamond structure (Fig. 3-10) is characteristic of group IV*B*—carbon (diamond), silicon, germanium, and gray tin. Tin exhibits polymorphism, also crystallizing in the tetragonal form (white tin), and hence is shown in Fig. 3-5 as belonging both to classes III and II. The obvious explanation of the type of crystal structure exhibited by the elements of class III is that the bonds are of the covalent type, and just as these elements form covalent molecules in the gas phase by sharing $(8 - N)$ electrons, so they crystallize by sharing $(8 - N)$ electrons with $(8 - N)$ nearest neighbors.

Class II Elements. Class II may be defined as comprising those elements whose crystal structures fall in neither class I nor III. Zinc and cadmium crystallize in a modification of an hcp arrangement in which the

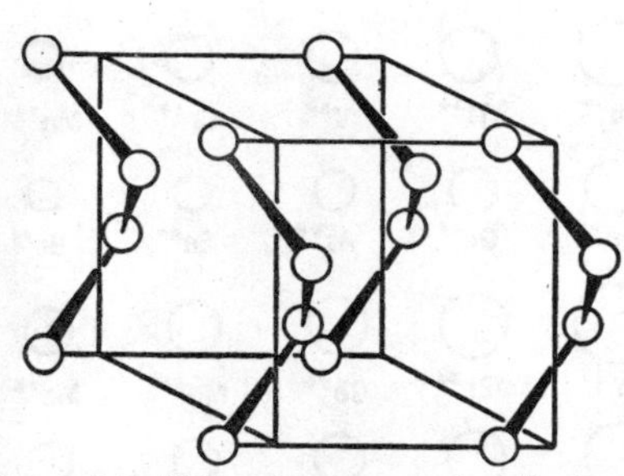

FIG. 3-9. The structure of tellurium. Atoms are arranged in spiral chains in which each atom has two nearest neighbors. (*From Barrett.*)

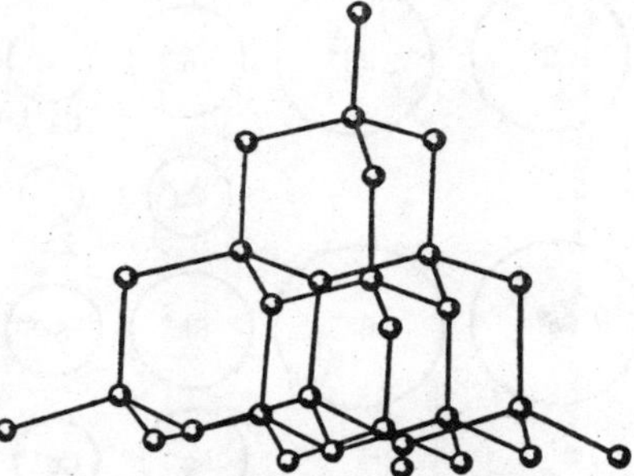

FIG. 3-10. The cubic structure of diamond. Each atom has four nearest neighbors occupying positions corresponding to the corners of a tetrahedron. (*From Barrett.*)

axial ratio is 1.856 and 1.886, respectively, instead of 1.633. Mercury crystallizes in a rhombohedral structure which may be regarded as a distorted simple cubic lattice; each atom has six close neighbors as might be anticipated from application of the $(8 - N)$ rule. The structure of indium is face-centered tetragonal, a structure whose unit cell is similar to a face-centered cube that has been slightly elongated along one of the major axes. Thallium and lead are included in this class by Hume-Rothery in spite of the fact that they crystallize in typical metallic structures; he considers that their rather large interatomic distance in conjunction with other evidence is sufficient to relegate them to class II. It will be noted that, had lead and thallium been included in class I, a rather awkward division of this class in the periodic chart would result.

ATOMIC RADIUS OF THE ELEMENTS

We have mentioned our adoption of the convention of atoms as hard spheres. Now the time has come to inspect this construct a little more closely. Several elements crystallize in more than one structure, *i.e.*, exhibit polymorphism. For example, iron crystallizes in the bcc lattice, the stable form of pure iron up to 910°C (and from 1400°C to the melting point), and also in the fcc lattice. This latter form, although stable from 910 to 1400°C for pure iron, is found at room temperature for several alloys, and the lattice constant at room temperature can be found by extrapolation. The relation between lattice parameter (length of the

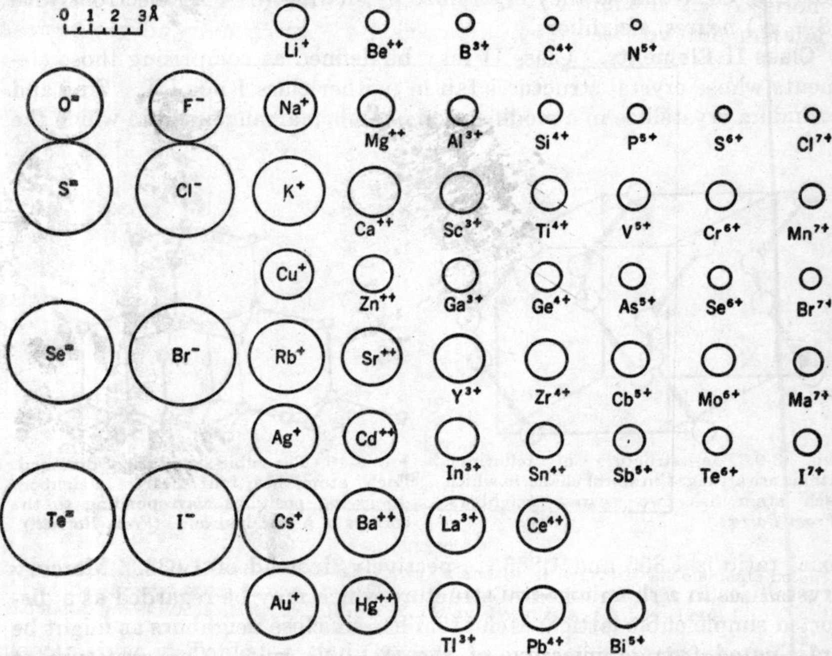

FIG. 3-11. The crystal radii of ions. (*From L. Pauling, "The Nature of the Chemical Bond," Cornell University Press, Ithaca, N.Y.*, 1940.)

unit cube) and the distance of closest approach of the atoms (equal to twice the radius of the hypothetical tangent atom spheres) is readily calculable from the geometry of the arrangement. For a body-centered cube this distance of closest approach is $a\sqrt{3}/2$, for a face-centered cube it is $a\sqrt{2}/2$, and for the close-packed hexagonal structure it is a.* The lattice constants determined by X-ray measurement for body-centered and face-centered iron are 2.86_1 and 3.56_4 A.† The interatomic distances

* For hcp structure with c/a unequal to 1.633 each atom has six (equatorial) neighbors at a distance a and six (in adjacent planes) at a distance $\sqrt{(a^2/3) + (c^2/4)}$.

† All atomic dimensions dependent upon X-ray measurement are here indicated in angstroms (A), as is customary. Actually the figures given are in kx units (the X-ray unit of length), the conversion being 1 kx = 1.00202 A.

calculated from these by use of the above relations are 2.47_8 and 2.52_0 A. It will be noted that there is a sort of compensation, in the sense that in the non-close-packed structure the atoms behave as though they were slightly smaller, thus giving rise to a smaller difference in density than would be observed if the interatomic distances were identical in the two cases.

Such direct comparison may be made for a number of elements. For others a comparison may be made by extrapolation of interatomic distances for alloys to obtain the interatomic distance for unstable crystalline forms. By considering numerous cases of this sort, Goldschmidt[1] found for a given metallic element that in general the interatomic distance is about 3 per cent less when the coordination number is 8 than when it is 12, 4 per cent less when 6 than when 12, and 12 per cent less when 4 than when 12. Twice the radius for CN 12 determined directly or in this manner is called the *Goldschmidt atomic diameter*.

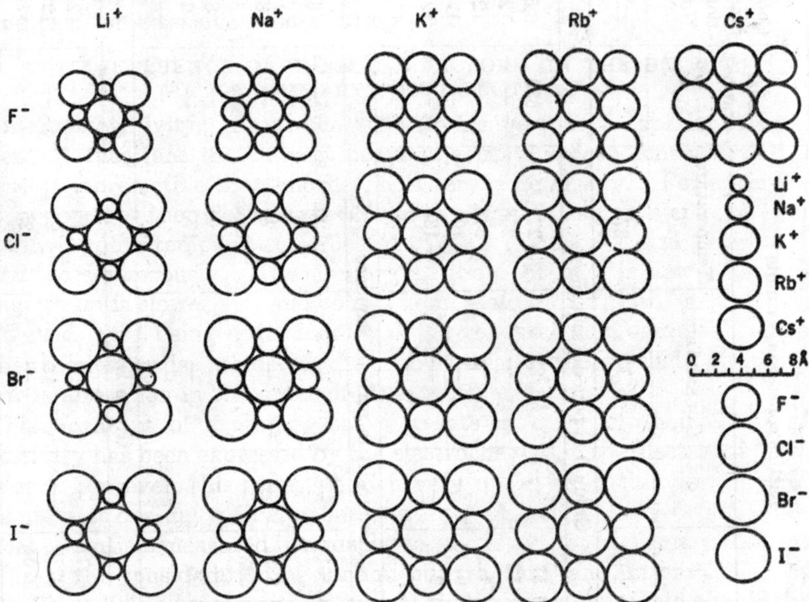

FIG. 3-12. The arrangement of ions in alkali-halide crystals with sodium chloride structure. (*From Pauling.*)

As already mentioned, the concept of the atoms as hard spheres, although crude, is found to be very useful; this will be demonstrated further in the next chapter. However, an atom behaves as though it had a fixed radius only under conditions which are similar; thus the metallic

[1] B. M. Goldschmidt, *Z. physik. Chem.*, **133**, 397 (1928).

TABLE 3-3. CRYSTALLOGRAPHY OF THE ELEMENTS*

Atomic No. and element	Crystal structure			Lattice constants, kx units				Axial ratio b/a, c/a	Coordination No.	Interatomic distances		Goldschmidt atomic diameter for CN12
	Allotrope	Type†	Space group	a	b	C or axial angle	Temperature, °C			d_1	d_2	
1 Hydrogen‡	—	⬡	?	3.75	—	6.12	−271	—	—	—	—	—
Group 0 in periodic sequence												
2 Helium‡	—	⬡	D_{6h}^4	3.57	—	5.83	−271.55	1.633	6,6	3.57	3.57	—
10 Neon‡	—	⊡	O_h^5	4.52	—	—	−268	—	12	3.19_6	—	—
18 Argon	—	⊡	O_h^5	5.4_2	—	—	−233	—	12	3.8_3	—	—
36 Krypton	—	⊡	O_h^5	5.6_8	—	—	−191	—	12	4.0_2	—	—
54 Xenon	—	⊡	O_h^5	6.2_4	—	—	−185	—	12	4.4_1	—	—
86 Radon	—	—	—	—	—	—	—	—	—	—	—	—
Group IA in periodic sequence												
3 Lithium‡	—	◫	O_h^9 or T_d^6§	3.501_7	—	—	20	—	8	3.032_6	—	3.13
11 Sodium	—	◫	O_h^9	4.282_0	—	—	20	—	8	3.708	—	3.83
19 Potassium	—	◫	O_h^9	5.333	—	—	Room	—	8	4.618	—	4.76
37 Rubidium	—	◫	O_h^9	5.62	—	—	−173	—	8	4.87	—	5.02
55 Cesium	—	◫	O_h^9	6.05	—	—	−173	—	8	5.24	—	5.40
87 Francium	—	—	—	—	—	—	—	—	—	—	—	—
Group IB												
29 Copper	—	⊡	O_h^5	3.607_4	—	—	18	—	12	2.55_1	—	2.551
47 Silver	—	⊡	O_h^5	4.0779	—	—	25	—	12	2.8835	—	2.883
79 Gold	—	⊡	O_h^5	4.0704	—	—	25	—	12	2.8782	—	2.878

Group II*A* in periodic sequence

4	Beryllium.....	$\alpha\|$	⬡	D_{6h}^4	2.2810_5	—	3.5771_4	18	1.5682_0	6,6	2.2211	2.2811	2.25
			⬡	—	7.1	—	10.8	Room	1.52	—	—	—	—
12	Magnesium...	—	⬡	D_{6h}^4	3.2030	—	5.2002	25	1.62354	6,6	3.1906	3.2030	3.20
20	Calcium......	$\alpha\|$	⊡	O_h^5	5.56	—	—	20	—	12	3.93	—	3.93
		β	—	—	—	—	—	Between 300–450	—	—	—	—	—
		γ	⬡	D_{6h}^4	3.98	—	6.52	450	1.638	6,6	3.98	3.99	3.93
38	Strontium.....	—	⊡	O_h^5	6.07_5	—	—	Room	—	12	4.29_6	—	4.29_6
56	Barium.......	—	⊡	O_h^9	5.01_5	—	—	Room	—	8	4.34_3	—	4.48
88	Radium.......	—	—	—	—	—	—	—	—	—	—	—	—

Group II*B*

30	Zinc.........	—	⬡	D_{6h}^4	2.6595	—	4.9368_5	25	1.8563_1	6,6	2.6595	2.9070	2.748
48	Cadmium......	—	⬡	D_{6h}^4	2.9731	—	5.6069_4	25	1.8858_8	6,6	2.9731	3.2872	3.042
80	Mercury......	—	▱	D_{3d}^5	2.999	—	$\alpha = 70°31.7'$	−46	—	6	2.999	—	3.10

Group III*A* in periodic sequence

5	Boron‡.......	—	▯	—	17.86	8.93	10.13	Room	—	—	—	—	—
		—	△	—	8.93	—	5.06	Room	—	—	—	—	—
13	Aluminum....	—	⊡	O_h^5	4.0414	—	—	25	—	12	2.8577	—	2.80 to 2.85_8
21	Scandium.....	$\alpha\|$	⊡	O_h^5	4.532	—	—	Room	—	12	3.2046	—	3.20
		β	⬡	D_{6h}^4	3.30	—	5.23	Room	1.585	6,6	3.23	3.30	3.27
39	Yttrium......	—	⬡	D_{6h}^4	3.66_3	—	5.81_4	Room	1.588	6,6	3.59_5	3.66_3	3.62_9
57	Lanthanum...	$\alpha\|$	⬡	D_{6h}^4	3.75_4	—	6.06_3	Room	1.613	6,6	3.72_7	3.75_4	3.741
		β	⊡	O_h^5	5.29_6	—	—	Room	—	12	3.745	—	3.745
89	Actinium.....	—	—	—	—	—	—	—	—	—	—	—	—

TABLE 3-3. CRYSTALLOGRAPHY OF THE ELEMENTS.*—(*Continued*)

Atomic No. and element	Crystal structure			Lattice constants, kx units				Axial ratio	Coordination No.	Interatomic distances		Goldschmidt atomic diameter for CN12
	Allotrope	Type†	Space group	a	b	C or axial angle	Temperature, °C	b/a, c/a		d_1	d_2	
Group III*B*												
31 Gallium	—	▯	V_h^{18}	4.5167	4.5107	7.6448	Room	b/a = 1:0.99868 c/a = 1:1.69257	—	2.43_7§	2.70_6§	2.7
49 Indium	—	△	D_{4h}^{17}	4.58_5	—	4.94_1	22	1.078	4,8	3.24	3.37	3.14
81 Thallium	α‖	⬡	D_{6h}^4	3.4496	—	5.5137	18	1.5984	6,6	3.4010	3.4496	3.42
	β	⊞	O_h^9	3.874	—	—	262	—	8	3.3550	—	—
Group IV*A* in periodic sequence												
22 Titanium	α‖	⬡	D_{6h}^4	2.95_3	—	4.72_9	Room	1.601	6,6	2.91_5	2.95_3	2.93
	β	⊞	O_h^9	3.32	—	—	900	—	8	2.87_5	—	—
40 Zirconium	α‖	⬡	D_{6h}^4	3.22_3	—	5.12_3	Room	1.589	6,6	3.16_6	3.22_3	3.19
	β	⊞	O_h^9	3.61	—	—	867	—	8	3.12_6	—	—
72 Hafnium	—	⬡	O_{6h}^4	3.20_0	—	5.07_7	Room	1.587	6,6	3.13_9	3.20_0	3.17
Group IV*B*												
6 Carbon, diamond	—	◇	O_h^7	3.5597	—	—	19	—	4	1.5414	—	—
Graphite	—	○	D_{6h}^4	2.4564	—	6.6906	19	2.7237	—	1.42	—	—
Graphite‡	—	○	D_{3d}^5	2.456	—	10.044	19	4.251	—	—	—	—
		▱	D_{3d}^5	3.635	—	α = 39.49°	—	—	—	—	—	—
14 Silicon	—	◇	O_h^7	5.417_3	—	—	20	—	4	2.3458	—	—

32	Germanium . . .	—	◇	O_h^7	5.64_7	—	—	Room	—	4	2.44_5	—	2.788
50	Tin, gray	α	◇	O_h^7	6.46	—	—	18	—	—	2.797	—	3.164
	White	β‖	△	D_{4h}^{19}	5.8194	—	3.1753	20	0.5456	4,2	3.0161	3.1753	—
82	Lead	—	⊡	O_h^5	4.9395	—	—	20	—	12	3.4927	—	3.49
	Group V*A* in periodic sequence												
23	Vanadium	—	◫	O_h^9	3.0338	—	—	25	—	8	2.6273	—	2.71
41	Niobium	—	◫	O_h^9	3.2941	—	—	20	—	8	2.8528	—	2.94
73	Tantalum	—	◫	O_h^9	3.2959	—	—	20	—	8	2.8543	—	2.95
	Group V*B*												
7	Nitrogen‡	α	□	T^4	5.66	—	—	−252	—	—	1.06	—	—
		β	○	?	4.03	—	6.59	−234	1.63_3	—	—	—	—
15	Phosphorus, black	—	□	V_h^{18}	3.31	4.38	10.50	Room	b/a = 1:1.32 c/a = 1:3.17	3	2.18	—	—
	yellow	‖	□	?	7.17	—	—	35	—	—	—	—	—
33	Arsenic	—	▱	D_{3d}^5	4.151	—	α = 53°49′	Room	—	3,3	2.508_6	3.146_5	—
51	Antimony	—	▱	D_{3d}^5	4.49762	—	α = 57°6′27″	25	—	3,3	2.898	3.357	3.228
83	Bismuth	—	▱	D_{3d}^5	4.7364	—	α = 57°14′13″	25	—	3,3	3.1050	3.4740	3.64
	Group VI*A* in periodic sequence												
24	Chromium	α‖	◫	O_h^9	2.878_6	—	—	17	—	8	2.492_9	—	2.57
		β	⬡	D_{6h}^4	2.71_7	—	4.41_8	Room	1.626	6,6	2.709	2.717	2.71
		γ	□	T_d^3	8.717	—	—	Room	—	—	1.04_3	2.04_3	—
42	Molybdenum‡.	—	◫	O_h^9	3.140_3	—	—	17	—	8	2.719_6	—	2.80
74	Wolfram¶	α‖	◫	O_h^9	3.1583	—	—	25	—	8	2.735	—	2.82
		β	□	O_h^3	5.038	—	—	20	—	—	2.519	2.816	—

TABLE 3-3. CRYSTALLOGRAPHY OF THE ELEMENTS.*—(*Continued*)

Atomic No. and element	Crystal structure			Lattice constants, kx units				Axial ratio b/a, c/a	Coordination No.	Interatomic distances		Goldschmidt atomic diameter for CN12
	Allotrope	Type†	Space group	a	b	C or axial angle	Temperature, °C			d_1	d_2	
Group VIB												
8 Oxygen‡	α	▯	?	5.50	3.82	3.44	−252	b/a = 0.694_1 c/a = 0.625	—	—	—	—
	β	▱	C_{3i}^2, C_{3v}^5 or D_3^7	6.19	—	$\alpha = 99.1°$	−238	—	—	—	—	—
	γ	□	Γ_h^6	6.83	—	—	−225	—	—	—	—	—
16 Sulfur, yellow‡	α‖	▯	V_h^{24}	10.48	12.92	24.55	Room	b/a = 1.233 c/a = 2.343	—	2.12	—	—
	β	▱	C_{2h}^5?	10.90	10.96	11.02 $\beta = 83°16'$	103	b/a = 1.101 c/a = 1.006	—	—	—	—
34 Selenium‡	‖	⬡	D_3^4 or D_3^6	4.3552	—	4.9494	20	1.136	2,4	2.32	3.46	—
	α	▱	C_{2h}^2	11.50	8.98	8.97_7 $\beta = 90°57'$	Room	—	—	—	—	—
	β	▱	C_{2h}^5	12.74	8.04	9.25, $\beta = 93°4'$	Room	—	—	—	—	—
52 Tellurium‡	—	⬡	D_3^4 or D_3^6	4.4469	—	5.9149	20	1.330	2,4	2.86	3.46	—
84 Polonium‡	—	▱	C_2^3	7.42	4.29	14.10, $\beta = 92°$	Room	—	—	3.4	—	—

					Group VIIA in periodic sequence								
25	Manganese....	α‖	□ **	T_d^3	8.894	—	—	Room	—	—	2.24	2.38§	—
		β	□ ††	O_7	6.29	—	—	Room	—	—	2.365	2.530§	—
		γ	⟁	D_{4h}^{17}	3.774	—	3.526	Room	0.934	8,4	2.582	2.669	—
43	Technetium‡..	—	—	D_{6h}^4	2.735	—	4.388	Room	1.604	6,6	2.698	2.735	—
75	Rhenium	—	⬡	D_{6h}^4	2.755_3	—	4.449_3	20	1.614_8	6,6	2.734_9	2.755_3	2.75
					Group VIIB								
9	Fluorine......	—	—	—	—	—	—	—	—	—	—	—	—
17	Chlorine‡.....	—	△	D_{4h}^{16}	8.56	—	6.12	−185	0.715	—	1.81	—	—
35	Bromine‡.....	—	▯	V_h^{18}	4.48	6.67	8.72	−150	—	—	2.27	—	—
53	Iodine‡.......	—	▯	V_h^{18}	4.7761	7.2501	9.7711	18	—	—	2.70	—	—
85	Astatine......	—	—	—	—	—	—	—	—	—	—	—	—
					Group VIII in periodic sequence								
26	Iron‡.........	α‖	⊞	O_h^9	2.8607	—	—	25	—	—	2.4775	—	—
		γ	⊡	O_h^5	3.56_4	—	—	Room	—	—	2.52_0	—	2.52
27	Cobalt........	α‖	⬡	D_{6h}^4	2.50_7	—	4.07_2	Room	1.624	6,6	2.49_9	2.507	2.50
		β	⊡	O_h^5	3.54_5	—	—	Room	—	—	2.507	—	2.507
28	Nickel‡.......	α	⬡	D_{6h}^4	2.49	—	4.08	—	1.64	6,6	2.49	2.49	2.49
		β‖	⊡	O_h^5	3.5168	—	—	25	—	12	2.4868	—	2.487
44	Ruthenium....	—	⬡	D_{6h}^4	2.6984	—	4.2730_5	20	1.5835	6,6	2.6442	2.6984	2.67
45	Rhodium......	α	□	O_h^3 or O_h^4	9.211	—	—	Room	—	—	—	—	—
		β‖	⊡	O_h^5	3.795_5	—	—	18	—	12	2.683_8	—	2.684
46	Palladium.....	—	⊡	O_h^5	3.882_4	—	—	20	—	12	2.745_3	—	2.745
76	Osmium.......	—	⬡	D_{6h}^4	2.7298	—	4.3104	20	1.5790	6,6	2.6700	2.7298	2.70
77	Iridium.......	—	⊡	O_h^5	3.831_2	—	—	18	—	12	2.709_1	—	2.709
78	Platinum.....	—	⊡	O_h^5	3.915_8	—	--	20	—	12	2.768_9	—	2.769

TABLE 3-3. CRYSTALLOGRAPHY OF THE ELEMENTS.*—(*Continued*)

Atomic No. and element	Crystal structure			Lattice constants, kx units				Axial ratio b/a, c/a	Coordination No.	Interatomic distances		Goldschmidt atomic diameter for CN12
	Allotrope	Type†	Space group	a	b	C or axial angle	Temperature, °C			d_1	d_2	
Rare-earth group												
58 Cerium.......	α	⬡	D^4_{6h}	3.65	—	5.91	Room	1.62	6,6	3.63	3.65	3.64
	β‖	⊡	O^5_h	5.14_3	—	—	Room	—	12	3.63_7	—	3.63_7
59 Praseodymium.	α‖	⬡	D^4_{6h}	3.662	—	5.908	Room	1.613	6,6	3.633	3.662	3.65
	β	⊡	O^5_h	5.151	—	—	Room	—	12	3.642	—	3.64
60 Neodymium...	—	⬡	D^4_{6h}	3.650	—	5.890	Room	1.613	6,6	3.621	3.650	3.63
61 Promethium...	—	—	—	—	—	—	—	—	—	—	—	—
62 Samarium.....	—	—	—	—	—	—	—	—	—	—	—	—
63 Europium.....	—	⊞	O^9_h	4.573	—	—	Room	—	8	3.960	—	4.08
64 Gadolinium...	—	⬡	D^4_{6h}	3.622	—	5.748	Room	1.587	6,6	3.554	3.622	3.59
65 Terbium......	—	⬡	D^4_{6h}	3.585	—	5.664	Room	1.580	6,6	3.508	3.585	3.54
66 Dysprosium...	—	⬡	D^4_{6h}	3.578	—	5.648	Room	1.579	6,6	3.499	3.578	3.54
67 Holmium......	—	⬡	D^4_{6h}	3.557	—	5.620	Room	1.580	6,6	3.480	3.557	3.52
68 Erbium.......	—	⬡	D^4_{6h}	3.532	—	5.589	Room	1.582	6,6	3.459	3.532	3.50
69 Thulium......	—	⬡	D^4_{6h}	3.523	—	5.564	Room	1.580	6,6	3.446	3.523	3.48
70 Ytterbium....	—	⊡	O^5_h	5.468	—	—	Room	—	12	3.866	—	3.87
71 Lutecium.....	—	⬡	D^4_{6h}	3.509	—	5.559	Room	1.584	6,6	3.439	3.509	3.47
Actinide Group												
90 Thorium......	—	⊡	O^5_h	5.07_7	—	—	Room	—	12	3.590	—	—
91 Protactinium..	—	—	—	—	—	—	—	—	—	—	—	—
92 Uranium......	—	▯	V^{17}_h	2.852	5.865	4.945	Room	b/a = 2.056 c/a = 1.734_1	—	2.76	2.85§	—

Footnotes to Table 3-3

* As tabulated by William Hume-Rothery in "Structure of Metals and Alloys," Monograph 1, Institute of Metals, London, 1945. Modified by references to literature to May, 1946, by H. C. Vacher, and to mid-1950 by others. Essentially as published in *Metal Progress*, **50**, 847 (1946), and reproduced by permission of the American Society for Metals.

† □ is cubic; ◫ is body-centered cubic; ⊡ is face-centered cubic; ⬡ is hexagonal; ⬢ is close-packed hexagonal; ◇ is cubic diamond; △ is tetragonal; ▱ is simple rhombohedral; ▯ is orthorhombic; ▯ is monoclinic; ▲ is face-centered tetragonal.

‡ Data for this element have been modified to correspond with recent publications.

§ Complex unit cell; dimensions are for smallest interatomic distances.

‖ The allotrope usually existing.

¶ Usual nomenclature indicates body-centered cubic form as βW, whose constant was reported as $a = 5.041 \pm 0.0005$ kx at 25°C by Rooksby, and as $a = 5.0408 \pm 0.0002$ kx at 18°C by Petch.

** Approximate where each lattice point is a cluster of 29 atoms; 1 type X, 5 type A, 12 type D_2 and 12 type D_1. The X atoms occupy the largest volume and the D_2 the smallest. Interatomic distances of neighboring atoms are as follows:

X atoms have:	A atoms have:	D_2 atoms have:	D_1 atoms have:
$12D_2$ at 2.71	$1X$ at 2.82	$1X$ at 2.71	$1A$ at 2.49 and 1 at 2.96
$4A$ at 2.82	$3D_1$ at 2.49 and 3 at 2.96	$2A$ at 2.69 and 1 at 2.89	$1D_2$ at 2.45, 2 at 2.51 and 2 at 2.66
	$3D_2$ at 2.69 and 3 at 2.89	$1D_2$ at 2.24 and 2 at 2.38	$6D_1$ at 2.67
		$1D_1$ at 2.45, 2 at 2.51 and 2 at 2.66	

†† Complicated structure with 20 atoms of 2 kinds in unit cell.

Each atom of the first kind has: 3 neighbors at 2.365; 3 neighbors at 2.530; 3 neighbors at 2.671; 3 neighbors at 2.675

Each atom of the second kind has: 2 neighbors at 2.530; 2 neighbors at 2.615; 4 neighbors at 2.659; 2 neighbors at 2.671; 2 neighbors at 2.675

radius may properly be applied only to a metal. In an ionic crystal, such as a metal chloride, an entirely different radius is to be expected. For example, the metallic radius of potassium is 2.35 A, whereas the crystal radius of K^+ is 1.33 A. Values of the Goldschmidt (metallic) atomic diameter for CN12 as well as other data relating to the crystalline elements are given in Table 3-3. The crystal radii of the ions are shown to scale in Fig. 3-11, and the arrangement of the ions in alkali halide crystals with sodium chloride structure is shown in Fig. 3-12.

PAULING'S THEORY OF ATOMIC RADIUS FOR METALS

Pauling[1] has proposed a rather fruitful elaboration of his concept of the nature of the chemical bond in the metallic state. His principal postulate, as already stated, is that the interatomic forces in metals are for the most part due to covalent bonds resonating among the available interatomic positions. The fundamental equation which he develops is based largely upon the carbon-carbon bond of organic chemistry, for which it is found that the distances for the single, double, and triple bonds are as follows:

Bond	Bond No., n	Interatomic distance, A	Atomic radius, A	Difference from single-bond radius, A
C—C	1	1.542	0.771	
C═C	2	1.330	0.665	0.106
C≡C	3	1.204	0.602	0.169

It will be noted that the ratio of the two radius differences listed in the last column is just equal to log 2:log 3. From this Pauling concludes that the relation between the atomic radius and the bond number n for such a case is $r(1) - r(n) = 0.353 \log n$, where $r(1)$ is used to designate the radius characteristic of a single bond, $r(n)$ that of an n-fold bond. It will be seen that this relation fits the data in the above table from which it was derived.

Resonance. The effect of resonance is illustrated by the fact that the carbon-carbon distance in benzene is not 1.418 A as given by this equation [with $n = \frac{3}{2}$ and $r(1) = 0.771$ A, as in the table] but is 1.390 A. This discrepancy of 0.028 A in interatomic distance or 0.014 in atomic radius is associated with resonance of the extra bond between the two alternative positions. In order to account for the effect of resonance of v

[1] L. Pauling, *Phys. Rev.*, **54,** 899 (1938); *J. Am. Chem. Soc.*, **69,** 542 (1947); *Proc. Roy. Soc.* (*London*), **A196,** 343 (1949).

valence bonds among N equivalent positions Pauling redefines the *bond number* n as v/N. By consideration of other substances as well as benzene, he finds the effect of resonance to be properly taken into consideration by slight alteration of the numerical constant[1] in the above equation, which then becomes:

$$r(1) - r(n) = 0.300 \log n \tag{3-1}$$

This simple equation, which was derived from consideration of nonresonating and resonating covalent bonds in simple nonmetallic substances of known structure, is now found to be remarkably useful in application to metallic crystals. It is to be regarded principally as an expression giving the bond number n as a function of the observed or predicted radius r.

Application of Pauling's Equation to Zirconium. It has already been mentioned that the closest packing of spheres is achieved in the fcc or the hcp arrangement with CN12 and that in the bcc arrangement each atom has eight nearest neighbors and six more at a distance only slightly greater. These latter distances are in the ratio $\sqrt{3}:2$; thus, the six farther neighbors are only 15.5 per cent farther than the nearest neighbors. Zirconium exists in two crystalline forms having bcc and hcp structures. We shall now calculate the radius of the zirconium atom in the hcp structure from the measured radius in the bcc structure. In this bcc structure the effective radius for the eight shortest bonds is 1.563 A and for the six longer bonds is 1.805 A, calculated from the lattice parameter of 3.61 A. The valence v of zirconium is 4; we shall find also how many of these four bonds are associated with the eight nearest neighbors and how many with the six farther neighbors. Let us designate by x the number of bonds associated with the eight nearest neighbors; the bond number for these is then $x/8$. Inserting these values in Eq. (3-1) we have:

$$r(1) - r\left(\frac{x}{8}\right) = 0.300 \log \left(\frac{x}{8}\right)$$

The remaining bonds, $4 - x$, are associated with six farther neighbors, the bond number in this case being $(4 - x)/6$. For these values the equation becomes:

$$r(1) - r\left(\frac{4-x}{6}\right) = 0.300 \log \left(\frac{4-x}{6}\right)$$

[1] In the case of benzene the correction in this numerical coefficient is 0.047, which is 0.014 (the change in atomic radius associated with resonance) divided by log 2, 2 being the number of positions between which the extra bond resonates.

Subtracting these two equations and evaluating the radius difference from the effective radii given above (1.805 − 1.563 = 0.242 A) we have:

$$0.242 = 0.300 \log \frac{6x}{8(4 - x)}$$

The solution of this is $x = 3.58$. Thus we find that 3.58 bonds are associated with the eight nearest neighbors and only 0.42 bond (about 10 per cent) with the six farther neighbors. Inserting this value of x in either of the above two simultaneous equations, $r(1)$ may be obtained.

For CN12 the bond number of zirconium is $\frac{4}{12} = \frac{1}{3}$. Hence from Eq. (3-1) we may write $r(1) - r(\frac{1}{3}) = 0.300 \log \frac{1}{3}$. Inserting the value for $r(1)$ just obtained it is found that the radius for CN12, $r(\frac{1}{3})$, is 1.601 A, which is 0.038 A greater than the shorter radius for the bcc lattice. The mean observed radius for the hcp structure is 1.597 A. The agreement here is astonishingly good.[1]

By the same method it is found for the hcp structures that the number of bonds associated with other than nearest neighbors is very small indeed, being about 0.001 in the case of titanium. This monopoly of the bonds of an atom by the nearest neighbors is in agreement with the well-established concept that atomic forces are of short range.

Pauling's Table of Metallic Valence and Metallic Radius. This table for both single bond, $r(1)$, and CN12, r(CN12), is here reproduced as Table 3-4. The metallic valences as given are in part the usual chemical valences, in particular those in the far left and far right quarters of the table. The value 5.78 for iron and other transition elements is derived from the fact that the saturation magnetic moment is 2.22 Bohr magnetons, indicating that there are 2.22 unpaired or nonbonding electrons. The remaining 5.78 electrons (8 − 2.22) outside the argonlike core are regarded as bonding electrons, it being noted that electrons in the incomplete inner shells (3*d*, 4*d*, and 5*d*) are included by Pauling among the bonding electrons for the metallic state. Some of the valences were obtained with the aid of Eq. (3-1), $r(1)$ being obtained by interpolation between the preceding and following element. All values of the valence are to be regarded as empirical in nature.

There is often disagreement between these valences and the corresponding ones resulting from the approximate quantum mechanical calculation, which gives, for example, 1 bonding electron for copper instead of 5.44 as tabulated. However, this tabulation, in which the valence or number of chemical bonds passes through a maximum in the center of both the short

[1] Had the four valence bonds been regarded as associated entirely with the eight nearest neighbors, the difference between radii for bcc and hcp structures would have been 0.053 instead of 0.038 A as calculated above or 0.034 A as observed.

TABLE 3-4. METALLIC VALENCE AND METALLIC RADIUS OF THE ELEMENTS*

	Li	Be	B											C	N	O	F
v	1	2	3											4	3	2	1
r(CN12)	1.549	1.123	0.98											0.914	0.88 0.92		
r(1)	1.225	0.889	0.80											0.771	0.70 0.74	0.66 0.74	0.64 0.72
	Na	Mg	Al											Si	P	S	Cl
v	1	2	3											4	3	2	1
r(CN12)	1.896	1.598	1.429											1.316	1.28	1.27	
r(1)	1.572	1.364	1.248											1.173	1.10	1.04	0.994
	K	Ca	Sc	Ti	V	Cr	Mn	Fe	Co	Ni	Cu	Zn	Ga	Ge	As	Se	Br
v	1	2	3	4	5	3 6	4 6	5.78	6	6	5.44	4.44	3.44	4	3	2	1
r(CN12)	2.349	1.970	1.620	1.467	1.338	1.357 1.267	1.306 1.261	1.260	1.252	1.244	1.276	1.379	[illegible]	1.366	1.39	1.40	
r(1)	2.025	1.736	1.439	1.324	1.224	1.176	1.171	1.165	1.162	1.154	1.173	1.249	1.245	1.223	1.21	1.17	1.142
	Rb	Sr	Y	Zr	Nb	Mo	Tc	Ru	Rh	Pd	Ag	Cd	In	Sn	Sb	Te	I
v	1	2	3	4	5	6	6	6	6	6	5.44	4.44	3.44	2.44 4	3	2	1
r(CN12)	2.48	2.148	1.797	1.597	1.456	1.386		1.336	1.342	1.373	1.442	1.543	1.660	1.620 1.542	1.59	1.60	
r(1)	2.16	1.914	1.616	1.454	1.342	1.296	1.271	1.246	1.252	1.283	1.339	1.413	1.497	1.412 1.399	1.41	1.37	1.334
	Cs	Ba	La	Hf	Ta	W	Re	Os	Ir	Pt	Au	Hg	Tl	Pb	Bi	Po	At
v	1	2	3	4	5	6	6	6	6	6	5.44	4.44	3.44	2.44	3	2	1
r(CN12)	2.67	2.215	1.871	1.585	1.457	1.394	1.373	1.350	1.355	1.385	1.439	1.570	1.712	1.746	1.70	1.76	
r(1)	2.35	1.981	1.690	1.442	1.343	1.304	1.283	1.260	1.265	1.295	1.336	1.440	1.549	1.538	1.52	1.53	
	Fr	Ra	Ac	Th	Pa	U	Np	Pu	Am	Cm							
v	1	2	3	4	5	5.78											
r(CN12)				1.795		1.516											
r(1)				1.652		1.421											
				Ce	Pr	Nd	Pm	Sm	Eu	Gd	Tb	Dy	Ho	Er	Tm	Yb	Lu
v				3.2	3.1	3.1		2.8	2	3	3	3	3	3	3	2	3
r(CN12)				1.818	1.824	1.818		1.85	2.084	1.795	1.773	1.770	1.761	1.748	1.743	1.933	1.738
r(1)				1.646	1.648	1.642		1.66	1.850	1.614	1.592	1.589	1.580	1.567	1.562	1.699	1.557

* From Linus Pauling, *J. Am. Chem. Soc.*, **69**, 542 (1947), and *Proc. Royal Soc. (London)*, **A196**, 343 (1949).

and long periods, is in accord with a great deal of experience. For exam ple, a large number of mechanical and physical properties such as com pressibility, coefficient of thermal expansion, heat capacity, melting poin tensile strength, and hardness exhibit a maximum or minimum (whe plotted against atomic number) for that element in each period which ha been assigned a maximum number of bonds. On the basis that it is th chemical bond which is responsible for these properties (*i.e.*, that mechan cal strength is the counterpart of the chemical bond strength), it woul seem mandatory that any satisfactory theory show such a correlatio between mechanical properties and bond strength. The tabulated rad are derived principally from measured lattice constants with the aid c Eq. (3-1).

The validity of Pauling's concept of the metallic bond as a resonatin covalent bond may be judged from a plot of the single-bond radius agains the atomic number. If the concept is correct, the calculated single-bon radii would be expected to lie on a common smooth curve with th observed normal covalent radii for the nonmetals of the same period This is indeed the case.

The Occurrence of the Maximum Valence in the Center of a Period The formation of each covalent (electron-pair) bond between two like atoms requires that each atom have (1) an electron to contribute and (2) an empty orbital to receive an electron. For each period in the periodic table it is seen that the maximum number of covalent bonds (the valence) is limited for the early elements by the number of electrons in the valence shell and for the late elements by the number of empty orbitals. Thus the maximum covalence is reasonably expected to occur in the center of the period. Since metallic behavior requires an extra orbital (for resonance), it is seen that the maximum metallic valence in the second half of the period must be less than the maximum (nonresonating) covalence. But in the beginning of each period there are plenty of extra orbitals and the metallic valence is equal to the covalence; *e.g.*, lithium has the same valence in the metallic state as in the gaseous Li_2 molecule.

Effect of Bond Character. Equation (3-1) expresses the metallic radius of an atom in terms of the bond number. It seems only reasonable to expect that the distance between two atoms in a crystal will depend not only on the number of bonds (electron pairs) but also on the type or character of the bond, *i.e.*, the subshell (s, p, or d) occupied by the pair or pairs. Thus, for Sn B, considered previously, bonds are of p type, whereas for Sn A one s and three p type bonds are formed. In cases such as this latter, it is assumed that the bond to any particular neighboring atom is sometimes p and sometimes s, and the bond is referred to as a *hybridized* sp^3 bond.

Pauling[1] has extended his interpretation of the relation between radius and bond number to include the character of the bonds. Thus the univalent radius for a pure *p* bond is taken about 0.030 A larger than for a hybridized sp^3 bond. The amount of *d* character of the bond is also regarded as important in determining the radii of the transition elements. Pauling also considers in some detail the change of radius from element to element (with atomic number) as related to the *d* character of the bonds. The interested student should consult this article. It should be noted that for a given element these considerations are of importance only in case of a change of the valence state or in dealing with alloys.

Structure of β-Manganese. Another interesting application of Pauling's equation is to the unusual structure of β-manganese. It will be recalled that manganese and tungsten are the only elements included in class I which crystallize in forms other than fcc, bcc, or hcp. According to the X-ray evidence, the unit cube of β-manganese contains 20 atoms. Crystallographically the atoms are of two kinds,[2] the unit cell containing eight atoms designated manganese I and twelve atoms designated manganese II. The manganese I and II have nearest neighbors as shown in the following table:

	Number	Type	Distance, A	Bond No.
Mn I	3	Mn I	2.36	0.92
	3	Mn II	2.53	0.48
	6	Mn II	2.67	0.28
Mn II	2	Mn I	2.53	0.48
	2	Mn II	2.60	0.37
	4	Mn II	2.66	0.28
	4	Mn I	2.67	0.28
	2	Mn II	3.24	0.03

The bond numbers given above are calculated from Eq. (3-1), using $r(1)$ as 1.168 A[3] as interpolated between chromium and iron. The sum of all the bond numbers gives $v = 3(0.92) + 3(0.48) + 6(0.28) = 5.88$ for

[1] L. Pauling, *Proc. Roy. Soc.* (*London*), **A196**, 343 (1949).

[2] When it is said that crystallographically there are two kinds of manganese atoms, this is not to be interpreted as implying necessarily that there is any difference in the manganese atoms themselves but rather a positional or configurational difference. X-ray investigation discloses that some of the manganese atoms have 12 nearest (or nearly nearest) neighbors and others 14. These are designated manganese I and II, respectively, solely because of this configurational difference.

[3] This example is taken from Pauling's earlier article [*J. Am. Chem. Soc.*, **69,** 542 (1947)], hence the slight discrepancy between this single-bond radius and the one tabulated.

manganese I, a value in good agreement with that, namely, 6, selected on the basis of magnetic measurements and recorded in Table 3-4. Similarly the sum of the bond numbers for manganese II gives $v = 4.00$, likewise in excellent agreement with the value tabulated, namely, 4, which was obtained with the aid of Eq. (3-1) from the parameters of γ-manganese. Hence it is concluded that β-manganese contains two kinds of atoms in the chemical sense (different valences) as well as in the crystallographic sense.

It is a geometrical fact that, although fcc and hcp structures exhibit the closest possible packing for spheres of equal radius, nevertheless there is a possible packing for CN12 which is even closer. Twelve spheres of equal radius may be close-packed about one with a radius 0.901 times that of the surrounding twelve. Thus if two different radii are permitted for manganese, corresponding to two valences according to Eq. (3-1), an even closer packing than that of fcc or hcp is possible. In fact the structure of β-manganese may be described as one in which each of the smaller atoms (manganese I) is surrounded by 12 atoms which approximate this closest possible packing and each of the larger atoms (manganese II) is surrounded by 14 atoms. Thus the anomalous crystallographic structure of β-manganese may now be interpreted as the closest possible packing consistent with the two tabulated valences, just as fcc and hcp are the closest packing for an element with a single valence. Unfortunately the complicated structure of α-manganese and that of one modification of tungsten cannot be interpreted so readily.

Pauling's Interpretation of the Class II Elements. Pauling[1] has introduced a further postulate to interpret the crystallography of metallic elements and alloys. This postulate is that the maximum amount of energy accompanying the resonance of bonds among the bond positions (corresponding to minimum energy of the crystal) occurs when the bond number n is equal to the ratio of small integers. Thus, in metallic structures maximum stability is attained with a bond number of $\frac{1}{2}$. This is followed in importance by $\frac{1}{3}$, $\frac{2}{3}$, $\frac{1}{4}$, $\frac{3}{4}$, . . . , etc. In accord with this postulate it might be expected that certain metals would crystallize in a structure corresponding to a deformation of one of the structures of the class I elements, or even in an alternative structure, if this deformed structure or alternative structure were particularly favorable to the especially preferred bond numbers $\frac{1}{2}$, $\frac{1}{3}$, etc.

By use of this postulate in conjunction with Eq. (3-1) and the metallic valences given in Table 3-4 Pauling is able to account for the structures of several of the class II elements. For example, zinc crystallizes in an

[1] L. Pauling, *Proc. Roy. Soc. (London)*, **A196**, 343 (1949); *J. Chem. Phys.*, **46**, 276 (1949).

hcp lattice with a c/a ratio of 1.856 rather than 1.633, the ratio for hexagonal close-packing of spheres. Each zinc atom has six equatorial neighbors at 2.660 A and six neighbors in adjacent planes at 2.907 A. In Table 3-4 the metallic valence of zinc is given as 4.44 or approximately 4.5; if each of the 12 neighbors were at an equal distance, the bond number would be 4.5/12 or $\frac{3}{8}$. However, by increasing the c/a ratio it is possible for zinc to assume a structure such that the six equatorial neighbors have a bond number of $\frac{1}{2}$ and the six others have a bond number of $\frac{1}{4}$. The observed c/a ratio for zinc is in good agreement with that calculated on this hypothesis. Pauling in his later articles finds even closer agreement by choosing bond numbers of $\frac{1}{2}$ and $\frac{1}{6}$, corresponding to a valence of 4. Similar agreement is found for cadmium.

The structure of mercury, which may be regarded as derived from an fcc lattice by compression along a 111 axis, may be interpreted in like manner. Six strong bonds with $n = \frac{1}{2}$ are directed to the atoms above and below the equatorial plane and the weak bonds with $n = \frac{1}{4}$ are directed to the atoms in the equatorial plane. The unique structure of white tin is accounted for by assuming that each atom has four stronger bonds, $n = \frac{1}{2}$, and two weaker ones, $n = \frac{1}{4}$, corresponding to an average valence of 2.5 as compared with 2.44 given in Table 3-4. Similarly, from the structure of gallium and Eq. (3-1) it is found that the bond numbers of the neighboring atoms are 1, $\frac{1}{2}$, and $\frac{1}{4}$. It will be noted that in this particular case each atom is linked to a single nearest neighbor by a bond of number unity, so that it is in one sense proper to describe this structure as a metallic packing of Ga—Ga diatomic complexes (molecules?). The structure of β-tungsten is also accounted for in terms of single, half, and quarter bonds, and that of α-uranium in terms of four single and eight quarter bonds.

Simplicity of Atomic Arrangement. The reader may already have inferred that crystal structure, as viewed in terms of the neighbors of a given atom, is relatively simple as compared with the same structure as viewed in terms of a unit cell. Pauling regards this simplicity as fundamental; his feeling here is perhaps somewhat akin to Einstein's faith in the "ultimate rationality of the universe."

Thus, the diamond structure is most simply described as a structure in which each atom is symmetrically surrounded by four atoms. The alternative description is that the unit cell is cubic (known as diamond cubic) of such nature that it may be regarded as an fcc structure to the interior of which four extra atoms have been added, one in the center of each of four of the eight subcubes. Similarly, the fcc and hcp structures seem more simply described in terms of the close packing of the 12 nearest neighbors of each atom. And again, the structure of β-manganese was previously described in terms of nearest neighbors, whereas the description in terms of the unit cell of 20 atoms is too formidable to undertake.

IMPERFECTION IN CRYSTALS

From the description of the crystalline elements which has been given so far, it might be concluded that such crystals are geometrically perfect. In fact, not many years ago it was the general opinion that nature was a perfect craftsman, not prone to the errors of man in construction. However, a great accumulation of evidence in the past two decades has led to the conclusion that most crystals, or grains as they are called in metals, exhibit departures from geometric perfection.

One of the most common types of departure from geometric perfection within a grain is readily understood by considering the manner of growth of a crystal during solidification of a liquid. One of the usual patterns followed is the dendritic method of growth, which may be described as producing a treelike structure. The main "trunk" forms first, "branches" grow from this, and then other branches in turn from these. Eventually the liquid between these freezes and the whole becomes a single grain. It will be noted that the slightest disturbance of any branch, *e.g.*, by convection of the liquid, will produce a misfit and thus the grain will be imperfect. Such misfits are multitudinous and are thought to produce a common type of irregularity known as *lineage structure*. The lines associated with this are sometimes called *Buerger's lineage pattern*, since Buerger[1] was the first to recognize them. The distance between these lines is of the order of 1 mm. Another type of irregularity, frequently called *block or mosaic structure*, occurs on a scale one-thousandth of this, *i.e.*, has dimensions of about one micron.

It is of some interest to note that the size of these blocks is of the same order of magnitude as the so-called *magnetic domains* in iron. However, the domains should not be associated directly with the mosaic blocks, since the domain pattern is variable. According to the generally accepted domain theory of ferromagnetism, each domain is always fully magnetized; *i.e.*, the electron spins within a domain are aligned in one direction. When the specimen is unmagnetized, the direction of magnetization of the domains is random; when the specimen is fully magnetized, the direction of magnetization of all the domains is the same. The domains may be made visible by placing upon a polished specimen a drop of a suspension of a very finely divided ferromagnetic substance (usually magnetite); this substance is attracted to the boundaries, outlining the domains. Although the departures from geometric perfection occasioned by accidents during growth may be regarded as errors which might be eliminated

[1] M. J. Buerger, *Z. Krist.*, **89**, 195 (1934).

by more careful procedure, it does not seem at present that the domains can be eliminated or grown to macroscopic dimensions.

Density as Evidence of Imperfections. Another discrepancy between the geometrically perfect crystal and the actual crystal is evidenced by a comparison of the measured density with that calculated from the lattice parameter and the atomic weight. Such a comparison is shown in Table 3-5, taken principally from Stockdale.[1] The disparity here, usually of

TABLE 3-5. COMPARISON OF MEASURED DENSITY (20°C) WITH THAT CALCULATED FROM X-RAY LATTICE PARAMETER

	Density, g/cm³		
	Measured	Calculated	Difference
Lead	11.343	11.346	0.003
Tin	7.282*	7.287*	0.005
Silver	10.499_7	10.503_9	0.004_2
Gallium	5.907	5.909	0.002
Zinc	7.137	7.139	0.002
Iron	7.874	7.873_4	0.000_6
Sodium chloride	2.1632	2.1634	0.0002
Diamond	3.514_2	3.5150	0.000_8
Calcite	2.7103_0	(2.7103_0) †	0.0000

* Values for tin are for 25°C.
† Arbitrarily taken as the standard

the order of a few parts in 10,000, has at times been blamed on the uncertainty of fundamental constants, on the lack of precision in the density measurements, on the lack of purity of the sample, and on porosity. However, the relatively good agreement in the case of the nonmetals listed would seem sufficient evidence to rule out any serious error in the numerical coefficient in the equation used to calculate the density.

[1] D. Stockdale, *J. Inst. Metals*, **66**, 287 (1940). Stockdale expressed the density as $1.65029\ \Sigma A/V$, where ΣA is the sum of the atomic weights of the atoms in a unit cell and V is the volume of the unit cell in (kx)³ units. The kx is the X-ray unit of length, formerly, but erroneously, referred to as the angstrom unit (A). The angstrom unit is defined as 10^{-8} cm, whereas recent agreement among physicists [E. A. Wood, *Phys. Rev.*, **72**, 436 (1947)] establishes the kx as 1.00202 A. On the basis of this figure the density ρ may be calculated from X-ray measurements by means of either of the equivalent formulas:

$$\rho = 1.66020 \frac{\Sigma A}{V} \qquad \text{for } V \text{ in (A)}^3$$

$$\rho = 1.65018 \frac{\Sigma A}{V} \qquad \text{for } V \text{ in (kx)}^3$$

The latter relation was used in obtaining the calculated values of Table 3-5.

Undoubtedly the other factors mentioned enter in the case of some of the metals; however, the extensive investigation of Stockdale on silver would appear to rule out such errors in the case of this metal. It will be observed that for metals, except iron, the calculated density exceeds the measured density. In spite of possible objections as to the reliability of the measured density, the data appear to indicate a real disparity between the observed density and the density of the perfect crystal.[1]

One interpretation which may be placed on the lower observed density is that it is due to empty lattice sites or holes of the size of an atom. Support has been added to this concept from an entirely different field—that of diffusion. The mechanism of the motion of an atom of a foreign element or isotope in the lattice seems most easily explained on the hypothesis that a small fraction of the lattice sites are empty.[2] It is to be emphasized that these holes are not necessarily regarded as defects, which could be removed by adequate annealing, but as an equilibrium phenomenon at elevated temperature where diffusion occurs. Even if they could be removed by some mechanism, it is postulated that the equilibrium number would again form in adequate time.

Wagner and Hantelmann[3] have devised and employed a clever method to evaluate the fraction of the total lattice sites which are empty in solid potassium chloride. Their method consists of determining the electrical conductivity and transference numbers for solid potassium chloride and the electrical conductivity of potassium chloride containing up to 0.8 per cent strontium chloride ($SrCl_2$) in solid solution. Because of the different valences, each Sr^{++} ion added to the KCl lattice is accompanied by a vacancy or hole. Knowing the increment to the conductivity occasioned by these added holes, it is possible to compute the equilibrium fraction of vacancies in pure potassium chloride. The results are as follows:

Temp., °C	*Fraction of lattice sites which are unoccupied*
600	0.8×10^{-5}
625	1.2
650	2.1
675	3.0
700	4.1
725	5.7

[1] As a 100-g sample can readily be weighed to $\frac{1}{10}$ mg, the realizable error in a density measurement due to weighing is of the order of 1 part in 100,000. By modern precision methods the other errors can be reduced to a comparable or smaller figure. The precision of lattice parameters determined by the best X-ray measurements is about the same.

[2] R. P. Johnson, *Phys. Rev.*, **56**, 814 (1939). H. B. Huntington, *ibid.*, **61**, 325 (1942). H. B. Huntington and F. Seitz, *ibid.*, **61**, 315 (1942). F. Seitz, *ibid.*, **74**, 1513 (1948).

[3] C. Wagner and P. Hantelmann, *J. Chem. Phys.*, **18**, 72 (1950).

The careful investigation of Maier[1] on the density of copper furnishes incontrovertible evidence that the density of polycrystalline copper at room temperature, determined to a precision of a few parts per million, varies rather disconcertingly, depending upon the previous history of the sample. Some of the data are given in the following table:

DENSITY OF COPPER AT 20°C

		g/cm^3
A	Single crystal	8.95285
B	Melted and solidified in vacuum	8.94153
C	Melted in vacuum and solidified in helium	8.94331
D	Commercial "soft-drawn" wire, annealed 970°C in vacuum for 12 hr	8.92426
	Sample D:	
	Reduced 34.82% in cross section by drawing	8.91414
	Reduced 67.90% in cross section by drawing	8.90526
	Reduced 96.81% in cross section by drawing	8.91187
E	Sample D, annealed 880°C in vacuum for 12 hr	8.93003
F	Sample D, annealed 995°C in vacuum for 12 hr	8.92438
G	Sample D, annealed 1035°C in vacuum for 12 hr	8.92763
H	Calculated from the lattice parameter* of Table 3-3 and the atomic weight	8.933

* Corrected to 20°C.

As indicated for sample *D*, the effect of cold work is first to decrease and finally to increase the density. In iron the effect is even more pronounced as shown in Fig. 3-13, also from Maier. If the large change in density occasioned by cold work is attributable to a "lack of fit" at the grain boundaries, one would expect the density to be restored by annealing above the recrystallization temperature. This frequently does not occur; annealing only partly restores the density to its original value as is evidenced by the variation of Maier's data on the density of copper and by data on the change of density of steel during annealing.[2]

X-ray evidence also gives some indication of a small structural unit within the crystal. Wood[3] makes the following observation, based on his own measurements: "The striking contrast on the one hand between the ease with which the grains are dispersed into crystallites immediately after deformation and, on the other, the subsequent impossibility even after the most severe deformation of breaking them down below about 10^{-5} centimeters . . . , indicates that the crystallite must be regarded as a fundamental unit of the metallic grain." Barrett[4] also found evidence of

[1] C. G. Maier, *Trans. AIME*, **122**, 121 (1936).
[2] K. Heindlhofer and F. L. Wright, *Trans. Am. Soc. Steel Treating*, **7**, 34 (1925).
[3] W. A. Wood, *Proc. Roy. Soc.*, **172**, 231 (1939).
[4] C. S. Barrett, *Trans. AIME*, **161**, 15 (1945).

localized cold-worked regions about 1 micron thick. Burgers[1] similarly comments, "Thus pure recovery [from cold work] leads to states intermediate between cold worked and completely annealed and which have considerable stability."

Bubble-raft Analogue of a Crystal. A very neat two-dimensional analogue of a close-packed arrangement of atoms in a crystal lattice has been presented by Bragg and Nye.[2] They prepared a raft of bubbles on

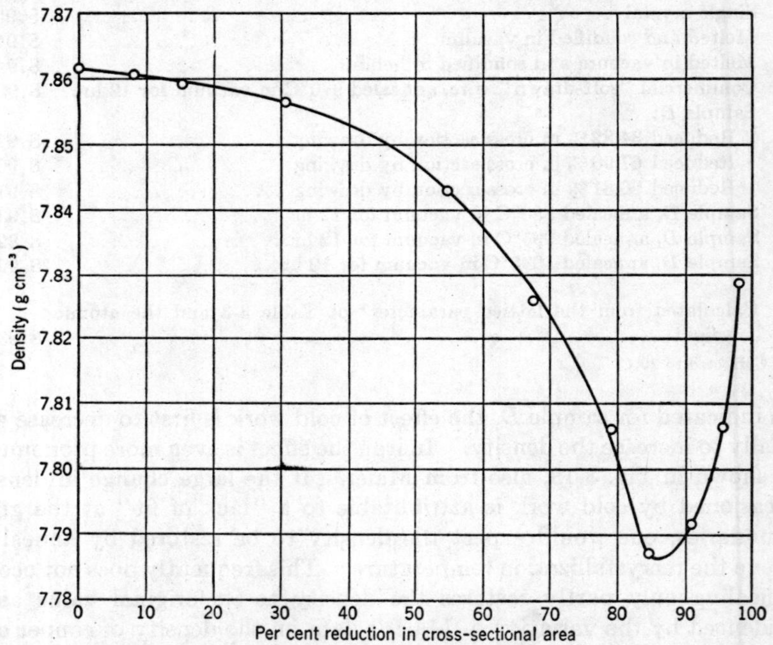

FIG. 3-13. Effect of cold work (drawing) on the density at 20°C of ingot iron wire, annealed at 1026°C prior to drawing. (*After Maier.*)

the surface of a soap solution by introducing air under controlled conditions through a jet beneath the surface of the solution. By this means they were able to produce bubbles of unusually uniform size. The various configurations assumed by the assemblage, or "raft," simulate the configurations of atoms in a crystal under various conditions. The perfect assemblage shown in Fig. 3-14 depicts a perfect lattice. "Grain boundaries" are shown in Fig. 3-15, a "dislocation" in Fig. 3-16, and another type of fault in Fig. 3-17. The lattice distortion accompanying the imperfections of Figs. 3-15 and 3-17 is seen to be confined to their

[1] W. G. Burgers, *Koninkl. Nederland Akad. Wetenschap. Proc.*, **50**, 452 (1947).

[2] L. Bragg and J. F. Nye, *Proc. Roy. Soc.* (*London*), **A190**, 474 (1947).

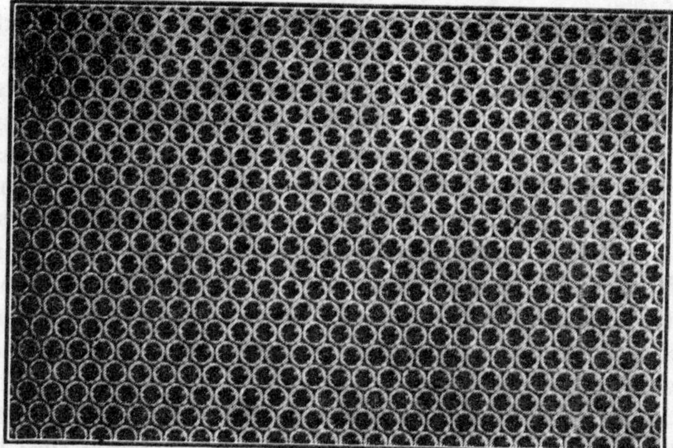

Fig. 3-14. Bubble raft simulating a perfect crystal lattice. [*From Bragg and Nye, Proc. Roy. Soc. (London)*, **A190**, 474 (1947).]

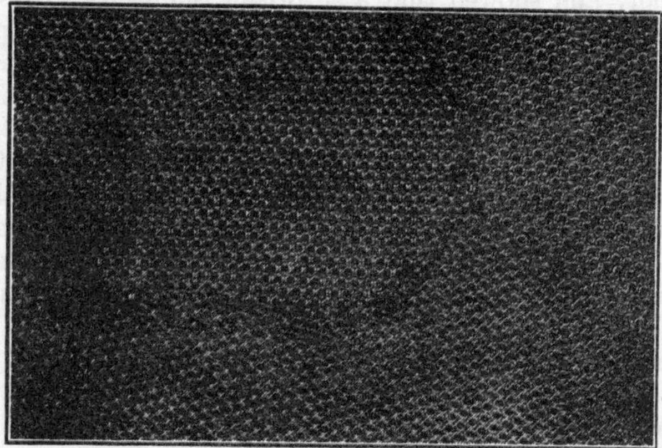

Fig. 3-15. Bubble raft depicting "grain boundaries." (*From Bragg and Nye.*)

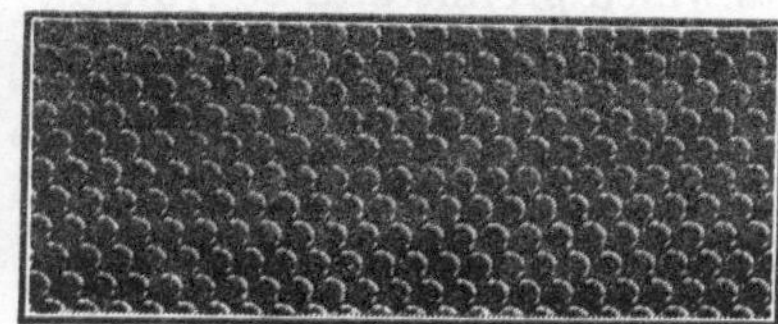

Fig. 3-16. Bubble raft showing a "dislocation." (*From Bragg and Nye.*)

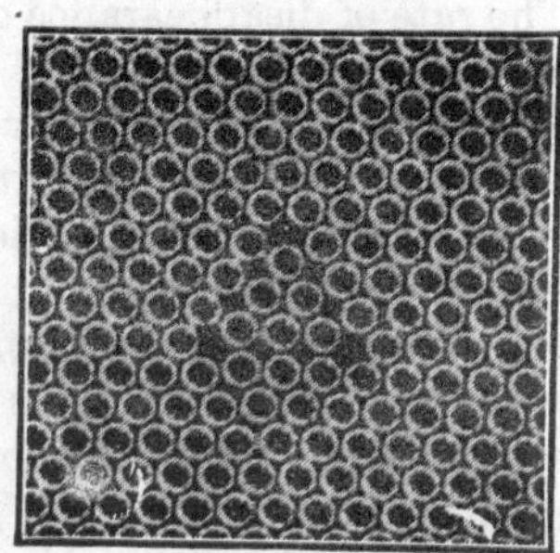

Fig. 3-17. Bubble raft with another type of fault. (*From Bragg and Nye.*)

immediate vicinity, whereas in the case of Fig. 3-16 the strain extends to a much greater distance.

It is found that a raft of bubbles has such properties as Young's modulus and Poisson's ratio and also exhibits many phenomena characteristic of metals such as grain growth, after being broken up into a mass of small "crystals," and slip, by the movement of dislocations when strained beyond the elastic limit. Since the relation between the attractive forces of surface tension and the repulsive pressure at the touching interfaces changes with size of bubbles, some control can be exerted over the "physical properties" of the crystal analogue. For example, the forces between copper atoms in metallic copper are found to correspond most closely to those between bubbles 1.2 mm in diameter.

The forces of surface tension are not identical in nature with those between the atoms of a crystal, and it is to be expected that the above analogue must be imperfect. Bragg and Lomer[1] investigated this point mathematically and showed that a surprising similarity exists between the two types of forces and that certain important conclusions can be drawn from experiments with bubble rafts. Plastic deformation proceeds by the motion of dislocations, which are produced either in pairs in the body of the raft or singly at its edge. The formation of dislocations (and consequent plastic yielding) can occur in an initially perfect lattice only at quite large shear strains, and the low shear strength of metallic single crystals is explainable only on the assumption that they are not perfect but already contain dislocations which move under very small stress.

In summary, it may be said that there is a considerable accumulation of evidence that actual metallic crystals or grains differ from idealized perfect geometric crystals. The nature of the difference is not yet fully understood. It seems fruitful to distinguish two types of departures: those which are stable, *e.g.*, the holes postulated to account for self-diffusion, and those which are unstable and tend to disappear on annealing. The rate of disappearance of the latter may sometimes be very slow as is indicated by the failure of a cold-worked metal to fully regain its original density on annealing for a reasonable length of time.

General Classification of Lattice Defects. H. B. Huntington has proposed a classification of lattice defects which promises to be fruitful. The types are as follows:

1. Point defects (cell defects). This class includes the vacancies or holes considered above as well as interstitial atoms. Either of these may provide a mechanism for diffusion.

[1] L. Bragg and W. M. Lomer, *Proc. Roy. Soc.* (*London*), **A196**, 171 (1949). W. M. Lomer, *ibid.*, **A196**, 182 (1949).

2. Line defects (dislocations). These provide the most widely accepted mechanism, or mechanisms, for plastic deformation and also are associated with internal friction. It will be noted that a dislocation is properly viewed as a line in the sense that its extent is much greater in one direction (normal to the plane of the paper in Figs. 3-3*b* and 3-16) than in other directions.

3. Surface and interface defects (grain boundaries and crystallite boundaries). The phenomena associated with surface and interface energy (or tension) fall in this class.

This classification implies that we have a chemistry not only of ordinary three-dimensional substances and of surfaces but also of lines and of points if we care to regard the matter in this light. Naturally the four are not independent of one another. Examples of the interrelations are numerous. Among the most important may be mentioned the accumulation of interstitial solute atoms (point defects) in the dislocations of cold-worked metal (*e.g.*, hydrogen in steel), the fact that grain boundaries serve as sources and sinks for dislocations and vacancies and that dislocations serve similarly as generators or absorbers for vacancies. It would seem reasonable to expect that the energy would decrease as defects of point type "coalesce" to line or interface type and as those of line type are converted to interface type.[1]

[1] This same reasoning would not necessarily apply to the free energy except when the number of defects greatly exceeds the equilibrium number. Thus, during recrystallization of cold-worked metal the dislocations migrate and "become" grain boundaries.

CHAPTER 4

SOLID SOLUTIONS AND INTERMETALLIC COMPOUNDS

If ether is added drop by drop to a vessel of water which is shaken between additions, it is found that the drops at first dissolve to give a homogeneous solution. On continued addition, however, a solubility limit is reached; no more ether dissolves in the water and an ether layer (containing some water) forms on top of the water. It was early recognized that a similar phenomenon occurs in many solid metallic systems, particularly at elevated temperature.

Thus, for example, if we consider a series of iron-copper alloys we find (near 1000°C) that the alloys with less than 10 weight per cent copper are homogeneous, containing but a single phase. However, alloys containing somewhat more than 10 per cent copper exhibit in addition a copper-rich phase. Although the lattice parameter may be altered in a regular manner, the crystal structure of the homogeneous low-copper alloys remains fcc, as that of the parent metal iron; this may be shown by X-ray diffraction. Such a solid solution, which is represented as one of the terminal regions on a temperature-composition equilibrium diagram, is known as a *primary solid solution*. Solid solution is common in metallic systems and frequent in other inorganic systems at elevated temperature. It is relatively uncommon in the traditional chemistry of aqueous solutions and in organic chemistry; in recent years, however, several instances of solid solubility of organic substances have come to light.

Types of Solid Solution. A question naturally arises as to the location of these "foreign" atoms in the crystal lattice of the metal. If the foreign atoms merely replace atoms of the parent metal on some of the lattice sites, the solution is known as a *substitutional solid solution*. This is the more common case. Sometimes, however, the atoms of the foreign element take up positions between the lattice sites, in the so-called interstitial positions, to form an *interstitial solid solution*. As might be expected, such an interstitial solid solution forms only when the atomic radius of the foreign element is small in comparison with that of the parent element. The metallic radii of hydrogen, boron, carbon, and nitrogen are all less than 1 A, and these elements frequently form interstitial solid solutions. Although the total number of known interstitial solid solutions is small compared with that of the substitutional type, one interstitial solid solution, that of carbon in γ-iron, known as *austenite*, is of great importance, since the heat-treatment of steel is based upon the transformation of this to other phases. Throughout the remainder of this chapter all mention of solid solutions refers to the substitutional type unless otherwise stated.

THE GEOMETRICAL LIMITATION IMPOSED BY THE SIZE FACTOR ON THE EXTENT OF PRIMARY SOLID SOLUTION

Bragg and Nye's[1] bubble-raft analogue of a crystal was discussed in the previous chapter. Another of their photographs, here reproduced as Fig. 4-1, shows the arrangement that results when a bubble of different size (an atom of a foreign element) is introduced. A widespread disrup-

FIG. 4-1. Bubble raft showing the extensive effect of an atom of impurity. A small bubble is to be seen below and to the right of the large bubble. [*From Bragg and Nye, Proc. Roy. Soc. (London)*, **A190**, 474 (1947).]

tion of the symmetry of the raft is seen to occur, suggesting a similar wide-spread effect of an atom of an impurity in a crystal. The extent of the effect has not been generally recognized in the past.

Further light on this phenomenon is shed by consideration of the change in volume, or lattice parameter, resulting from the introduction of an alloying constituent to form a substitutional solid solution. Suppose that one atom in a given volume element of a pure crystal is replaced by a foreign atom of larger radius. If the disturbance caused by the disparity in size is purely local, *i.e.*, if the adjacent atoms are compressed

[1] Bragg and Nye, *Proc. Roy. Soc.* (*London*), **A190**, 474 (1947).

to compensate for the disparity, then the volume element will contain the same number of atoms as it did originally and the mean lattice parameter will remain unaltered. However, if the disturbance occasioned by the foreign atom extends to a great distance, *i.e.*, to the boundary of the volume element, then the lattice parameter will be altered. In other words, if a plot of lattice parameter against concentration of alloying element exhibits a horizontal tangent at low concentration, then we would seem justified in concluding that the disorganizing effect of the foreign

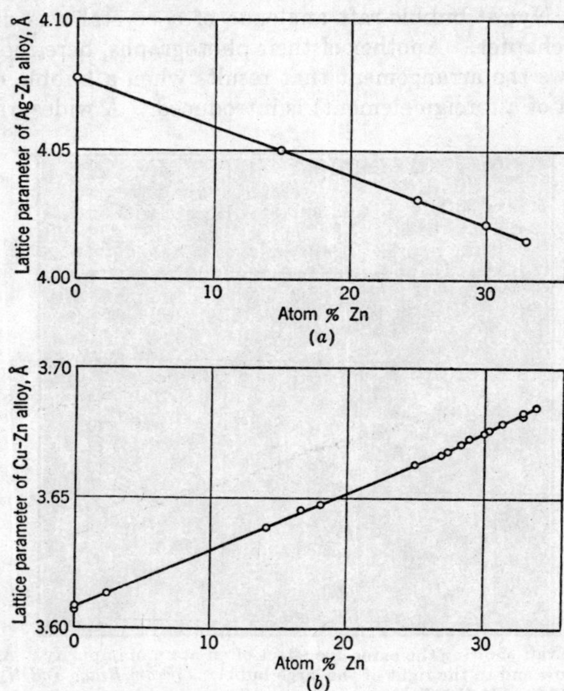

FIG. 4-2. (*a*) Effect of Zn addition on the lattice parameter of Ag (20°C). (*b*) Effect of Zn addition on the lattice parameter of Cu (room temperature).

atom is local; whereas if the tangent is not horizontal, the conclusion would seem inescapable that the disorganizing effect is, at least in part, widespread, as indicated by the soap-bubble analogy. Actually it has been known for some time that the lattice parameters of primary solid solutions vary nearly linearly with the atomic percentage of the solute element; this is known as *Vegard's law*.[1] As typical examples, the change in the lattice parameter of silver occasioned by the addition of zinc is

[1] L. Vegard, *Z. Physik*, **5**, 17 (1921). L. Vegard and H. Dale, *Z. Krist.*, **67**, 148 (1928).

shown in Fig. 4-2a* and that of copper by the addition of zinc in Fig. 4-2b.† There is no indication of any flattening of the curve in the vicinity of zero zinc content in either case. Another example is to be found in the iron-aluminum system, shown later in Fig. 4-17. Hence, in accord with the foregoing reasoning, it may be concluded that the disorganization of the lattice accompanying the introduction of a foreign atom of different radius is widespread and not confined to the immediate vicinity of the foreign atom.

Mechanical Analogue of Size Factor. It is of some interest at this point to consider a model illustrating the effect of the introduction of a foreign atom into a crystal of a pure metal. The model taken here as the analogue of the crystal is one which conforms to classical elastic theory. It is shown in Fig. 4-3 as a sphere of radius b.‡ At the center of this there is a small spherical cavity of radius a, to be filled with an incompressible fluid representing the foreign atom. The spherical hole is connected to the outside by a small capillary through which the incompressible fluid may be injected.

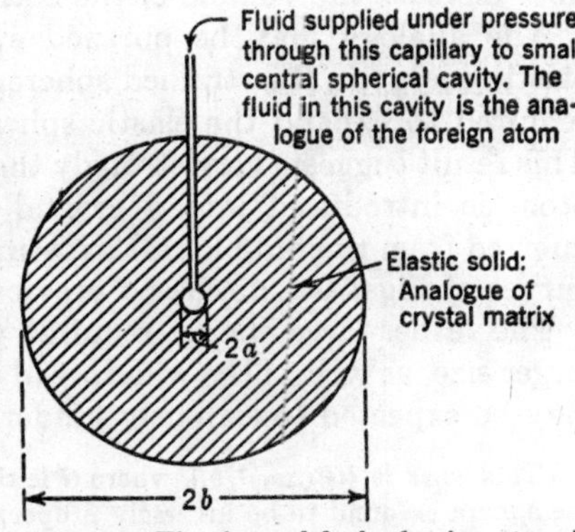

FIG. 4-3. Elastic model of a foreign atom in a crystal.

The behavior of this elastic model may readily be calculated§ from equations of the elastic theory. It is found that if, after filling the cavity at atmospheric pressure (its volume then being v_1), an additional volume of the incompressible fluid Δv_1 is forced in, the volume change of the sphere Δv_2 then is related to Δv_1 by the equation

$$\frac{\Delta v_2}{\Delta v_1} = \frac{3(1 - \mu)}{1 + \mu}$$

valid when b is very large compared with a. μ is *Poisson's ratio,* which is the ratio of the change in length in the direction perpendicular to the

* D. Stockdale, *J. Inst. Metals,* **66,** 287 (1940).

† E. O. Kirkendahl, *Trans. AIME,* **147,** 104 (1942).

‡ The large sphere may be visualized as being composed of rubberlike material, the capillary as a hypodermic needle, and the incompressible fluid as mercury. The whole assemblage is enclosed in a dilatometer, the intervening space being connected to the outside by a small tube and filled with a liquid such as mercury, so that any change in the total volume of the sphere (as the incompressible fluid is forced into the cavity) is indicated by a rise of the liquid in the tube.

§ The authors are indebted to K. Heindlhofer for performing this calculation. Similar reasoning has been followed by A. W. Lawson, *J. Chem. Phys.,* **15,** 831 (1947).

applied force to that in the direction of this force. For an incompressible substance Poisson's ratio is 0.5, and in this case $\Delta v_2/\Delta v_1 = 1$, as expected. For many metals Poisson's ratio is about 0.3; hence $\Delta v_2/\Delta v_1 = 1.5$, a rather surprising result—the volume increase of the outer sphere is greater than that of the inner sphere. It will be noticed that the behavior of a foreign atom in the crystal lattice does indeed appear to be in qualitative accord with this model, in that the introduction of an atom of larger size does increase the volume of the entire crystal.

The analogy may be pursued even further by investigation of the elastic energy in the strained sphere. The result found is that the work[1] required to expand the elastic sphere increases rather rapidly with Δv_1. This result suggests very strongly that such is also the case when a foreign atom is introduced into a crystal of a pure metal—a result already inferred from the bubble-raft experiments as well as the finite slope of the curves of Figs. 4-2, discussed previously.

The rather large energy required to introduce an atom of appreciably larger size, as well as the widespread disorganization of the crystal lattice, may be expected to seriously hinder the extent of solid solubility of one

[1] This work is $\frac{2}{3}G[(\Delta v_1)^2/v_1]$, where G is the shear modulus. The energy density in the sphere is found to be inversely proportional to the sixth power of the radial distance. The elastic energy per gram atom at infinite dilution, designated W, can also be written in the form

$$W = \frac{2}{3}GV\left(\frac{V' - V}{V}\right)^2 = \frac{2}{3}GV\left(\frac{V'}{V} - 1\right)^2 = \frac{2}{3}GV\left[\left(\frac{r'}{r}\right)^3 - 1\right]^2$$

where V is the molal volume, r is the atomic radius, and the prime designates a property of the foreign species. In Chap. 13 it is shown that a solution usually verges on immiscibility if the excess partial molal free energy (at infinite dilution) exceeds $2RT$; pronounced limitation of relative solubility (to order of magnitude 1 atom per cent) occurs when this quantity is $4RT$. The corresponding limit imposed by elastic energy is thus seen to occur when

$$W = \frac{2}{3}GV\left[\left(\frac{r'}{r}\right)^3 - 1\right]^2 = 4RT$$

or when

$$\frac{r'}{r} = \left(1 \pm \sqrt{\frac{6RT}{GV}}\right)^{\frac{1}{3}}$$

The product GV does not vary much either with temperature or from metal to metal. Taking an average value of 60,000 cal for GV, it is found, at 1000°K, that

$$\frac{r'}{r} = (1 \pm \sqrt{0.2})^{\frac{1}{3}} \cong 1 \pm 0.15.$$

This evaluation thus seems to furnish a very understandable basis for Hume-Rothery's empirical rule that substantial solid solution cannot be expected if the atomic radii differ by more than 15 per cent. It will be noted, however, that classical elastic theory could hardly be expected to apply for such a large displacement. The agreement must be regarded as partly fortuitous.

element in another when the atomic radii are unequal. This type of limitation on the extent of solid solution is frequently referred to as *size factor*. Hume-Rothery found that the extent of primary solid solution is seriously hindered whenever the disparity in atomic radii exceeds 15 per cent. We may regard this as a first rule governing the extent of primary solid solution. It should be carefully noted that the inverse of this rule is not necessarily so; it is not to be inferred that primary solid solution will be extensive whenever the size factor is favorable, for other factors, to be discussed immediately, may operate to limit the solid solubility. In other words, a favorable size factor is a necessary but not a sufficient condition for extensive solid solution.

THE INFLUENCE OF CHEMICAL FACTORS ON THE EXTENT OF PRIMARY SOLID SOLUTION

Electronegativity. From the time of his very first experience with chemical reactions the chemist becomes accustomed to considering one element more active than another, in a very general way. For example, let us consider the relative rates of reaction of the various elements with water at room temperature. Sodium reacts vigorously, and potassium even more vigorously. The similar reaction of calcium is slower, that of magnesium is slower yet, and so on through aluminum, iron, tin, copper, and gold, gold being quite inert. If the reaction with some substance other than water is considered, the chemist might expect a slight but not a serious alteration in the order of these elements.

This concept of arranging the elements in order of their chemical reactivity has been expressed by the electrochemist in the electrochemical series. However, this series, though very useful from the viewpoint of the chemist dealing with aqueous solutions (since it is derived from electrical potentials measured in aqueous solutions), is not of primary interest to us in the consideration of nonaqueous reactions and, in particular, of metallic solid solutions and intermetallic compounds. This should be clear from the fact that in aqueous solutions many ions form hydrates and other complexes which are not of direct interest in consideration of the reactions between the elements themselves.

It has already been mentioned that the two common types of bonds, the covalent and the ionic, do not usually occur in pure form; most of the bonds in chemical compounds may properly be interpreted as partly ionic and partly covalent. Pauling introduces the assumption that the energy of the normal covalent bond is independent of the elements bonded. From this it would follow that for a reaction involving only covalent bonds the energy change (or heat of reaction) will be nil. For example, if all the bonds of reactants and products of the reaction

$$\text{H—H} + \text{Cl—Cl} = 2\text{H—Cl}$$

were purely covalent, it would then be expected that the heat of the reaction would be zero, it being noted that two bonds are broken and two formed. Such is not the case, however, for the H—Cl bond is partly ionic in character, whereas consideration of the symmetry of the Cl_2 and H_2 molecules leads to the conclusion that the bonds involved here have very little ionic character.

It might on first thought seem possible to arrange the elements in a series and assign to each a number such that the difference between the numbers assigned to any two elements would give the heat evolved when a chemical bond is formed between them. This cannot quite be done, but Pauling did find that arrangement on a numbered scale was possible in such a manner that the square of the difference in the numbers assigned to any two elements is roughly proportional to the heat accompanying the formation of a bond between these elements. The square-root scale used seems not unreasonable in view of the fact that the predicted heats must therefore have a positive sign, in accord with experience (with a few exceptions), discussed later. This scale is known as the *electronegativity scale.* The electronegativity of an element is represented by the symbol x; its units are the square root of electron volts per bond. An extended table of electronegativities given by Gordy* is reproduced as Table 4-1.

Gordy pointed out that the electronegativity is quite well expressed in terms of the number of valence electrons n'† and the single-bond covalent radius r by the relation‡

$$x = 0.31\left(\frac{n' + 1}{r}\right) + 0.50 \tag{4-1}$$

* W. Gordy, *Phys. Rev.*, **69,** 604 (1946).

† Gordy originally used the symbol n for the number of valence electrons. We designate this n' to guard against confusion with the symbol used for the bond number in the previous chapter.

‡ Gordy arrived at this relation by consideration of the possibility of redefining the electronegativity of a neutral atom as the potential Ze/r, where Ze is the effective nuclear charge of the atom acting on a valence electron (of charge e) at a distance r from the nucleus equal to the single-bond covalent radius of the atom. By making the simplifying assumption that all the electrons in the closed shells below the valence shell exert their full screening power and that the electrons in the valence shell have a screening constant of 0.5, Z is found to be equal to $0.5(n' + 1)$, where n' represents the number of electrons in the incompletely filled valence shells. Substituting this value of Z in the preceding expression for the potential, we obtain the following expression for electronegativity:

$$x = k\left(\frac{n' + 1}{r}\right)$$

where k is a constant. As the origin of the electronegativity scale is arbitrary, a constant may be added to this equation.

TABLE 4-1. ELECTRONEGATIVITY, COVALENT RADIUS, AND NUMBER OF VALENCE ELECTRONS FOR SEVERAL ELEMENTS (FROM GORDY)

Atom	Single-bond covalent radius, r	Valence electrons, n'	Electro-negativity,* x	$0.31\left(\frac{n'+1}{r}\right)+0.50$
Ag	1.53	1	1.9	0.91
Al	1.26	3	1.5	1.48
As	1.21	5	2.0	2.04
Au	1.50	1	3.1	0.92
B	0.88	3	1.9	1.91
Ba	2.17	2	0.9	0.93
Be	1.06	2	1.45	1.38
Bi	1.40	5	1.8	1.83
Br	1.14	7	2.75	2.68
C	0.77	4	2.55	2.52
Ca	1.78	2	1.0	1.03
Cd	1.48	2	1.1	1.13
Cl	0.99	7	2.97	3.00
(Cr)	1.25	6	(2.2)	2.24
Cs	2.25	1	0.75	0.78
Cu	1.35	1	2.2	0.96
F	0.72	7	3.95	3.94
Ga	1.26	3	1.4	1.48
Ge	1.22	4	1.7	1.77
H	0.37	1	2.13	2.17
Hg	1.50	2	1.0	1.12
I	1.33	7	2.45	2.36
In	1.44	3	1.4	1.36
K	1.96	1	0.80	0.82
(La)	1.86	3	(1.2)	1.17
Li	1.34	1	0.95	0.96
Mg	1.40	2	1.2	1.16
(Mn)	1.18	7	(2.6)	2.60
(Mo)	1.36	6	(2.1)	2.09
N	0.74	5	2.98	3.01
Na	1.54	1	0.90	0.90
(Nb)	1.43	5	(1.8)	1.76
O	0.73	6	3.45	3.47
P	1.10	5	2.1	2.19
Pb	1.46	4	1.5	1.56
(Po)	1.46	6	(2.0)	1.98
Rb	2.11	1	0.78	0.79
S	1.04	6	2.53	2.58
Sb	1.41	5	1.8	1.82
Sc	1.61	3	1.3	1.27
Se	1.17	6	2.4	2.35
Si	1.17	4	1.8	1.82
Sn	1.40	4	1.7	1.61
Sr	1.93	2	1.0	0.98
(Tc)	1.31	7	(2.4)	2.39
Te	1.37	6	2.1	2.08
Ti	1.45	4	1.6	1.57
Tl	1.47	3	1.3	1.34
(V)	1.30	5	(1.9)	1.93
Y	1.75	3	1.3	1.21
Zn	1.31	2	1.2	1.21
Zr	1.58	4	1.6	1.48

*** Values given in parentheses are taken from the final column; others are derived in a manner independent of Gordy's equation.**

Values calculated by this equation are given in the last column of Table 4-1. The constants in this equation were determined from a plot of $(n' + 1)/r$ against electronegativity, shown in Fig. 4-4. The conformity of the points to a straight line is excellent except for copper, silver, and gold. It will be noted, however, that there are minor discrepancies between the single-bond covalent radii given here and in Table 3-4. The number of valence electrons n' used in this relation differs from the valence v, particularly in the case of the nonmetals; for sodium n' and v both have the value 1, whereas for oxygen n' is 6 and v is 2. A serious discrepancy between Pauling's concept, as expressed by the metallic

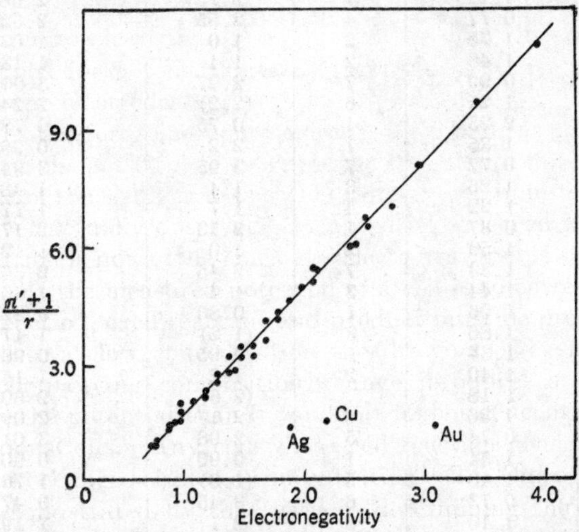

FIG. 4-4. The relation between electronegativity x and $(n' + 1)/r$, for the elements listed in Table 4-1 (except those in parentheses). r is the single-bond covalent radius, and n' is the number of valence electrons. [*From W. Gordy, Phys. Rev.*, **69**, 604 (1946).]

valences given in Table 3-4, and Gordy's, as expressed by the number of valence electrons given in Table 4-1, arises in the vicinity of the transition elements. For example, for copper, silver, and gold Pauling gives a metallic valence of 5.44 whereas Gordy sets n' equal to 1. The wide discrepancy in Fig. 4-4 for these elements is greatly diminished if n' is set equal to Pauling's metallic valence v. In fact it might be suggested, in view of this, that, in using Gordy's equation to calculate the electronegativity, Pauling's value for v be used in place of n' for the class I elements. The validity of Pauling's concept of metallic valence has received further confirmation from Hume-Rothery and Christian,[1] who found that the form of the equilibrium phase diagram of iron with tita-

[1] W. Hume-Rothery and J. W. Christian, *Phil. Mag.*, **36**, 835 (1945).

nium, vanadium, chromium, manganese, cobalt, or nickel (elements immediately preceding and following iron in the periodic table) can be understood if Pauling's ideas are combined with the already existing theory of alloy structure.

The numerical factor to convert from electron volts per atom to calories per gram atom is 23,060. Hence the heat of formation Q of a gas molecule containing single bonds from elements in normal covalent states[1] is

$$Q = 23{,}060\Sigma(x_a - x_b)^2 \tag{4-2}$$

where the summation is to be carried over all the bonds in the molecule; $(x_a - x_b)$ is the difference in electronegativity of the elements joined by the single bond under consideration. For nitrogen and oxygen the normal gaseous standard states $N_2(g)$ and $O_2(g)$ are much more stable than they would be if the molecules involved only single covalent bonds. This extra stability amounts to 110,200 cal/mole for N_2 and 48,400 cal/mole for O_2; hence for molecules involving nitrogen and oxygen the foregoing equation becomes

$$Q = 23{,}060\Sigma(x_a - x_b)^2 - 55{,}100n_N - 24{,}200n_O \tag{4-3}$$

in which n_N and n_O are the numbers of atoms of nitrogen and oxygen in the molecule. This rather astonishing simplification of thermochemical data is brought about largely, as Pauling states, "through the corrections for the two elements nitrogen and oxygen, which are alone among the elements in having their standard states much different in stability from single-bonded states."

Since the heat of formation reflects the formation of bonds with at least partial ionic character, we should expect a relationship between the amount of ionic character and the difference in electronegativity of the two elements joined by the bond. This relationship is shown in Fig. 4-5. By use of this figure and the table of electronegativity, Table 4-1, we see, for example, that the bond in HCl is about 20 per cent ionic and those in SiO_2 50 per cent ionic.

In the foregoing, then, a semiquantitative theory has been developed correlating differences in electronegativity of elements, amount of ionic character of the bond, and heat of reaction. From this it is clear that, if the difference in electronegativity of two elements under consideration is large, the heat of compound formation is also large and exothermic, or, in general, the compound is very stable.[2]

[1] Such states are $H_2(g)$, C (diamond), $F_2(g)$, Si(c), $P_4(g)$, $S_8(g)$, $Cl_2(g)$, Ge(c), $As_4(g)$, $Se_8(g)$, $Br_2(g)$, and $I_2(g)$.

[2] For the present we overlook the distinction between energy and free energy and tentatively assume that, as in a mechanical system, the state of lowest energy is the stable state. This is a crude approximation, but sufficient for the inference made.

It also follows, in case the difference in electronegativity is large, that the formation of a primary solid solution, similar to that of a compound, may be expected to be rather highly exothermic, since bonds between atoms of different electronegativity are formed in either case. However, it frequently happens that there is a crystallographic arrangement different from that of the primary solid solution, such that each atom of one element forms more bonds with the other element and fewer with its own kind. Such an arrangement constitutes an intermetallic compound and will be discussed further in the next section. The intermetallic compound tends to form at the expense of the primary solid solution when the difference in electronegativity is large. In view of the fact that a

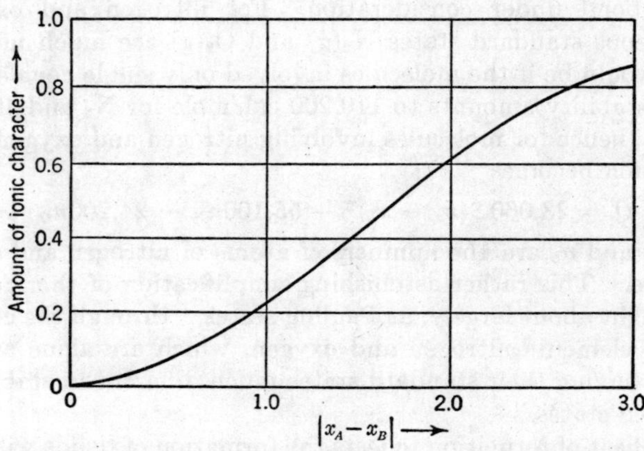

FIG. 4-5. Curve relating the amount of ionic character of a bond $A—B$ to the difference in electronegativity of the atoms, $x_A - x_B$. *(From L. Pauling, "The Nature of the Chemical Bond," Cornell University Press, Ithaca, N.Y.,* 1940.)

difference in electronegativity adds stability to both the primary solid solution and the intermetallic compound, it is difficult to assign a general value to the difference in electronegativity above which extensive primary solid solution is hindered by the tendency toward compound formation. It may be stated, however, that this difference is usually in the vicinity of one-half unit on the electronegativity scale.

Valence. In addition to the effect of difference in electronegativity, which is a measure of bond energy, upon the extent of primary solid solution, we must also consider the effect of valence or number of bonds formed. The total energy involved in the substitution of an atom of element B in place of an atom of A in an A lattice is proportional not only to the square of the electronegativity difference but also to the valence of B. This is true also for any compound formed. Hence the stability of such a compound increases with the valence, electronegativity and size being constant. It would thus be expected that a given difference in electronegativity would limit the extent of solid solution much more in the case of the higher valence elements. That this is indeed so is shown by Fig. 4-6, where the solid solubility in copper of the immediately subsequent elements in the periodic table is shown. The solidus curves also are shown in this figure.

Let us now reconsider Gordy's relation [Eq. (4-1)], in particular the functional relationship, namely, that the electronegativity is a function of

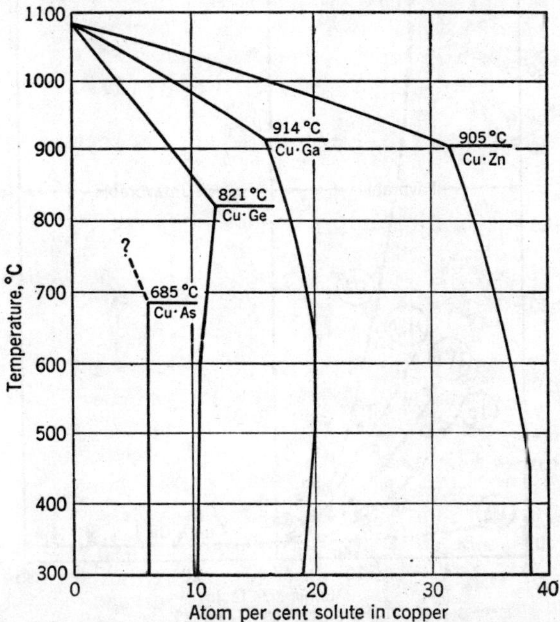

FIG. 4-6. Solidus and solid solubility curves for binary systems involving copper and the elements immediately following it in the periodic table. (*From W. Hume-Rothery, "Structure of Metals and Alloys," Institute of Metals, London,* 1944.)

the single-bond covalent radius and the valence. As factors determining the extent of primary solid solution, we have so far considered (1) size factor, (2) electronegativity, and (3) valence. However, if any relationship between these, such as that proposed by Gordy, is valid as it seems to be from Fig. 4-4, then these three factors governing the extent of primary solid solution are not independent; only two may properly be so regarded. Clearly any two may arbitrarily be selected as independent, since if x is a function of n' (or v) and r, then n' is a function of x and r and r is a function of x and n'. Thus it would seem that the extent of solid solution of various elements in a given element could be expressed as a function of the radius difference and the valence of the solute. That this is so in the case of magnesium is shown by Carapella,[1] whose plot of maximum solubility vs. percentage difference in radius, r(CN12), for various alloying elements is reproduced as Fig. 4-7. It will be observed that the solubility is negligibly small whenever the radius difference exceeds 15 per cent and that, when this difference is less than 15 per cent, the solubility for elements of the same valence (same subgroup in the

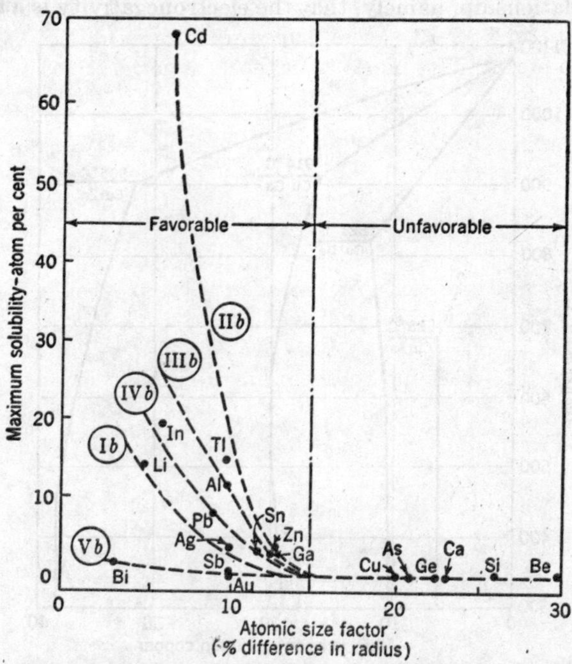

Fig. 4-7. Influence of atomic size on the solubility in magnesium of metals from various *B* groups. [*From L. A. Carapella, Metal Progress*, **48**, 297 (1945).]

periodic table) plots on a reasonably smooth curve. Thus it would seem fruitful to consider that the limitation on primary solid solubility is imposed by two factors, one geometrical in nature and the other chemical.

The simultaneous effect of size factor and electronegativity in limiting solid solution may be illustrated by a plot whose abscissa is r(CN12), from Table 3-4 and whose ordinate is the electronegativity, Table 4-1. On such a map each element is represented by a point, and it might reasonably be expected that points representing elements with high mutual solubility would lie close together. This is illustrated in Fig. 4-8 in

[1] L. A. Carapella, *Metal Progress*, **48**, 297 (1945).

which elements whose maximum solubility in magnesium is over 5 atom per cent are shown as full circles and those whose maximum solubility is known to be less than 5 per cent are shown as open triangles. It will be noticed that there is a pronounced grouping of points representing the more soluble elements. In fact, all these points are included within an ellipse whose maximum width is ± 15 per cent of the value of r(CN12) for magnesium and whose maximum height is about ± 0.4 unit in electronegativity. A similar chart for aluminum (Fig. 4-9) shows, as before,

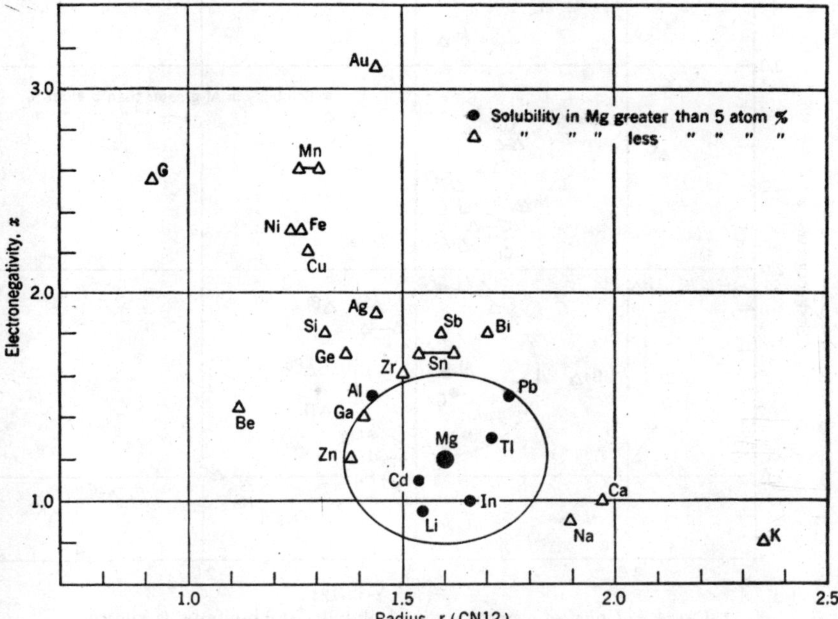

Fig. 4-8. Chart of elements whose solubility in magnesium is known.

elements with maximum solubility greater than 5 atom per cent as full circles, those with less as open triangles. The grouping in this case is not quite so good as for magnesium.

A similar chart for silver is shown in Fig. 4-10. In this case we find that, although there is pronounced grouping of the more soluble elements, the point representing silver itself does not fall in the center or even within the group. The key to this rather paradoxical situation, which occurs even more pronouncedly in the case of copper, is probably closely associated with the fact that these elements do not follow the general relationship between electronegativity, valence, and size factor shown in Fig. 4-4. This in turn seems to be closely related to the fact that the metallic valence of silver and copper (5.44, Table 3-4) differs markedly from the more familiar valence manifested in salts, from the heat of formation of which the value of the electronegativity was obtained. In fact it will be observed that, if copper and silver are arbitrarily assigned electronegativities such that these elements fall near the centers of their respective groups ($x_{Cu} = 1.5$ and $x_{Ag} = 1.4$), then the points corresponding to these elements fall in the expected positions in Fig. 4-10.

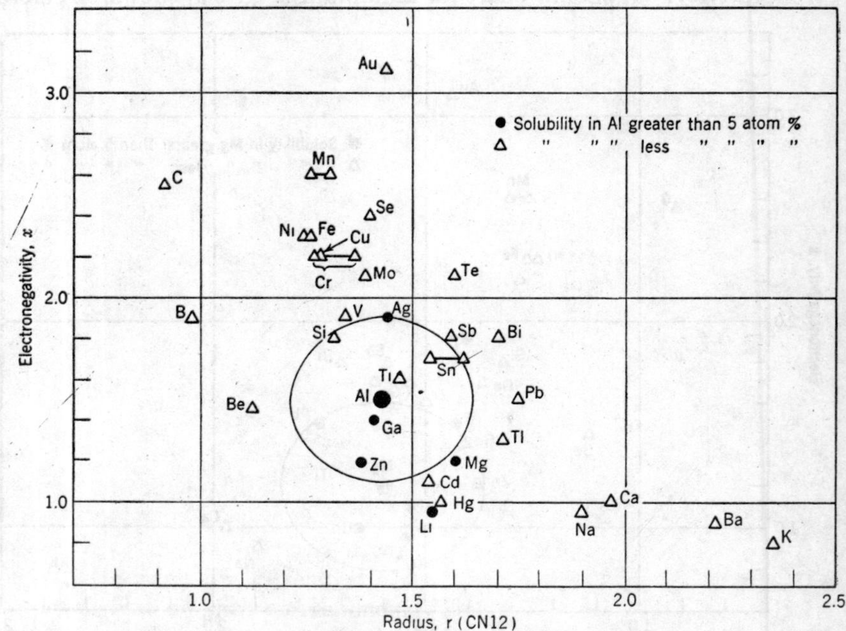

FIG. 4-9. Chart of elements whose solubility in aluminum is known.

The Relative Valency Effect. In considering whether the tendency for two metals to form solid solution is reciprocal, Hume-Rothery and coworkers[1] found that, other things being equal, a metal of lower valence is more likely to dissolve one of higher valence than vice versa. For example, the solubility of silicon in copper is 14 atom per cent whereas that of copper in silicon is less than 2 atom per cent, the size factor being favorable. In the system copper-magnesium where the size factor is unfavorable, the solubility of magnesium in copper is 6.5 atom per cent and of copper in magnesium is 0.01 per cent. However, it will be noted that, in this case and in all other cases involving copper or silver, the

[1] W. Hume-Rothery, G. W. Mabbott, and K. M. Channel-Evans, *Phil. Trans. Roy. Soc.*, **233**A, 44, (1934).

success of the rule hinges on assigning a valence of 1 rather than Pauling's value of 5.44 to these elements. The class III elements, crystallizing in accord with the $(8 - N)$ rule, seldom take into solid solution any appreciable quantity of a group I element. However, an element of class III sometimes forms extensive solid solution with another element of the same group in the periodic table; *e.g.*, bismuth forms a continuous series of solid solutions with antimony.

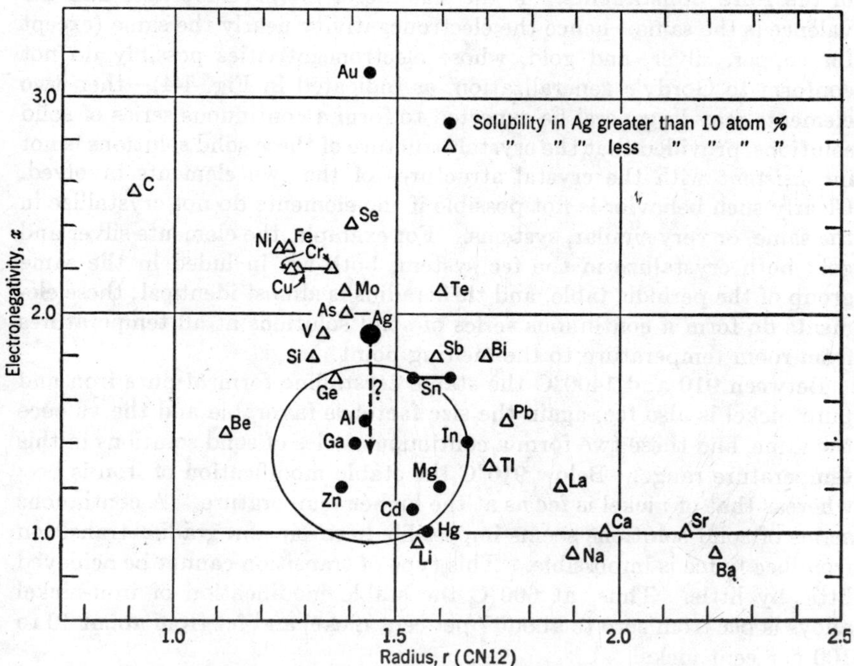

Fig 4-10. Chart of elements whose solubility in silver is known.

Referring again to Fig. 4-7, it will be noticed that the solubility of lithium (valence 1) in magnesium (valence 2) is much lower than might be expected were the curve labeled I*B* absent; this curve obviously does not fall in sequence with the others. This might lead us to infer a generalization of the relative valency rule, namely, that a disparity in valence is conducive to lower solubility and that this disparity has an especially pronounced effect when the valence of the solute is less than that of the solvent.

Incidentally, it should be noted that the position of silver and gold in Fig. 4-7 is in general accord with either the valence 1 assigned by Hume-Rothery or 5.44 designated by Pauling. However, the extensive earlier work on copper and silver alloys by Hume-Rothery and coworkers led to very fruitful results under the assumption of unit valence for these elements. They showed that most divalent or polyvalent elements are more soluble in copper or silver than is copper or silver in these other elements.

Limitation Imposed upon Extent of Solid Solution by Crystal Nature of the Pure Constituents. If the size factor is very favorable and the valence is the same—hence the electronegativity nearly the same (except for copper, silver, and gold, whose electronegativities possibly do not conform to Gordy's generalization, as indicated in Fig. 4-4)—then two elements may in general be expected to form a continuous series of solid solutions, provided that the crystal structure of these solid solutions is not inconsistent with the crystal structures of the two elements involved. Clearly such behavior is not possible if the elements do not crystallize in the same, or very similar, systems. For example, the elements silver and gold both crystallize in the fcc system, both are included in the same group of the periodic table, and their radius is almost identical; these elements do form a continuous series of solid solutions at all temperatures from room temperature to the melting point.

Between 910 and 1400°C the stable crystalline form of pure iron and pure nickel is also fcc, again the size factor is favorable and the valence the same, and these two form a continuous series of solid solutions in this temperature range. Below 910°C the stable modification of iron is bcc, whereas that of nickel is fcc as at the higher temperature. A continuous series of solid solutions seems impossible here, since a gradual transition from bcc to fcc is impossible. This type of transition cannot be achieved little by little. Thus, at 600°C the stable modification of iron-nickel alloys is bcc from zero to about 7 per cent nickel and fcc from about 10 to 100 per cent nickel.

There are certain types of crystalline arrangements between which transition can be made gradually, *e.g.*, between hcp structures with different values of c/a. Similarly, the face-centered-tetragonal structure may be regarded as an elongation, along one axis, of the fcc structure. Manganese crystallizes in face-centered-tetragonal form over a limited range of temperature in the vicinity of 1200°C. There is some question as to whether it forms a continuous series of solid solutions with iron which is fcc in this temperature range.

Polarizability or Ionic Deformation. So far, the principal factors considered as determining the role which an ion or atom plays in a crystal structure have been (1) its size and (2) its charge (or valence). Goldschmidt[1] in his original work, now regarded as the foundation of modern

[1] V. M. Goldschmidt, Geochemical Distribution Laws VII and VIII, Oslo, 1926.

crystal chemistry, considered the polarization properties to be of equal importance to these other two. In the interpretation of some properties the polarization is of minor importance; this is true if one restricts himself to (1) crystals of high symmetry, (2) crystals composed of ions of low charge, or (3) properties which do not involve asymmetric force fields. Polarizability (ability to form an electric dipole) plays a pronounced role, perhaps even more important than the other two factors, in the structure-sensitive properties and in surface chemistry. The particularly important role of this property in catalysis and surface chemistry has recently been emphasized by Weyl.[1]

In pure metals the high symmetry and covalent nature of the bonds in the interior lead to a low polarization. At the surface this is no longer true, for the metal is highly polarizable by virtue of the high mobility of the free electrons or electron pairs. Surface atoms, having fewer nearest neighbors and hence less opportunity for resonance, behave less metallically. It has been shown[2] that nickel films of a thickness below 400 A have a negative temperature coefficient of electrical resistance (*i.e.*, the nickel has become a semiconductor); deviations from Ohm's law were also found.

It has also been found[3] that the conductivity of platinum foil (20,000 A thick) depends on the gaseous atmosphere. For instance, helium (a gas of low polarizability) increased the conductivity 0.1 per cent over that in vacuo. Argon and oxygen, which are more polarizable, were found to be three and five times as effective as helium, respectively. This illustrates that the surface effect must penetrate many atom layers (several hundred at least) and that the abrupt transition (metal-vacuum) may be made less abrupt by a polarizable layer.

The polarizability (A^3) of some ions is as follows:[4]

Li^+	11.7	Mg^{++}	11.4	F^-	25.8	Cu^+	27.2
Na^+	20.2	Ca^{++}	34.9	Cl^-	62.8	Ag^+	39.4
K^+	42.3	Sr^{++}	49.1	Br^-	81.0	Tl^+	90.1
Rb^+	58.1	Ba^{++}	70.6	I^-	109.4		
Cs^+	71.3			O^{--}	30.0		
				S^{--}	($\sim$100)		

Here, perhaps, is an interpretation of the high surface activity of sulfur, so important metallurgically (as in hot shortness of steel).

In ionic crystals the surface may be expected to be formed principally of the ions with the higher polarizability; these may adjust their electron

[1] W. A. Weyl, *Trans. N.Y. Acad. Sci.*, II, **12**, 245 (1950).

[2] Von Itterbeek and de Greve, *Nature*, **158**, 100 (1946).

[3] W. Braunbeck, *Z. Naturforsch.*, **3a**, 216 (1948).

[4] S. Roberts, *Phys. Rev.*, **76**, 1215 (1950); *ibid.*, **77**, 258 (1950).

distribution the more and thus decrease the space field and lower the free energy. It will be noted that ions exert forces (at normal crystalline distances) which correspond to fields of several million volts per centimeter. Because of the symmetry these nearly cancel in the interior of an ionic crystal, but not at the surface. A systematic treatment of polarization has been given by Fajans and coworkers.[1]

The effect of polarizability may be expected to be found in the thermodynamic properties of multicomponent alloy systems where pronounced asymmetry of atoms in the first coordination shell may be encountered. There is, however, no available evidence on this subject.

ORDER-DISORDER TRANSFORMATION AND THE SUPERLATTICE

Long-range Order. For a long time it was supposed that the replacement of atoms of one element by those of another to form a primary substitutional solid solution took place in a purely random manner, *i.e.*, that

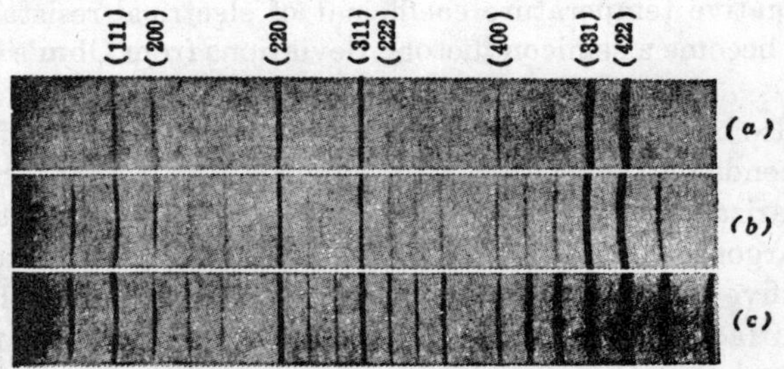

Fig. 4-11. Powder X-ray diffraction pattern of the superlattice $AuCu_3$: (*a*) disordered, (*b*) partially-ordered, (*c*) highly ordered. [*From C. Sykes and H. Evans, J. Inst. Metals*, **58**, 255 (1936).]

the resulting solution, in whatever way it happened to be formed, was composed of a random arrangement of solute and solvent atoms upon the lattice sites. The strongest evidence that this is not always the case is furnished by the occurrence of superstructure lines on X-ray diffraction patterns. For example, in the copper-gold system, the X-ray diffraction pattern of a quenched alloy containing 25 atom per cent gold, shown in Fig. 4-11*a*, is very similar to that of both pure copper and pure gold. However, if the same alloy is slowly cooled, its diffraction pattern is observed to contain not only these lines but also numerous additional lines as shown in Fig. 4-11*c*. The interpretation placed upon these additional lines is that they are occasioned by a superlattice or superstructure. In order to understand this phenomenon it must be recalled that the

[1] K. Fajans, "Chemical Forces and Optical Properties," McGraw-Hill Book Company, Inc., New York, 1931; other references in *Chem. Eng. News*, **27**, 900 (1948).

scattering factor (reflectivity) of atoms of different elements increases directly with the atomic number, and in the alloy under consideration, if the atoms are ordered in a manner similar to that shown in Fig. 4-12, then the two X-ray beams shown to the right in this figure, whose paths differ by one-half of the wave length λ, will not cancel as they would if all atoms were alike but produce an additional line on the X-ray plate. The unit cube of the ordered structure for this composition is shown in Fig. 4-13.

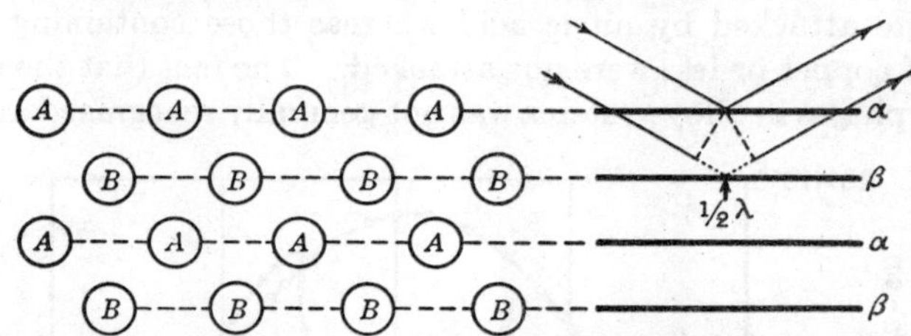

Fig. 4-12. Explanation of the origin of a superstructure line. [*From Nix and Shockley, Rev. Modern Phys.*, **10**, 1 (1938).]

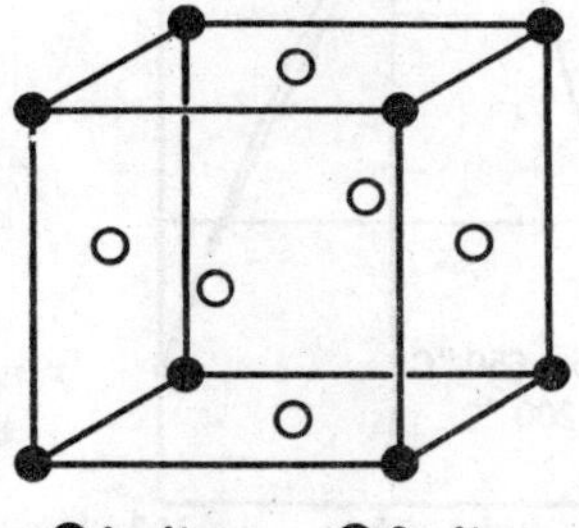

Fig. 4-13. The superlattice of $AuCu_3$ (cubic). (*From C S. Barrett, "Structure of Metals" McGraw-Hill Book Company, Inc., New York*, 1952.)

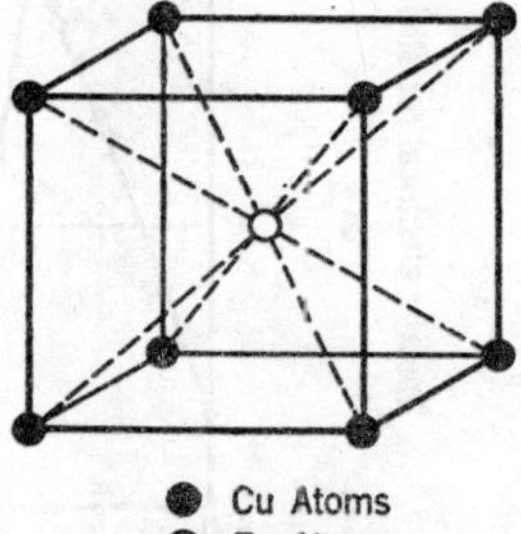

Fig. 4-14. The superlattice of β-brass (cubic, CsCl type). (*From Barrett.*)

Figure 4-11*b* is a diffraction pattern of the partially ordered alloy produced by cooling at an intermediate rate.

In the gold-copper system there is also another ordered arrangement which occurs in annealed specimens of composition 47 to 53 atom per cent. The ordering in this case consists in the formation of alternate layers of copper and gold atoms on the 001 planes; this ordering distorts the lattice so that it is not fcc but face-centered tetragonal with an axial ratio of 0.93.

The earliest superstructure lines were observed in this system by Bain[1] in 1923. Since that time ordered arrangements have been found in many alloy systems. The structure shown in Fig. 4-13 for $AuCu_3$ has also been

[1] E. C. Bain, *Chem. & Met. Eng.*, **28**, 65 (1923).

observed in the following alloys: $PtCu_3$, $PdCu_3$, $FeNi_3$, $MnNi_3$. The type of superlattice shown in Fig. 4-14 is characteristic of alloys of the following compositions: CuZn, CuBe, CuPd, AgMg, AgZn, AgCd, AuNi, NiAl, FeCo.

It is interesting to note that in 1919 Tammann[1] postulated the existence of such ordered alloys from their chemical behavior. He observed that annealed copper-gold alloys containing more than 50 atom per cent copper were attacked by nitric acid whereas those containing 50 atom per cent of copper or less were not attacked. The fact that the existence of ordered phases in alloy systems was not generally recognized until fairly

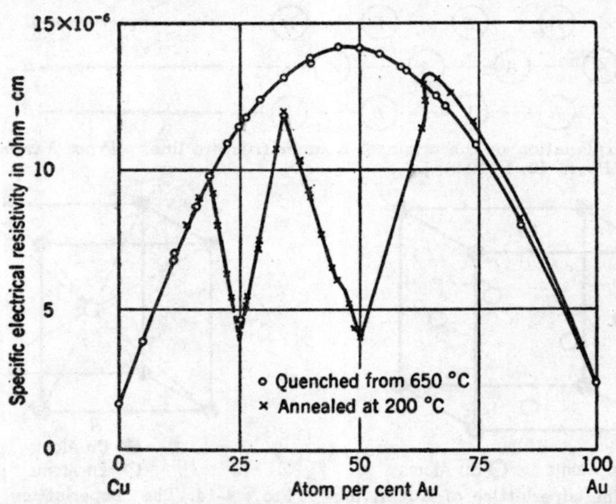

FIG. 4-15. The variation of electrical resistivity with composition for the copper-gold system. Ordering is indicated in alloys annealed at 200°C but not in those quenched from 650°C. (*From Barrett.*)

recently should not be interpreted as indicating that the change in properties accompanying the order-disorder transition is small, for such is indeed not the case. For example, the electrical resistivity of annealed and quenched copper-gold alloys is compared in Fig. 4-15. The resistivity is lowest at points corresponding to the ordered compositions mentioned previously.

At elevated temperature the effect of thermal agitation may become sufficiently great to disrupt the order. Accordingly, on heating, the superlattice disappears sharply, or nearly sharply, at a certain temperature. The disordered lattice which prevails above this temperature may be retained at lower temperature by quenching, as indicated previously. The disappearance of the superlattice at a critical point on heating bears a

[1] G. Tammann, *Z. anorg. allgem. Chem.*, **107**, 1 (1919).

strong resemblance to the disappearance of the lattice itself on melting. As will be pointed out in the next chapter, however, the lattice structure is not completely eradicated on melting, and in the same way the superlattice does not disappear at this critical point. To continue the analogy, just as energy (heat of fusion) must be supplied to melt a crystal, so must energy be supplied to destroy the superlattice. Although the major part of the superlattice structure disappears isothermally on heating, just as the major part of the lattice structure disappears at the melting point, the disappearance of the superlattice at the critical point is not accompanied by an isothermal heat absorption analogous to the heat of fusion. Instead, this heat absorption is spread out over a range of temperature below the critical temperature and appears as an increase in the heat capacity in this range; this is shown in Fig. 4-16 for β-brass (CuZn), where it will be noted that the heat capacity just below the critical temperature is well over twice the Dulong and Petit value of 6 cal/(gram atom)(deg). Such a phase change, where the heat capacity suffers a discontinuity but the heat content does not, is referred to as a *phase change of the second kind*. The magnetic transformation of iron at the Curie point (760°C) is another example of this type of phase change; it will be recalled that here, too, the change involves a disordering, but in this case a disordering of the magnetic domains.

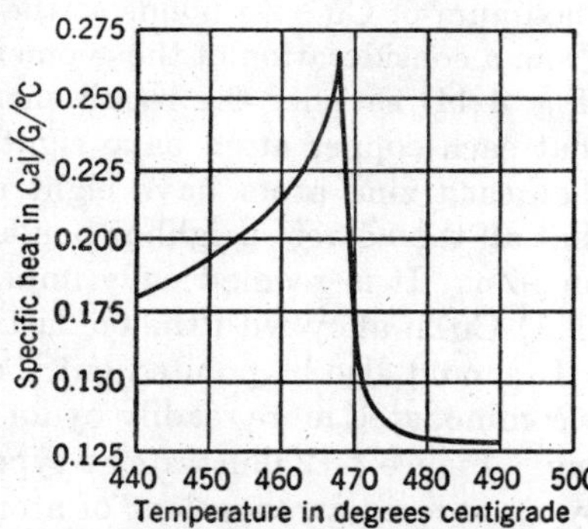

Fig. 4-16. The specific heat of β-brass (50.4 per cent Zn) in the temperature range where order disappears. (*From Barrett.*)

Mechanical properties also change when an order-disorder reaction occurs. Hardness, tensile strength, and elastic limit are usually greater for the ordered phase than for the disordered. Plastic deformation, or cold work, frequently destroys order and produces the disordered phase. The difference between the behavior of such an alloy and a pure metal is pronounced: Annealing softens a pure metal but hardens an alloy capable of ordering; in the latter case the observed effect produced by tempering is similar to a precipitation hardening.

The type of order discussed in the preceding paragraphs is known as *long-range order*. Frequently such an ordered phase is of the same crystal type as the parent elements; this is true of the ordered phases in the copper-gold system. It also happens that such a phase is of different crystallographic structure from the parent metals; such is the case for β-brass. In either case these phases are frequently referred to as *inter-*

metallic compounds. This use of the word "compound" should be interpreted without prejudice, as there is no intention to imply any meaning other than has been expressed in the preceding paragraphs of this section.

In the case of β-brass of composition CuZn, which is bcc whereas copper is fcc, the effect of the zinc addition in favoring the change from fcc to bcc is relatively easy to understand. Reference to Table 4-1 shows a rather large difference in electronegativity between copper and zinc ($\Delta x = 1.0$), in other words, the Cu—Zn bond is relatively stable as compared to either the Cu—Cu or Zn—Zn bond. It is reasonable, therefore, to expect that that structure will be stable which is capable of providing a maximum of Cu—Zn bonds at the expense of Cu—Cu or Zn—Zn bonds. From a consideration of the geometry it is clear that the bcc arrangement (Fig. 4-14) satisfies the requirement, for the ordered structure at least, that each copper atom have eight zinc atoms as nearest neighbors and that each zinc atom have eight copper atoms as nearest neighbors, so that all the nearest neighbor bonds are Cu—Zn and none are Cu—Cu or Zn—Zn. It is geometrically impossible to satisfy this condition for the (1:1) CuZn alloy with the fcc lattice.

It should also be pointed out that a difference in atomic radius can be accommodated more readily by an ordered structure than by a disordered one. Figure 3-12 illustrates a type of close packing that can be achieved by an ordered arrangement of atoms with large disparity in size. Note, however, that an ordered bcc arrangement cannot tolerate too great a size disparity. Resorting to the hard-sphere analogy, if the central sphere is too small, the eight corner spheres will be in mutual contact and the central sphere cannot be tangent to all the others. This point is reached when the radius ratio equals $\sqrt{3} - 1$ or 0.73. Experimentally the bcc β-phase is not found when the radius ratio is less than this.

As just pointed out, the bcc arrangement is particularly suited to an ordered phase of 1:1 composition, as CuZn, in that all nearest neighbors are unlike and the size disparity may be easily accommodated.[1] It is of interest to note that the fcc arrangement is particularly favorable to an ordered phase of 3:1 composition, as Cu_3Au, as may be seen from Fig. 4-13.[2] In this arrangement each atom of the minor constituent (the

[1] It is true that the copper-gold system forms an fcc ordered phase at the composition CuAu; however, each atom has as neighbors eight like and four unlike atoms. Apparently the tendency of both copper and gold to crystallize in the fcc system predominates here.

[2] It should be noted that ordered fcc phases of 3:1 composition seldom, if ever, occur except when both parent metals crystallize in this system. Ordered phases corresponding to the compositions Cu_3Al and Cu_3Ga are bcc, contrary to what might be inferred from the foregoing statements.

black circles) has 12 nearest neighbors of the major constituent (the white circles), and each atom of the major constituent has 12 nearest neighbors of which 4 are the minor constituent and 8 the major. Clearly at the composition 3:1 an arrangement in which each atomic species has only the other as nearest neighbors is not to be expected.

The iron-aluminum system involves a very interesting and somewhat more complicated set of superlattice arrangements. The superlattices in

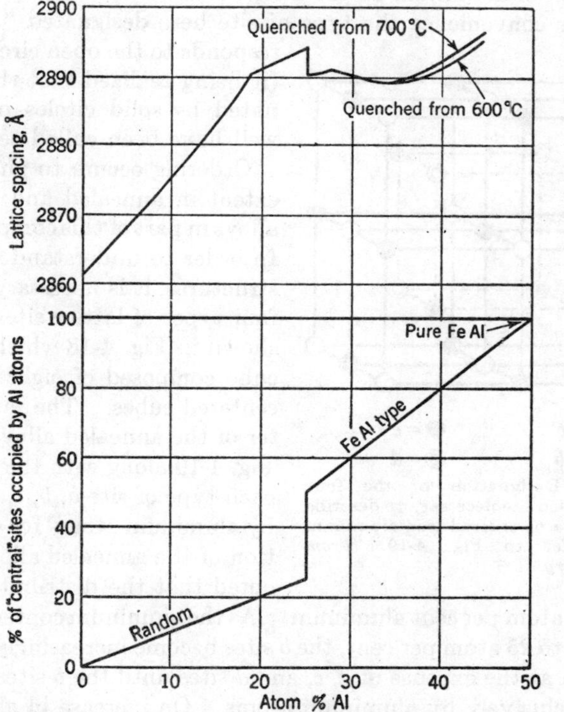

FIG. 4-17. The lattice spacing and crystal structure of quenched iron-aluminum alloys.

this system have been thoroughly investigated by Bradley and Jay,[1] whose work constitutes one of the highest peaks yet reached in the application of X-ray crystal analysis to the study of alloys. Let us first consider a series of alloys containing 0 to 50 atom per cent aluminum quenched from 600 and 700°C. The lattice parameters are shown in Fig. 4-17. Bradley and Jay found that in these quenched alloys the distribution of the aluminum atoms was random in the range 0 to 25 atom

[1] A. J. Bradley and A. H. Jay, *J. Iron Steel Inst.*, **125**, 339 (1932); *Proc. Roy. Soc. (London)*, **A136**, 210 (1932).

per cent aluminum; *i.e.*, no superlattice was detected. However, in the range 25 to 50 atom per cent a superlattice was found even though the alloys were quenched. This superlattice is of the same type as that of β-brass shown in Fig. 4-14. It will be noted that at 25 atom per cent where, upon increasing the aluminum content, the structure changes from disordered to ordered, there is also a discontinuous decrease in the lattice parameter. Here is direct evidence that the ordered arrangement provides a better fit of the atoms and hence a more dense structure. The distribution of aluminum on the lattice is shown in the lower half of Fig. 4-17. For convenience the type of site here designated "central" corresponds to the open circle of Fig. 4-14 (it being realized that the sites designated by solid circles might equally well have been called central).

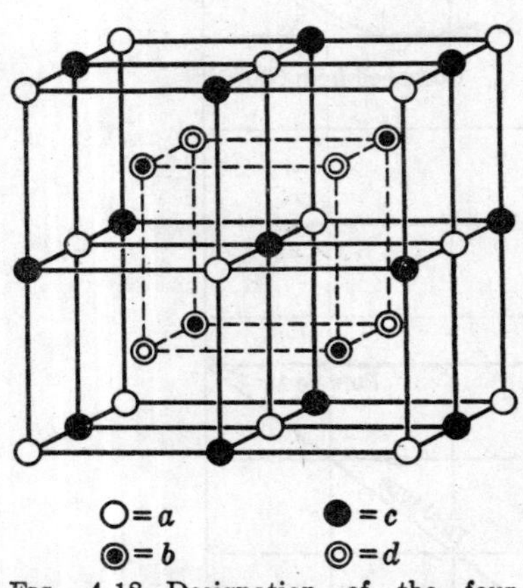

FIG. 4-18. Designation of the four types of lattice site necessary to describe the structure of ordered iron-aluminum alloys. Refer to Fig. 4-19. (*From Hume-Rothery.*)

Ordering occurs to an even greater extent in annealed and slowly cooled alloys in part of this composition range. In order to understand these ordered structures, it is necessary to designate four types of lattice sites: *a*, *b*, *c*, *d*, as shown in Fig. 4-18 which represents a cube composed of eight simple body-centered cubes. The lattice parameter of the annealed alloys is shown in Fig. 4-19 along with the percentage of each type of site *a*, *b*, *c*, or *d* occupied by aluminum atoms for each composition of the annealed alloy. It will be noted that the distribution is random up to 18.5 atom per cent aluminum. As the aluminum content is increased from 18.5 to 25 atom per cent, the *b* sites become increasingly preferred by aluminum at the expense of *a*, *c*, and *d* sites until the *b* sites are occupied almost exclusively by aluminum atoms. On increase in aluminum content up to about 30 atom per cent, additional aluminum atoms occupy *d* sites leaving the *b* sites nearly filled, on further increase the *b* and *d* sites tend to become equally preferred, and at about 38 atom per cent the structure becomes the same as that of the quenched alloys in which *b* and *d* sites are equally occupied.

In connection with the rather extensive ordering found in the iron-aluminum system, it is interesting to recall the fact that, although the coordination number for bcc is 8, each atom has six more near neighbors which are only slightly farther away than the eight nearest neighbors.

On the basis of Pauling's equation, about one-tenth of the total number of bonds is associated with these six next nearest neighbors. This deduction that an appreciable fraction of the bonds is associated with other than the nearest neighbors is in accord with the rather complicated superstructure, for it will be noticed that the ordered structure of the 25

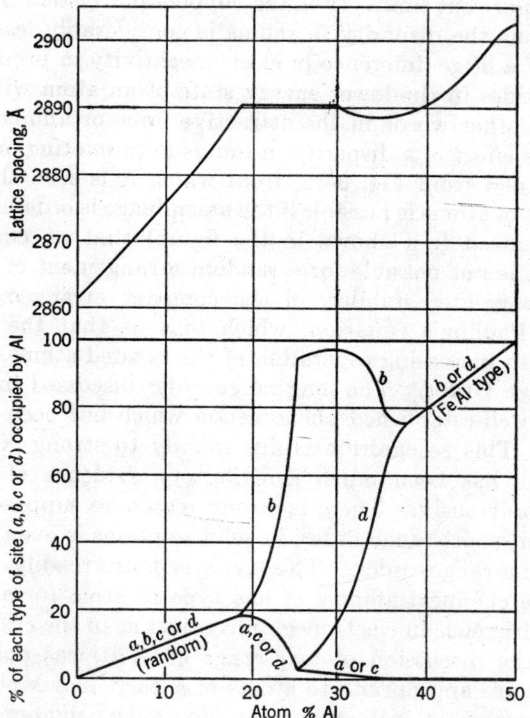

Fig. 4-19. Lattice spacing and crystal structure of annealed iron-aluminum alloys. The letters refer to types of lattice site shown in Fig. 4-18.

atom per cent aluminum alloy is such that none of the 14 nearest and next nearest atoms to an aluminum atom is another aluminum atom. This fact is sometimes described by saying that an aluminum atom has no aluminum neighbor in either the first or second *coordination zones*. The bond number n of the atoms in the second coordination zone is much smaller than the bond number for the first coordination zone; hence the forces responsible for the complicated superstructure of the annealed alloys are much weaker than those responsible for the simpler ordering of the quenched alloys.

It is only reasonable to inquire as to the nature of the difference of the iron-aluminum system from other systems in which the rather complicated ordering of the iron-aluminum system has not been observed. This question may be resolved by reference to Table 4-1 which shows a fairly large difference in electronegativity between iron and aluminum, $\Delta x = 0.8$, and to Table 3-4 which shows a disparity in radius which is near the 15 per cent limit. For the copper-zinc system Δx is about the same (1.0) but the disparity in radius is considerably less (8 per cent). The effect of a large difference in electronegativity in promoting ordered structure resides in the lower energy state of an atom with a dissimilar neighbor, in other words in the attractive force or affinity of dissimilar atoms. The effect of a disparity in radius in promoting order is perhaps best understood from Fig. 3-12, from which it is seen that a compact arrangement of atoms is possible if the assemblage is ordered, whereas it is readily visualized (not shown in the figure) that an equally compact arrangement is not possible for a random arrangement of different-sized atoms. The greater stability of the compact arrangement is in part implied by Pauling's equation, which tells us that the bond number decreases with increasing separation of the bonded atoms.

Short-range Order. The long-range order discussed in the foregoing section is a well-established phenomenon which has been recognized for some time. This recognition is due mainly to strong X-ray evidence, though there has been much contributory evidence of other nature, mentioned previously. There is strong reason to suppose that another type of order, short-range order, in solid solutions is even more common than the long-range order. This type is most readily understood in terms of the chemical affinity of one type of atom for another, *i.e.*, in terms of a difference in electronegativity, just as in the case of long-range order. In the discussion of long-range order it was pointed out that certain sites are appropriate to atoms of a particular kind, the arrangement being such that each atom has more unlike neighbors than it would in a purely random arrangement. This same state of affairs can also prevail, though perhaps not to so great an extent, even when there is no long-range order. For example, in a solution of sodium chloride each sodium atom has statistically more Cl^- than Na^+ ions in its immediate vicinity, although there is not even a lattice in this case.

Similarly it does not seem at all unreasonable to suppose that β-brass has some order of this type even above the transformation temperature at which the long-range order disappears. The difference in electronegativity still persists, and the affinity of copper atoms for zinc is not altered by the fact that the superlattice is destroyed by thermal agitation above the transformation temperature. Evidence of the persistence of some short-range order above this temperature is found in the specific heat curve of Fig. 4-16, which shows that the specific heat is abnormally high in this range. The specific heat of iron is abnormally high and decreases with rising temperature for 150°C above the Curie point. The high specific heat in the case of β-brass and other ordered structures reflects the thermal energy required to break down the ordered arrangement and replace unlike neighbors by like.

Since the different natures of different chemical elements commonly lead to a greater affinity between unlike atoms, as has already been made clear under the discussion of electronegativity, it seems only reasonable to expect that some extent of short-range order should be the rule rather than the exception for solid solutions. Size difference may be considered as a second factor also favoring some extent of short-range order. Furthermore, cases are known where the concept of electronegativity seems definitely misleading. For example, the solubility of silver in iron and of iron in silver is exceedingly limited in both the solid and the liquid states. These unlike atoms certainly seem to have less affinity than do like atoms of either kind. This sort of repulsion, if so it may be interpreted, may also be reasonably expected to produce a mild ordering in which more like pairs occur than in a random arrangement. In fact, any type of chemical dissimilarity of the components of a binary system may be expected to give rise to some sort of departure from random arrangement of the two types of atoms on the lattice.

Since such short-range order is more subtle and more difficult to demonstrate experimentally than the long-range type, it is not so generally recognized by metallurgists. Various theoretical treatments of both types have been reviewed in an excellent manner by Nix and Shockley.[1] The hypothesis of long- and short-range order thus expanded into a rather involved statistical theory may be regarded as established by experimental verification of the consequences or predictions thereof, just as Dalton's atomic hypothesis and the kinetic theory of gases were established, not by actual sense perception of the molecules and their thermal agitation, but by experimental verification of the relations derived mathematically from these postulates.

By this statement we do not mean to imply that the statistical mechanics of the order-disorder transformation has already been completely developed in a form adapted to direct practical application. However, rapid strides are being made in such mathematic development; Onsager[2] and Wannier[3] are among those who have made significant contributions.

[1] F. C. Nix and W. Shockley, *Rev. Modern Phys.*, **10**, 1 (1938).

[2] L. Onsager, *Phys. Rev.*, **65**, 117 (1944).

[3] G. H. Wannier, *Rev. Modern Phys.*, **17**, 50 (1945).

An approximate treatment by Birchenall and Mehl[1] has shed considerable light upon the connection between thermodynamic properties and short-range order in the copper-zinc system. Darken and Smith[2] have shown that the interstitial solution of carbon in γ-iron (austenite) is also characterized by short-range order in the sense that carbon atoms tend to avoid adjacent interstitial positions; in this case also the thermodynamic properties are interpreted quantitatively in terms of this short-range order.

INTERMEDIATE PHASES IN METALLIC SYSTEMS

In many systems composed of two or more metallic elements there appear under some conditions of composition and temperature phases other than primary solid solutions. These are commonly known as *intermediate phases.* From the discussion in the preceding section it is already realized that the difference between a primary solid solution and an intermediate phase need not be great. We may consider the following as a logical sequence of steps between the primary solid solution and the intermediate phase.

1. Primary solid solution with short-range order (α solution of zinc in copper).
2. Primary solid solution with long-range order (copper-gold and iron-aluminum systems).
3. Ordered phase with crystal structure different from α solution (β phase in copper-zinc system).

It will be recognized that the third step represents an intermediate phase. β-brass is distinguishable from α-brass not only crystallographically (by means of X rays) but also metallographically and in addition by means of a variety of physical properties.

As already pointed out, the β solid solution of zinc in copper exists over a range of composition; this range broadens somewhat with increasing temperature. In other cases the range of composition of both the intermediate phase and the primary solid solution may be much smaller. In the copper-aluminum system, the maximum solubility of aluminum in the α solid solution of aluminum in copper is about 20 atom per cent, and the range covered by the intermediate phase of approximately the composition corresponding to $CuAl_2$, known as the θ phase, is about 1 atom per cent. As the difference in electronegativity of the two elements under consideration increases, both composition ranges usually decrease.

Ordinary Valence Type Compounds. In the magnesium-tin system the maximum solubility of tin in magnesium is about 5 atom per cent, the intermediate phase has the composition Mg_2Sn, and its composition

[1] C. E. Birchenall and R. F. Mehl, *Trans. AIME,* **171,** 143 (1947).

[2] L. S. Darken and R. P. Smith, *J. Am. Chem. Soc.,* **68,** 1172 (1946) (Appendix).

range is very small. In the case of the magnesium-antimony system, where the atomic radii are nearly identical but the difference in electronegativity is about 0.6 unit, the composition range of both the α solid solution and the intermediate phase Mg_3Sb_2 is very small. It will be noted that in the last two systems mentioned the compounds formed, Mg_2Sn and Mg_3Sb_2, are essentially ionic and of the ordinary valence type. The system sodium-chlorine may be considered as the final step in this sequence; here the α solid solution of chlorine in sodium and of both sodium and chlorine in sodium chloride is small.[1] In such a case where the intermediate phase has a very small composition range and is practically completely ordered, it is commonly called a *compound;* and if,

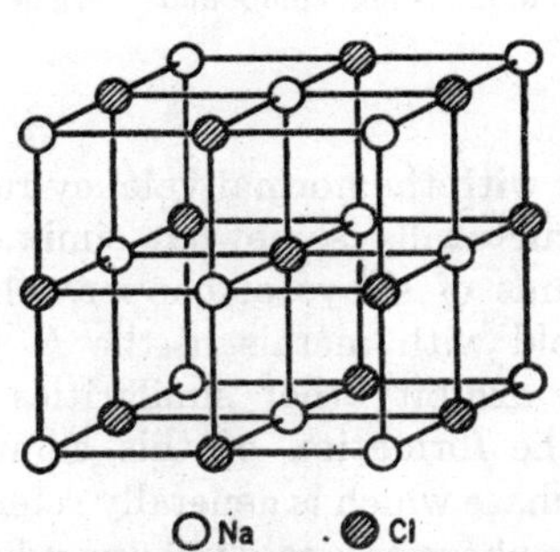

Fig. 4-20. The sodium chloride structure, exhibited by the following compounds: MgSe, CaSe, SrSe, BaSe, CaTe, SrTe, BaTe MnSe, SnTe, PbSe, and PbTe.

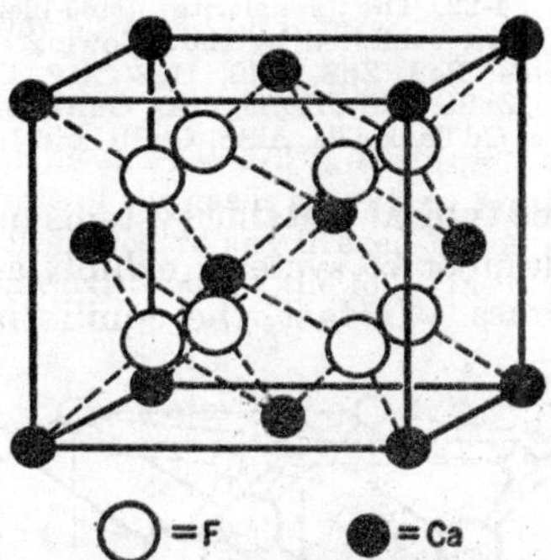

Fig. 4-21. The calcium fluoride structure, exhibited by the following compounds: Mg_2Si, Mg_2Ge, Mg_2Sn, Mg_2Pb, and Cu_2Se.

in addition, the two elements are regarded as metallic (members of Hume-Rothery's class I or II), the intermediate phase is commonly called an *intermetallic compound.* It will be noted in Fig. 3-12 that even here there is no implication of molecular structure. The metals form normal valence compounds only with the elements of Hume-Rothery's class III (including tin and also with lead).

The common crystallographic types of ionic intermetallic compound are shown in Figs. 4-20 to 4-24. Examples of each type are listed under the corresponding figure. It will be noticed that in all cases the arrangement is such that nearest neighbors are dissimilar atoms.

Intermediate Phases Conforming to Certain Electron-Atom Ratios. In addition to those intermediate phases which crystallize in one of the

[1] In the case of KCl, the ratio of potassium to chlorine atoms can be made to exceed unity by about one part per thousand by heating the salt in the presence of potassium vapor. H. Rogener, *Ann. Physik,* **29,** 386 (1937); F. Seitz, *Rev. Modern Physics,* **18,** 384 (1946).

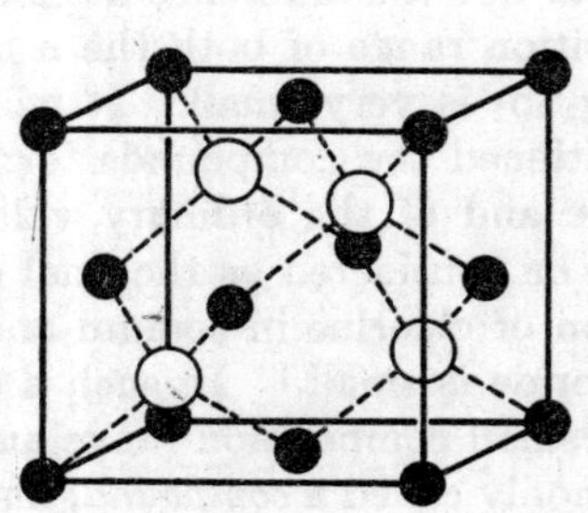

FIG. 4-22. The sphalerite (zinc-blende) structure, exhibited by the following compounds: BeS, ZnS, CdS, HgS, AlP, GaP, BeSe, ZnSe, CdSe, HgSe, AlAs, GaAs, BeTe, ZnTe, CdTe, HgTe, AlSb, GaSb, and InSb.

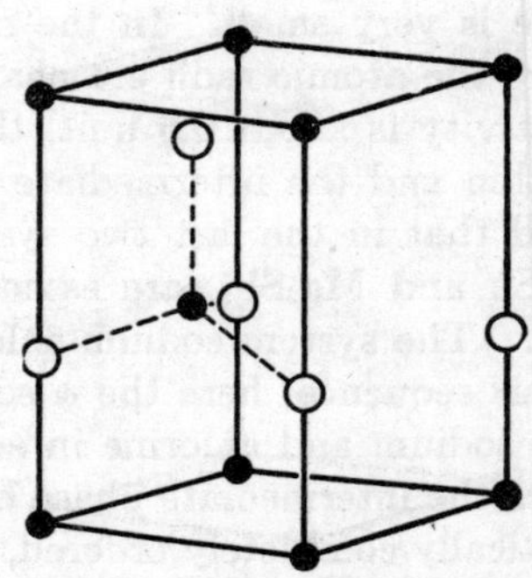

FIG. 4-23. The wurtzite structure, exhibited by the following compounds: MgTe and CdSe.

three typical metallic systems or in accord with the normal valency rules, a number of systems exhibit a phase structurally somewhat similar to γ-brass. In fact, the equilibrium diagrams of alloys of copper, silver, and gold with metals of the *B* subgroups exhibit other similarities besides the formation of this complex cubic phase which is generally referred to as a γ phase or is said to have γ-brass structure. The unit cell of γ-brass may be considered as made up of 27 unit cells of bcc β-brass with 2 atoms removed and the other 52 atoms shifted somewhat in position.

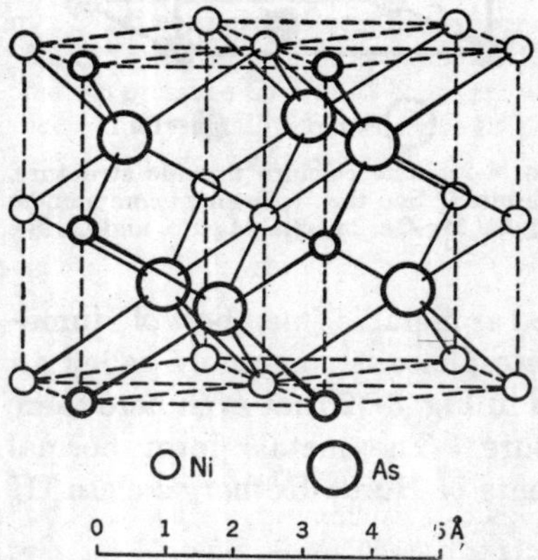

FIG. 4-24. The nickel arsenide structure, exhibited by the following compounds: CrS, CoS, FeS, NiS, CoSe, FeSe, NiSe, CrSe, CoTe, FeTe, NiTe, CrTe, MnTe, PdTe, PtTe, FeSn, AuSn, CuSn, PtSn, NiSn, CoSb, FeSb, NiSb, CrSb, MnSb, PdSb, PtSb, MnAs, NiAs, and NiBi.

Hume-Rothery points out that the ratio of electrons to atoms in these γ phases is 21:13. This is rather remarkable in that the idealized composition differs considerably for different systems as may be seen from Table 4-2. Hume-Rothery also notes that a number of intermediate phases crystallizing with bcc structure, known as β structure by analogy to β-brass, have an electron-atom ratio of 3:2, as do also some intermediate phases which crystallize with the same structure as β-manganese (complex cubic). Others, known as ε phases, crystallize in the hcp structure; in this case the electron-atom ratio is 7:4. The approximate composition of a number of such phases is given in Table 4-2. Phases approximating a 1:1 atomic ratio which crystallize in the bcc system (first column of Table 4-2) are readily understood in terms of principles already discussed as well as in terms of the electron-atom ratio. The other three phases in this column are not readily understood except on

TABLE 4-2. ELECTRON-ATOM RATIO AND STRUCTURE OF A NUMBER OF INTERMEDIATE PHASES OF IDEALIZED COMPOSITION*

Electron-atom ratio 3:2		Electron-atom ratio 21:13	Electron-atom ratio 7:4
β Structure (bcc)	β-Manganese structure (cubic, complex)	γ-Brass structure (cubic, complex)	ϵ-Brass structure (hcp)
AgCd	Ag_3Al	Ag_5Cd_8	$AgCd_3$
AgMg	Au_3Al	Ag_5Hg_8	Ag_3Sn
AgZn	Cu_5Si	Ag_5Zn_8	Ag_5Al_3
AuCd	$CoZn_3$	Au_5Cd_8	Ag_5In
AuMg		Au_5Zn_8	$AgZn_3$
AuZn		Cu_5Cd_8	$AuCd_3$
CuBe		Cu_5Hg_8	$AuZn_3$
CuZn		Cu_5Zn_8	Au_3Sn
Cu_3Al		Cu_9Al_4	Au_3Hg
Cu_3Ga		Cu_9Ga_4	Au_5Al_3
Cu_5Sn		Cu_9In_4	$CuCd_3$
CoAl		$Cu_{31}Si_8$	$CuZn_3$
FeAl		$Cu_{31}Sn_8$	Cu_3Ge
NiAl		Co_5Zn_{21}	Cu_3Si
		Fe_5Zn_{21}	Cu_3Sn
		Ni_5Zn_{21}	$FeZn_7$
		Pd_5Zn_{21}	
		Rh_5Zn_{21}	
		Pt_5Zn_{21}	
		Ni_5Cd_{21}	
		$Na_{31}Pb_8$	
		$Cu_7Zn_4Al_2$	

* From C. S. Barrett, "Structure of Metals," McGraw-Hill Book Company, Inc., New York, 1952.

the latter basis. Similarly, the phases with approximately 3:1 atom ratio, given in the last column, are easily interpreted as being very similar to ordered phases of this composition with fcc structure, previously discussed; again, however, several will be seen to have a different composition. It should be noted that in the application of Hume-Rothery's *electron-atom rule*, calling for the ratio of electrons to atoms of 3:2, 21:13, or 7:4, the number of electrons is to be taken as the group number in the periodic table except that the number of electrons assigned to the transition elements, group VIII, is zero. The actual observed range of composition for several of these phases is shown in Fig. 4-25.

The relation between the observed composition shown in Fig. 4-25 and the idealized composition will be better understood in terms of the free-energy relationships which will be discussed later. However, at this time it might be pointed out that the extension of the observed composition on either side of the idealized composition may be interpreted in terms of

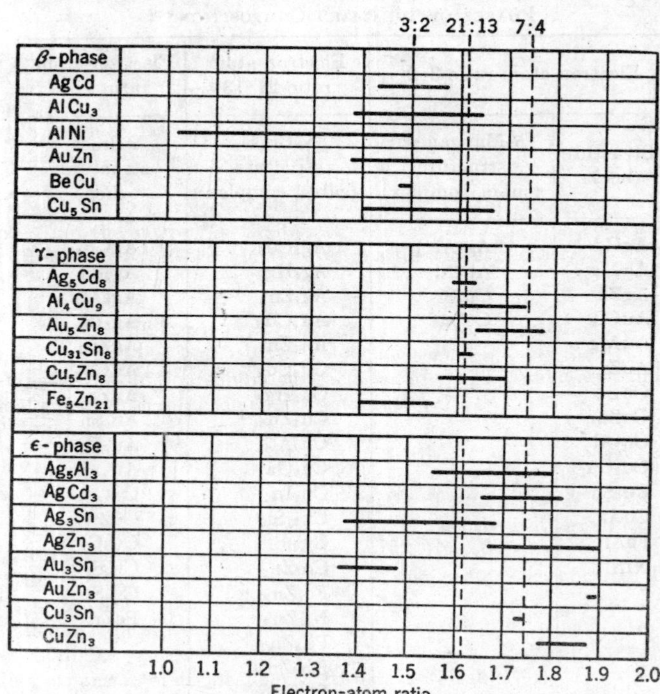

FIG. 4-25. Observed composition range for several β, γ, and ϵ phases and comparison with the Hume-Rothery electron-atom ratios. (*From Barrett.*)

the substitution of one type of atom for another. Ranges of observed composition which do not correspond at all with the idealized composition, such as the so-called γ phase of the iron-zinc system, may be tentatively viewed as associated with thermal agitation. In some cases the validity of the atomic ratio corresponding to the idealized composition, for example, Cu_5Zn_8 for γ-brass, is not definitely established. Norbury[1] has suggested, on the basis of considerable evidence, the formula $Cu_{21}Zn_{31}$ which gives an electron-atom ratio of 83:52 instead of 21:13. This is compatible with his idea that the unit cell of this structure may be

[1] A. L. Norbury, *J. Inst. Metals*, **65**, 355 (1939).

regarded as derived from 27 bcc unit cells from which two atoms have been removed and the rest displaced slightly.[1] It would perhaps seem best to regard the stability of these complex cubic structures as arising from the fact that such structures offer an opportunity for the greatest number of dissimilar nearest neighbors, for a particular system in the observed composition range, consistent with the disordering influence of thermal agitation.

The Structure of Cementite, Fe_3C. On the basis of available experimental evidence the composition range of cementite in the iron-carbon system is negligibly small.[2] Cementite is metastable in that it tends to decompose into iron and graphite, though this decomposition usually is slow. Its crystal structure is orthorhombic and of such nature that the iron atoms are reasonably close-packed,[3] some having 12 neighbors at the average distance 2.62 A and others 11 at the average distance 2.58 A. Each carbon atom occupies a position in the center of a trigonal prism of six iron atoms, the iron-carbon distance being 2.01 A. Thus, crystallographically, there are the two different kinds of iron atoms which may be designated Fe I and Fe II, respectively. Pauling[4] very fruitfully applied his equation to this intermediate phase. Using the value of $r(1)$ from Table 3-4 and $r(n)$ from the above observed interatomic distances, direct solution of Pauling's equation, $r(1) - r(n) = 0.300 \log n$, gives the following values for the bond number n:

		Bond No., n	Σn
Fe I	12Fe at 2.62 A	0.33	5.48
	2C at 2.01 A	0.76	
Fe II	11Fe at 2.58 A	0.38	5.70
	2C at 2.01 A	0.76	
C	6Fe at 2.01 A	0.76	4.56

The summation of the bond numbers given in the final column corresponds to the valence. Thus, for iron, Σn does not differ greatly from the valence given in Table 3-4.

[1] The number of atoms per unit cell, in this case 52 as determined from X-ray diffraction data, may be considered as one of the fundamentals in determining the idealized composition. If the composition is represented in the customary way, Cu_5Zn_8, then the sum of the subscripts must be 52 or an integral factor thereof. In some cases the unit cell of the γ structure is seen to contain more than 52 atoms; since it is cubic, the number of atoms contained is 52 multiplied by a factor which is the cube of an integer. Thus the unit cell for the composition $Cu_{31}Sn_8$ contains 52×3^3 atoms.

[2] E. D. Campbell, *Am. Chem. J.*, **18**, 836 (1896).

[3] H. Lipson and N. J. Petch, *J. Iron Steel Inst.*, **142**, 95 (1940).

[4] L. Pauling, *J. Am. Chem. Soc.*, **69**, 542 (1947).

However, there is a strong presupposition that the valence of carbon is 4 and not 4.56. Assuming the valences as given in Table 3-4 and the bond number given in the second column of the following table, Pauling found by trial and error that his equation leads to the interatomic distances in the third column. These are in reasonably good agreement with the observed distances given in the last column.

	Bond No., n	Interatomic distance, A	
		Calculated	Observed
Fe–C	0.67	2.04	2.01
Fe I–Fe I, Fe I–Fe II	0.37	2.59	2.60 (mean)
Fe II–Fe II	0.43	2.55	

The slight disparity between the observed and calculated distances, which is different in sign for Fe—C and Fe—Fe, may be regarded as lack of exact fit and hence as resulting in a strain of such nature that the carbon atoms are forced into somewhat undersized holes and the iron frame work is slightly expanded.

In a later article[1] Pauling emphasizes the relation between bond character and radius and points out that the above slight disparity may be interpreted in terms of an increase in the amount of *d* character in the orbitals of the iron atoms involved in Fe—C bonds and a corresponding decrease in the amount of *d* character in the orbitals involved in forming Fe—Fe bonds. This effect of bond character was mentioned in Chap. 3.

Hägg's Rule. A systematic study of carbides, hydrides, borides, and nitrides led Hägg[2] to the general conclusion that these compounds are metallic in nature—having high luster and conductivity—if formed from transition metals but nonmetallic in character if formed from other metals. On the basis of X-ray data he observed that the ratio of the radii of the constituent atoms was a factor strongly influencing the structure. For a metalloid-metal radius ratio less than about 0.59 the observed arrangement of *metal* atoms was found to be simple, usually fcc or hcp but sometimes bcc or simple hexagonal ($c/a = 1$), while for a radius ratio greater than 0.59 the structure was complicated, though in many cases the phase was still metallic in nature. The statement that the structure of these metalloid-metal compounds shall be simple or complex according as the radius ratio is less than or greater than 0.59 has become known as *Hägg's rule*.

[1] L. Pauling, *Proc. Roy. Soc.* (*London*), **A196,** 343 (1949).
[2] G. Hägg, *Z. physik. Chem.*, **B6,** 221 (1930); **B12,** 33 (1931).

More recent data[1] indicate that the borides, whether of simple or complex structure, have radius ratios both above and below the 0.59 limit; they therefore do not follow the Hägg rule. This is explained on the basis of a tendency of boron atoms to form chains or other networks with each other.

Considering the atoms as hard spheres, Hägg concluded from an analysis of the crystal structures that a metalloid atom can occupy an interstitial position in a metal lattice only when the dimensions are such as to permit contact with the surrounding metal atoms. This principle provides geometrically a lower limit for the metalloid-metal radius ratio for an interstitial structure. For example, in the fcc lattice the largest interstitial position, the octahedral position where the metalloid atom would have six nearest neighbors, can be occupied only by a metalloid atom such that the metalloid-metal radius ratio is greater than 0.41 (this figure corresponding to the size of the interstitial position for a lattice of hard spherical atoms). When the ratio is less than 0.41, the metalloid atom can maintain contact with metal atoms only if it occupies the tetrahedral interstitial position between one corner and three face-center metal atoms (four nearest metal neighbors) where contact can be maintained with a radius ratio down to 0.23. A more modern approach to this whole subject of the interpretation of structure in terms of atomic size is on the basis of Pauling's resonating-bond theory, as discussed above.

Molecules (?). It is well to recall that in all the foregoing discussion of intermediate phases in metallic systems there has been no mention of molecules or of molecular structure. In fact in all the systems discussed it is very difficult, if not impossible, to give meaning to the term "molecule" unless it be applied to the whole crystal or grain. In organic crystals and in some inorganic crystals such as H_2O and CO_2, there are local groupings of atoms to which the primary valence bonds are restricted. In such cases it is quite proper to refer to these groups as molecules. For example, in the case of water ice each oxygen atom is surrounded by four hydrogen atoms, but two of these are much closer (corresponding to a greater bond number) than the other two and may reasonably be considered as forming a molecule with the oxygen atom. However, no parallel of this situation exists in the metallic systems considered here. For example, in β-brass each zinc atom is surrounded by eight copper atoms symmetrically located. None of the eight copper atoms may be said to be associated with a given zinc atom more than the other seven. There seems to be no justification for speaking of molecules in intermediate metallic phases.

[1] R. Kiessling, *Acta Chem. Scand.*, **4**, 209 (1950).

CHAPTER 5

LIQUIDS

The most quantitative means at our disposal for describing the structure of a liquid are on the basis of experimental results and theories in the field of X-ray diffraction. Although X-ray diffraction patterns of liquids and supercooled liquids and glasses were obtained by Friedrich, Debye and Scherer, and Keesom and de Smedt in the decade following 1913, the interpretation of these patterns awaited the theoretical advances of Zernicke and Prins[1] and of Debye and Menke,[2] who completed the theoretical analysis and applied the results to liquid mercury.

THE STRUCTURE OF LIQUIDS AS DEDUCED FROM X-RAY DIFFRACTION DATA

The experimental technique used in obtaining the X-ray diffraction pattern for a liquid is similar in principle to the Debye-Scherer technique used for a polycrystalline solid. Monochromatic radiation is used. The pattern does not contain lines as does that produced by a crystalline material but shows merely gradations of intensity with angle of diffraction θ. Figure 5-1, from the comprehensive paper by Gingrich,[3] shows the variation in intensity of the diffracted beam for liquid zinc with $(\sin\theta)/\lambda$, λ being the X-ray wave length in kx units. A similar graph for liquid mercury is shown in Fig. 5-3.[4] It should be noted that in this case, as in others where distribution curves are given for several temperatures, the origin is shifted vertically to preclude overlapping.

The intensity of the diffracted beam is interpreted, as first suggested by Debye, in terms of the fact that every pair of atoms serves as the source of an interference pattern, and hence, if the distribution of atoms in the liquid were known, the pattern would be calculable. The inverse problem, namely, that of calculating the distribution of atoms from the X-ray pattern, is considerably more complicated, although involving the same principles, and was solved, as mentioned previously, by Zernicke and Prins and by Debye and Menke. Further contributions were made by

[1] F. Zernicke and J. Prins, *Z. Physik*, **41**, 184 (1927).
[2] P. Debye and H. Menke, *Ergeb. Tech. Rontgenk*, II (1931).
[3] N. S. Gingrich, *Rev. Modern Phys.*, **15**, 90 (1943).
[4] J. H. Hildebrand, *J. Chem. Phys.*, **11**, 330 (1943).

Warren and Gingrich[1] and Warren.[2] The calculation of atomic distribution is based on a minimum of hypotheses[3] in the case of liquid elements. The calculation is more complex and involves more hypotheses for compounds or solutions, and for this reason we shall restrict ourselves principally to the consideration of liquid elements. The theory leads from X-ray intensity graphs such as Figs. 5-1 and 5-3 to atomic distribution

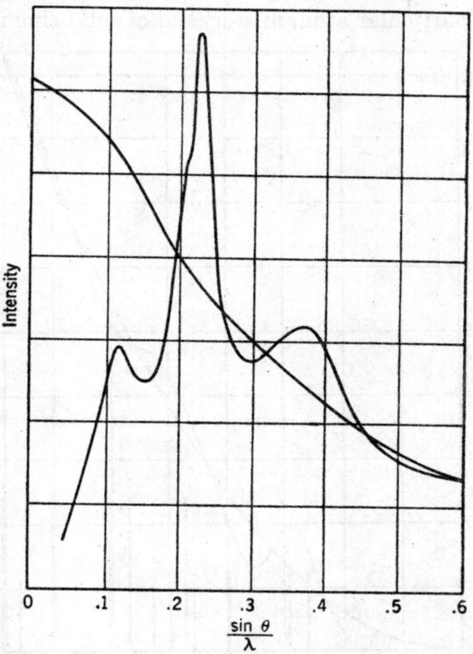

Fig. 5-1. X-ray diffraction pattern for liquid zinc at 460°C. [*From N. S. Gingrich, Rev. Mod. Phys.*, **15**, 90 (1943).]

charts such as Figs. 5-2 and 5-4. To understand the meaning of these atomic distribution plots, let us consider as the origin any particular atom and investigate the number of atoms in a spherical shell of thickness dr at distance r from this particular atom. If the average number of atoms per unit volume at any particular point is called the density ρ (commonly written $\rho(r)$ to indicate that density is a function of radius), then the number of atoms in the spherical shell is $4\pi r^2\rho\, dr$. If $4\pi r^2\rho$ is

[1] Warren and Gingrich, *Phys. Rev.*, **46**, 368 (1934).

[2] B. E. Warren, *J. Applied Phys.*, **8**, 645 (1937).

[3] The special assumptions involved are (1) that the liquid is monatomic or quasi-monatomic, (2) that the liquid is "molecularly homogeneous," exhibiting no large fluctuations of density or of degree of order.

plotted against r, the area under the curve between any two values of r is numerically equal to the number of atoms contained in the corresponding spherical shell.

Liquid Zinc. In the case of zinc, integration of Fig. 5-2 shows that there is a total of 11 neighboring atoms within a sphere of radius 3.6 kx units from any particular atom, the position of the first minimum in the atomic distribution curve. It will be recalled that the number of neighbors of any particular atom in a crystal of solid zinc within a sphere

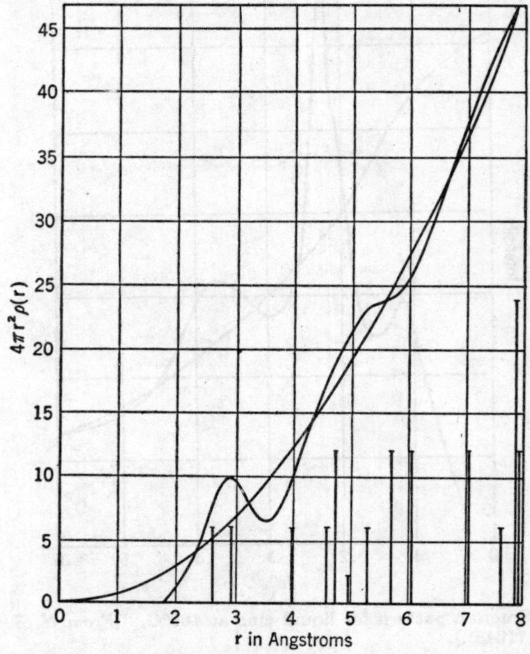

FIG. 5-2. Atomic distribution chart for liquid zinc at 460°C. Distribution for solid shown by vertical lines. (*From Gingrich.*)

of the same radius is 12 (6 at 2.659 kx units and 6 at 2.906, the positions indicated by vertical lines in Fig. 5-2). Also shown is a parabola representing the purely random distribution characteristic of an ideal monatomic gas; thus the atomic distribution for solid, liquid, and ideal gas are compared in this figure.

Liquid Mercury. The atomic distribution chart for liquid mercury (Fig. 5-4) shows very clearly the progressive approach with rising temperature of the atomic distribution in the liquid to that in the gas. In the vicinity of the melting point the maxima and minima of the curve for the liquid indicate a pronounced similarity to the solid. On the other hand the atomic-density distribution of the liquid becomes more smooth and approaches that of the gas in the vicinity of the boiling point, although even here the maxima and minima have not entirely disappeared.

Liquid Yellow Phosphorus. In contrast to these atomic distribution curves for liquid metals, that for a typical class III or nonmetallic ele-

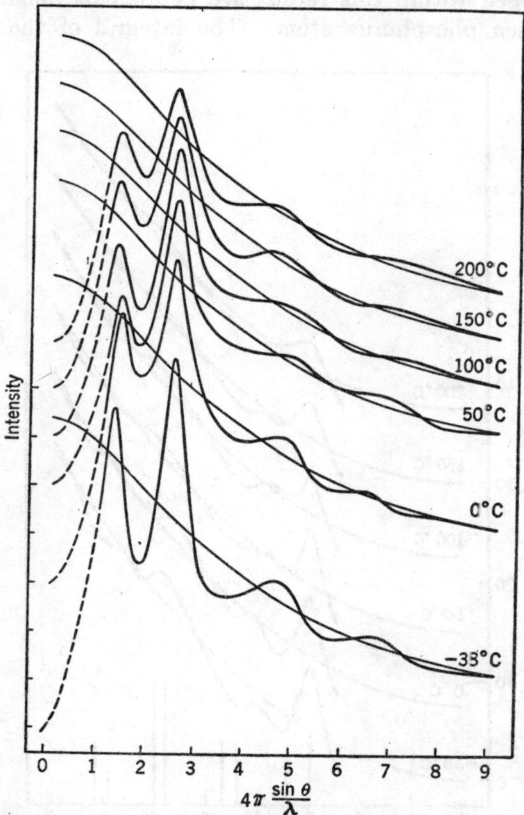

FIG. 5-3. X-ray diffraction pattern for liquid mercury. [*From Campbell and Hildebrand, J. Chem. Phys.*, **11**, 330 (1943).]

ment, liquid yellow phosphorus, is shown in Fig. 5-5.[1] Here a new phenomenon is seen. The atomic density is zero, or very nearly so, at about 3 A. Although the chart affords information only on the time average of the density, it is clear, since the density can never be less than zero, that, if the time average of the density is zero, then the density itself

[1] This atomic distribution chart, as well as those for chlorine, sulfur, argon, aluminum, potassium, and tin, to follow, are from N. S. Gingrich, *Rev. Modern Phys.*, **15**, 90 (1943).

is zero at all times in the vicinity of this radial distance. From this it follows that the atoms within a sphere of radius about 3 kx units are permanently trapped, for if they were occasionally to traverse the boundary, then the density here would not be zero. In other words the atoms trapped within this radius are permanent neighbors of the arbitrarily chosen phosphorus atom. The integral of the area up to

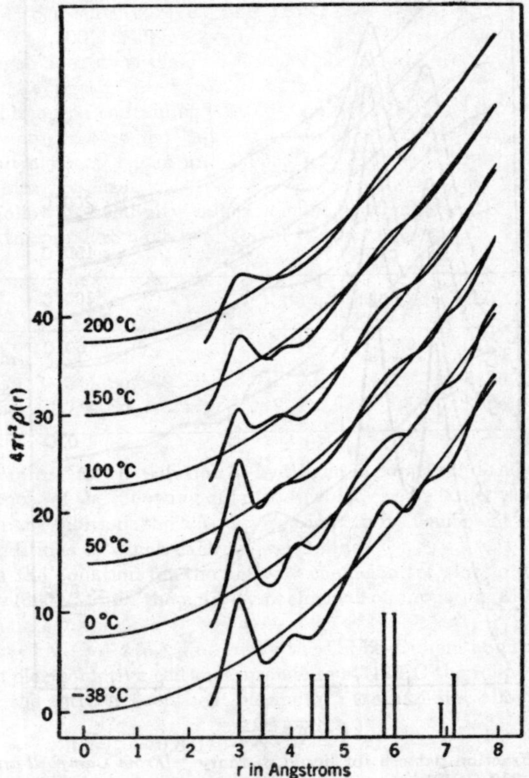

Fig. 5-4. Atomic distribution chart for liquid mercury. Distribution for solid shown by vertical lines. *(From Campbell and Hildebrand.)*

$r = 3$ kX units is 3, indicating that each phosphorus atom has three permanent neighbors. In a case such as this it is clear that there is indeed good reason to speak of *molecules in the liquid state;* in particular, any arbitrarily chosen phosphorus atom along with its three permanent neighbors constitute a P_4 molecule.

Liquid Chlorine. The atomic distribution curve for liquid chlorine, shown in Fig. 5-6, exhibits a feature similar to that for phosphorus in that

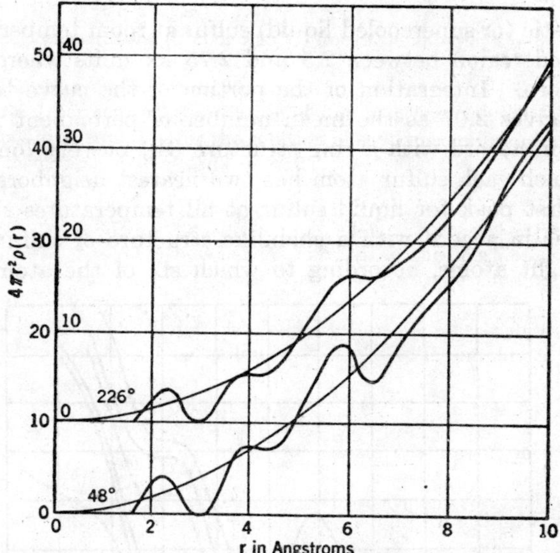

FIG. 5-5. Atomic distribution chart for liquid yellow phosphorus. *(From Gingrich.)*

the atomic density passes through a zero value—in this case at a distance of about 2.4 kx units. This vicinity is again a prohibited region, and the atom within a sphere of this radius is a permanent neighbor of the chlorine atom under consideration. The area under the curve to the left of this point corresponds to a single atom; hence it appears quite proper to speak of liquid chlorine as having the structure Cl_2, each atom with its single permanent neighbor forming a molecule.

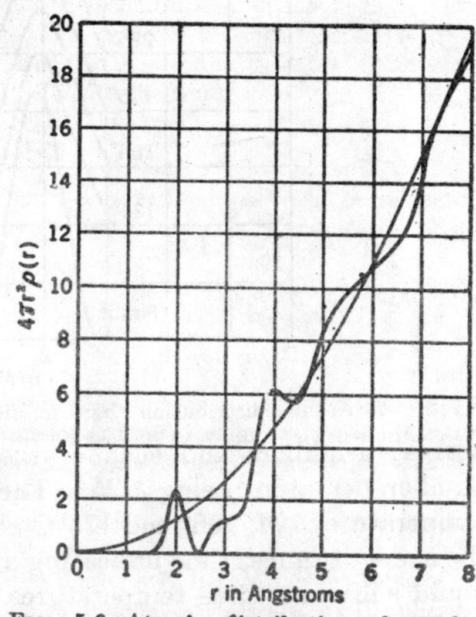

FIG. 5-6. Atomic distribution chart for liquid chlorine at 25°C. *(From Gingrich.)*

Liquid and Plastic Sulfur. The transition with temperature from a well-defined molecular state such as P_4 or Cl_2 to a state where the meaning, if any, of the term "molecular" becomes hazy is well illustrated by the case of sulfur, for which the atomic distribution curves at a series of temperatures are shown in Fig. 5-7. The lowest curve in this figure, that for plastic (or supercooled liquid) sulfur at room temperature, shows a well-defined region between 2.5 and 2.75 kx units where the atomic density is zero. Integration of the portion of the curve to the left of this region gives 2.07 as the mean number of permanent neighbors of each atom, in accord with a ring structure (S_8) or very long chains for either of which each sulfur atom has two nearest neighbors. The area under the first peak for liquid sulfur at all temperatures shown is 1.7. This value is in accord with a chainlike structure of sulfur with chain length of eight atoms, according to which six of the atoms have two

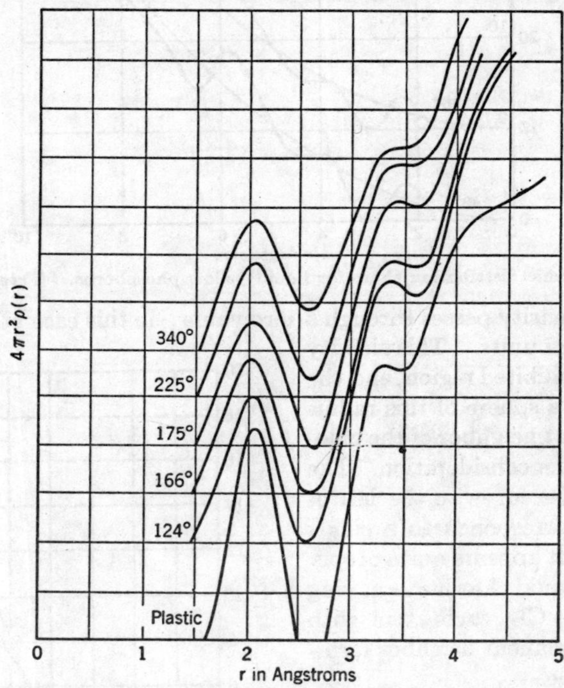

FIG. 5-7. Atomic distribution chart for liquid and plastic sulfur. (*From Gingrich.*)

nearest neighbors and each of the end two has but one—the mean number per atom being 1.75. The three curves corresponding to the temperatures 124, 166, and 175°C all show zero atomic density at about the same distance, thus indicating a well-defined molecular structure for liquid sulfur at these temperatures. However, the curves for 225 and 340°C do not exhibit this phenomenon, the first minimum in the density being significantly greater than zero and occurring at a progressively greater density the higher the temperature. The exact meaning of the departure of the minimum density from a value of zero cannot be fully explained in this case on account of the complex molecular structure of this liquid. It may reflect an extensive dissociation of S_8 molecules into some other species;[1] at any rate it is clear that X-ray evidence of molecular constitution of liquid sulfur tends to disappear at elevated temperature.

Liquid Argon. The atomic distribution curves for liquid argon at a series of temperatures are shown in Fig. 5-8. The approach to the inert gas state with rising temperature is apparent.

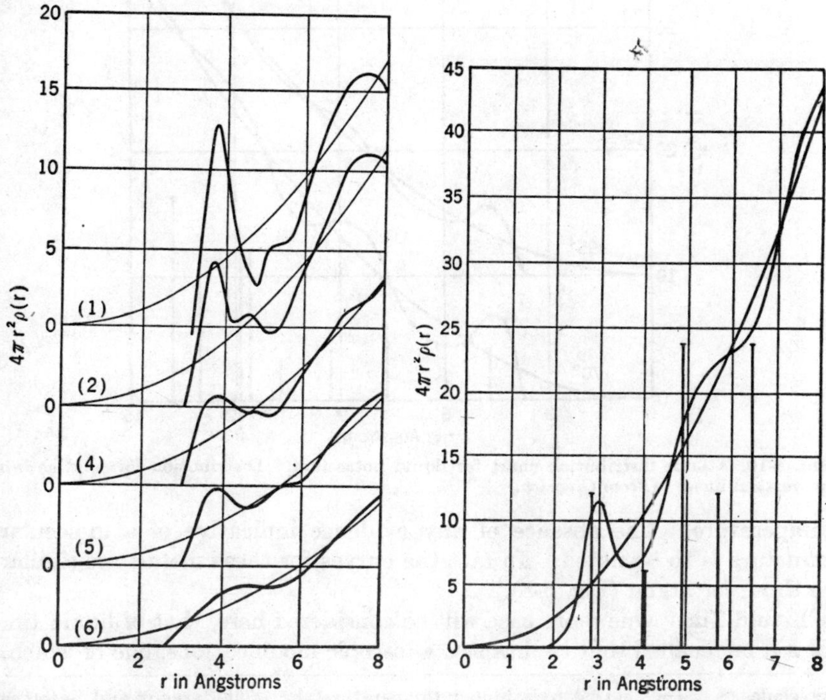

FIG. 5-8. Atomic distribution chart for liquid argon at (1) 84.4°K and 0.8 atm, (2) 91.8°K and 1.8 atm, (4) 126.7°K and 18.3 atm, (5) 144.1°K and 37.7 atm, (6) 149.3°K and 46.8 atm. (*From Gingrich.*)

FIG. 5-9. Atomic distribution chart for liquid aluminum at 700°C. Distribution for solid shown by vertical lines. (*From Gingrich.*)

Liquid Aluminum and Potassium. To proceed to class I elements, atomic distribution charts for aluminum, fcc, and potassium, bcc, are shown in Figs. 5-9 and 5-10, respectively. Of particular interest is the

[1] The generally accepted viewpoint [as given, for example, by G. P. Ford, and V. K. LaMer, *J. Am. Chem. Soc.*, **72**, 1959 (1950)] is that the mobile liquid, known as Sλ, which is stable at temperatures slightly above the melting point, is composed of ring

fact that the atomic distribution in the liquid is similar to that in the solid except that it is "smeared." The conclusion seems practically inescapable that the structure of a liquid element is related to that of the solid, the relation being closest at the melting point and less at higher

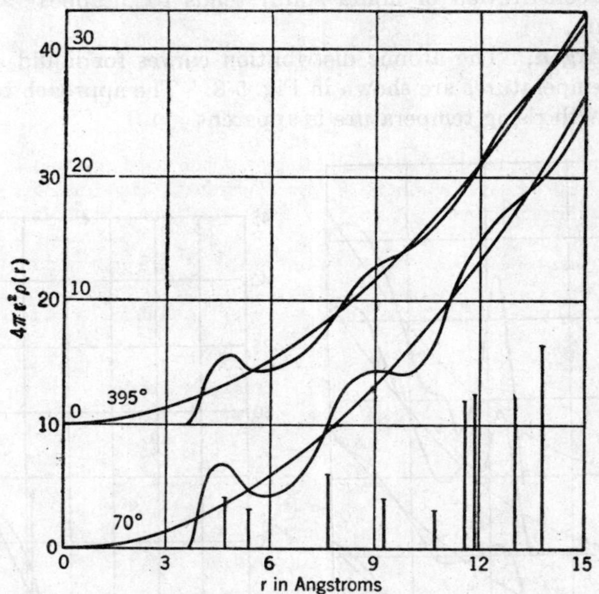

FIG. 5-10. Atomic distribution chart for liquid potassium. Distribution for solid shown by vertical lines. (*From Gingrich.*)

temperature. The absence of any evidence indicative of a molecular structure is to be noted. In fact the curves for these metals are similar to those for argon (Fig. 5-8).

Liquid Tin. One more case will be considered here, that of liquid tin. It will be recalled that tin has two allotropic modifications, one of which,

molecules S_8. On heating to a higher temperature the melt darkens and becomes more viscous; this liquid, known as $S\mu$, is usually presumed to contain chains. The transition appears to be nearly reversible on slow heating or cooling. The X-ray data considered above are apparently not in perfect agreement with this viewpoint unless the disparity between the observed area (1.7 atoms) and the expected area (2 atoms) for a ring structure is attributed to experimental error. However, the disappearance of the zero minimum (Fig. 5-7) between 175 and 225°C occurs in the temperature range where chain molecules are believed to become pronouncedly more predominant. Ford and LaMer give the amount of $S\mu$ (chain molecules) as 3.6 per cent at 120°C and 8.5 per cent at 150°C. At 225° and above we can readily imagine that the making and breaking of chains have become so rapid that the minimum of Fig. 5-7 no longer has a zero value.

gray tin (with a diamond-type lattice), is included by Hume-Rothery among the class III elements, and the other, white tin (tetragonal), is metallic in nature and is included among the class II elements by Hume-Rothery (see Fig. 3-5). The atomic distribution of liquid tin is compared with that of these two solids in Fig. 5-11, the distribution for gray tin being shown by dotted vertical lines and for white tin by full vertical

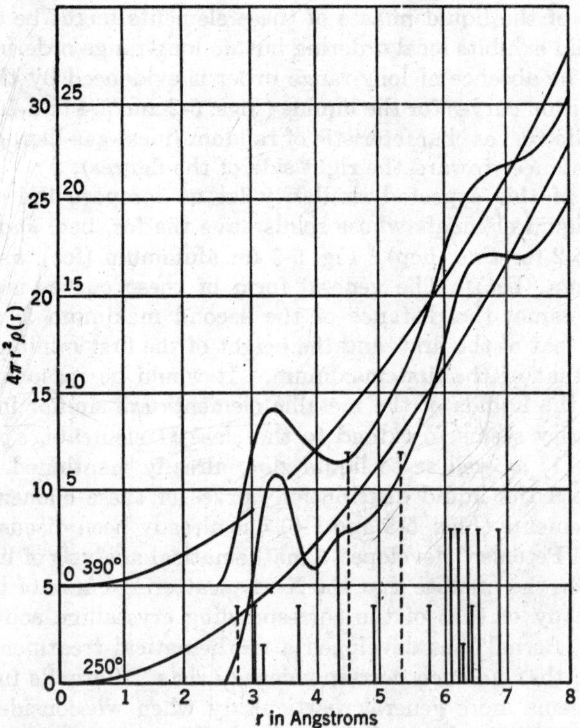

Fig. 5-11. Atomic distribution chart for liquid tin. Distribution for white tin shown by full vertical lines, that for gray tin by dotted vertical lines. (*From Gingrich.*)

lines. The similarity of the structure of the liquid to that of white tin rather than gray tin is quite pronounced. This similarity is in accord with the fact that liquid tin, like white tin, exhibits metallic luster and conducts electricity.

The Structure of the Liquid Class I Elements. Although we may now regard the general relationship between the structure of the solid and the liquid to be established by the foregoing evidence, it is well to consider the limitations of this relationship. It will be recalled that in the discussion of the class I solids it was found that the distinction between the three crystalline forms (bcc, fcc, and hcp) was not great from the viewpoint of the atom; *i.e.*, the atoms in both fcc and hcp crystals are close-packed with CN12; bcc crystals, though not quite so closely packed, have 8 nearest neighbors and 6 next nearest, at a distance only slightly greater, making a total of 14 near neighbors. In view of this similarity of fcc, bcc, and hcp crystals from the local viewpoint of an atom it might seem reasonable to expect that the structural similarities of the liquid phases of these elements might be even closer, since a liquid exhibits local ordering but no long-range ordering as does a crystal. The absence of long-range order is evidenced by the fact that the distribution curves for the liquids (Figs. 5-2 and 5-4 to 5-11) approach the parabolic curves characteristic of random (ideal-gas-like) distribution at great distance (toward the right side of the figures).

In view of this expected similarity let us compare the distribution curves of liquid elements whose solids have the fcc, bcc, and hcp structure: Fig. 5-2 for zinc (hcp),[1] Fig. 5-9 for aluminum (fcc), and Fig. 5-10 for potassium (bcc). The general form of these curves is seen to be nearly the same; the distance of the second maximum is a little less than twice that of the first, and the height of the first minimum is about six-tenths that of the first maximum. It would be reasonable to conclude that the liquids of the metallic elements are similar in structure. This similarity seems to extend to the class II elements, *e.g.*, to liquid tin (Fig. 5-11) as well as to liquid zinc, already mentioned. The contrast between the liquid distribution curves of these elements and the class III elements (Figs. 5-5 and 5-6) has already been discussed.

Prins and Petersen[2] developed a mathematical analysis of liquid structure based on the premise that the X-ray scattering curve of the liquid is a blurred copy of that of the corresponding crystalline solid. On the other hand, Bernal[3] has developed a mathematical treatment based on the premise that no such correspondence exists. Bernal's fundamental postulate seems more general, particularly when we consider elements such as iron which exist in different crystalline metallic forms. Bernal considers that three principal variables are required to describe liquid structure: (1) the mean radius of the coordination shells, (2) the number of atoms (or molecules) in each shell, and (3) the irregularity or scatter of the distribution in each shell. Actually only the nearest neighbors (first coordination shell) are considered. Bernal's treatment, while of con-

[1] Although the structure of zinc is hexagonal close-packed, the axial ratio is 1.856 instead of 1.633. It is taken as an example because no distribution curve is available for the liquid phase of an element with an ideal hcp structure in the solid.

[2] J. A. Prins and H. Petersen, *Physica*, **3**, 147 (1936).

[3] J. D. Bernal, *Trans. Faraday Soc.*, **33**, 27 (1937).

siderable interest, involves further assumptions as to the representation of the energy and entropy; it becomes mathematically cumbersome and is too lengthy to reproduce here.

X-ray Diffraction Investigation of Liquid Alloys. Intensity patterns, such as those of Figs. 5-1 and 5-3 for liquid elements, have been published for only a few liquid alloys. Hendus[1] included three molten Au-Sn alloys in his investigation of several molten elements (Pb, Tl, In, Au, Sn, Ga, Bi, Ge). The interpretation for alloys is difficult, and the radial distribution function ($4\pi r^2 \rho$) has not been calculated. It will be noted that several such functions would be required to represent the distribution of each of the two kinds of atoms. Hendus' Au-Sn alloys with 15 and 30 atom per cent Sn gave patterns which are substantially indistinguishable from that of pure Au. However his third alloy, 50 atom per cent Sn, exhibited a double maximum at the first peak. Hendus interpreted this (qualitatively) in terms of local ordering, *i.e.*, a similarity to the structure of the intermediate solid phase AuSn. No such phenomenon was observed by Glocker and Richter[2] in their investigation of Sn-Pb, Sn-Bi, and Pb-Bi alloys. Latin[3] has included molten alloys in his review of the structure of liquid metals.

The Structure of Liquid Salts. Considerable light has been shed on the structure of liquid salts by the X-ray investigations of Harris, Wood, and Ritter[4] on aluminum chloride, $AlCl_3$, and of Wood and Ritter[5] on indium iodide, InI_3, and tin iodide, SnI_4. Their method of investigation and interpretation is similar to that described above for liquid elements. The analysis of the data is, of course, considerably more complicated.

The radial distribution curve for aluminum chloride was found to exhibit a zero minimum, indicating a molecular structure as in the case of elemental phosphorus, Fig. 5-5, or chlorine, Fig. 5-6. The coordination number of aluminum was found by integration to be 4.7. This is interpreted, in view of the experimental error, as consistent with a coordination number of 4 or 5 but not with a coordination number of 3, as would be the case if the liquid were composed of molecules of $AlCl_3$. The ratio of the first two peak distances in the radial distribution curve, $3.60/2.20 = 1.636$, is very nearly $\sqrt{\frac{8}{3}} = 1.633$ which is the ratio of edge length to center-apex distance of a regular tetrahedron. This also strongly suggests a structure based on tetrahedral coordination of chlorine atoms about an aluminum atom. Such coordination is not

[1] H. Hendus, Z. *Naturforsch.*, **2A**, 505 (1947).
[2] R. Glocker and H. Richter, *Naturwissenschaften*, **31**, 236 (1943).
[3] A. Latin, *J. Inst. Metals*, **66**, 177 (1940).
[4] R. L. Harris, R. E. Wood, and H. L. Ritter, *J. Am. Chem. Soc.*, **73**, 3151 (1951).
[5] R. E. Wood and H. L. Ritter, *J. Am. Chem. Soc.*, **74**, 1760 (1952).

possible for a molecule containing only one aluminum atom but is quite probable for Al_2Cl_6—the common gaseous species. Further analysis of the curve indicates that the interatomic distances in the molecule of liquid aluminum chloride agree well with those previously reported for gaseous Al_2Cl_6.

It is concluded that liquid aluminum chloride is composed of molecules of Al_2Cl_6. The atomic arrangement is in the form of a double tetrahedron with a common edge, the aluminum atoms being at the two centers and the chlorine atoms at the apices. In solid aluminum chloride the structure may be described as based on a regular spherical close packing of chlorine atoms. The aluminum atoms occupy some of the octahedral spaces and thus each aluminum atom has six nearest-neighbor chlorine atoms. The structure of the liquid obviously differs considerably from that of the solid.

Liquid tin iodide is also found to be of molecular structure—SnI_4 of regular tetrahedral arrangement. The solid may be viewed as composed of symmetrical SnI_4 molecules arranged in such a way that the iodine atoms fall into a face-centered cubic arrangement. Thus SnI_4 has the same molecular structure in the gaseous, liquid, and solid states.

Molecules of SnI_4 may well have been expected in the liquid state in view of their existence in both the gas and the solid. The structure of liquid aluminum chloride is perhaps a little surprising in that it resembles the gas more than the solid. One is not tempted to believe this molecular behavior to be typical of fused salts, especially at high temperature. The amphoteric nature of aluminum (that is, its tendency to form groups similar to $SO_4^=$, NO_3^-, and $SiF_6^=$) may offer some clue to the structure of its liquid chloride. Certainly it seems very likely that liquid sodium chloride, for example, would, like the ionic solid, exhibit no semblance of molecular structure. Unfortunately, such data are not available.

Solid cadmium iodide has a nonmolecular (ionic) structure in which each cadmium atom has as nearest neighbors six iodine atoms. Liquid cadmium iodide might reasonably be expected to have a structure similar to either that of the solid or that of the gas (CdI_2 molecules). Unpublished work of H. L. Ritter indicates that in liquid cadmium iodide, each cadmium atom has six iodine neighbors, as in the solid, but that two of these are slightly closer than the other four. Thus, in this case, the liquid strikes a compromise.

The results described above pertain to temperatures slightly above the respective melting points: 144°C for SnI_4, 190°C for $AlCl_3$, and 388°C for CdI_2. Extension of this work, particularly to higher temperature, would be very valuable in furnishing insight as to the nature of molten salts and slags.

THE "HOLE" THEORY OF LIQUID STRUCTURE

The principal observable difference between a liquid and a solid is, of course, the fluidity of the former, *i.e.*, its ability to take the form of the confining vessel, or, in more precise terminology, its inability to support a static shear force. Other outstanding differences are the higher energy content of the liquid as evidenced by the heat of fusion, the higher compressibility and greater thermal expansion of the liquid, and the higher diffusivity in the liquid.[1] A satisfactory general theory of the liquid state should, of course, account for these differences. The "hole" theory of liquids, proposed by Eyring,[2] does indeed account semiquantitatively for them. Although this theory was proposed originally for molecular liquids, it will be applied here, with slight modification and extension, to atomic liquids of the class I and class II elements.

According to the hole theory the structure of a liquid is regarded as similar to that of the corresponding solid except that some of the "lattice sites" are empty, the whole "lattice" being distorted thereby. These empty "sites" are called *holes* or *vacancies*. The coordination number is accordingly smaller. To a rough approximation we may say that the liquid corresponding to a close-packed solid has a coordination number of 11 instead of 12; *i.e.*, each atom of the liquid has on the average 11 atoms and one hole as nearest neighbors. This hypothesis is in general accord with the X-ray evidence previously considered. The coordination number for a liquid element is equal to the area under the first peak of Fig. 5-2 or Figs. 5-4 to 5-11. In the case of the metallic elements where the first minimum does not have an ordinate of zero, the exact area to be evaluated is somewhat indefinite; however, it seems reasonable to include the area up to the radius of the first minimum. The area so evaluated by Gingrich for several metallic elements is given in Table 5-1.

Fluidity. This picture of liquid structure is readily seen to account for the fluidity; the inability of the liquid to support a static shearing force is now interpreted as arising from the ease with which an atom moves into an adjacent hole. From a slightly different viewpoint we might say that the dislocations, in terms of which we interpreted slip in crystalline metals, are so numerous in liquids that the resistance to shearing decreases almost to the vanishing point. On this picture it would be expected, then, that the viscosity of liquids would rise with the application of pressure. This is indeed the case. Bridgman[3] found that

[1] This higher diffusivity in the liquid is evidenced both by the self-diffusivity (determined by the aid of isotopes) and the diffusivity of other substances therein.

[2] H. Eyring, *J. Chem. Phys.*, **4**, 283 (1936).

[3] P. W. Bridgman, "The Physics of High Pressure," The Macmillan Company, New York, 1931; *Rev. Modern Phys.*, **18**, 1 (1946).

TABLE 5-1. COORDINATION NUMBER OF SEVERAL LIQUID METALLIC ELEMENTS (Evaluated as the area up to the radius of the first minimum of the atomic distribution curve)

Element	*Atoms*
Na	10.6
Li	9.8
Al	10.6
K	8.
Zn	10.8
Cd	8.3
In	8.4
Sn	10. at 250°C
	8.9 at 390°C

the viscosity of many liquids increases tenfold to one hundredfold when the pressure is increased from 1 to 10,000 atm.

Compressibility. The hole theory of liquid structure also obviously accounts for the observed fact that the compressibility of liquids is generally considerably greater than that of solids. The comparison is made for several metallic elements in Table 5-2.

TABLE 5-2. CHANGE IN VOLUME, COMPRESSIBILITY, AND THERMAL EXPANSIVITY ON MELTING FOR SEVERAL METALLIC ELEMENTS

Element	Volume change, %	Coefficient of compressibility, atm^{-1}			Coefficient of thermal expansion, $°C^{-1}$	
		Solid		Change at m.p.	Solid (room temp.)	Change on melting
Cs	1.4	50×10^{-6}	(50°C)	$23. \times 10^{-6}$	97×10^{-6}	
K	2.7	31	(45°C)	4.7	83	75×10^{-6}
Na	2.8	15	(30°C)	3.3	71	89
Bi	−0.3	2.8	(25°C)	0.057	13.3	80

Thermal Expansion. Table 5-2 includes some data on the change in the coefficient of thermal expansion on melting. The higher coefficient of thermal expansion commonly associated with liquids may also be readily interpreted upon considering that the holes originate from thermal agitation. It will be recalled that in the discussion of the nature of crystals a small number of vacant lattice sites or holes was postulated even for crystals (perhaps one lattice site in 10^3 to 10^6 being vacant) and that their number increased with temperature. Upon melting, there is a large increase in the number of holes and a further slow increase with rising temperature. On the basis of the present hypothesis the increase of number of holes with temperature gives rise to a greater increase in volume than would otherwise occur, and accordingly the coefficient of thermal expansion is greater for the liquid than for the solid.

Diffusivity. Perhaps the most satisfactory mechanism yet proposed to account for diffusion in solid metals is based upon the hole theory. It will suffice here to say that this theory "permits" an atom to move only into a vacant lattice site. Since the hole theory calls for only a relatively small number of holes in a crystalline metal (perhaps one for 10^3 to 10^6 atoms, even near the melting point), whereas for liquids it requires one hole for about every ten atoms, the diffusivity of a liquid might reasonably be expected to be 10^2 to 10^5 times that for the corresponding solid.[1] Although few measurements of diffusivity in liquid metals have been made, the above figures appear to be correct, at least as to order of magnitude.

Heat of Fusion. The hole theory of liquid structure also accounts for the heat of fusion. If on the average one of the twelve nearest neighbors of each atom in a close-packed solid is replaced (on melting) by a hole, then one-twelfth of the total chemical bonds are broken and the corresponding energy must be supplied. This reasoning may be extended, in a semiquantitative way, as follows: In the evaporation of a liquid to a monatomic vapor, each atom loses all eleven of its nearest neighbors, so that the heat of vaporization might be expected to be eleven times the heat of fusion (except for the effect of the difference between these two temperatures). That this is so may be seen by comparing Trouton's rule, $\Delta H_v/T_v = 21$ cal/deg, with Richard's rule, $\Delta H_f/T_f = 2$ cal/deg, where ΔH_v and ΔH_f are the heats of vaporization and fusion, respectively. These rules represent average experimental values. The ratio 21:2 again neglecting the effect of temperature, is indeed very nearly the anticipated ratio 11:1. This coincidence should not mislead us into thinking that the ratio of atoms to holes is exactly 11:1, for this ratio probably varies considerably with the element under consideration as well as with temperature and pressure.

Change in Density on Fusion. The density change accompanying fusion is also easily interpreted in terms of the hole theory. The quantitative check is, however, not quite so good in this case. If one-twelfth of the total number of atoms is replaced by holes, the density would be expected to decrease on fusion by about 8 per cent, providing the holes are of the same size as the atoms replaced. The average density change of metals on fusion is about 3 per cent; allowing for the fact that this

[1] The diffusivity of manganese in liquid iron at 1500°C is 6×10^{-5} cm²/sec, whereas that of manganese is γ-iron at the same temperature (extrapolated) is 4×10^{-9}. The ratio 1.5×10^4 agrees with the above estimate. This comparison is quite crude; if another temperature were selected, the ratio would be quite different.

average includes metals whose crystals are not close-packed, the average volume change of a close-packed metal does not exceed 5 per cent. If we reason from this figure alone, we should conclude either that the number of holes is slightly smaller than previously stated or that their size is slightly smaller than the atoms. All in all, a better quantitative agreement can hardly be expected in view of the present lack of knowledge as to the exact structure of liquids.

The only elements known to be exceptions to the general rule that the solid is more dense than the liquid at the melting point are bismuth,[1] crystallizing according to the $(8 - N)$ rule, and gallium (rhombohedral). As both of these solids have a rather open structure, it might seem reasonable to conclude that the liquid state has a greater tendency to be close-packed (with the exception of the holes) than the solid. This conclusion is also in accord with X-ray evidence which indicates, as mentioned previously, somewhat similar structure for all molten class I and class II elements.

TABLE 5-3. VOLUME CHANGE DURING FUSION OF ALKALI METALS

Metal	$\Delta V/V$		
	bcc → liquid observed	fcc → bcc calculated*	fcc → liquid calculated†
Na	0.0279	0.016	0.044
K	0.0268	0.024	0.051
Rb	0.0185	0.025	0.043
Cs	0.0136	0.030	0.044

* By aid of Eq. (3-1).

† The sum of the preceding columns.

Change in Density on Fusion of a Close-packed Crystal. If it is correct to postulate that liquid elements of classes I and II are nearly close-packed, then we should expect the fractional volume change on fusion to be more uniform for close-packed crystals than for others. As precise data for such elements are not plentiful, let us consider the fractional change in volume $(\Delta V/V)$ on fusion of the bcc alkali metals, as shown in the second column of Table 5-3. It will be noted that this quantity changes by a factor of nearly 2 in proceeding from sodium to cesium.

[1] Bismuth is also an exception as regards the change of electrical conductivity on melting. The class I and II elements exhibit an increase in electrical resistivity on melting, a fact readily interpreted in terms of the wavelike nature of the electron and the loss of periodicity upon fusion. The opposite behavior of bismuth would suggest an increase in the coordination number leading to a resonance of bonds characteristic of the metallic state, as well as an increase in density, as observed.

The corresponding quantity for the fusion of the metastable fcc form may be computed if $\Delta V/V$ is known for the transformation fcc $\rightarrow$ bcc; this latter may be computed from Pauling's relation [Eq. (3-1)] and is listed in the third column. The sum of these two quantities, corresponding to $\Delta V/V$ for the fusion of the fcc crystal, is given in the last column. This is seen to be nearly constant, corresponding to a volume change of 4 to 5 per cent accompanying the fusion of the close-packed structure. This near constancy seems to support the postulate stated at the beginning of this paragraph.

Unfortunately, no equally precise data seem to be available for close-packed class I elements; available data indicate that the fractional volume change on fusion is 0.027 for lead and 0.050 for copper and is between these values for many other fcc and hcp class I elements.[1]

Surface Tension. By assuming a quasi-crystalline model for liquids Belton and Evans[2] were able to derive an expression for the surface tension. Experimental values of surface tension are in good agreement with their relation, thus giving further evidence of the fruitfulness of the postulate of similarity in structure of the solid and the liquid state.

The surface tension of liquid metals is very high compared with that of molecular liquids. Many organic liquids have a surface tension between 20 and 40 ergs/cm^2, that of water is 70 ergs/cm^2, and fused salts usually have a surface tension of about 100 ergs/cm^2. The surface tension of several metals near the melting point is given in Table 5-4.

Refinement of the Hole Theory. Eyring and coworkers[3] have developed this hole, or vacancy, theory of liquids by means of their theory of absolute reaction rates; reasonably good agreement exists between the results of their extended mathematical treatment and the observed behavior, in particular as regards the viscosity and diffusivity. For example, they find the following simple relation between the diffusivity D and the viscosity η:

$$D\eta = \frac{d_1}{d_2 d_3} kT$$

where k is Boltzmann's constant, d_1 is the smallest dimension of the molecule or atom, and $d_2 d_3$ is its largest area. For a spherical molecule or atom $d_1/d_2 d_3 = 1/d$ or $1/(2r)$, r being the radius. In this extended theory they consider that the number of lattice sites is about 1.54 times

[1] The value is approximately 0.029 for both γ-iron and zinc.

[2] Belton and Evans, *Trans. Faraday Soc.*, **37**, 1 (1941).

[3] H. Eyring, *J. Chem. Phys.*, **4**, 283 (1936). J. Hirschfelder, D. Stevenson, and H. Eyring, *ibid.*, **5**, 896 (1937). F. Cernuschi and H. Eyring, *ibid.*, **7**, 547 (1939). D. Frisch, H. Eyring, and J. F. Kincaid, *J. Applied Phys.*, **11**, 75 (1940). R. E. Powell, W. E. Roseveare, and H. Eyring, *Ind. Eng. Chem.*, **33**, 430 (1941). W. E. Roseveare, R. E. Powell, and H. Eyring, *J. Applied Phys.*, **12**, 669 (1941).

TABLE 5-4. SURFACE TENSION OF SEVERAL LIQUID METALS

Metal	Temp, °C	σ, ergs/cm²	Reference
Zn	419	754	
Sn	232	531	
Pb	232	444	Semenchenko and Pokrovski*
Cd	320	630	
Bi	269	378	
Hg	−39	467	
Sn	1000	497	
Pb	1000	401	
Sb	1000	355	
Zn	640	761	Sauerwald†
Cu	1131	1103	
Ag	995	923	
Au	1120	1128	
Cu	1150	1034	Becker, Harders, and Kornfeld‡
Fe	1600	1360	

* V. K. Semenchenko and N. L. Pokrovski, "Progress in Chemistry," Vol. VI, Book 6, p. 777, 1937.
† F. Sauerwald and coworkers, *Z. anorg. allgem. Chem.*, **154**, 79 (1926); **162**, 301 (1927); **181**, 355 (1929); **213**, 310 (1933).
‡ G. Becker, F. Harders, and H. Kornfeld, *Arch. Eisenhüttenw.*, **20**, 363 (1949).

the number of atoms for an atomic liquid. This is calculated from the fact that the heat of fusion divided by the absolute temperature of melting is usually about 2 cal/deg and corresponds to a substantially larger number of holes than we have previously considered. However, Frisch, Eyring, and Kincaid[1] have shown that the volume of one of these holes required to interpret flow phenomena (viscosity) is about one-sixth that of a molecule or atom; thus the total vacant space and the coordination number are about the same as previously considered.

Auluck and Kothari,[2] by application of the hole theory of liquids, have been able to give an essentially correct account of the variation of viscosity with pressure. In a sense, they, as others, attribute a sort of physical substantiability to the vacant spaces in liquids.

MISCELLANEOUS EVIDENCE OF RELATION BETWEEN LIQUID AND SOLID

The differential relation between the vapor pressure p of a liquid or solid and the temperature T is given by the expression

$$\frac{d \ln p}{d(1/T)} = \frac{-\Delta H}{R}$$

[1] D. Frisch, H. Eyring, and J. F. Kincaid, *J. Applied Phys.*, **11**, 75 (1940).
[2] Auluck and Kothari, *Nature*, **153**, 777 (1944).

where ΔH is the molal heat of vaporization or sublimation and R is the gas constant. Normally ΔH shows little variation with temperature, and in this event the integration is easily performed giving

$$\ln p = \frac{-\Delta H}{RT} + I$$

indicating a linear relationship between $\ln p$ or $\log p$ and $1/T$. At the melting point, liquid, solid, and vapor are all in equilibrium, and hence the vapor pressure of the solid is equal to that of the liquid. Thus if $\log p$ is plotted against $1/T$ for solid and liquid, two nearly straight lines are obtained, as shown in Fig. 5-12; these intersect at the melting point. The older viewpoint was that, for any particular substance, these two lines reflected the separate properties of the solid and the liquid and that

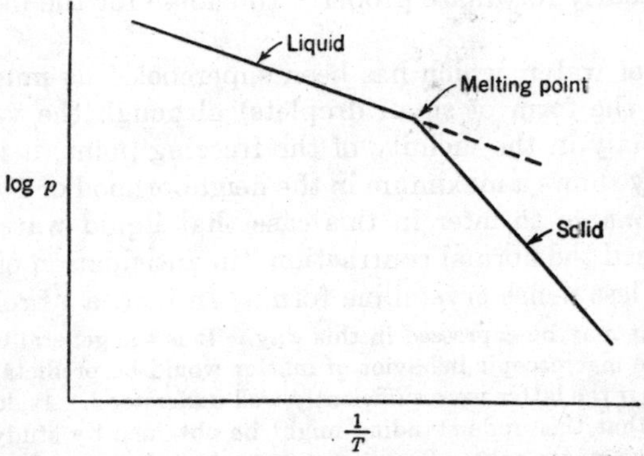

Fig. 5-12. Schematic plot showing variation of vapor pressure of solid and liquid with temperature.

the melting point was simply the point where the two lines happened to intersect. This point of view was given some support by the fact that many liquids can be supercooled, *i.e.*, cooled perhaps 10°C below the freezing point (or even 100 if care is taken to eliminate nuclei) without crystallizing. Vapor pressures of such supercooled liquids have been measured and the corresponding points found to fall on the smooth extension of the curve above the melting point as shown by the dotted portion of Fig. 5-12.

The old argument based on this phenomenon of supercooling is then as follows. The liquid does not "know" it is about to freeze, since its properties (in particular the vapor pressure) change smoothly in the vicinity of the melting point and even below in just the same way as they did above. This argument may be continued as follows. Two straight, nonparallel lines in a plane always have some point of intersection; the intersection of the two "independent" lines, the melting point, can never be determined by consideration of one alone. The pitfall represented by the above reasoning is no longer as deluding as it was only a decade ago. In the first place, the two lines are, in general, not completely independent; the difference in their slope, corresponding to the heat of fusion, is approximated by Richard's rule, referred to previously, stating that the heat of fusion (the difference between the heats of sublimation and vaporization) is equal to RT_f. Second, although many liquids may be rather easily supercooled, there is no known instance of superheating of a crystalline metal, as though the melting point were a singular point on the curve for the solid—even without reference to the curve for the liquid.[1] Third, the temperature range in which supercooling has actually been achieved is, with few exceptions, small.[2] The vapor pressure is not a particularly fortunate property to choose for the investigation of structure.

In the case of water, which has been supercooled as much as 20°C in bulk (38°C in the form of small droplets) although the vapor pressure shows no anomaly in the vicinity of the freezing point, it is well known that the density shows a maximum in the neighborhood of 4°C. It would not be unreasonable to infer in this case that liquid water, on cooling here, has reversed the normal contraction "in anticipation of" its crystallization to the less dense crystalline form. In fact, a "broken-down ice

[1] This argument may be expressed in this way. It is the general opinion of most scientists that the macroscopic behavior of matter would be predictable in terms of atomic properties if the latter were sufficiently well understood. It does not seem at all unreasonable that this understanding might be obtained by study of crystalline and (ideal) gaseous states only. If such were so, then the properties of the liquid, as well as the melting point, would be predictable from these atomic properties, or in other words, the properties of the liquid would be predictable from those of the solid and gas with the aid of atomic constructs. This argument depends on a "faith in the ultimate rationality of nature."

[2] D. Turnbull [*J. Applied Phys.*, **20**, 817 (1949) also *J. Chem. Phys.*, **18**, 198 (1950)] has recently pointed out that the common limitation to supercooling lies in the preservation of nuclei. By subdividing the liquid into small drops a few nuclei can no longer start crystallization of the whole. A fine dispersion of mercury in alcohol (with added sodium oleate or iodine) was found to supercool 46°C, and a similar dispersion of gallium, 70°. Small droplets of tin in aluminum may be supercooled 100°C. It is interesting to note that in each case this maximum degree of supercooling is one-quarter to one-fifth the absolute temperature of the melting point. J. C. Fisher, J. H. Hollomon, and D. Turnbull [Science, **109**, 168 (1949)] show that the experimental range of maximum supercooling (in the absence of foreign nuclei) is rather small—a few degrees. This is in accord with the fact that the rate of production of nuclei by thermal agitation changes very rapidly with temperature—by a factor of about 10^{10} in 5°C. Hence it may be said that there exists a lowest temperature to which a liquid may be supercooled; this temperature depends only slightly on the time scale of the experiment.

structure" for liquid water is supported by both X-ray[1] and Raman[2] spectra. The concept of "micro-icebergs" is applied quantitatively by Frank[3] and is found fruitful in interpreting the properties of aqueous solutions.

A Kinetic Theory of Liquids. Born and Green[4] in the third of a series of papers on a comprehensive "General Kinetic Theory of Liquids" develop general equations for fluids. The expression for the energy flux as well as that for viscosity and that for the thermal conduction consists of two parts, one corresponding to the thermal motion and the other to the atomic forces (or molecular forces, for a molecular liquid). The first part is found to be predominant for a gas, and the second for a liquid. Since this theory is found to be in good agreement with experiment in many verifiable respects, it may be said to substantiate further the viewpoint adopted here that the similarity of liquid (near freezing point) to gas is not particularly impressive or fruitful.

LIQUID ALLOYS

It seems reasonable to extend the general theory of the liquid state, developed in the foregoing sections, to include solutions, in particular liquid alloys. The general nature of a molten alloy we should expect to be characterized by a similarity to the corresponding solid alloy, except for the holes. This similarity does not include, of course, a regular crystalline lattice of many atoms but applies only to the immediate vicinity of any particular atom. Thus β-brass has a well-ordered structure below 460°C. Above this temperature the long-range order disappears, but short-range order still persists in the sense that each zinc atom has a preponderance of copper atoms as nearest neighbors and vice versa. This was interpreted in terms of the difference in electronegativity as well as the size factor. These two factors still exert their influence in the liquid state, though perhaps that of the size factor may be considerably diminished. Hence there is every reason to believe that the short-range order still exists in the liquid state of the 50-50 copper-zinc alloy, although it undoubtedly diminishes slowly with rising temperature.

It will be noted that this picture leaves no room for the concept of a molecule in such a molten alloy. Indeed it would seem absurd to assume that a nonmolecular solid, such as β-brass, should become molecular on

[1] Bernal and Fowler, *J. Chem. Phys.*, **1**, 515 (1933). S. Katzoff, *ibid.*, **2**, 841 (1934).

[2] Cross, Burnham, and Leighton, *J. Am. Chem. Soc.*, **59**, 1134 (1937).

[3] H. S. Frank, *J. Chem. Phys.*, **13**, 478, 493, 507 (1945).

[4] M. Born and H. S. Green, *Proc. Roy. Soc. (London)*, **A190**, 455 (1947). All three papers are now published in a book, "A General Kinetic Theory of Liquids," Cambridge University Press, London, 1949.

melting. The increased thermal agitation in the liquid state would, on the contrary, tend to break down any molecular structure if such had existed in the solid. Characteristic metallic properties, such as luster and electrical conductivity, still persist in the liquid state, and in particular, the liquid consists of the same atoms with the same atomic properties as the solid. If we accept Pauling's theory of resonating bonds for crystalline metals, there would seem to be no reason to discard it for liquid metals. In other words, the whole tenor of the modern evidence seems to leave the word "molecular" bereft of meaning as applied to liquid metals.

Solubility of Gases—Sievert's Law. The solubility of the inert gases (He, A, etc.) in metals is immeasurably small. Other gases, particularly the bimolecular gases of class III elements, such as hydrogen, nitrogen, and oxygen, frequently dissolve to a significant extent in a solid or liquid metal. The nature of the metallic solution resulting from solution of a gas may be partially determined from investigation of the equilibrium between the gas and the metal. Since such solutions are usually dilute, the mass law may be expected to apply. Thus, if hydrogen dissolved molecularly, as H_2, in a metal the equilibrium would be written

$$H_2(g) = \underline{H_2}$$

where the underscore denotes solution in the metal; the mass law constant is then $k = c_{H_2}/p_{H_2}$, c_{H_2} being the concentration of hydrogen in the gaseous phase. If the gas dissolved molecularly, the concentration of dissolved hydrogen would be proportional to the pressure of hydrogen gas. Experimentally this is not so; rather it is found that the concentration of dissolved hydrogen (or other bimolecular gas) is proportional to the square root of the partial pressure of hydrogen in the gaseous phase. This square-root relationship, commonly known as *Sievert's law*, is readily interpretable if it is assumed that the gas is dissolved atomically in the metal; thus the equilibrium may be written: $H_2(g) = 2\underline{H}$. The mass law constant[1] is then $k = c_H^2/p_{H_2}$, whence it follows that $c_H = \sqrt{kp_{H_2}}$, concentration being thus proportional to the square root of the observed pressure. Here again is evidence of the atomic rather than the molecular nature of metallic solutions.

Solubility of Gases in Alloys. Some further insight into the nature of binary alloys may be gained from the solubility of gases therein. Liang,

[1] This "constant" may be regarded as truly constant only for a particular gas in a particular metal at a particular temperature at pressures sufficiently low that the concentration of the gas in the metallic phase is very small. Normally the restriction as to pressure is not serious, inappreciable error being encountered up to a pressure of perhaps 100 atm (provided no third phase forms).

Bever, and Floe[1] investigated the solubility of hydrogen at a fixed pressure in a series of iron-silicon alloys at 1500 and 1650°C. A plot of their results (Fig. 5-13) shows that there is a nearly linear relation between the logarithm of the hydrogen solubility and the composition of the alloy from 0 to 50 and from 50 to 100 atom per cent silicon. The abrupt change in slope at the 50-50 composition is in accord with the presumption that this alloy is highly ordered—somewhat like β-brass; we might say that the affinity of iron and silicon for each other is so strong that a relatively inert atom such as hydrogen is squeezed out. However, if the

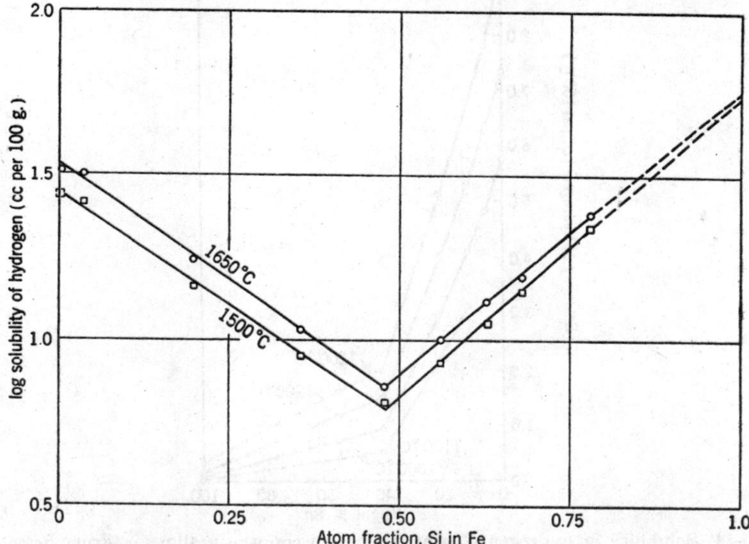

FIG. 5-13. Solubility of hydrogen (1 atm) in molten iron-silicon alloys at 1500 and 1650°C.

affinity were not sufficiently high to produce a rather pronounced degree of order in the solution, the plot of the logarithm of the hydrogen solubility vs. composition might reasonably be expected to be a smooth curve with a minimum rather than a discontinuity in slope at the 50-50 composition. The discontinuity in Fig. 5-13 occurs at the composition of one of the intermediate solid phases of the iron-silicon system. The liquid has an extraordinarily large heat of formation, thus furnishing direct evidence of the strong affinity of iron for silicon.

The solubility of hydrogen in liquid copper-tin alloys is plotted in Fig. 5-14.[2] Again a discontinuity in slope is evident—in this system at an atom ratio of 3:1 copper to tin. Here also, the break in the solubility

[1] Hung Liang, M. B. Bever, and C. F. Floe, *Trans. AIME*, **167**, 395 (1946).
[2] M. B. Bever and C. F. Floe, *Trans. AIME*, **156**, 149 (1944).

curve occurs at the composition of an intermediate solid phase. A somewhat similar phenomenon, observed for the solubility of graphite in copper-manganese alloys, is shown in Fig. 5-15.[1] In this case the significance of the composition at which the break occurs is not clear. The mere fact that the break does exist seems rather convincing evidence of an ordered rather than a random structure in these liquids.

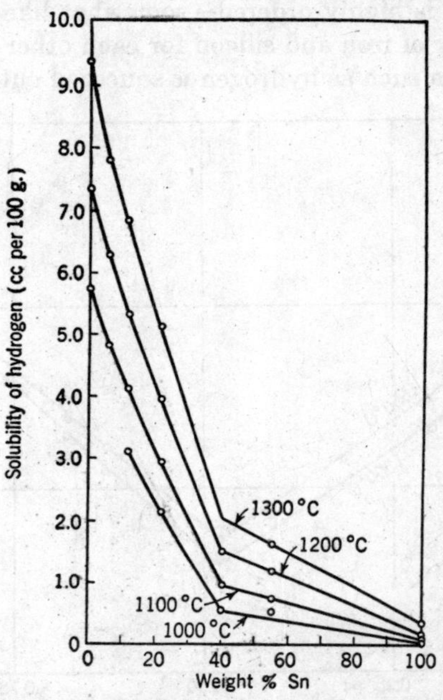

FIG. 5-14. Solubility of hydrogen (1 atm) in molten copper-tin alloys. [*From Bever and Floe, Trans. AIME,* **156,** 149 (1944).]

Oxides in Liquid Iron. When iron and iron oxide (or oxygen) are brought together at a temperature above the melting point of iron, *e.g.*, 1600°C, two phases, metal and oxide, are formed. In view of the rather large difference in electronegativity (between iron and oxygen) it is not surprising that these are not miscible. It is found that under these circumstances some oxygen, about 0.23 per cent, is dissolved in the liquid iron. There is no reason to suppose that this alloy is structurally different from the other alloys we have considered. Each of the relatively few oxygen atoms probably has in its first coordination shell several iron atoms. If the bonds to the oxygen are of the resonating covalent type typical of the metallic state, then the radius of the oxygen atom from Table 3-4 is seen to be similar to that of carbon (about two-thirds that of

[1] J. R. Anderson and M. B. Bever, *Trans. AIME,* **171,** 119 (1947).

the iron atom), whence from the analogy with carbon and the geometry of the situation it is seen that about six iron atoms surround each oxygen atom. From this it would seem difficult to attach significance to the presence of FeO as a molecular species in liquid iron.

If a small amount of chromium is added to the solution of oxygen in iron, it might reasonably be expected, from the general picture of liquids as presented in this chapter and from the fact that chromium has a higher affinity for oxygen than does iron,[1] that there would be more

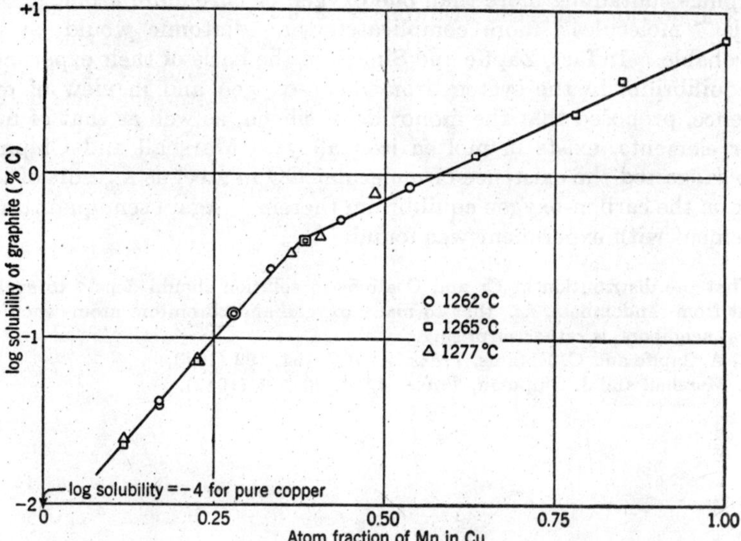

FIG. 5-15. Solubility of graphite in molten copper-manganese alloys in the vicinity of 1270°C.

oxygen in the first coordination shell of chromium than of iron. Equilibria in this system have been investigated by Hsin-Min Chen and Chipman.[2] They found that, at a given low chromium content, the oxygen content of the liquid alloy in equilibrium with solid chromium oxide was higher than anticipated on the basis of a random distribution of chromium and oxygen in the liquid alloy. They interpreted their results in terms of molecular CrO dissolved in liquid alloy; combination of this hypothesis with the mass law gave semiquantitative agreement with the experimental observations. These two interpretations of the structure of the solution are not quite so different as they might at first

[1] This is evidenced by the fact that the free energy of formation of chromium oxide is less than that of iron oxide and is in general accord with the chemical nature of chromium; *e.g.*, it is a mild deoxidizer for steel.

[2] H. M. Chen and J. Chipman, *Trans. ASM*, **38**, 70 (1947).

seem. If a chromium and an oxygen atom in its first coordination shell be arbitrarily called a "molecule" of CrO, the two pictures would be identical.[1] The mathematical treatment in terms of the mass law and CrO molecules is simpler but is probably nearly equivalent to that corresponding to the picture first suggested.

If it is correct, as suggested in the foregoing, that these two concepts of the constitution of liquid ternary alloys are equivalent, then, on the basis of the laws of probability, it would be exceedingly unlikely to encounter groupings containing more than one oxygen or chromium atom; in other words, "molecules" more complicated than diatomic would be very improbable. In fact, Zapffe and Sims,[2] on the basis of their experiments on equilibrium in the system iron-silicon-oxygen and in view of other evidence, proposed that the monoxide of silicon, as well as that of many other elements, exists in molten iron alloys. Marshall and Chipman[3] have suggested the existence of molecular CO in steel as a result of their work on the carbon-oxygen equilibrium therein. Again semiquantitative agreement with experiment was found.

[1] That the distribution of Cr and O atoms in solution should depart to such an extent from randomness, *i.e.*, that so many oxygen and chromium atoms should be nearest neighbors, is rather surprising.

[2] C. A. Zapffe and C. E. Sims, *Trans. AIME*, **154**, 192 (1943).

[3] S. Marshall and J. Chipman, *Trans. ASM*, **30**, 695 (1942).

CHAPTER 6

THE FIRST LAW OF THERMODYNAMICS

A *system* is any portion of the universe selected for consideration. Since the difficulty of understanding a system increases with its complexity, it is customary to choose as simple a system as possible consistent with the aim in view; hence the parts of a system are usually connected in some manner mechanically, thermally, physically, chemically, or logically.

The concept of *state* has already been introduced and discussed. At this point, however, it seems advisable to repeat that this concept is one of the most important, both historically and systematically, in the development of thermodynamics. The state (or macrostate) of a system is defined at any instant by giving all its state properties; this, of course, does not include atomic properties or arrangements but does include temperature, pressure, volume, viscosity, vapor pressure, surface tension, etc. If attention is focused, as it is here, upon the chemical behavior of a system, the geometrical form is not usually considered, since it has no effect on chemical or thermodynamic properties unless the surface energy related thereto is also under consideration.

From experience it has been learned that a complete listing of all the properties of a state is superfluous and that all the chemical and thermodynamic properties are implied when only a few are explicitly given. All these properties, values of which are implied when a few are explicitly given, are called *state properties*. As an example, the state of a system consisting of 10 g of water may be defined further by saying that it is at 25°C and 1 atm pressure, not too finely subdivided and not under the influence of electric, magnetic, gravitational, or other kind of external field. All other properties are thereby implicitly fixed and may be found by experiment, or by reference to the literature.

The properties of a system may conveniently be divided into two classes: those which are additive, the *extensive properties*, and those to which a value may be assigned at each point in the system, the *intensive properties*. A few common extensive properties are mass, volume, and energy; it will be noted that the total volume of a system is the sum of the volumes of the component parts. A few common intensive properties are temperature, pressure, density, refractive index, and molal energy.

The extensive properties of a homogeneous system are proportional to the total mass; since the ratio of any two is therefore independent of the total mass and may be assigned a value at any point, this ratio is an intensive property. All the so-called specific properties, such as the specific volume (volume per unit mass, the reciprocal of the density), and the molal properties, such as the molal volume or molal energy, fall in this category. In common usage the word *property* refers to an intensive property.

Of the various interrelations between state properties one of the most important is the *equation of state*, defined as a relation between pressure, volume, and temperature. It is common to assume that all pure substances obey an equation of state, and this assumption will be made in the subsequent thermodynamic development. Although there is usually little doubt as to the validity of this assumption for gases and liquids, the state of solids, particularly metals, is sometimes difficult of definition, and a property of a given mass of a metal, at a given temperature and pressure, may not be so definite as it was for water, considered above.

For example, the density of iron or copper is a function of the extent and manner of previous cold work and heat-treatment, *i.e.*, its history. Although it may be felt that this is explainable in terms of dislocations or other departures from perfect crystallinity, no state variable has as yet been invented to describe this behavior adequately. The concept of state clearly calls for a definition of the system in terms of present properties rather than past history, and to be fruitful the number of properties required for adequate definition must be small. The lack of a definitive state for a substance such as cold-worked iron may give rise to only an insignificant error in many properties (*e.g.* heat of combustion) but may be very important in others (*e.g.* the solubility of hydrogen). It is usually assumed that an adequately annealed metal has a definite, reproducible density; however, this has yet to be demonstrated by precise experiment. This whole field seems a promising one for future investigation.

P. W. Bridgman[1] has recently contributed to the application of thermodynamics to nonequilibrium states. He considers such a state to have a definite internal energy which is recovered when and if the state is recovered after a finite excursion; thus an equilibrium energy is associated with the state. This leads to a description of the thermodynamic properties of the nonequilibrium state in terms of the parameters of an equilibrium state.

For example, Bridgman describes the state of an alloy possessing a nonequilibrium degree of order at a given temperature in terms of the tem-

[1] P. W. Bridgman, *Rev. Modern Phys.*, **22**, 56 (1950).

perature at which this degree of order would exist at equilibrium. Also, he expresses the thermal effects of an irreversibly strained body in terms of the properties of the unstrained body. In general, the thermodynamic properties of a closed system which is out of internal equilibrium in one or more respects are represented as functions of the existing pressure (or stress) and temperature and, in addition, of one or more other parameters, each such parameter being, for example, a temperature, a volume, or a pressure (or stress) at which the system would be in equilibrium in one of these respects. Thus the macroscopic method characteristic of classical thermodynamics is applied to irreversible systems and processes.

ENERGY AND THE FIRST LAW

Development. The concept of energy is one of the most important in all modern science. It is an exceedingly fruitful concept whose use has been developed over a relatively long period of time. A large part of scientific thinking is now done in terms of energy rather than in terms of force, which was more common up to a century ago. For example, in the foregoing chapters we found that the energy associated with chemical bonds, rather than the interatomic forces, led to the concept of electronegativity which provided a semiquantitative formulation of the observed phenomena.

The fact that its development extended over such a long period of time is evidence that the concept of energy is not obvious. Insight as to the source of its fruitfulness may readily be gained by historical considerations. Only the briefest historical outline can be given here. The first use of this concept was in simple kinetic systems in which fields and frictional forces were absent and collisions perfectly elastic. In such systems it was recognized long ago that the *kinetic energy* $\frac{1}{2}mv^2$ for a single rigid body or $\Sigma\frac{1}{2}mv^2$ for a system of many bodies, called at the time *vis viva*, remains constant. In a potential field, such as a gravitational field, this is no longer true. However, conservation laws—statements that certain quantities remain constant under certain conditions—are fruitful as well as appealing, and the term *potential energy*, defined as the negative integral of the force over the distance, was invented because it was found that the sum of the kinetic and potential energy, known as the *dynamic energy*, remains constant in a *conservative system*, *i.e.*, one in which the forces are functions only of the coordinates. This constancy, which can be deduced from Newton's laws of motion, is stated as the law of conservation of dynamic energy. A system in which frictional forces are operative is not conservative, and the law does not apply.

Most actual processes do not fully satisfy the conditions under which the law of conservation of dynamic energy holds. Frictional forces may be present, the bodies in the system may be deformed (in particular, compression or expansion may occur), and most important from the chemical viewpoint, dynamic energy may be created (or destroyed) chemically or thermally. Thus the chemical and thermal "forces" can hardly be considered conservative on the macroscopic scale under consideration. Again it appeared necessary, in order to save the conservation law in such cases, to broaden the concept of energy further, beyond the limits of dynamic energy.

The concept of *heat* as an additional form of energy dates from the classic work of Count Rumford (1798), who observed that the heat generated in the boring of cannon was proportional to the amount of work done or the amount of dynamic energy expended. Joule's famous determination of the mechanical equivalent of heat (1845) followed. The *latent heat* of fusion, vaporization, or chemical reaction was recognized by Black (1762) as analogous to potential energy in the sense that a latent heat may be converted to sensible heat just as potential energy may be converted to kinetic energy. At a considerably later date (1905) Einstein pointed out that mass may also be converted to energy, thus necessitating another addition to the known forms of energy required to preserve the conservation law.

Energy. We have, as yet, given no general definition of the term "energy." In 1851 Lord Kelvin offered a definition which is difficult to improve: "The energy of a material system is the sum, expressed in mechanical units of work, of all the effects which are produced outside the system when the system is made to pass in any manner from the state in which it happens to be to a certain arbitrarily fixed initial (standard) state." Since the energy of a system is a function of an arbitrarily chosen standard state, there is no such thing as *absolute energy* but only *relative energy;* only energy differences can ever be measured. From the viewpoint of the chemist and metallurgist, kinetic energy (associated with sensible motion) is of little interest—in fact it seldom appears in the systems in which he is interested. The total energy other than the kinetic—sometimes called the *internal energy* of the system—is designated E. Henceforth we shall consider only this internal energy, referring to it simply as energy, on the assumption that the kinetic energy of the system is negligibly small or sensibly constant. A kinetic-energy term could easily be included in all subsequent equations but is omitted, since it will have no purpose here. Similarly, we do not include energy arising or disappearing from mass-energy transformations according to Einstein's relation.

The transition of a system from one state to another is called a *process* or a *reaction.* The energy change ΔE associated with any process is, from the definition of energy,[1]

$$\Delta E = q - w$$

where q is the amount of *heat absorbed* by the system from the surroundings and w is the *work done by the system* on the surroundings. For an infinitesimal change

$$dE = \delta q - \delta w \tag{6-1}$$

These relations, which are now seen to be a consequence of the definition of energy, are usually referred to as statements of the *law of conservation of energy* or the *first law of thermodynamics*.

To refer to a definition as a "law" is obviously somewhat misleading. To reassure the reader of the fact that the foregoing equations are in fact definitions of energy change, we need merely remind him that the only way of measuring the energy change of a system is by measuring the heat it absorbs from its surroundings and the work it does upon its surroundings. In other words, there is no such thing as an energy meter, and the only way the energy change of a system can be determined is by measuring all the heat and work exchanges with the surroundings. In principle this is done by posting sentries (measuring instruments) all about its periphery; their duty is to report heat and work exchanges with the surroundings. Most experiments are designed to minimize the number of measuring instruments required by making the conditions at various points on the boundary of the system as much alike as possible, *e.g.*, by making the pressure the same throughout the system.

Energy a State Property. The question now arises as to whether the energy so defined is a state property of the system, *i.e.*, whether a system caused to pass from some initial state through any sequence of other states and back to the same initial state has then the same energy it had initially. The answer to this question must be determined by appeal to experiment. The experiment, in principle, is of the following nature: Measurements are made of all the heat and work effects outside the system while the system passes from the initial state, through the intermediate states, and back again to the initial state. In symbolic language, the sentries posted all along the boundary keep accounts—all of which are combined in one total at the end of the experiment. If at the end it is

[1] We use the symbol Δ to express a finite change in a function (such as E) of the *state* of a system and d to indicate the corresponding infinitesimal change, an exact differential. A finite exchange of heat or work, not being a function of the state of a system, is represented simply by q or w, and an infinitesimal exchange by δq or δw. This symbolism is in accord with common, although not universal, usage and emphasizes the distinction between the exact and the inexact differential, which is fundamental to the understanding of thermodynamics.

found that the sum of all the heat and work effects produced outside the system during this cycle is zero, then, from the definition of energy, the energy change of the system is zero and the system has the same energy content at the end of the cycle as at the beginning. Experiments of this nature, when carefully performed, do uniformly yield a zero energy change whenever any system is carried through any cycle. Herein is the meaning of the first law, namely, that *the energy of a system is a function of its state*, and whenever the system is in a certain state, it has a definite energy, and whenever it passes from one state to another, the energy change is a fixed quantity independent of the course followed or the particular sequence of states chosen.

That the second part of this statement follows directly from the first may be illustrated with the help of Fig. 6-1. Let us first consider the passage of the system from state A to state B by course 1, designated $A1B$, and its return to state A via the same course, $B1A$. Since this constitutes a cycle, though a special case, the energy change involved is zero.

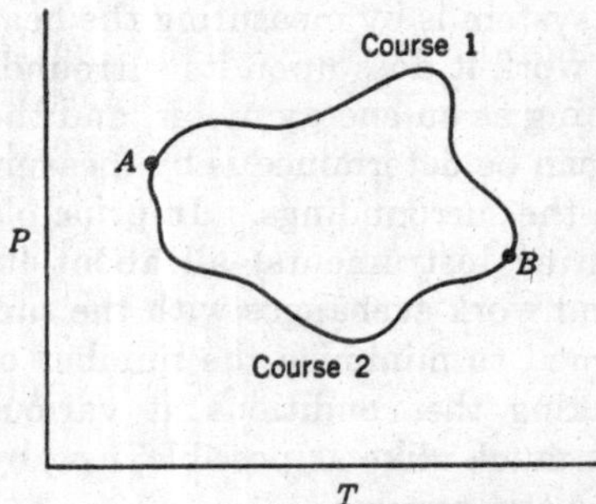

Fig. 6-1. Schematic diagram leading to the conclusion that E is a state property.

$$\Delta E_{A1B} - \Delta E_{B1A} = \Delta E_{\text{cycle}} = 0$$

or

$$\Delta E_{A1B} = -\Delta E_{B1A}$$

In general the energy change involved in passing from any state A to any state B is the negative of the energy change from state B to state A.

We next consider the passage of the system around the complete cycle of Fig. 6-1—from A via course 1 to B, then from B via course 2 to A. Designating the energy change of the system in the first part of the cycle as ΔE_{A1B} and in the second as ΔE_{B2A}, we have

$$\Delta E_{A1B} + \Delta E_{B2A} = \Delta E_{\text{cycle}} = 0$$

Substituting from above,

$$-\Delta E_{B1A} + \Delta E_{B2A} = 0$$

or

$$\Delta E_{B1A} = \Delta E_{B2A}$$

Thus, from the experimental fact that the energy change of a system in passing through any cycle is zero, it has been demonstrated that the energy change in passing from any state A to any state B is independent of the particular course taken and hence that energy may be regarded as a state property.

The reader who is not thoroughly familiar with the historical and logical development of the first law may be somewhat puzzled by the emphasis placed upon the demonstration that the energy of a system is a function only of its state, *i.e.*, that E is a state variable. He might think back over the better known properties associated with systems and conclude that they all are functions of state and hence not be surprised that the energy of a system likewise is a function of state. The fallacy of this train of thought lies in the fact that it is the useful variables which are functions of state. Variables which are not functions of state are difficult if not impossible to handle by the usual mathematical procedures. For example, the differentials of such variables are not exact in terms of state variables. If y represents a property which is a function of the state of a closed[1] homogeneous system, then we may write (in the absence of gravitational, magnetic, or electrical fields; surface effects; etc.)

$$dy = \left(\frac{\partial y}{\partial P}\right)_T dP + \left(\frac{\partial y}{\partial T}\right)_P dT$$

which is simply the fundamental equation for partial differentiation expressing the complete differential of a function in terms of its partial derivatives. If, however, y is not a function of state, in particular of pressure and temperature, then such an equation has no meaning. It will be recalled from the very beginning of the differential calculus that the four-step rule for differentiation starts: "let us consider y, a function of x; $y = f(x)$." If y is not a function of x, it cannot be differentiated with respect to x. It is obvious, therefore, that, if a property of a system is not a function of the state variables, the quantitative (mathematical) treatment thereof is very much hampered.

Ability to Absorb Heat or Perform Work Not a Function of State. At this point the uninitiated reader may well state that he fully understands the desirability of this functional relationship in any property we might wish to treat mathematically, but he may wonder why it is necessary to use such strict tests before admitting energy as a state property when temperature, pressure, density, and others are admitted rather informally. The answer to this question lies in the fact that energy, or energy change, in accord with Lord Kelvin's definition, is compounded of two terms, namely, heat effects and work effects produced outside the system, neither of which can be regarded as uniquely related to the state, or change in the state, of the system.

[1] By a *closed system* is meant one to which or from which no mass is transferred during the course of the processes under consideration. An *open system* is, of course, one which suffers a change in mass.

For example, let us consider a system composed of an ideal gas. In view of the ideal-gas equation, the isothermal relation between P and V is hyperbolic as shown by the light curves of Fig. 6-2. The work done in passing from state A to state B is $\int P\,dV$ between these two states. There is a variety of ways by which the system may undergo this transition: It may do so isothermally, in which case the work corresponds to the area under curve 1, or the gas may be cooled and reheated during the process so that the work is the area under curve 2, or it may be heated and recooled (curve 3). Thus the work done by the system in passing from state A to state B may arbitrarily be made very large or very small or made to assume any intermediate value. Therefore the ability of the system to do work in passing between states A and B is not a function of these two states alone but is a function of the arbitrary nature of the course taken. Similarly the ability of the system to absorb heat in passing between states A and B is not a function of these states alone. Hence one may well be surprised to find that, although neither of these two quantities is a state variable, their difference is.

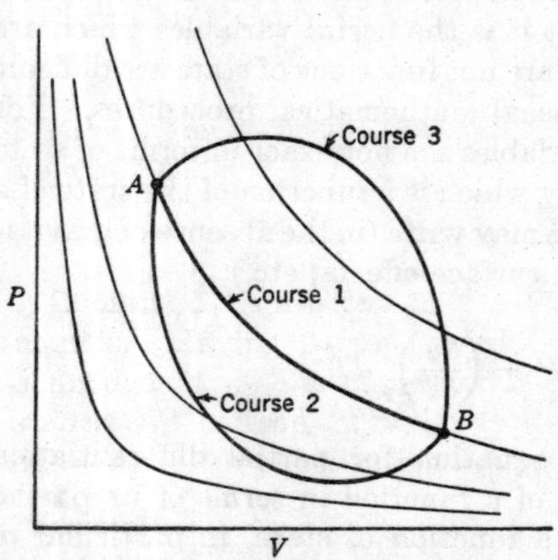

Fig. 6-2. Schematic diagram leading to the conclusion that q and w are not state properties.

THE CONCEPT OF EQUILIBRIUM

In the preceding section we discussed $\int P\,dV$, the work done by the system under consideration upon the surroundings during a volume change. From the definition of energy, the pressure involved in this term is that exerted by the surroundings upon the system. In order to be able to evaluate the work by a single term as above, it is necessary that this pressure be uniform over the entire boundary of the system. If such is not the case and the pressure varies over the boundary, then the work term may be expressed as the sum or integral of several such contributions. If, further, the force is not normal to the boundary, *e.g.*, if both the system and the surroundings are solids capable of supporting a shear at the boundary, the work is not properly expressed by such a term alone. It should be noted that by proper choice of the surroundings the work may be expressed by $P\,dV$ terms only, even though the system itself may be subject to shearing stress. For example, a shearing stress within a cube of metal may be produced by applying pressure to two opposite faces and none to the other four; the work done here is calculable as the integral of the applied pressure over the volume traversed by the two faces under pressure.

Although the pressure to be used in the work term is that of the immediate surroundings of the system, it is convenient, when possible, to think of pressure as being a property of the system itself rather than of the surroundings. This is proper only when a uniform pressure and no shear prevails within the system; under these conditions the pressure of the system is equal to that of the surroundings upon the system. Pressure, in accord with its definition as force per unit area, can be measured only by the introduction of suitable instruments to determine this force. The pressure, so measured, has experimentally verifiable meaning only at the boundary of a system unless the measuring instrument itself is considered part of the system. Nevertheless, if the measuring instrument, *i.e.*, the pressure gage, indicates a uniform pressure in a variety of test positions, it is common and fruitful to regard this pressure as a property associated with the interior as well as the boundary of the system. A system in which the pressure is at all points the same (in the absence of a field of force) is said to be in pressure equilibrium or *mechanical equilibrium.* On any small volume element of such a system the resultant mechanical force is zero, and hence no motion can arise therefrom.

The concept of mechanical equilibrium which arises from the above and other mechanical considerations is found to have counterparts thermally and chemically. Just as matter does not move when the pressure is uniform, so heat does not flow when the temperature is uniform. A system in which there are no temperature gradients is said to be in *thermal equilibrium.* This subject will be considered in detail in the next chapter in the treatment of the second law of thermodynamics. *Chemical equilibrium* is said to prevail when there is no further tendency for the substances concerned to react, or in terms of the theory of reaction rates, when the rates of reaction in the forward and reverse directions are equal. Just as equality of pressure is the criterion of mechanical equilibrium and equality of temperature is the criterion of thermal equilibrium, the equality of the Gibbs chemical potential throughout all phases is the criterion of chemical equilibrium. The development of this concept will appear in Chap. 11. However, it may be stated here that chemical equilibrium prevails when the chemical or atomic forces do not tend to move any constituent from one phase to another or from one molecular species to another, if such exist in the reaction under consideration.

Complete *thermodynamic equilibrium* is said to prevail when the system under consideration is in equilibrium in all three of the foregoing respects. A true pressure must prevail and must be uniform throughout the system, the temperature must be uniform, and the chemical potential of each of the components must be constant throughout, no matter whether the system is composed of one or many phases.

Reversibility. The concept of reversibility is very closely related to that of equilibrium. The term *equilibrium* is applied to a system, whereas the term *reversible* is applied to a process involving a system which is displaced from equilibrium only to an infinitesimal extent. Thus a reversible process may be defined as the hypothetical passage of a system through a series of equilibrium states. In principle the direction of a reversible process can be reversed by an infinitesimal change of the imposed conditions. Although no going process is, in a strict sense, reversible, it can be regarded as substantially reversible providing the forces are balanced, the temperature is uniform, and chemical equilibrium prevails—all within the experimental error. Some processes are inherently irreversible: the flow of heat from a hot to a cold body; the mixing of two gases or liquids to form a solution; processes involving friction; and the flow of electric current through a resistance.

Frequently it happens that a system undergoes a process which is substantially reversible and yet the immediate surroundings undergo changes that are irreversible. For example, if the gas in a cylinder expands slowly the process may be reversible as far as the gas is concerned, but the friction of the piston on the cylinder wall introduces an irreversible element into the larger system gas-cylinder-piston. Similarly, if a well-stirred solution in a beaker is heated by a bunsen flame, the heating may be substantially reversible as far as the solution is concerned but irreversible for the system fuel-air-beaker-solution.

Partial Equilibrium. Although the criteria just mentioned are the conditions necessary for complete equilibrium, we often wish to consider a system which is not in complete equilibrium but which is in equilibrium only in certain aspects. For example, in the vicinity of room temperature, the P-V-T relations of an H_2-O_2 gaseous mixture may be investigated in spite of the fact that the system is not in equilibrium with respect to the reaction by which water is formed. In fact, the simultaneous solubility of a mixture of these gases in water might be investigated and equilibrium attained with respect to the solution process, but not with respect to the formation of water, at least in the absence of catalysts. Similarly, if a system contains a semipermeable membrane, as discussed in Chap. 2, the chemical potential of the component to which the membrane is permeable will ultimately become equal on both sides of the membrane, whereas the chemical potential of the component to which the membrane is impermeable will be unequal on the two sides. The system may then be said to be equilibrated with respect to the permeable constituent but not with respect to the impermeable constituent.

Thus, if oxygen and nitrogen are separated by a membrane of silver, which is permeable to oxygen but not to nitrogen at elevated temperature, p_{O_2} will approach in reasonable time the same value on both sides whereas p_{N_2} will not. It is of interest to note, further, that, although this system can easily be brought to thermal equilibrium, or will spontaneously approach thermal equilibrium if isolated (*i.e.*, insulated thermally from the surroundings), it will not in reasonable time approach equality of pressure on the two sides of the membrane. p_{O_2} will equalize, but p_{N_2} will not, except by the exceedingly slow process of diffusion of nitrogen through the silver; and hence the total pressure $P = p_{O_2} + p_{N_2}$ will not be equal on the two sides of the membrane. The membrane must support this inequality of pressure. Such partial equilibria may be treated thermodynamically.

It may be noticed from the foregoing example that the criterion of equilibrium depends upon the time scale of the experiment. In the usual period of time allowed for an experiment, the passage of nitrogen through a sheet of silver is insignificant, so that we are justified in treating the experimental results as though the oxygen had equilibrated and no passage of nitrogen had occurred. On the other hand, if the experiment had been conducted over an exceedingly long period of time (a "geologic" time), it would be found that the very slight permeability of silver to the nitrogen would be sufficient to bring about equalization of p_{N_2} as well as of p_{O_2}; thus complete equilibrium would prevail and again the case could readily be treated by thermodynamics.

Let us now consider an intermediate case where the time allowed for the process is long compared with the diffusion of oxygen but short compared with the "geologic" scale. In this case, the partial pressure of oxygen will be equilibrated on the two sides of the diaphragm, but the diffusion of nitrogen must be considered as a going process. On this intermediate time scale the system may by no means be regarded as being in equilibrium. A very similar situation may prevail in the heat-treatment of steel, since carbon diffuses rapidly and other elements relatively slowly. In this case iron acts as the diaphragm which is relatively more permeable to carbon than to the other elements, *e.g.*, manganese, molybdenum, or chromium. In reasonably short treatments it is permissible to consider the carbon as equilibrated and the other elements as static, and in very long processes a complete equilibrium may be approached, but in processes of intermediate duration it is not permissible to regard the system as being in equilibrium. Frequently the time allotted in commercial heat-treatment falls in this intermediate range, so that a strict thermodynamic treatment is impossible.

It is generally found, in conducting an experiment, that certain processes are relatively rapid and others relatively slow. If the more rapid ones are fast enough to approach equilibrium, within the desired limit of error, and the slow ones sufficiently slow that they can be ignored, again within the desired limit of error, then it is quite proper to treat the system as being at equilibrium with respect to the rapid processes alone. For example, there is reason to believe that, under certain circumstances during the refining period of the open-hearth steelmaking process, the transfer of sulfur, manganese, and phosphorus between the metal and the slag is rapid. The volatilization of iron from the bath is appreciable at this time, but the reaction of oxygen and this iron with the silica roof proceeds slowly, though continuously. Clearly, at complete equilibrium the entire roof would be dissolved. This fact, however, need not deter us from considering the foregoing slag-metal equilibria. Another factor in this case is the reaction of the carbon in the metal with oxygen which crosses the slag-metal interface, producing gaseous oxides of carbon—a reaction which does not approach equilibrium very closely. Thus a going reaction takes place at the same interface where the sulfur, manganese, and phosphorus equilibria prevail. That this is essentially the case had to be demonstrated experimentally and could by no means be deduced from thermodynamic considerations. Quite the contrary is usually the case, it being commonly found that a going process does influence other equilibria. The phenomenon of thermal transpiration, cited previously in Chap. 2, serves as an illustration of this. The thermoelectric effect (Peltier effect) is another example, the flow of heat giving rise to an electrical potential. The Soret effect, or thermal diffusion, a further example of this general phenomenon, includes many cases where a composition difference arises from a temperature difference (flow of heat).[1]

The Object of Chemical Thermodynamics. The main object which the chemist has in mind in the development of chemical thermodynamics is the development of a systematic mathematical treatment of chemical equilibria. He desires a mathematical system to treat the variations of chemical equilibria with state variables, especially temperature and pressure, and to show the relationship between different equilibria. In other words, he wishes to catalogue known quantitative information concerning equilibria and to be able to make quantitative predictions, if possible, about equilibria which have not been directly investigated

[1] This branch of thermodynamics dealing with irreversible processes promises to become very important. An excellent treatment is given by S. R. de Groot in "Thermodynamics of Irreversible Processes," Interscience Publishers, Inc., New York, 1951.

experimentally. Since in this cataloguing and predicting the emphasis is upon the quantitative aspect, it is highly desirable that the quantities dealt with should bear a functional relationship, so that they can be readily handled by conventional mathematics. It is the fact that the energy is a function of the state variables that makes it so valuable to us in our efforts to formulate mathematically the relations and interrelations pertaining to chemical equilibria.

Expression of Work in Terms of State Variables. Let us now reconsider the differential relation defining an energy change:

$$dE = \delta q - \delta w \tag{6-1}$$

This will henceforth be referred to as a statement of the first law of thermodynamics for a closed system, in accord with convention. It is well to note in passing that the δq term occurs here with a plus sign and the δw term with a minus sign because, conventionally, δq refers to the heat absorbed by the system whereas δw refers to the work done by, rather than the work done on, the system. Let us now divide the work term into two parts: the work done against an external pressure, $P\,dV$, and all other forms, designated $\delta w'$. $\delta w'$ includes electrical work; surface work (work accompanying a change in amount of surface); work against external electrical, magnetic, or gravitational fields; etc. The equation then becomes

$$dE = \delta q - P\,dV - \delta w' \tag{6-2}$$

It is not unusual for the system to do no work other than that against pressure; hence we shall consider for the present, and for the future unless otherwise stated, that $\delta w'$ is zero. Thus

$$dE = \delta q - P\,dV \tag{6-3}$$

It will be noted that the differentials occurring here are exact except for δq, or in other words, the variables except for q are functions of state. Because of the presence of this inexact differential the equation is still not in its most fruitful mathematical form. However, this is the best that can be done with the first law alone. The casting of this equation into a form such that all variables are functions of state must await the introduction of the second law of thermodynamics.

***dE* in Terms of Partial Derivatives.** We have already discussed the equation of state, which we assume generally to exist for any homogeneous substance. This may be written in the form[1]

$$f(P,V,T) = 0$$

[1] f is used in this section to designate a functional relationship. The use of the same symbol f in different equations does not, of course, imply that the same function is involved.

or, equally well, in any one of the following forms: $P = f(V,T)$, $V = f(P,T)$, and $T = f(P,V)$. For a closed homogeneous system it is found from experience that any variable which is a function of state may be represented as a function of three state variables only; these may be chosen conveniently as P, V, and T, so that the energy of such a system is a function of P, V, and T; $E = f(P,V,T)$. If the system obeys an equation of state, then P, V, or T in the functional expression for energy can be eliminated by aid of one of the foregoing equations of state. Thus we may write $E = f(P,V)$, $E = f(P,T)$, or $E = f(V,T)$. Furthermore, by use of the fundamental theorem of partial derivatives, dE may be expressed in any of the following forms:

$$dE = \left(\frac{\partial E}{\partial P}\right)_V dP + \left(\frac{\partial E}{\partial V}\right)_P dV \tag{6-4}$$

$$dE = \left(\frac{\partial E}{\partial P}\right)_T dP + \left(\frac{\partial E}{\partial T}\right)_P dT \tag{6-5}$$

$$dE = \left(\frac{\partial E}{\partial V}\right)_T dV + \left(\frac{\partial E}{\partial T}\right)_V dT \tag{6-6}$$

HEAT CAPACITY

Rewriting the first law in the form $\delta q = dE + P\,dV$ and substituting for dE from Eq. 6-6, we obtain

$$\delta q = \left(\frac{\partial E}{\partial T}\right)_V dT + \left[\left(\frac{\partial E}{\partial V}\right)_T + P\right] dV \tag{6-7}$$

The *heat capacity* of a substance is the amount of heat required to raise its temperature one degree. More strictly it is the limiting ratio of the heat increment to the temperature rise as both approach zero; thus the heat capacity, designated C, is $\delta q/dT$. The term *specific heat* refers to the heat capacity of unit mass of the substance under consideration. Unless otherwise specified, the term heat capacity will be used in this book to designate the heat capacity in calories per degree centigrade of one gram atom or one gram mole. However, as noted previously, δq is not an exact differential in terms of state variables and is determinate for a given process only if the course is specified. Hence $\delta q/dT$ and C, to be definite, should always carry a subscript designating the particular manner in which the temperature is raised. For example, in an adiabatic process, one during which the system exchanges no heat with the surroundings (approximated by the rapid expansion or compression of gas in a pump or engine), δq, and hence $\delta q/dT$, is zero. It is not common practice to refer to this as a heat capacity.

Experimentally it is most convenient to determine the heat capacity under the condition of constant pressure (usually atmospheric); this *heat capacity at constant pressure* is designated C_P. Theoretical consideration of the subject is usually found easier under the condition of constant volume; the *heat capacity at constant volume* is called C_V. C_V may readily be expressed in terms of the energy by use of the preceding equation, which reduces to $\delta q = (\partial E/\partial T)_V\, dT$ for a process during which there is no volume change. Dividing by dT and expressing the constancy of volume by the subscript: $(\delta q/\partial T)_V = (\partial E/\partial T)_V$. Since the quantity on the left side of this equation is, by definition, C_V, we have

$$C_V = \left(\frac{\partial E}{\partial T}\right)_V \tag{6-8}$$

In order to evaluate C_P in terms of energy, Eq. (6-5) is combined with Eq. (6-3) to give

$$\delta q - P\,dV = \left(\frac{\partial E}{\partial P}\right)_T dP + \left(\frac{\partial E}{\partial T}\right)_P dT$$

Under the condition of constant pressure the first term on the right is zero; dividing by dT and transposing, we find for C_P, defined[1] as $(\delta q/\partial T)_P$,

$$C_P \equiv \left(\frac{\delta q}{\partial T}\right)_P = P\left(\frac{\partial V}{\partial T}\right)_P + \left(\frac{\partial E}{\partial T}\right)_P \tag{6-9}$$

It will be observed that $P(\partial V/\partial T)_P$ may also be written $(\partial PV/\partial T)_P$ and hence

$$C_P = \left(\frac{\partial(E + PV)}{\partial T}\right)_P \tag{6-10}$$

A relation between C_P and C_V may be obtained by combining Eqs. (6-8) and (6-9) with the following mathematical modification of Eq. (6-6).

$$\left(\frac{\partial E}{\partial T}\right)_P = \left(\frac{\partial E}{\partial T}\right)_V + \left(\frac{\partial E}{\partial V}\right)_T \left(\frac{\partial V}{\partial T}\right)_P$$

Combination of these three gives

$$C_P - C_V = \left[P + \left(\frac{\partial E}{\partial V}\right)_T\right]\left(\frac{\partial V}{\partial T}\right)_P \tag{6-11}$$

This expression will be put into a more useful general form after considera-

[1] The identity sign $\equiv$ is used instead of the equality sign to distinguish equalities which follow from definition or mathematical identity, *i.e.*, those equalities which depend upon no experimental finding whatever.

tion of the second law in the next chapter.[1] It may be applied in the present form to an ideal gas, which we previously defined as one which obeys Boyle's law and for which the energy is a function only of temperature. Combination of Boyle's law with the definition of temperature in terms of the ideal gas gives the ideal-gas law $PV = RT$. Since

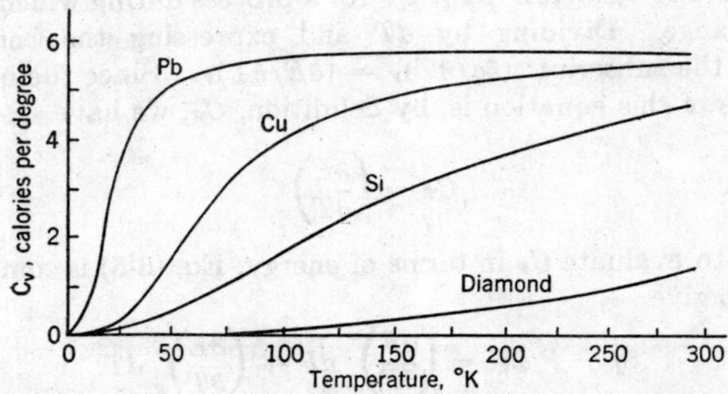

FIG. 6-3. The atomic heat capacity of Pb, Cu, Si, and diamond. (*From Eastman and Rollefson, "Physical Chemistry," McGraw-Hill Book Company, Inc., New York,* 1947.)

the energy of such a gas is a function only of temperature, it does not change with volume at constant temperature and $(\partial E/\partial V)_T = 0$. By partial differentiation of the ideal-gas law, $(\partial V/\partial T)_P = R/P$. By substitution of these values in Eq. (6-11) it is found for an ideal gas that

$$C_P - C_V = R \tag{6-12}$$

Heat Capacity as a Function of Temperature—Debye's Theory. A number of solid elements at low temperature follow the same general pattern in the variation of C_V with temperature. C_V for a few elements is shown in Fig. 6-3. Lewis and Gibson[2] first observed that, if C_V is plotted against $\log T$, these curves are nearly identical except for a horizontal displacement. This similarity may be shown by plotting C_V against $\log T/\theta$, where θ is a constant, characteristic of the element. The original plot of Lewis and Gibson is shown in Fig. 6-4. It will be noted that the value of C_V approached at higher temperature is $3R$ [5.97 cal/(gram atom)(deg)], in accord with the older *law of Dulong and Petit.* Debye[3] was able to derive theoretically the form of this curve from consideration of waves (vibrations due to thermal agitation) in a homo-

[1] In Chap. 7 [Eq. (7-20)] it will be demonstrated that Eq. (6-11) may be reduced to the form $C_P - C_V = \alpha^2 VT/\beta$, where α is the coefficient of thermal expansion $(\partial \ln V/\partial T)_P$ and β is the coefficient of compressibility, $-(\partial \ln V/\partial P)_T$.

[2] Lewis and Gibson, *J. Am. Chem. Soc.*, **39**, 2554 (1917).

[3] P. Debye, *Ann. Physik*, **39**, 789 (1912).

geneous elastic continuum.[1] He obtained the formula

$$\frac{C_V}{3R} = \frac{12}{x^3}\int_0^x \frac{y^3\,dy}{e^y - 1} - \frac{3x}{e^x - 1}$$

where $x = \theta/T$ and y may be regarded as an integration variable which, it will be noted, disappears when the integration is performed and the limits substituted. The above expression gives C_V as a function of θ/T; tables of this function are available.[2]

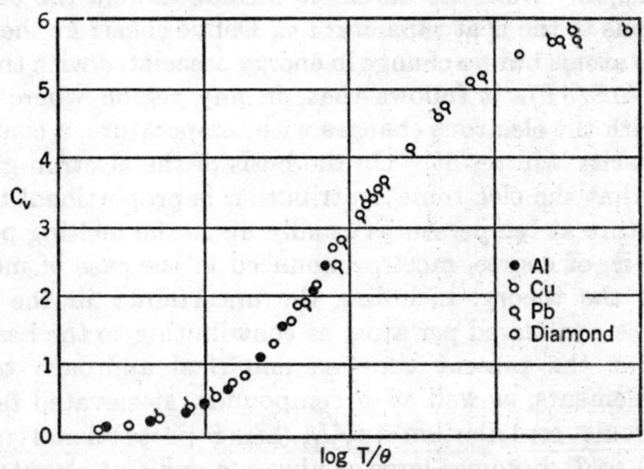

FIG. 6-4. The variation of C_V with log (T/θ) for Al, Cu, Pb, and diamond. *(From Eastman and Rollefson.)*

Two facts regarding this function are particularly noteworthy: (1) At high temperature, corresponding to a low value of θ/T, C_V approaches $3R$ in accord with the law of Dulong and Petit, and (2) at very low temperature the expression reduces to

$$\frac{C_V}{3R} = 77.927\left(\frac{T}{\theta}\right)^3 \quad \text{or} \quad C_V = 464.5\left(\frac{T}{\theta}\right)^3 \quad \text{cal/(mol)(deg)}$$

indicating a rapid approach of C_V toward a zero value as T approaches zero.

Debye's equation is found to be most useful in the low-temperature range, in particular at temperatures below those at which the heat capacity has been measured experimentally. It will be noted that C_P rather than C_V is usually measured experimentally whereas Debye's

[1] Debye assumed an upper limit imposed upon the spectrum of frequencies by the atomistic structure of an actual solid. Moreover, the N atoms of a solid are assumed to have $3N$ degrees of freedom, so that the total number of frequencies is $3N$.

[2] F. H. MacDougall, "Thermodynamics and Chemistry," p. 481, John Wiley & Sons, Inc., New York, 1939.

theory relates to C_V. In order to interconnect theory and experiment the difference between C_P and C_V may be determined from the previously mentioned relation $C_P - C_V = \alpha^2 VT/\beta$, which will be derived in the next chapter [Eq. (7-20)].

Electronic Contribution to C_V. At a temperature much above room temperature it is not unusual to find that C_V exceeds the Dulong and Petit value; Debye's theory is not applicable in such a case. The deficiency in Debye's theory at elevated temperature is generally believed to arise principally from the failure to include therein the contributions of the electrons to the heat capacity; *i.e.*, Debye considers the vibrational energy of the atoms but no change in energy associated with the electrons. Since $C_V = (\partial E/\partial T)_V$ it follows that, in any region where the energy associated with the electrons changes with temperature, a contribution to the heat capacity will result. On the basis of the electron-gas theory it is indicated that the electronic contribution is proportional to the absolute temperature at temperatures usually up to the melting point. This contribution is, of course, most pronounced in the case of metals. The crudeness in the theory, including the uncertainty in the number of electrons to be considered per atom as contributing to the heat capacity, necessitates at the present time an empirical approach to the heat capacity of elements, as well as of compounds, at elevated temperature.

The electronic contribution to C_V, being proportional (at least in simple cases) to T, becomes large in absolute value at elevated temperature. However, at low temperature it again becomes large as compared with the lattice contribution, since the latter is small (proportional to T^3) in this range. This fact is illustrated by the low-temperature specific heat of zinc and of tungsten shown in Figs. 6-5 and 6-6, taken from the work of Silvidi and Daunt.[1] In the low-temperature range shown it will be noted that the curves are not cubic parabolas as called for by the Debye theory but appear to approach a straight line (with nonzero slope) at 0°K. The curves of these two figures are adequately represented by the following equations.

For Zn:

$$C_V = 464.5\left(\frac{T}{291}\right)^3 + 1.50 \times 10^{-4}T \qquad \text{cal/(mole)(deg)}$$

For W:

$$C_V = 464.5\left(\frac{T}{169}\right)^3 + 51.1 \times 10^{-4}T \qquad \text{cal/(mole)(deg)}$$

In these equations the first term on the right represents the Debye term or lattice contribution as discussed in the preceding subsection and the

[1] A. A. Silvidi and J. G. Daunt, *Phys. Rev.*, **77**, 125 (1950).

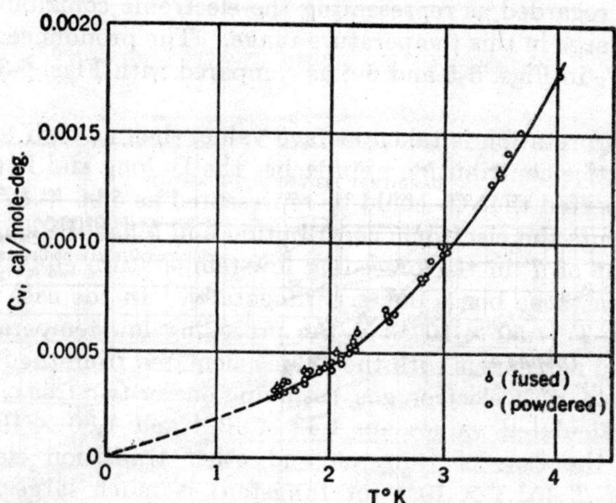

FIG. 6-5. The low-temperature heat capacity of zinc. [*From Silvidi and Daunt, Phys. Rev.*, **77**, 125 (1950).]

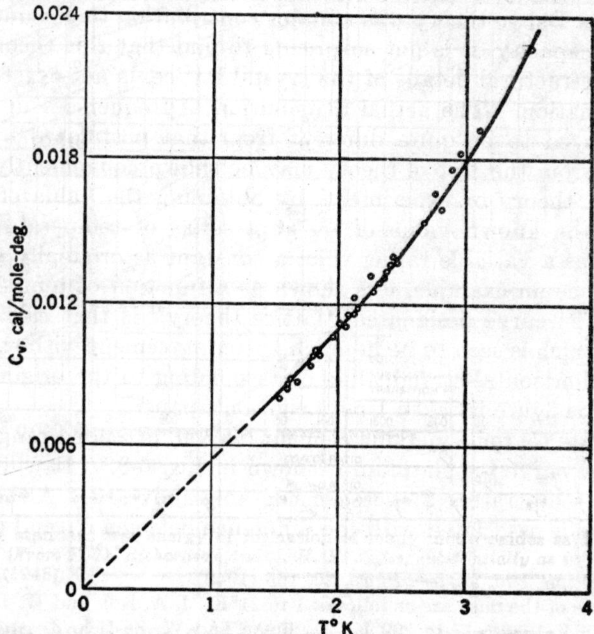

FIG. 6-6. The low-temperature heat capacity of wolfram. (*From Silvidi and Daunt.*)

final term is regarded as representing the electronic contribution, which is relatively large in this temperature range. The pronounced difference in scale for C_V in Figs. 6-5 and 6-6 as compared with Figs. 6-3 and 6-4 is to be noted.

If this interpretation is taken at face value, then at high temperature where the lattice contribution approaches the Dulong and Petit value, it would be expected that C_V could be represented as $5.96 + bT$, where bT again represents the electronic contribution and b has the same value as the coefficient of T in the preceding low-temperature equations. This does not seem to be borne out experimentally. In the case of zinc the coefficient of T (1.50×10^{-4}) in the preceding low-temperature equation is in good agreement with the value calculated from the Sommerfeld formula for a perfect electron gas, assuming one or two free electrons per atom (the calculated values are 1.43×10^{-4} and 1.80×10^{-4}, respectively). In the case of tungsten and other transition elements the coefficient of T (51.1×10^{-4} for tungsten) is much larger than that required to fit either the high-temperature heat capacity or existing theory. This matter is discussed in some detail by Silvidi and Daunt.

Extended Theory of Lattice Vibration Contribution to Heat Capacity. Although the Debye theory was a major contribution to our understanding of heat capacity, it is not surprising to find that this theory, which ignores the structural details of the crystal lattice, is not exact but only an approximation. The actual distribution of frequencies in a crystal is now believed to be quite different from that postulated by Debye. Departures from the Debye theory may be shown conveniently, if C_V is known from theory or experiment, by obtaining the value of θ which reproduces the known value of C_V at a series of temperatures—thus regarding θ as a variable rather than a constant as originally supposed. Using silver as an example, θ is shown as a function of temperature in Fig. 6-7. The curve designated "lattice theory" is that calculated by Leighton,[1] which is seen to be in much better agreement with experiment than is the horizontal straight line corresponding to the original Debye theory. This figure is taken from Leighton's paper.

To illustrate C_P for a metallic element over an extended range of temperature, the curve for platinum is shown in Fig. 6-8.[2] It will be noted that at high temperature C_P rises considerably above $3R$. Although part of this excess is associated with the difference between C_P and C_V, which

[1] R. B. Leighton, *Rev. Modern Phys.*, **20**, 165 (1948).

[2] The sources of the data are as follows: 1 to 21°K., J. A. Kok and W. H. Keesom, *Physica*, **3**, 1035 (1936); 17 to 209°K., F. Simon and W. Zeidler, *Z. physik. Chem.*, **123**, 383 (1926); 100 to 1600°C., F. M. Jaeger and E. Rosenbohm, *Proc. Acad. Sci. Amsterdam*, **33**, 457 (1930).

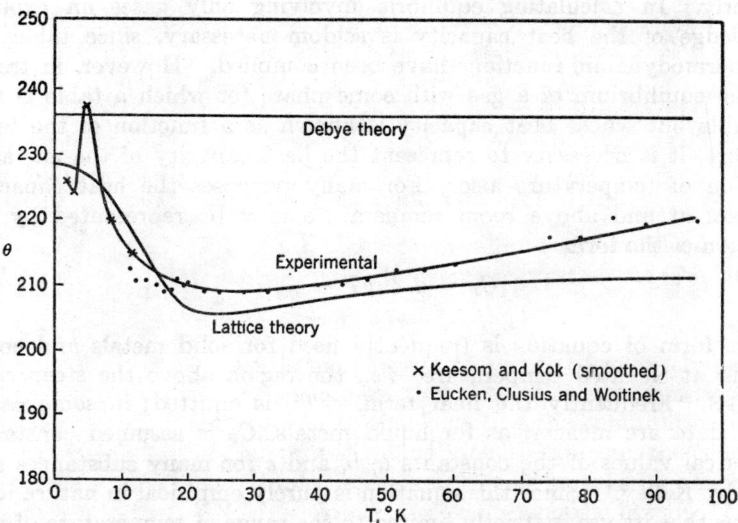

FIG. 6-7. The variation with temperature of θ for silver. [*From R. B. Leighton, Rev. Mod. Phys.*, **20**, 165 (1948).]

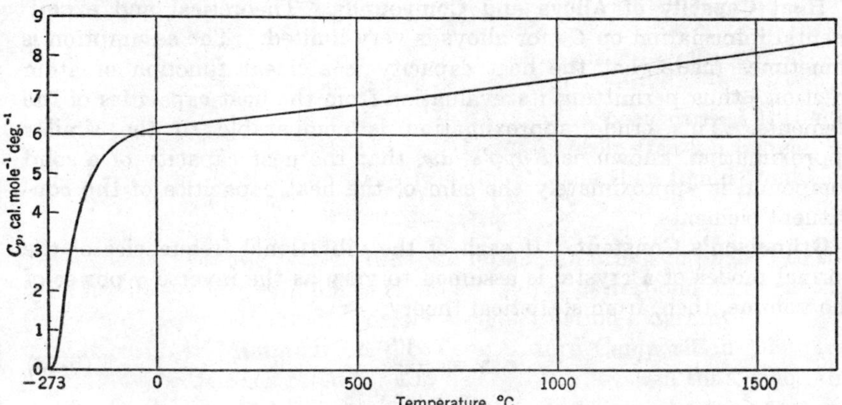

FIG. 6-8. The heat capacity of platinum.

is difficult to estimate at high temperature, it seems reasonable to assume that the excess is attributable in large part to the electronic contribution and possibly to other factors not considered by Debye's theory. Regardless of the explanation, the heat capacity of platinum in the high-temperature region is seen to be a reasonably smooth function of temperature.

Empirical Representation of C_P. The thermodynamic functions of gases, including the heat capacity, are at the present time usually based on spectroscopic measurements rather than upon thermal data, as formerly. In calculating equilibria involving only gases an explicit knowledge of the heat capacity is seldom necessary, since tables of the thermodynamic functions have been compiled. However, in treating the equilibrium of a gas with some phase for which a table is not available but whose heat capacity is known as a function of the temperature, it is necessary to represent the heat capacity of the gas as a function of temperature also. For many purposes the heat capacity of gases at and above room temperature may be represented by an equation of the form

$$C_P = a + bT - \frac{c}{T^2}$$

This form of equation is frequently used for solid metals and compounds at elevated temperature—*i.e.*, the region above the steep rise, Fig. 6-8. Frequently the final term, c/T^2, is omitted; in some cases where data are meager, as for liquid metals, C_P is assumed constant. Numerical values of the constants a, b, and c for many substances are given by Kelley.[1] Since this equation is purely empirical in nature it is obvious that its use is strictly limited to the range of temperature of the data from which it was derived.

Heat Capacity of Alloys and Compounds. Theoretical and experimental information on C_P for alloys is very limited. The assumption is sometimes made that the heat capacity is a linear function of atom fraction—thus permitting its evaluation from the heat capacities of the elements. This crude approximation is comparable to the similar approximation, known as *Kopp's law,* that the heat capacity of a solid compound is approximately the sum of the heat capacities of the constituent elements.

Grüneisen's Constant. If each of the vibrational frequencies of the normal modes of a crystal is assumed to vary as the inverse γ power of the volume, then, from statistical theory,

$$\gamma = \frac{\alpha V}{\beta C_V} \tag{6-13}$$

where α is the coefficient of thermal expansion $(1/V)(\partial V/\partial T)_P$ and β is the coefficient of compressibility $-(1/V)(\partial V/\partial P)_T$. This relation was first derived by Grüneisen, who found empirically that γ, known as Grüneisen's constant, has a value between 1.5 and 2.5 for many crystalline solids.[2] Also, on the basis of Debye's theory of specific heats and the assumption that Poisson's ratio is independent of the volume, an

[1] K. K. Kelley, Contributions to the Data on Theoretical Metallurgy X, *U.S. Bur. Mines Bull.* 476, 1949.

[2] E. Grüneisen, "Handbuch der Physik," Vol. 10, Springer-Verlag, Berlin, 1926.

expression for γ can be derived involving only β and its change with volume:

$$\gamma = -\frac{1}{6} + \frac{1}{2}\frac{d \ln \beta}{d \ln V} \tag{6-14}$$

Theoretically the two methods of computing γ should give approximately the same result, and a useful relation is thus provided between the various experimental quantities.

A comparison of the value of γ at room temperature as calculated by the two methods is shown in Table 6-1. The agreement is seen to be fair

TABLE 6-1. VALUES OF GRÜNEISEN CONSTANT AT ROOM TEMPERATURE FOR SEVERAL METALS AND ALKALI HALIDES

Substance	Grüneisen constant γ*	
	Eq. (6-13)	Eq. (6-14)
LiF	1.34	3.02
LiCl	1.52	2.11
LiBr	1.70	2.06
NaCl	1.63	1.85
NaBr	(1.56)	1.75
KF	1.45	2.26
KCl	1.60	1.59
KBr	1.68	1.62
KI	1.63	1.54
RbBr	1.37	1.46
RbI	(1.41)	1.50
Li	1.17	0.63
Na	1.25	1.83
K	1.34	2.55
Al	2.17	1.27
Au	3.03	8.6
Pb	2.73	2.42
Fe	1.68	1.60
Co	2.1	1.87
Ni	2.2	1.88
Cu	1.9	1.96
Pd	2.4	2.23
Ag	2.5	2.40
W	1.7	1.62
Pt	3.3	2.54

* Data for substances in the first two groups from J. C. Slater, "Introduction to Chemical Physics," McGraw-Hill Book Company, Inc., New York, 1939; data for the last group from J. C. Slater, *Phys. Rev.*, **57**, 744 (1940).

for metals as well as for ionic crystals. Because of the assumptions involved in the two derivations an exact check is not to be expected.

ENTHALPY

The quantity $E + PV$ appearing in Eq. (6-10) occurs so frequently in the treatment of systems at constant pressure that it is convenient to designate it by a single symbol H and to give it a special name, *enthalpy*. The term *heat content* is sometimes used, and the symbol H is derived therefrom.

$$H \equiv E + PV \tag{6-15}$$

From the definition of the enthalpy of a system it follows that the enthalpy, like the energy, is a function of the state of a system, since it is defined in terms of the energy, a function of state, and the state variables P and V. Hence the enthalpy change for any cyclic process involving a thermodynamic substance is zero.

The first law may be expressed in terms of the enthalpy instead of the energy. To do this the definitional equation for H is differentiated,

$$dH = dE + P\,dV + V\,dP$$

and this expression is added to Eq. (6-2), giving

$$dH = \delta q + V\,dP - \delta w'$$

For a process in which no work is done other than that against pressure, $\delta w' = 0$ and

$$dH = \delta q + V\,dP \tag{6-16}$$

If, further, the process is carried out at constant pressure, Eq. (6-16) reduces to $dH = \delta q$ and hence, for such a process,

$$\Delta H = q \tag{6-17}$$

Thus for a process at constant pressure the heat exchanged between the system and the surroundings is seen to be the difference between the initial and final enthalpy of the system itself and to be independent of the particular process by which the final state is reached. This conclusion follows from Eq. (6-17) and the previously established fact that the enthalpy of a system is a function only of its state. It is to be emphasized that Eq. (6-17) applies only to a system at constant pressure, for, as has been shown, q is in general a function not only of the initial and final states of the system but also of the path followed.

To illustrate further the mathematical simplification resulting from the introduction of the enthalpy, it is worth noting at this point that the equivalent of Eq. (6-10) follows immediately from Eq. (6-16) under the condition of constant pressure.

$$C_P \equiv \left(\frac{\delta q}{\partial T}\right)_P = \left(\frac{\partial H}{\partial T}\right)_P \tag{6-18}$$

Heat Balance. For a system which undergoes a process or reaction at constant volume, as in the bomb calorimeter, it follows from Eq. (6-3) that $dE = \delta q$ and hence that $\Delta E = q$. In view of the first law, the energy change for any process is zero if we include the energy changes of the system and the surroundings. Hence for a process at constant volume, sometimes called isometric, the energy increase of the system must equal the heat lost by the surroundings, or conversely the energy lost by the system must equal the heat gained by the surroundings. Thus for any such process a heat balance may be prepared in which the increments of energy gained by the system are entered on one side and the heat lost by the surroundings on the other. By this means the adequacy of the approximations necessary in the computation for any actual process can be checked by comparing the two totals, or if there is only one principal unknown term, this can be computed by "forcing" the heat balance.

Processes taking place at constant volume are few compared to those at constant pressure, usually atmospheric. As already pointed out, under the condition of constant pressure $\Delta H = q$. In this case, then, the enthalpy increase of the system must equal the heat lost by the surroundings; the latter constitutes the externally observed heat effect. Hence in this case also a heat balance may be prepared in which the increases in enthalpy of the system are tabulated in one column and the losses of heat by the surroundings in the other. Any lack of balance of the two columns is due to experimental error.

Heat of Reaction. Let us consider the reaction at constant pressure of two substances to form a third, *e.g.*, the reaction of iron and gaseous oxygen to form hematite (Fe_2O_3). This process may be carried out in two ways: (1) isothermally, and (2) the reactants may be heated to some elevated temperature, the reaction carried out at this elevated temperature, and the products cooled to the initial temperature. The initial and final states of the system are the same for the two processes, and thus the enthalpy change of the system is the same. Suppose now that the isothermal process is carried out in two steps: The iron is oxidized to magnetite (Fe_3O_4), and the magnetite then oxidized to hematite. Since the initial and final states are as before, the enthalpy change for the over-all process is seen to be independent of the number or nature of the steps involved.

In general, the heat of a chemical reaction may be represented as in Fig. 6-9. The heat content of the reactants at constant pressure is represented as a function of temperature by the curve *ABC*, and that of the products by the curve *DEFG*. Let us consider first the course *ABCEF*, which corresponds to heating the unmixed reactants to T_2, mixing and allowing the reaction to take place at this temperature, then heating the products to T_F. The alternate course *ABDEF* consists of heating the reactants to T_1, mixing and reacting at T_1, and further heating to T_F. It is obvious from the figure that the enthalpy of the system at

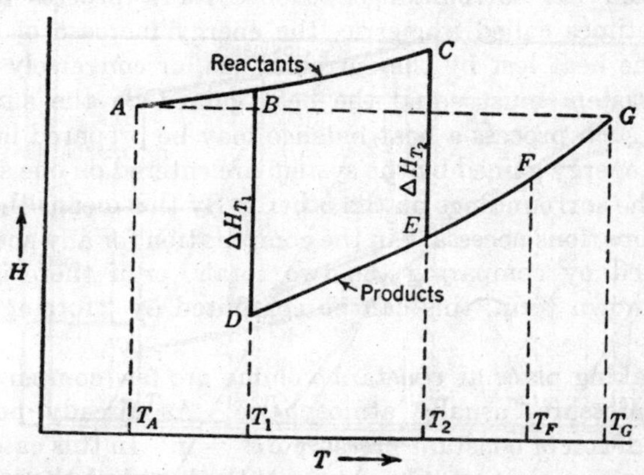

Fig. 6-9. Schematic diagram illustrating change of heat of reaction with temperature.

T_F is identical whether either of these courses or any other course was pursued, and the enthalpy change in passing from state *A* to state *F* is also independent of the course.

As previously shown, the heat capacity C_P of a system is $(\partial H/\partial T)_P$; thus the heat capacity of the reactants C'_P is equal to the slope of the curve *ABC* at any particular temperature, and that of the products C''_P is equal to the slope of curve *DEFG*. The enthalpy change of the reactants during heating from T_1 to T_2 is $\int_{T_1}^{T_2} C'_P\,dT$, and that of the products during heating through the same temperature interval is $\int_{T_1}^{T_2} C''_P\,dT$. The enthalpy change of the reactants during cooling through the same temperature interval is $-\int_{T_1}^{T_2} C'_P\,dT$. As shown on the diagram the enthalpy change of the reaction at T_1 is designated ΔH_{T_1} and that at T_2 is designated ΔH_{T_2}. If the reaction is caused to proceed in the reverse direction at T_2, the enthalpy change is $-\Delta H_{T_2}$. Let us now sum up the enthalpy changes when the system is made to pass through the cycle *BDECB* and equate this sum to zero:

$$\Delta H_{T_1} + \int_{T_1}^{T_2} C''_P \, dT - \Delta H_{T_2} - \int_{T_1}^{T_2} C'_P \, dT = 0 \quad \text{or}$$

$$\Delta H_{T_2} - \Delta H_{T_1} = \int_{T_1}^{T_2} (C''_P - C'_P) \, dT$$

The difference of the heat capacities of the products and reactants $C''_P - C'_P$ is commonly designated ΔC_P, whence

$$\Delta H_{T_2} - \Delta H_{T_1} = \int_{T_1}^{T_2} \Delta C_P \, dT \tag{6-19}$$

It will be noted that this same equation may be derived by applying Eq. (6-18) to the products and to the reactants. Thus $(\partial H''/\partial T)_P = C''_P$ and $(\partial H'/\partial T)_P = C'_P$, H'' and H' being the enthalpies of the products and reactants, respectively. Subtracting, it is found that

$$\left[\frac{\partial(H'' - H')}{\partial T}\right]_P = C''_P - C'_P$$

Since $H'' - H'$ is the enthalpy change of the reaction ΔH and $C''_P - C'_P$ has already been designated ΔC_P,

$$\left(\frac{\partial \Delta H}{\partial T}\right)_P = \Delta C_P \tag{6-20}$$

This expression is known as *Kirchhoff's law*. Writing this as a differential equation, valid at constant pressure, $d\,\Delta H = \Delta C_P \, dT$, and integration from T_1 to T_2 gives Eq. (6-19). Thus it is seen that the change of the heat of a reaction with temperature is directly related to the difference between the heat capacity of the products and that of the reactants. If this difference is zero, the heat of reaction is independent of temperature. Such a fortuitous occurrence is rare, and we must in general treat the heat of reaction as a function of temperature.

Let us return again to Fig. 6-9 and consider an adiabatic reaction at constant pressure. If the system is initially in state *A*, consisting of reactants at T_A, and the reaction is allowed to proceed in an adiabatic manner, then, since H cannot vary, the temperature will rise and the reaction will proceed (in the case illustrated) to completion, the system then consisting wholly of products in state *G*. The temperature attained during the course of the reaction is calculable if the following are known: ΔH at any temperature (such as T_1), C'_P in the range T_A to T_1, and C''_P in the range T_1 to T_G. The enthalpy change of the system in following the course *ABDEFG* is $\int_{T_A}^{T_1} C''_P \, dT + \Delta H_{T_1} + \int_{T_1}^{T_G} C''_P \, dT$. T_G may be found by equating this expression to zero (adiabatic process) and solving.

Hess' Law. As already stressed, the energy change and the enthalpy change of a system upon passing from one state to another are independent of the particular course followed. In the previous section we considered paths of different temperature; let us now consider isothermal paths which differ chemically. As mentioned above, iron may be oxidized to hematite in one step according to the equation[1]

$$2Fe(\alpha) + \tfrac{3}{2}O_2(g) = Fe_2O_3(\text{hem.}) \qquad \Delta H_A \qquad (A)$$

or in two steps:

$$3Fe(\alpha) + 2O_2(g) = Fe_3O_4(\text{mag.}) \qquad \Delta H_B \qquad (B)$$

$$2Fe_3O_4(\text{mag.}) + \tfrac{1}{2}O_2(g) = 3Fe_2O_3(\text{hem.}) \qquad \Delta H_C \qquad (C)$$

[1] As pointed out in Chaps. 3 to 5, most of the substances with which the metallurgist deals are *not* molecular in structure. Since the original introduction of chemical equations to most of us was in terms of molecules, the question might be raised as to the precise meaning of equations such as the above. To answer this question let us consider the nature of the evidence of a chemical reaction such as the first above. Primarily, it is observable that metallic iron with a certain structure and lattice parameter disappears; similarly gaseous oxygen disappears; and an oxide of iron with certain physical characteristics, and in particular with a certain structure and parameter revealed by X rays, appears. From these statements it is clear that our primary interest is focused upon the appearance and disappearance of certain phases. Hence the equation is intended to represent the disappearance of 2 gram atoms of iron from crystalline α-iron and of 3 gram atoms of oxygen from gaseous oxygen and the appearance of the same atoms in the crystalline structure known as hematite. With this in mind, the importance of adequately designating the phases involved in a chemical equation becomes apparent. The phases involved should be designated in every equation unless they are specified in the context or are otherwise unambiguous. Thus in Eq. (*A*), bcc iron is designated $Fe(\alpha)$ and the two atoms of iron and three atoms of oxygen are designated Fe_2O_3(hem.). This is necessary, in general, since we might want to consider two atoms of iron and three atoms of oxygen in some entirely different phase, *e.g.*, in a melt or, perhaps, in magnetite. The latter, though commonly represented by the formula Fe_3O_4, $FeO{\cdot}Fe_2O_3$, has at elevated temperature a considerable range of solid solution with oxygen content in excess of that represented by Fe_3O_4 and thus may be considered a solution of Fe_2O_3 in Fe_3O_4.

The traditional chemist may wish to add to the observed phenomena concerned in the foregoing equation that the composition of the phases is determinable and is important. This is indeed so. He might be inclined to consider, further, that the criterion of hematite is its composition, Fe_2O_3. Actually the composition of hematites does conform very closely to this formula. However, a similar situation does not prevail for magnetite, and in the case of the lower oxide of iron, wüstite, the discrepancy is great. It is doubtful if a wüstite of exactly the composition FeO has ever been prepared; the ratio of oxygen to iron in stable wüstites is appreciably greater than 1. If a solid phase is not designated explicitly, it is usually to be taken for granted that the phase intended is the stable phase which is nearest to the composition corresponding to the written formula. It is very important to bear in mind that the properties, including the enthalpy, are frequently very sensitive to the phase.

Taking two-thirds of reaction (*B*) as written and one-third of reaction (*C*),

$$2\text{Fe}(\alpha) + \tfrac{4}{3}\text{O}_2(\text{g}) = \tfrac{2}{3}\text{Fe}_3\text{O}_4(\text{mag.}) \qquad \tfrac{2}{3}\Delta H_B \qquad (D)$$

$$\tfrac{2}{3}\text{Fe}_3\text{O}_4(\text{mag.}) + \tfrac{1}{6}\text{O}_2(\text{g}) = \text{Fe}_2\text{O}_3(\text{hem.}) \qquad \tfrac{1}{3}\Delta H_C \qquad (E)$$

it is seen that the sum of these two chemical equations is identical with Eq. (*A*); *i.e.*, the sum of the two processes represented by Eqs. (*D*) and (*E*) has in common with the process represented by Eq. (*A*) the same initial and final states. Since the enthalpy change is a function only of the initial and final states, it follows that the enthalpy change corresponding to the process represented by Eqs. (*D*) and (*E*) is identical with that corresponding to the process represented by Eq. (*A*), or

$$\tfrac{2}{3}\Delta H_B + \tfrac{1}{3}\Delta H_C = \Delta H_A$$

The same type of reasoning may be applied to any set of chemical equations; the enthalpy change of the reaction corresponding to the sum of the chemical equations is equal to the sum of the enthalpy changes corresponding to the individual equations. This statement is sometimes known as *Hess' law* and was found empirically by Hess before it was realized to be derivable from the first law. It should be borne in mind that the processes represented by the equations to be added must be truly additive. For example, the magnetite produced by process (*D*) must be identical in state with that consumed by process (*E*); temperature and pressure as well as crystalline form must be the same. The additivity relation is most frequently and easily used under the condition that all initial and final temperatures and pressures are identical; thus ΔH_A, ΔH_B, or ΔH_C refers to heat absorbed by the system from the surroundings when the reaction is carried out at constant temperature and pressure.

This same additivity holds also for changes in any other state property, *e.g.*, volume or energy. It is most fruitfully applied to energy change under the condition that temperature and volume remain constant, for under this condition, as noted previously, the energy change is the heat absorbed by the system.

CHAPTER 7

THE SECOND LAW OF THERMODYNAMICS; ENTROPY AND FREE ENERGY

It is well to recall at this point that our general aim in developing the thermodynamics of chemical equilibrium is to achieve general relations between the compositions of the various phases and the temperature and pressure of a system. Even a casual survey of elementary chemistry leaves no doubt that the chemical behavior of a substance is highly specific; the chemical behavior of potassium, for example, can be inferred only in a general way from that of sodium. This specific type of behavior is what is usually loosely classified as chemical and is in contrast to the general type of behavior traditionally classed as physical. Introductory physics courses deal with processes involving gravitational and electric fields, etc., where the behavior of a particle depends upon its mass or charge, etc., and further knowledge of the properties is unnecessary.

In view of the highly specific nature of the properties with which we must deal, we cannot expect to develop a general relation involving only temperature, pressure, and composition of phases at equilibrium which will be valid for all systems (all substances). Hence the general equilibrium relations will contain, in addition to temperature, pressure, and composition, at least one other variable, which will be specific in nature in the above-mentioned sense. The most useful form in which this variable may appear is as a thermal quantity, *i.e.*, one which may be determined calorimetrically, in particular C_P or ΔH.

For a homogeneous substance any change in energy may be represented by $dE = \delta q - \delta w$, where, according to the first law, the energy E is a function of state. If in the processes under consideration no work is done except that against pressure, this relation becomes $dE = \delta q - P\,dV$. If other types of work are also performed, an additional term is required for each type; each such term, however, may readily be expressed in terms of state variables.[1] The equation still contains one differential, δq, which is inexact, *i.e.*, which is not expressed in terms of state variables. The mathematical treatment is seriously hampered and, in fact, fruitful mathematical treatment is well-nigh impossible unless δq can be expressed

[1] For example, if electrical work is performed, this term may be expressed as the product of the electric charge and the differential of the electric potential.

in terms of state variables. Thus, in accord with the first law, energy is a function of state, but in order to make full use of this fact it must be so expressed, at least in differential form.

In order to express δq in terms of state variables another experimental law must be invoked. Since this problem deals primarily with the transfer of heat between a system and its surroundings, it seems proper to inquire first into the nature of this transfer. It is a matter of common experience that heat flows spontaneously from a hot body to a cold body, but never the reverse. This observation is intimately connected with the definition of temperature and the idea of thermal equilibrium discussed earlier. Although heat never flows directly from a body at lower temperature to one at higher, we must consider the possibility that an engine or other mechanism might be devised to accomplish this result indirectly. Many people have tried; their success is attested by, among other things, the electric or gas refrigerator. Their experimental findings are summed up in the *second law of thermodynamics*, which may be stated as follows: Heat cannot in any way be transferred from a reservoir[1] at lower temperature to one at higher temperature without producing a permanent change in the surroundings. With the aid of this law we are able, for a reversible process, to express δq in terms of state variables. This is done in the subsequent sections. An ideal gas is considered first, and the reasoning later extended to include any thermodynamic substance.

THE CARNOT CYCLE—ENTROPY

Application of the First Law to an Isothermal or Adiabatic Compression (or Expansion) of an Ideal Gas. When 1 mole of an ideal gas is expanded or compressed isothermally, the work done may be found by substituting $P = RT/V$ from the ideal-gas law in the general expression for work, $\int P\,dV$. Since RT is constant,

$$\text{Isothermal work done} = \int_{V_i}^{V_f} P\,dV = RT\int_{V_i}^{V_f}\frac{dV}{V}$$
$$= RT\int_{V_i}^{V_f} d\ln V = RT\ln\frac{V_f}{V_i} \qquad (7\text{-}1)$$

where the subscripts i and f refer to the initial and final states, respec-

[1] A heat reservoir may be defined as a system in complete equilibrium, whose temperature, pressure (or volume), and masses remain constant throughout the process considered. Such a reservoir usually consists of a homogeneous thermodynamic substance, *i.e.*, one which obeys an equation of state, and may be conveniently visualized as a common water thermostat which is sufficiently large or adequately controlled so that the addition or withdrawal of heat due to processes under consideration produces no more than an essentially infinitesimal temperature change.

tively. Since the energy of an ideal gas is a function of temperature only, its energy change for any isothermal process is zero.

$$\text{Isothermal energy change} = 0 \tag{7-2}$$

The heat absorbed per mole during isothermal expansion or compression is hence found, from the first law, to be equal to the work done, as given by Eq. (7-1).

$$\text{Isothermal heat absorbed} = RT \ln \frac{V_f}{V_i} \tag{7-3}$$

An adiabatic process is one during which no heat is exchanged between the system and the surroundings.

$$\text{Adiabatic heat absorbed} = 0 \tag{7-4}$$

The expression for the first law then becomes $dE = -\delta w$. Expressing dE in terms of its partial derivatives at constant temperature and at constant volume, $dE = (\partial E/\partial T)_V\, dT + (\partial E/\partial V)_T\, dV$, and setting $(\partial E/\partial T)_V = C_V$, it is found for an adiabatic process that

$$C_V\, dT + \left(\frac{\partial E}{\partial V}\right)_T dV = -\delta w \tag{7-5}$$

Since the energy of an ideal gas is a function of temperature only, it does not vary with volume at constant temperature, or $(\partial E/\partial V)_T = 0$, and

$$C_V\, dT = -\delta w$$

For simplicity it will be assumed that C_V is constant; this is strictly so only for the rare gases helium, argon, etc., but is approximately true for other gases over a short range of temperature.

The integration may now be performed, and we find that

$$\text{Adiabatic work done} = -C_V(T_f - T_i) \tag{7-6}$$

Since $dE = -\delta w$, it follows that

$$\text{Adiabatic energy change} = C_V(T_f - T_i) \tag{7-7}$$

It is now desirable to determine the relation between volume and temperature for an adiabatic process involving an ideal gas. Returning to Eq. (7-5), again setting $(\partial E/\partial V)_T = 0$ and equating $\delta w = (RT/V)\, dV$, we have

$$C_V\, dT = -\frac{RT}{V}\, dV$$

The variables may be separated by dividing the equation by T. Dividing also by R,

$$d \ln V = -\frac{C_V}{R} d \ln T$$

Integrating between limits,

$$\ln \frac{V_f}{V_i} = -\frac{C_V}{R} \ln \frac{T_f}{T_i} = \ln \left(\frac{T_f}{T_i}\right)^{-C_V/R}$$

whence

$$\frac{V_f}{V_i} = \left(\frac{T_f}{T_i}\right)^{-C_V/R} \tag{7-8}$$

This equation could also be written, upon integration without limits, in the form $V = IT^{-C_V/R}$, where I is an integration constant. The latter is the equation for the adiabatic expansion of an ideal gas. Eliminating T by means of the ideal-gas law this equation becomes $PV^{1+(R/C_V)} = I'$, I' being another constant. Thus it is seen that the adiabatic relation between pressure and volume is somewhat similar to the isothermal relation $PV = I''$, from the ideal-gas law. The plot of V against P rises more sharply for the adiabatic process.

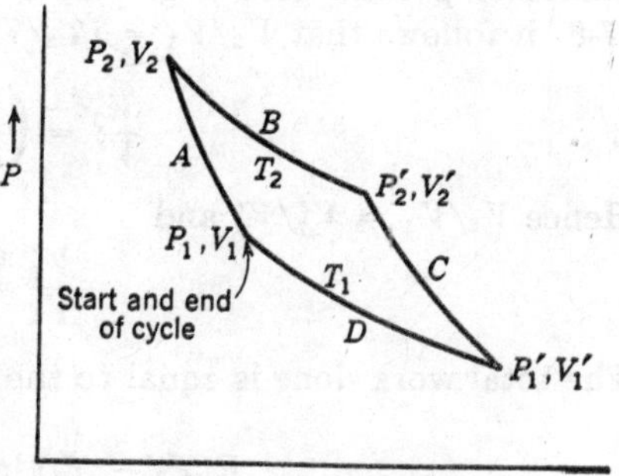

FIG. 7-1. Schematic diagram illustrating the steps of the Carnot cycle.

Carnot Cycle for an Ideal Gas. Let us now consider the work done by, the heat absorbed by, and the energy change of an ideal gas during a Carnot cycle. The Carnot cycle (after Sadi Carnot, 1824) consists of two isothermal and two adiabatic steps, all reversible, and may be visualized in the following manner. The working substance, *e.g.*, the ideal gas under consideration, is enclosed in a cylinder fitted with a piston. There are available two large heat reservoirs or thermostats at different temperatures and suitable insulating material, so that the gas can be compressed or expanded either isothermally (in contact with one of the reservoirs) or adiabatically (insulated). The sequence of steps to be followed may be traced in Fig. 7-1. The cylinder containing the medium is placed first in contact with the thermostat or heat reservoir at the lower temperature T_1, where it assumes the volume V_1 and pressure P_1. Step A: It is now insulated and compressed adiabatically until it reaches the temperature T_2 of the hotter reservoir. Step B: The insulation is removed, the cylinder is placed in contact with the reservoir at T_2, and the gas is expanded isothermally by withdrawing the piston until it reaches some volume V'_2 and pressure P'_2. Step C: The cylinder is removed from the reservoir and insulated, and the medium is further expanded adiabatically by continued withdrawal of the piston until the initial temperature T_1 is reached, the volume and the pressure being now V'_1 and P'_1. Step D: The insulation is again removed, the system is placed in contact with the cold reservoir, and the medium is compressed isothermally until the original volume V_1 and pressure P_1 are reached, thus completing the cycle.

The work done by, the heat absorbed by, and the energy change of 1 mole of the ideal gas during each step of such a cycle are readily calculated by use of the relations developed in the preceding section. They are given in Table 7-1. The sum of each of these quantities obtained by addition for all steps is also given. The revaluation of the sums of the heats and works shown in the final row is obtained from Eq. (7-8) for an adiabatic process (see Fig. 7-1) in the following manner. From Eq. (7-8) it follows that $V_2/V_1 = (T_2/T_1)^{-C_V/R}$ and also that

$$\frac{V'_2}{V'_1} = \left(\frac{T_2}{T_1}\right)^{-C_V/R}$$

Hence $V_2/V_1 = V'_2/V'_1$ and

$$\frac{V'_2}{V_2} = \frac{V'_1}{V_1} \tag{7-9}$$

The total work done is equal to the total heat absorbed,

$$R(T_2 - T_1) \ln \frac{V'_2}{V_2} = q_2 + q_1$$

and since, from the heat absorbed in step B, $\ln (V'_2/V_2) = q_2/RT_2$, it follows that

$$\frac{R(T_2 - T_1)q_2}{RT_2} = q_2 + q_1$$

Cancellation of R and division by q_2 give

$$\frac{q_2 + q_1}{q_2} = \frac{T_2 - T_1}{T_2} \tag{7-10}$$

Since $q_2 + q_1$ is equal to the total work done (see Table 7-1), Eq. (7-10) evaluates the *thermodynamic efficiency* of this reversible heat engine. The efficiency is defined as the ratio of the total work done to the heat absorbed at the higher temperature. The efficiency is thus seen to be equal to the difference between the two operating temperatures divided by the higher temperature (in degrees Kelvin).

TABLE 7-1. WORK DONE BY, HEAT ABSORBED BY, AND ENERGY CHANGE OF 1 MOLE OF IDEAL GAS DURING EACH STEP OF CARNOT CYCLE

Step	Work done	Heat absorbed	Energy change
A (adiabatic)	$-C_V(T_2 - T_1)$ [Eq. (7-6)]	0 [Eq. (7-4)]	$+C_V(T_2 - T_1)$ [Eq. (7-7)]
B (isothermal at T_2)	$RT_2 \ln \frac{V_2'}{V_2}$ [Eq. (7-1)]	$q_2 = RT_2 \ln \frac{V_2'}{V_2}$ [Eq. (7-3)]	0 [Eq. (7-2)]
C (adiabatic)	$+C_V(T_2 - T_1)$ [Eq. (7-6)]	0 [Eq. (7-4)]	$-C_V(T_2 - T_1)$ [Eq. (7-7)]
D (isothermal at T_1)	$RT_1 \ln \frac{V_1}{V_1'}$ [Eq. (7-1)]	$q_1 = RT_1 \ln \frac{V_1}{V_1'}$ [Eq. (7-3)]	0 [Eq. (7-2)]
Sum	$RT_2 \ln \frac{V_2'}{V_2} - RT_1 \ln \frac{V_1'}{V_1}$ or, using Eq. (7-9), $R(T_2 - T_1) \ln \frac{V_2'}{V_2}$	$q_2 + q_1 = RT_2 \ln \frac{V_2'}{V_2} - RT_1 \ln \frac{V_1'}{V_1}$ or, using Eq. (7-9), $R(T_2 - T_1) \ln \frac{V_2'}{V_2}$	0

Carrying out the indicated division in Eq. (7-10),

$$1 + \frac{q_1}{q_2} = 1 - \frac{T_1}{T_2}$$

whence

$$\frac{q_1}{q_2} = -\frac{T_1}{T_2}$$

or

$$\frac{q_1}{T_1} + \frac{q_2}{T_2} = 0 \tag{7-11}$$

Thus we obtain an important fact by consideration of the Carnot cycle, namely, that the summation of the reversible heat exchange divided by the temperature of that exchange is zero for this cyclic process, as expressed by Eq. (7-11). It will be noted that only the heats exchanged during the isothermal steps *B* and *D* need be considered, since the heats for the adiabatic steps are zero. Hence for an ideal gas undergoing this particular cyclic process the summation of q/T is found to be zero.

Let us next consider any arbitrary reversible cycle through which a mole of ideal gas may be carried. It is obvious that the values of the heat absorbed, work done, and energy change for this cycle can be evaluated by considering that this arbitrary cycle is the equivalent of a large number of suitably chosen infinitesimal Carnot cycles; a corresponding large number of heat reservoirs are assumed to be available. Hence for any arbitrary reversible cycle through which an ideal gas may be carried, it follows that $\sum_i (q_i/T_i) = 0$ or, in the language of the calculus,

$$\oint(\delta q/T) = 0,$$

where the symbol $\oint$ refers to integration over a cycle. This expression is usually written $\oint(\delta q_r/T) = 0$, the subscript r denoting that this equation is valid only for a reversible cycle. From our previous consideration of properties which are functions of state, it is clear that any property whose initial and final values are the same for any arbitrary cycle is a state property, since its value for any particular set of state variables is a function of state variables only. The differential of such a state property is said to be exact in terms of the state variables. It follows from this and the foregoing [Eq. (7-11)] that for an ideal gas $\delta q_r/T$ is exact and that its integral between any two states is a function of these two states only and is not a function of the particular path followed in passing from one state to the other. It is convenient to define a new function S such that $dS = \delta q_r/T$, it being noted that the property S of a system, thus defined is a state variable. This function S is known as the *entropy*. It i invented only for mathematical convenience in that it is a function of state and hence its derivative may be represented in terms of partials in the usual way.

Carnot Cycle for Any Thermodynamic Substance. Let us now consider any thermodynamic substance, *i.e.*, any substance—gas, liquid, or solid—obeying an equation of state, the energy of which is a function of pressure and temperature[1]—and, using it as a medium in an engine, subject it to a Carnot cycle, as was done previously for an ideal gas. It is convenient to select this cycle so that the temperatures of the two reservoirs are identical with those for the ideal gas just considered and the total work done per cycle is the same as before. This equality of work

[1] Although, in accord with the first law, the energy is a function of state, it does not necessarily follow for a given substance that the energy is a function of pressure and temperature, even though the substance obeys an equation of state. For example, a mixture of hydrogen and chlorine at room temperature occupies essentially the same volume at 1 atm whether the molecular constitution is H_2 and Cl_2 or HCl. However, it is obvious that the state of the system is different in these two cases and also that the energy is correspondingly different.

The working substance, or medium, considered here may be any substance whose state is completely specified by its pressure and temperature (or by its volume and temperature or by its pressure and volume). Thus it may be a substance which undergoes chemical reaction (or phase change) in the course of the cycle providing that the extent of the reaction can be expressed as a function of pressure and temperature, as is the case when the reaction proceeds to equilibrium at each stage in the cycle. For example, the mixture of hydrogen and chlorine just considered would be a suitable working substance at a temperature sufficiently elevated or in the presence of a catalyst, so that the amounts of H_2, Cl_2, and HCl are functions of temperature and pressure and are not functions of previous history. Also a mixture of H_2 and Cl_2 would be a suitable medium if the cycle were carried out without catalyst at such a low temperature that the reaction did not occur to any measurable extent at any stage of the cycle during the actual time required. However, such a mixture would not be a suitable working substance (thermodynamic substance) at any intermediate temperature such that reaction took place to some extent but not to equilibrium for in this case the molecular composition would depend upon the manner of approach to any particular temperature and pressure.

It should be noted that the extent of chemical reaction depends not only upon the temperature range of the cycle but also upon the time allotted. Thus at any particular temperature the mixture considered will behave as a suitable thermodynamic substance if the time allowed is very short so that no appreciable reaction occurs or very long so that equilibrium is essentially attained. If, however, the cycle extends over an intermediate time period so that some reaction occurs but equilibrium is not attained, then the mixture is not a thermodynamic substance and will not serve as a suitable medium in the cycle we wish to consider.

The working substance may equally well be a solid instead of a fluid. In this case the container must be so designed that a true pressure or tension (components equal in all directions) is exerted upon the substance. This solid substance may also undergo chemical reactions, including phase changes, with the qualifications just considered.

may be achieved by appropriate adjustment of the extent of compression and expansion during the isothermal steps. Let us imagine that these two engines, one the ideal-gas engine and the other with any other thermodynamic substance as medium, are mechanically connected in such a way that the expansion of one medium takes place at the same time as the compression of the other.[1] We shall consider first that the ideal-gas engine operates with the same sequence of steps as before and that the other engine operates with the reverse sequence. Thus the work w done by the ideal-gas engine in operating through a cycle is done upon the other substance, and hence the work w' done by the other substance is negative. We shall designate the heat absorbed by the ideal gas at the higher temperature as q_2 and that at the lower temperature as q_1 as in Table 7-1; it will be noted that q_1 is negative, heat being transferred from the gas to the reservoir at the lower temperature. The corresponding heat exchanges of the other substance will be designated q'_2 and q'_1, respectively, q'_2 being negative and q'_1 positive.

We now consider the two substances in the coupled engines as one system. When the coupled system passes through a complete cycle, its energy change is zero, in accord with the first law, both substances being restored to their initial states. Also the total work done by the coupled system in a complete cycle is zero, since the cycle of the other substance was selected so that the work received would be the same as that done by the ideal gas. By the usual statement of the first law, $\Delta E = q - w$, the total heat exchange between this compound system and the surroundings is zero, or

$$q_1 + q'_1 + q_2 + q'_2 = 0$$

We now define $q_2 + q'_2$ as a, noting that this is the total heat absorbed by the compound system from the hot reservoir in this case. From the above equation by transposition it is seen that

$$a = q_2 + q'_2 = -q_1 - q'_1$$

That is, the total heat absorbed from the hot reservoir is equal to the total heat given up to the cold reservoir.

Let us change our viewpoint for a moment and regard the reservoirs as the system under consideration, the ideal gas and the other substance being the surroundings which undergo no net change during the cycle. In accord with our fundamental statement of the second law, heat cannot by any mechanism whatsoever be transferred from a cold reservoir to a

[1] In order to have these two coupled engines operate smoothly, it is, in general, necessary to postulate that appropriate gears or other device are provided to compensate for the difference in stroke.

hot reservoir unless a change is produced in the surroundings. Hence the heat a transferred from the hot reservoir to the cold reservoir cannot be a negative quantity and must be zero or a positive quantity, that is, $a \geq 0$. Let us now consider the situation when both engines are reversed, *i.e.*, when the sequence of steps is taken in the opposite order. The signs of all heat exchanges are now changed, and hence we find also that $a \leq 0$. The only value of a that is consistent with both these relations is zero, and from the above equation,

$$q_2 + q_2' = 0$$

and

$$q_1 + q_1' = 0$$

Since, as was shown in Eq. (7-11) of the preceding section,

$$\frac{q_1}{T_1} + \frac{q_2}{T_2} = 0,$$

it follows from the above relations that

$$\frac{q_1'}{T_1} + \frac{q_2'}{T_2} = 0 \tag{7-11a}$$

Recalling that q_1' is the heat absorbed by any thermodynamic substance at the temperature T_1 and q_2' is that absorbed at T_2 and that the heat absorbed during the adiabatic steps is zero, it follows that, when any thermodynamic substance is carried reversibly through a Carnot cycle, $\Sigma q_r/T = 0$. Now it remains only to note that any arbitrary cycle through which any thermodynamic substance is carried may be regarded, in so far as heat and work effects are concerned, as the summation or integral of a large number of very small (infinitesimal) Carnot cycles, and hence for any thermodynamic substance we can write, as we did for an ideal gas, $\oint(\delta q_r/T) = 0$. Thus we have demonstrated from the fundamental statement of the second law that $\delta q_r/T$ is an exact differential for *any* thermodynamic substance. This exactness is such an important property that the function S is defined so that dS is identical with $\delta q_r/T$. The function S, called the entropy, is seen by the same token to be a function of state variables; *e.g.*, in the absence of electrical, magnetic, gravitational, etc., fields, it is a function of P and T.

Entropy Change for Various Processes. From the definitional relation

$$dS = \frac{\delta q_r}{T}$$

we see that for any reversible process taking place at constant temperature the entropy change of the system is

$$\underline{\Delta S} = \frac{q_r}{T}$$

If the process takes place at constant pressure as well as constant temperature, then $\Delta S = \Delta H/T$. For a thermodynamic substance heated or cooled at constant pressure δq_r is $C_P\,dT$, whence

$$dS = \frac{C_P}{T}\,dT = C_P\,d\ln T \tag{7-12}$$

Similarly for a thermodynamic substance heated or cooled at constant volume,

$$dS = \frac{C_V}{T}\,dT = C_V\,d\ln T \tag{7-12a}$$

Combined Expression of the First and Second Laws of Thermodynamics. Having demonstrated by means of the second law that dS, defined as $\delta q_r/T$, is an exact differential for any thermodynamic substance, we can now rewrite the expression for the first law, namely, $dE = \delta q - \delta w$, substituting for δq the term $T\,dS$.

$$dE = T\,dS - \delta w \tag{7-13}$$

Or if the only work done is that against pressure,

$$dE = T\,dS - P\,dV \tag{7-13a}$$

This expression applies only to a reversible process, since it is only for a reversible process that $dS = \delta q/T$; the subscript r has been used to indicate this fact.

Let us now consider the general irreversible process. Such a process could be made reversible if heat could be made to flow from a cold reservoir to a hot one without requiring a change in anything else.[1]

[1] This statement is to be regarded as a corollary of the second law. The reader may convince himself of its truth by consideration of the means by which any particular irreversible process could be made reversible if the second law were false and heat could be made to flow from a cold to a hot reservoir. For example, a bar of copper initially hot at one end and cold at the other actually tends to equalize its temperature—an irreversible process. The initial temperature gradient could be restored by immersing the ends in heat reservoirs of appropriate temperature. If heat could only be made to flow by some other mechanism from the cold to the hot reservoir without any change in the surroundings, the gradient in the bar could be restored and maintained indefinitely and the process would become reversible.

As another example, the spontaneous reaction

$$Pb + PbO_2 + 2H_2SO_4 = 2PbSO_4 + 2H_2O$$

can be made to proceed in the opposite direction by forming the $PbSO_4$ into the

Hence any irreversible process may be regarded as equivalent to a number of steps all of which are reversible except one, this one involving only the flow of heat from a hot reservoir at T_2 to a cold reservoir at T_1. The entropy change of the system plus its surroundings during all the equivalent reversible steps is zero, since the heat absorbed by the system is exactly equal to that lost by the surroundings and the temperature is the same. The net entropy change of the system plus the surroundings during this process is determined by the entropy change occurring during the single irreversible step, namely, during the transfer of heat from the hot to the cold reservoir. In this step the entropy change of the hot reservoir occasioned by loss of the heat q is $-q/T_2$,* that of the cold reservoir during gain of the same heat is q/T_1, and the sum of these two quantities, the total entropy change ΔS, is $q(1/T_1 - 1/T_2)$. Since T_1 is less than T_2, it is seen immediately that ΔS is positive. Hence for any actual (or irreversible) process the net entropy change of the system and surroundings is positive. If we consider the system and its surroundings as an isolated system, then it may be said of this system that for any actual process the entropy change is positive. By *isolated system* is meant one which exchanges no heat with and does no work on the surroundings, *i.e.*, one which is surrounded by a rigid, adiabatic envelope.

In accord with the above reasoning we may write for any isolated system $\Delta S \geq 0$ or, in differential notation, $dS \geq 0$. The equality sign pertains to any reversible process, *i.e.*, one at equilibrium at all stages; the inequality sign pertains to any going, actual, or spontaneous process, *i.e.*, one which is not reversible.†

---

electrodes of a storage battery and passing the appropriate quantity of electricity (charging the battery). If now the heat evolved by the original reaction at, say, 25°C could be transferred to a reservoir at a higher temperature (*e.g.*, 100°C), then an engine could operate on a Carnot cycle between these two temperatures; this engine could drive a generator which would supply the current to charge the battery, thus restoring the initial conditions without any permanent change in the surroundings. In a similar way, any irreversible process could be imagined as convertible to a reversible one by suitable steps if heat could be made to flow from a cold to a hot reservoir without producing a permanent change in the surroundings.

* Since the entropy change of a reservoir depends only on its change of state, the entropy change occasioned by the addition or subtraction of the heat q does not depend upon whether this heat is exchanged reversibly or irreversibly.

† E. A. Guggenheim ("Thermodynamics," Interscience Publishers, New York, 1949) presents a somewhat different viewpoint of entropy and the second law. His statement of the second principle of thermodynamics is that there exists a function S (the entropy) of the state of a system which has the following properties:

1. The total entropy of the system is the sum of the entropies of the parts.
2. The entropy of a system may change either (*a*) by interaction with the surroundings or (*b*) by virtue of changes taking place within the system. The former is

The use of the entropy as a criterion of equilibrium is seriously hampered experimentally by the rather awkward conditions required to apply such a test. In order to use the fact that the entropy change for a given reversible process is zero, it is necessary to consider the entropy changes not only of the system but also of the surroundings, which together constitute an isolated system. Since our interest, as chemists and metallurgists, usually centers upon the system and not upon the surroundings, it seems abnormally burdensome to be called upon to evaluate entropy changes of the surroundings as well as those of the system. In fact, this procedure is not only tedious but particularly liable to error, since one must be careful to include in the surroundings *all* substances outside the system which undergo change during the process under consideration. We shall find that the application of the second law is much easier and more fruitful after the introduction of the function known as the free energy in a subsequent section.

SOME CONSEQUENCES OF THE SECOND LAW

The Thermodynamic Temperature Scale. The temperature T was defined previously in terms of an ideal gas. As pointed out by Lord Kelvin, an identical scale may be obtained by virtue of the Carnot cycle using any thermodynamic substance. It was shown in an earlier section [Eq. (7-11*a*)] that $q_1'/T_1 + q_2'/T_2 = 0$, or that

$$\frac{T_2}{T_1} = -\frac{q_2'}{q_1'}$$

where the primes signify that the working substance is any thermodynamic substance. Thus the ratio of absolute temperatures may in principle be determined by experimental measurement of the heats absorbed and liberated in carrying any substance through a Carnot cycle for which the hot and cold reservoirs are held at these two temperatures. The relative size of the degree at two temperatures is seen to be determined by the foregoing; the absolute size of the degree is, of course, arbitrary, as is the size of the unit of length, *e.g.*, the centimeter or inch.

Thermodynamic Equations of State. Having demonstrated by means of the second law that the entropy is a function of state and hence that

---

equal to q/T; the latter is positive for all natural changes, is zero for all reversible changes, and is never negative.

This method of presentation has advantages, in particular that cycles need never be considered. However, to the authors of this book the advantages seem outweighed by the remoteness between these statements and direct experience. The ideas expressed, whether regarded as fundamental (Guggenheim) or derived (as in this text), are of great importance.

dS is exact, let us now proceed immediately to find what use can be made of this fact. The general form [Eq. (7-13)] of the combined statement of the first and second laws becomes, for any thermodynamic substance which does work only against pressure, $dE = T\,dS - P\,dV$. Inserting herein the expression for dE in terms of its partials, namely,

$$dE = \left(\frac{\partial E}{\partial T}\right)_V dT + \left(\frac{\partial E}{\partial V}\right)_T dV$$

and rearranging and collecting terms, it is found that

$$dS = \frac{1}{T}\left(\frac{\partial E}{\partial T}\right)_V dT + \frac{1}{T}\left[\left(\frac{\partial E}{\partial V}\right)_T + P\right] dV$$

This equation is of the form $dz = M\,dx + N\,dy$, for which, provided dz is exact and M and N are functions of x and y, $(\partial M/\partial y)_x = (\partial N/\partial x)_y$.* Applying this fact,

$$\left\{\frac{\partial}{\partial V}\left[\frac{1}{T}\left(\frac{\partial E}{\partial T}\right)_V\right]\right\}_T = \left\{\frac{\partial}{\partial T}\left[\frac{(\partial E/\partial V)_T + P}{T}\right]\right\}_V$$

Carrying out the indicated differentiation,

$$\frac{1}{T}\frac{\partial^2 E}{\partial V\,\partial T} = \frac{1}{T}\left[\frac{\partial^2 E}{\partial T\,\partial V} + \left(\frac{\partial P}{\partial T}\right)_V\right] - \frac{1}{T^2}\left[\left(\frac{\partial E}{\partial V}\right)_T + P\right]$$

Multiplying by T and canceling the second derivatives of the energy,*

$$0 = \left(\frac{\partial P}{\partial T}\right)_V - \frac{1}{T}\left[\left(\frac{\partial E}{\partial V}\right)_T + P\right]$$

* If z is a function only of x and y, then dz is exact in x and y and may be written

$$dz = \left(\frac{\partial z}{\partial x}\right)_y dx + \left(\frac{\partial z}{\partial y}\right)_x dy$$

whence, by comparison with the relation $dz = M\,dx + N\,dy$, $M = (\partial z/\partial x)_y$ and $N = (\partial z/\partial y)_x$. By differentiation,

$$\left(\frac{\partial M}{\partial y}\right)_x = \left[\frac{\partial}{\partial y}\left(\frac{\partial z}{\partial x}\right)_y\right]_x = \frac{\partial^2 z}{\partial y\,\partial x}$$

and

$$\left(\frac{\partial N}{\partial x}\right)_y = \left[\frac{\partial}{\partial x}\left(\frac{\partial z}{\partial y}\right)_x\right]_y = \frac{\partial^2 z}{\partial x\,\partial y}$$

Since the order of differentiation is immaterial, $\partial^2 z/\partial y\,\partial x = \partial^2 z/\partial x\,\partial y$. It follows that

$$\left(\frac{\partial M}{\partial y}\right)_x = \left(\frac{\partial N}{\partial x}\right)_y$$

which may be rewritten in the form

$$P = T\left(\frac{\partial P}{\partial T}\right)_V - \left(\frac{\partial E}{\partial V}\right)_T \qquad (7\text{-}14)$$

This relation, which is obeyed by any thermodynamic substance, is known as a *thermodynamic equation of state*, since it involves the variables P, V, and T generally used to describe the state of a substance.

From the primary combined equation of the first and second laws for a system doing work only against pressure, namely, $dE = T\,dS - P\,dV$, and the derivative of the definitional equation of H, namely,

$$dH = dE + P\,dV + V\,dP$$

it follows that

$$dH = T\,dS + V\,dP \qquad (7\text{-}15)$$

This expression may be regarded as an alternate combined expression of the first and second laws. By treating in a manner analogous to that of the preceding paragraph, it may be shown that

$$V = T\left(\frac{\partial V}{\partial T}\right)_P + \left(\frac{\partial H}{\partial P}\right)_T \qquad (7\text{-}16)$$

This is also a thermodynamic equation of state.

Relation between C_P and C_V. In the preceding chapter, by means of the first law only, it was shown [Eq. (7-11)] that

$$C_P - C_V = \left[P + \left(\frac{\partial E}{\partial V}\right)_T\right]\left(\frac{\partial V}{\partial T}\right)_P$$

By Eq. (7-14) the expression in brackets is equal to $T(\partial P/\partial T)_V$, which may also be written $-T(\partial V/\partial T)_P/(\partial V/\partial P)_T$.* The foregoing equation

* If z is a function of x and y only, then, as noted previously,

$$dz = \left(\frac{\partial z}{\partial x}\right)_y dx + \left(\frac{\partial z}{\partial y}\right)_x dy$$

Dividing by dx, $dz/dx = (\partial z/\partial x)_y + (\partial z/\partial y)_x(dy/dx)$. This expression is true in general and hence is true in the particular case that z is constant or that $dz/dx = 0$; in this case the constancy of z is expressed in the equation itself by replacing (dy/dx) with $(\partial y/\partial x)_z$. Thus, $0 = (\partial z/\partial x)_y + (\partial z/\partial y)_x(\partial y/\partial x)_z$, or

$$\left(\frac{\partial y}{\partial x}\right)_z = -\frac{(\partial z/\partial x)_y}{(\partial z/\partial y)_x}.$$

Setting P for y, T for x, and V for z, this becomes $(\partial P/\partial T)_V = -(\partial V/\partial T)_P/(\partial V/\partial P)_T$.

This type of transformation, based upon the fundamental equation of partial differentiation, will be used frequently. The student not well versed therewith will

thus becomes

$$C_P - C_V = -\frac{T(\partial V/\partial T)_P^2}{(\partial V/\partial P)_T} \tag{7-17}$$

It is convenient to express this equation in terms of the coefficient of thermal expansion[1] α, which is defined

$$\alpha \equiv \frac{1}{V}\left(\frac{\partial V}{\partial T}\right)_P \tag{7-18}$$

and in terms of the coefficient of compressibility[1] β, defined

$$\beta \equiv -\frac{1}{V}\left(\frac{\partial V}{\partial P}\right)_T \tag{7-19}$$

In terms of these coefficients Eq. (7-17) becomes

$$C_P - C_V = \frac{\alpha^2 VT}{\beta} \tag{7-20}$$

Most theoretical considerations lead to an expression for C_V, whereas C_P rather than C_V is usually evaluated experimentally; consequently, the foregoing thermodynamic relation furnishes a convenient link between theoretical and experimental values of the heat capacity.

---

find it advantageous to familiarize himself with a number of alternate forms, such as the following:

$$\left(\frac{\partial P}{\partial T}\right)_V = -\left(\frac{\partial V}{\partial T}\right)_P\left(\frac{\partial P}{\partial V}\right)_T \qquad \left(\frac{\partial V}{\partial T}\right)_P = -\frac{(\partial P/\partial T)_V}{(\partial P/\partial V)_T}$$

Such equations may be easily written by bearing in mind that they must be dimensionally correct, that each subscript appears but once, that any given letter cannot appear as a subscript to a partial containing it, and that a minus sign is always involved.

[1] The coefficients of thermal expansion and compressibility defined here are sometimes called volume coefficients to distinguish them from the linear coefficients, which are perhaps more frequently used. The coefficients used here correspond to the fractional change in volume per degree or per atmosphere and are three times as great as the corresponding linear coefficients. This fact may readily be deduced by considering a cube, initially with edge length l, whose size is altered by infinitesimal variation of temperature or pressure so that the edge length becomes $l + dl$. The volume, originally l^3, becomes $(l + dl)^3$ or $l^3 + 3l^2\,dl$. The fractional change in cube edge is $[(l + dl) - l]/l$ or dl/l, and the fractional change in volume dV/V is

$$\frac{[(l^3 + 3l^2\,dl) - l^3]}{l^3}$$

or $3dl/l$, three times the fractional change in edge length. Thus, for example, the coefficient of thermal expansion, $\alpha \equiv (1/V)(\partial V/\partial T)_P$, adopted here, is three times the linear coefficient of thermal expansion, which is $(1/l)(\partial l/\partial T)_P$. Occasionally the linear coefficient is defined as $(1/l_0)(\partial l/\partial T)_P$, where l_0 is the length under standard conditions. The use of l or l_0 makes little difference in the numerical value unless large changes in length are involved.

Reconsideration of the Definition of an Ideal Gas. It will be recalled that an ideal gas was originally defined as one which obeys Boyle's law and whose energy is a function of temperature only; temperature was defined in terms thereof. By means of Eqs. (7-14) and (7-16), which follow from the second law, it will now be shown that for any substance obeying the relation $PV = RT$ the energy is a function of temperature only. Directly from this relation it is found that $(\partial P/\partial T)_V = R/V = P/T$ and $(\partial V/\partial T)_P = R/P = V/T$. Substitution of these values in Eqs. (7-14) and (7-16), respectively, gives $(\partial E/\partial V)_T = 0$ and

$$\left(\frac{\partial H}{\partial P}\right)_T = \left[\frac{\partial(E + PV)}{\partial P}\right]_T = \left[\frac{\partial(E + RT)}{\partial P}\right]_T = \left(\frac{\partial E}{\partial P}\right)_T = 0$$

Hence the energy of such a substance does not vary with either volume or pressure at constant temperature and is a function of temperature only (energy changes due to gravitational, electric, or other fields being excluded). Thus, by virtue of the second law, it is no longer necessary to include in the definition of an ideal gas the fact that its energy is a function of temperature only. Obedience to the ideal-gas law may be considered necessary and sufficient evidence for the ideality of a gas.

Application of a Thermodynamic Equation of State to the Elastic Compression or Expansion of a Solid. When a metal—or, in general, a solid—is subjected to a hydrostatic pressure, it undergoes a change in volume which is significant if the pressure is sufficiently high. It will be noted that this is a special case of an elastic strain. The work done upon the solid is commonly referred to as the strain energy; however, it is not proper, thermodynamically, to refer to the work done as an energy change except in the case that no heat is exchanged. Let us now investigate the heat exchanged between a solid and a thermostat and the energy change of the solid when the solid is isothermally compressed. First let us rewrite Eq. (7-14) for an isothermal process; this is done by removing the subscript T from $(\partial E/\partial V)_T$. (We now state in words that the process is isothermal and need not state it in the equation itself.) Thus,

$$P = T\left(\frac{\partial P}{\partial T}\right)_V - \frac{dE}{dV}$$

Multiplying by dV and rearranging terms,

$$dE = T\left(\frac{\partial P}{\partial T}\right)_V dV - P\,dV$$

Comparing this with the statement of the first law, $dE = \delta q - P\,dV$, it follows that

$$\delta q = T\left(\frac{\partial P}{\partial T}\right)_V dV \tag{7-21}$$

Expressing the partial derivative in terms of the coefficients of thermal expansion and compressibility,

$$\delta q = T\frac{\alpha}{\beta}\,dV$$

The ratio of the heat absorbed by the solid from the thermostat to the work done upon the system, $-P\,dV$, is

$$\frac{\delta q}{-P\,dV} = -\frac{T}{P}\frac{\alpha}{\beta}$$

At pressures in the neighborhood of atmospheric this ratio is very high for metals; for iron at room temperature and atmospheric pressure it is $-17{,}000$. Hence the heat effect is large in comparison with the work done during the elastic strain of a metal. The ratio of the energy change of the metal to the elastic work done upon it by a hydrostatic pressure is readily calculable as $(\delta q - P\,dV)/(-P\,dV)$, which is $1 - (T/P)(\alpha/\beta)$, a large negative quantity, in general.

Thus we see that, when a metal or other solid is isothermally compressed by a moderate pressure, it gives up a relatively large quantity of heat to the thermostat and hence loses energy even though work is done upon it. Of course, this energy is regained from the thermostat when the pressure is released and is not usually considered in the treatment of elastic strain. Although in the foregoing the high ratio of the liberated heat to the work done upon the solid has been stressed, it should be noted that the actual work done in the compression of a solid by moderate pressure is very small, so that even when multiplied by a large factor to give the heat, the result is still small by the usual standards of heat measurement. Thus, if this heat were added to the metal itself, the temperature rise normally would not exceed a degree.

The Thermoelastic Effect. In the preceding section it was shown [Eq. (7-21)] that, isothermally, $\delta q = T(\partial P/\partial T)_V\,dV$. It follows that

$$\left(\frac{\delta q}{\partial V}\right)_T = T\left(\frac{\partial P}{\partial T}\right)_V \quad \text{or} \quad \left(\frac{\delta q}{\partial P}\right)_T\left(\frac{\partial P}{\partial V}\right)_T = T\left(\frac{\partial P}{\partial T}\right)_V$$

Applying the general relation $(\partial y/\partial x)_z = -(\partial z/\partial x)_y/(\partial z/\partial y)_x$ to $(\partial P/\partial T)_V$ and to $(\delta q/\partial P)_T$,

$$\left(\frac{\delta q}{\partial T}\right)_P\left(\frac{\partial T}{\partial P}\right)_q\left(\frac{\partial P}{\partial V}\right)_T = T\left(\frac{\partial P}{\partial V}\right)_T\left(\frac{\partial V}{\partial T}\right)_P$$

Noting the cancellation and setting $(\delta q/\partial T)_P = C_P$ and $(\partial V/\partial T)_P = \alpha V$,

$$\left(\frac{\partial T}{\partial P}\right)_q = \frac{T\alpha V}{C_P} \tag{7-22}$$

thus evaluating $(\partial T/\partial P)_q$, the adiabatic temperature rise or thermoelastic effect for unit pressure change. For unidimensional tension σ the expression becomes (approximately)

$$\left(\frac{\partial T}{\partial \sigma}\right)_q = -\frac{T\alpha_l V}{C_P} \tag{7-23}$$

where $\alpha_l = \alpha/3$, the coefficient of linear thermal expansion. The thermoelastic effect (in tension) has recently been measured[1] for iron and for nickel. The results are in good agreement with Eq. (7-23) at temperatures where α has been reliably determined and are used to evaluate α at higher temperature (up to 750°C for iron).

FREE ENERGY

Having found a combined expression of the first and second laws [Eq. (7-13)] applicable to any thermodynamic system, the problem is now to cast this in more fruitful form. For a closed system at equilibrium and in the absence of any work except that against pressure this expression, as noted previously, becomes

$$dE = T\,dS - P\,dV \tag{7-24}$$

In a differential equation such as this, the variables whose derivatives appear on the right-hand side are regarded as the independent variables. Thus the mathematician would regard this differential equation as derived from or leading to an expression in which the energy E appears as a function of the entropy S and the volume V. From the viewpoint of the chemist or metallurgist such an expression is indeed a very awkward form, for the independent regulation of the volume V (particularly of liquids and solids) is difficult and that of the entropy S is usually impossible without prior knowledge of the thermodynamic behavior of the system.[2] In the actual conduct of an experiment the two variables most easily independently controlled are the temperature, by means of a thermostat or furnace, and the pressure, frequently at 1 atm, achieved simply by exposing the system to normal atmospheric pressure.

[1] R. Rocca and M. B. Bever, *Trans. AIME*, **188**, 327 (1950).

[2] If previous experiments had established the entropy as a function of temperature and volume, the entropy of the system could be set at any arbitrary value by appropriately fixing the temperature for any preassigned volume.

The Helmholtz Free Energy A. Occasionally, particularly with gases, the temperature and volume are easily controlled. Let us consider a modification of Eq. (7-24) which is convenient in this case. If we define a new function A, known as the Helmholtz free energy or sometimes as the *isothermal work content*, by the following relation

$$A \equiv E - TS \tag{7-25}$$

then the derivative $dA = dE - T\,dS - S\,dT$ may be added to Eq. (7-24) to give the expression

$$dA = -S\,dT - P\,dV \tag{7-26}$$

which is valid for a thermodynamic system at equilibrium. Thus for any such system at constant temperature and volume $dA = 0$; at constant temperature $dA = -P\,dV$, the negative work against pressure, giving rise to the name "isothermal work content." The negative sign reflects the fact that, as the system does work, it loses the ability to do work. It will be noted that A is a function of state, since it is defined in terms of variables which are themselves functions of state, and the maximum work which may be obtained as a system passes *isothermally* from a given initial state to a given final state is equal to the negative change in the A function of the system between these two states.

The Gibbs Free Energy F. Although the Helmholtz free energy described in the previous section is a function whose derivative may be expressed explicitly in terms of T and V, it is our ultimate object to express the first and second laws explicitly in terms of T and P, which are experimentally most easily controlled. The function by means of which the combined statement of the first and second laws may be so cast is the Gibbs free energy, which we shall designate as F. This function, to which we shall subsequently refer simply as the free energy,[1] is sometimes also called the Lewis free energy and was called by Gibbs himself the thermodynamic potential.

F is defined by the relation

$$F \equiv E + PV - TS \tag{7-27}$$

or, since $H = E + PV$,

$$F = H - TS \tag{7-28}$$

Differentiation of Eq. (7-27) and union with the combined statement of the first and second laws [Eq. (7-24)] give another form of this combined

[1] Gibbs used the Greek symbol ζ (zeta) for this function, and the international symbol is G in honor of Gibbs. However, American usage was established by G. N. Lewis and the classic text of Lewis and Randall, "Thermodynamics and the Free Energy of Chemical Substances," McGraw-Hill Book Company, Inc., New York, 1923. We shall adhere to the nomenclature of Lewis and use F for the free energy and A for the Helmholtz free energy.

statement:

$$\begin{aligned} dF &= dE + P\,dV + V\,dP - T\,dS - S\,dT \\ dE &= -P\,dV + T\,dS \\ \hline dF &= V\,dP - S\,dT \end{aligned} \qquad (7\text{-}29)$$

This is, perhaps, the most fruitful mathematical form for the combined statement of the first and second laws, since P and T appear as the independent variables. It should be borne in mind that this applies to a system of constant mass which does no work other than that against pressure and all parts of which are in equilibrium with one another.

Maxwell's Relations. Let us now review the four forms of the combined statement of the first and second laws, all of which apply to a system of constant composition and mass at equilibrium.

$$dE = T\,dS - P\,dV \qquad (7\text{-}24)$$

$$dH = T\,dS + V\,dP \qquad (7\text{-}15)$$

$$dA = -S\,dT - P\,dV \qquad (7\text{-}26)$$

$$dF = V\,dP - S\,dT \qquad (7\text{-}29)$$

If work other than that against pressure is to be considered, then another term, $-\delta w'$, should occur on the right-hand side of each equation.

P, V, and T are state variables by their very nature. E is a function of state according to the first law, and S is a function of state according to the second law. H, A, and F are functions of state, since they are defined in terms of these variables which themselves are functions of state [Eqs. (6-15), (7-25), and (7-27)]. Since this functional relationship exists (*i.e.*, the differentials are exact), the fundamental relation of partial differentiation, discussed previously, may be applied to each of the foregoing equations. It will be recalled that, if $z = f(x,y)$, then

$$dz = M\,dx + N\,dy,$$

where $M = (\partial z/\partial x)_y$ and $N = (\partial z/\partial y)_x$, and from this it was shown (footnote, page 179) that $(\partial M/\partial y)_x = (\partial N/\partial x)_y$. Applying this to each of the four above equations [(7-24), (7-15), (7-26), and (7-29)], it is found that

$$\left(\frac{\partial T}{\partial V}\right)_S = -\left(\frac{\partial P}{\partial S}\right)_V \qquad (7\text{-}30)$$

$$\left(\frac{\partial T}{\partial P}\right)_S = \left(\frac{\partial V}{\partial S}\right)_P \qquad (7\text{-}31)$$

$$\left(\frac{\partial S}{\partial V}\right)_T = \left(\frac{\partial P}{\partial T}\right)_V \qquad (7\text{-}32)$$

$$\left(\frac{\partial V}{\partial T}\right)_P = -\left(\frac{\partial S}{\partial P}\right)_T \qquad (7\text{-}33)$$

These four equations are known as *Maxwell's relations*. The last two are particularly useful in evaluating the change of entropy with volume and pressure, respectively, from an equation of state. The Clausius-Clapeyron equation may be derived from the third [Eq. (7-32)], though we shall find it more convenient, in a later section, to derive it in a different manner.

Free Energy as a Criterion of Equilibrium. Equation (7-29), by integration, gives the free-energy change for a system of constant mass which does no work other than that against pressure, providing all its parts are in equilibrium. It is now of interest to consider a going (non-equilibrium) process in a system of constant mass which does work only against pressure. We restrict our consideration to a process which takes place at constant temperature and pressure. This we may imagine as being broken up into a series of steps of the following nature: Let us first allow the spontaneous process to occur to a small extent in an isolated manner; *i.e.*, the system is enclosed in a rigid adiabatic envelope. After this small change has occurred in the isolated system, let the rigid envelope be removed and the system be returned to its initial temperature and pressure by bringing it into contact with a thermostat-barostat. The system is now again enclosed in the rigid adiabatic envelope, and the cycle repeated. The spontaneous departure from the initial temperature and the return are illustrated schematically in Fig. 7-2. There is postulated, of course, a similar spontaneous and simultaneous departure and forced return to a uniform pressure.[1]

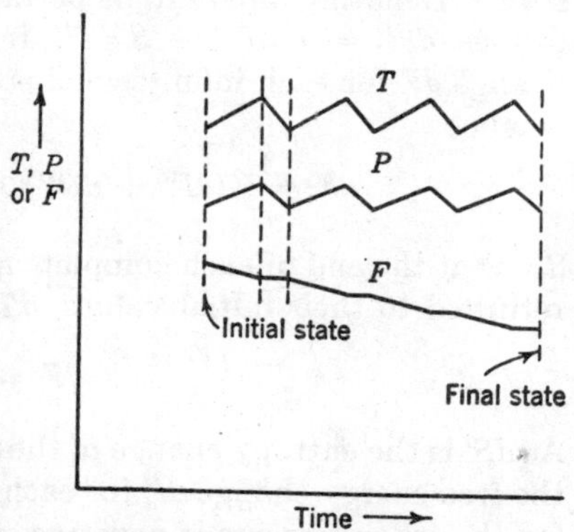

FIG. 7-2. Schematic plot leading to the conclusion that the free-energy change of a spontaneous process is negative.

It will now be seen that the extent of reaction, and hence the extent of

[1] For example, in the irreversible reaction of copper with nitric acid at constant temperature and pressure, we can imagine that in the first part of each step the copper and nitric acid (in contact) are enclosed in a rigid adiabatic envelope and the reaction is allowed to proceed to a small extent. The copper is then lifted out of the acid to stop the reaction. In the second part of each step, the copper, acid, and gaseous phase are returned reversibly to the initial temperature and pressure.

It is obvious that, as the steps are shrunk to infinitesimal size, the over-all effect of a series of such steps (each restoring the system to the initial temperature and pressure) is identical with that of a process carried out at constant temperature and pressure.

temperature or pressure change, in each step can be reduced to an infinitesimal without affecting the nature of the over-all reaction. In each infinitesimal step the entropy change dS', during the first part when the system is adiabatic and isometric, is positive (second law). The general free-energy change from the definitional equation is

$$dF = dE + P\,dV + V\,dP - T\,dS - S\,dT$$

Since the system is isolated, dE' is zero, and since it is isometric, dV' is zero; hence $dF' = V\,dP' - T\,dS' - S\,dT'$, where the prime indicates a change during the first part of each step. During the second part of each step the system is returned reversibly to its initial temperature and pressure. Denoting differentials pertaining to this part of the step by double primes, $dF'' = V\,dP'' - S\,dT''$, from Eq. (7-29). The total free-energy change dF for each infinitesimal step is obtained by adding dF' and dF'', whence

$$dF = V(dP' + dP'') - S(dT' + dT'') - T\,dS'$$

Since at the end of each complete step both temperature and pressure are returned to their initial values, $dT' = -dT''$ and $dP' = -dP''$, so that

$$dF = -T\,dS'$$

As dS' is the entropy change of the system when isolated, it is positive and the free-energy change dF for each complete step is negative. Hence ΔF for the whole process is negative.

Since the free-energy change in passing from the initial to the final state is independent of the course followed, the free-energy change for the above process is the same as if the temperature and pressure had been maintained constant throughout. In fact, the free-energy change is the same as it would have been for any other possible course, even one involving large variations in pressure and temperature, as long as the pressure and temperature were finally brought back to the initial values. Thus we may conclude that, during any process which proceeds spontaneously and in such a manner that the pressure and temperature of the products are the same as for the reactants, the free energy of the system decreases. In a slightly more restricted sense we may say that the free-energy change of any system which undergoes spontaneous reaction at constant temperature and pressure is negative.

It already has been shown [Eq. (7-29)] that the free-energy change of a system undergoing a reversible process at constant temperature and pressure and doing no work other than that against pressure is zero.

From this and the final statement of the preceding paragraph we may write in general for any system of constant mass at constant temperature and pressure which does no work other than that against pressure

$$dF \leq 0 \tag{7-34}$$

Any process for a system of constant mass at constant temperature and pressure for which $\Delta F < 0$ is said to be thermodynamically possible. It is usually considered that under these conditions the process does take place, but the resistance to reaction (activation energy) may be so high that the rate may be very slow, or even imperceptible, in the absence of a suitable catalyst. However, it may be said with certainty from the foregoing reasoning that, if $\Delta F > 0$, the reaction will not take place.[1] For example, the free-energy change accompanying the freezing of a pure liquid below its melting point is negative; however, we cannot say with certainty that such a liquid will spontaneously freeze during the time ordinarily allotted to such an experiment. We can only appeal to experience and note that large degrees of supercooling are not common—silicate, phosphate, and borate systems being notable exceptions. However, above the melting point the free-energy change of fusion is positive, and it can be said with certainty that freezing will not take place.

The free-energy criterion of equilibrium and spontaneous processes is usually the most convenient one, since temperature and pressure are the variables most easily controlled experimentally in most cases. There are, however, several other criteria—all equivalent in that they stem from the second law. The entropy criterion has already been discussed at length; for a system of constant energy and volume $\Delta S \geq 0$. By reasoning similar to that of the preceding paragraphs, the following may also be shown for a system doing work only against pressure: For a system at constant entropy and volume, $\Delta E \leq 0$; for a system at constant entropy and pressure $\Delta H \leq 0$; for a system at constant temperature and volume $\Delta A \leq 0$. In all cases the equality sign pertains to a reversible process and the inequality sign to an irreversible process. It is found from practical experience that the free-energy criterion of the direction in which a reaction will proceed is more useful than any of the others, and the equality $\Delta F = 0$ for a process at equilibrium is one of the most fruitful relations derived from the second law.

[1] It is of historical interest to note that the first approach to this problem was the supposition that a process would not occur spontaneously if ΔH (rather than ΔF) were positive; this erroneous idea is known as the Berthelot-Thomson (1867 and 1853) principle.

The Partial Derivative of the Free Energy. The partial derivative with respect to P or T may be obtained directly from Eq. (7-29), the combined statement of the first and second laws.

$$\left(\frac{\partial F}{\partial P}\right)_T = V \tag{7-35}$$

$$\left(\frac{\partial F}{\partial T}\right)_P = -S \tag{7-36}$$

It frequently is convenient to eliminate S from Eq. (7-36). At constant pressure, using the definitional relation $F = H - TS$, this equation may be written

$$dF = \left(\frac{F - H}{T}\right) dT$$

or

$$T\,dF - F\,dT = -H\,dT$$

Upon dividing by T^2 the left member is $d(F/T)$; so that, at constant pressure, $d(F/T) = Hd(1/T)$ or

$$\left(\frac{\partial(F/T)}{\partial(1/T)}\right)_P = H \tag{7-37}$$

This relation is particularly useful as will become evident in Chap. 9.

CHAPTER 8

THE THIRD LAW OF THERMODYNAMICS

It follows from the first and second laws of thermodynamics that the energy and the entropy of a system are functions of its state. Since these laws deal only with *changes* in energy and entropy, respectively, the zero level for both is arbitrary, just as sea level or a laboratory bench top may be chosen as the zero level for gravitational potential energy. Let us now inquire as to whether there is any particular choice for the zero level of energy or entropy which might be regarded as a natural choice rather than an arbitrary one. Since, as deduced from the behavior of ideal gases and verified by the second law, there is an absolute zero to the temperature scale and since the energy and entropy of a body decrease with falling temperature, the absolute zero of temperature suggests itself as a candidate for the point of zero energy and zero entropy.

In the case of energy, however, a few simple reflections immediately indicate that the energy of all pure substances cannot be assigned a value of zero at the absolute zero. For example, if the energies of H_2 and O_2 at 0°K are assigned zero values, then that of H_2O may be assigned a zero value only if the energy of formation of water is zero at 0°K. Experimentally, from heat capacities and the observed energy of formation ΔE, at some temperature above 0°K, it is found that this is not so. It will be noted that there is no particular trouble in evaluating $C_V\,dT$ as long as C_V does not become infinite between absolute zero and the temperature at which ΔE is known. It is quite general that the energy of reaction does not approach zero at 0°K; hence if the energy of each element at absolute zero is assigned a zero value, the energy of a compound at absolute zero is, in general, not zero. The assignment of zero energy for each element at 0°K is sometimes made; this is entirely arbitrary and can in no sense be called a law.

ΔS AT ABSOLUTE ZERO

As noted above, the choice of absolute zero as the zero point for the energy of the elements is not a particularly fruitful choice, since the energy of compounds at absolute zero does not show any unusual regularity; *i.e.*, the energy change accompanying chemical reactions is not zero at absolute zero, nor does it show any regularity more than that at any other temperature. However, the choice of the absolute zero for the zero point of entropy is fruitful and leads to the third law of thermodynamics.

Early work leading to the third law was done by T. W. Richards* (1902), who found that for many reactions ΔS and ΔC_P approach zero at low temperature. Nernst† (1906) generalized these findings in the statement that *for all reactions involving substances in the condensed state ΔS is zero at the absolute zero.* This is known as the *Nernst heat theorem* and constitutes one statement of the *third law of thermodynamics.* Thus for the chemical reaction $A + B = AB$, where A and B are elements and AB a compound, the entropy change ΔS which equals $S_{AB} - S_A - S_B$ is, according to the Nernst heat theorem, zero at the absolute zero. If the entropies of the elements S_A and S_B are assigned zero values at the absolute zero, then the entropy of the compound S_{AB} is also zero.

Nernst did not originally state his law in exactly this form. He said that $(\partial \Delta H/\partial T)_P$ and $(\partial \Delta F/\partial T)_P$ both approach zero at the absolute zero. However, it is seen immediately from Eq. (7-29), which takes the form $d\,\Delta F = \Delta V\,dP - \Delta S\,dT$ for a reaction, that $(\partial \Delta F/\partial T)_P = -\Delta S$ and hence that the statements $(\partial \Delta F/\partial T)_P = 0$ and $\Delta S = 0$ are equivalent. Nernst's statement that $(\partial \Delta H/\partial T)_P$, and hence ΔC_P, is zero at the absolute zero follows from the second law and the preceding statement that $\Delta S_0 = 0$.‡ This may be seen by differentiating the definitional equation for the free-energy change of a chemical reaction, $\Delta F = \Delta H - T\,\Delta S$. Partial differentiation with respect to T at constant P gives

$$\left(\frac{\partial \Delta F}{\partial T}\right)_P = \left(\frac{\partial \Delta H}{\partial T}\right)_P - T\left(\frac{\partial \Delta S}{\partial T}\right)_P - \Delta S$$

Since $(\partial \Delta F/\partial T)_P = -\Delta S$ (shown above from the second law) and $(\partial \Delta H/\partial T)_P = \Delta C_P$,

$$\Delta C_P = T\left(\frac{\partial \Delta S}{\partial T}\right)_P$$

and thus at $T = 0$ it follows that $\Delta C_P = 0$ unless $(\partial \Delta S/\partial T)_P$ is infinite.

It may be shown in a different way that $\Delta C_P = 0$ at the absolute zero, provided that $\Delta S_0 = 0$. Rearranging the definitional equation for ΔF in the form $(\Delta F - \Delta H)/T = -\Delta S$, it follows from the third law that

* T. W. Richards, *Z. physik. Chem.*, **42**, 129 (1902).

† W. Nernst, *Nachr. kgl. Ges. Wiss. Göttingen, Math-physik. Klasse*, vol. 1, 1906.

‡ A numerical subscript to a thermodynamic function is used to denote the temperature under consideration, expressed on the Kelvin scale (°K).

$\lim_{T\to 0} (\Delta F - \Delta H)/T = 0$. Hence, by L'Hôpital's theorem,*

$$\left(\frac{d\Delta F}{dT}\right)_{T=0} - \left(\frac{d\Delta H}{dT}\right)_{T=0} = 0.$$

Since $(\partial\Delta F/\partial T)_P = -\Delta S$, which is zero at $T = 0$, it follows that

$$\left(\frac{\partial\Delta H}{\partial T}\right)_{P,T=0} = \Delta C_{P,T=0} = 0.$$

It will be recalled from our previous discussion of heat capacity that C_P for many pure crystalline substances is observed to approach zero at the absolute zero. In fact, according to Debye's law, the heat capacity of a pure crystalline substance obeys the relation† $C_V = aT^3$ in the vicinity of the absolute zero. It might seem that the foregoing reasoning might be reversed, and from the fact that $\Delta C_P = 0$ at absolute zero it might be shown that $\Delta S = 0$. Although this cannot quite be done as the student may verify by attempting to reverse the order of the steps in the preceding paragraph, it may definitely be regarded as one of the clues suggestive of the third law.

$S = 0$ AT ABSOLUTE ZERO

As a result of further experimental work and particularly of contributions of the quantum mechanics, the *third law* may now be stated more precisely: *The entropy of any homogeneous substance which is in complete internal equilibrium may be taken as zero at* 0°K.‡ It will be noted that the law thus applies to solids, liquids, and gases.§ As the crystalline form is the stable one at sufficiently low temperature, it is apparent that liquids and gases are metastable in the vicinity of the absolute zero. This fact does not affect the applicability of the law to these substances, however, provided they are in complete internal equilibrium.

As was previously pointed out, the absolute value of the entropy, similar to that of the energy, is arbitrary. The third law, however,

* L'Hôpital's theorem may be stated as follows. An indeterminate form 0/0 or ∞/∞ may be replaced by the quotient of the derivatives of numerator and denominator.

† The Debye expression includes only the contribution to C_V of the harmonic vibrations of the atomic nuclei. The electronic contribution at low temperature is usually represented by a small term in aT. This does not alter the above argument.

‡ As the entropy of most heterogeneous substances is the sum of the entropies of the component homogeneous parts, it follows that a similar statement may be made for such a heterogeneous substance. It should be noted that only homogeneous equilibrium is required, not heterogeneous.

§ However, the ideal-gas law is not applicable as the absolute zero is approached.

makes it clear that one particular arbitrary choice, namely, that the entropy of all internally equilibrated chemical elements is zero at the absolute zero, is especially fruitful, since it leads through the third law to the consequence that the entropy of all internally equilibrated compounds is also zero at the absolute zero. It should be borne in mind that the assignment of a zero value to the entropy of crystalline elements is purely a convention and that the assignment of a zero value to the entropy of most compounds is a consequence of convention as well as the third law.

Taking by convention the entropy of all elements at the absolute zero to be zero, it then follows that the entropy of most compounds is also zero at the absolute zero. It is fitting to inquire as to the limitation imposed by the inclusion of the word "most" in the preceding statement. The *third law* is sometimes stated in the following form: *The entropy of any phase whose quantum states and atomic arrangement correspond to a unique lowest energy state at the absolute zero is zero.* To understand the meaning of "a phase of unique lowest energy," it is desirable to know something of statistical mechanics as well as something of the structure of crystalline compounds as described in the earlier chapters of this book. From the viewpoint of modern statistical mechanics, any phase (not necessarily the stable one) when cooled to the absolute zero under equilibrium conditions will approach a unique lowest energy state. For example, in a crystal the atoms on the lattice sites will approach a unique most orderly arrangement and each atom will approach a unique lowest energy quantum state.

Lack of Internal Equilibrium. In actual practice it may, and frequently does, happen that, as a phase is cooled to the vicinity of the absolute zero, the more random atomic arrangement characteristic of higher temperature is "frozen" and the unique most ordered arrangement characteristic of the lower temperature is not actually realized. For example, when solid carbon monoxide is cooled to a very low temperature, near absolute zero, there is a very strong presumption that the molecules do not arrange themselves in the most orderly fashion consistent with the crystal structure, corresponding to rows of the type —C–O—C–O—C–O—. Instead, the structure of solid carbon monoxide at absolute zero probably corresponds more nearly to a random arrangement of CO molecules on the lattice sites, *e.g.*, one in which C–O and O–C occur thus —C–O—O–C—C–O—C–O—. The entropy of the usual crystalline solid carbon monoxide, arranged in this random order, is not zero, but the entropy of the idealized completely ordered crystalline carbon monoxide is presumed, with considerable evidence, to be zero. Similar considerations apply to nitrous oxide and to nitric oxide.

A phase may lack internal equilibrium in ways other than that just mentioned. In the case of water it is found that the value of $S_{298} - S_0$ obtained calorimetrically (by integrating the low-temperature heat capacities and adding the entropy of fusion) is in error by 0.83 EU [cal/(mole)(deg)]. This error is established by the fact that $S_{298} - S_0$ as determined from spectroscopic data is in agreement with that determined from equilibrium data [*e.g.*, for the reaction $MgO + H_2O = Mg(OH)_2$]. This failure of the direct method is attributed to lack of internal equilibrium in ice at low temperature; specifically, it is believed that at low temperature the hydrogen bonds still possess the random arrangement characteristic of high temperature rather than the perfectly ordered arrangement characteristic of the low-temperature equilibrium state. In fact the entropy attributable to this source has been calculated and found equal to the above figure. Thus when ice is cooled to the vicinity of the absolute zero, it fails to approach the lowest energy (equilibrium) state and does not liberate the heat it otherwise would; accordingly the observed heat capacity is lower than that for the equilibrium state, and $S_{298} - S_0$ as calculated therefrom is low (by 0.83 EU).

Moreover, for hydrogen, which may exist in two different sets of quantum states (ortho and para), it is found that the transition from one set to another occurs readily at high temperature but very slowly if at all at low temperature. This effect gives rise at the absolute zero to an extra entropy [1.64 cal/(mole)(deg) observed as the discrepancy between the calorimetric and spectroscopic values for $S_{298} - S_0$] which also is in fair agreement with the calculated value. The effect involves nuclear spin and is associated with the low mass of hydrogen (or deuterium); it seems unlikely to be encountered elsewhere.

In principle the third law is presumably applicable to *solutions* and *glasses* which are in complete internal equilibrium. As noted in Chap. 4 many solid solutions transform to an ordered state on slow cooling. If this state is completely ordered, its entropy may be taken as zero at the absolute zero, as discussed more completely later in this chapter. However, such complete order (which is regarded as characteristic of any solution in complete internal equilibrium at 0°K) can very seldom if ever be achieved experimentally. It is perhaps more common to find a solid solution, *e.g.*, the solid solution KCl-KBr, which is nearly completely random at all temperatures of experimental investigation. The extra entropy corresponding to this randomness, calculated as $R(N_1 \ln N_1 + N_2 \ln N_2)$ assuming the solution to be ideal, is found to be in good agreement with the observations in this case.

Liquid glycerin readily supercools and becomes a glass at low temperature. The entropy difference between the glassy state and the crystalline state at 0°K has been determined and is discussed in detail by Jones and Simon.[1] The more slowly the glass is cooled, the smaller this difference becomes. Jones and Simon conclude "although a practical limit is very soon reached it is safe to predict that by an infinitely slow approach to the absolute zero one would approach vanishing entropy differences." Thus it is not the fact that a glass is metastable (relative to crystal) which gives rise to its measured entropy at the absolute zero but rather the fact that it is a supercooled *disordered* phase. This disorder becomes "frozen in" during cooling through some particular temperature range (easily recognized by the change in C_P), and thus the glass does not further approach the ordered state of internal equilibrium as the temperature is further lowered.

EXPERIMENTAL VERIFICATION OF THE THIRD LAW

Direct Verification. The third law of thermodynamics, like any other scientific law, is based upon experiment. The second law was based upon the common experience that heat will not flow from a cold to a hot body, and we did not have to appeal to any special experiments to be convinced of this. The third law is not so obvious; hence it seems desirable to describe some of the evidence by which it was established. If it is required to find the entropy change of a reaction at absolute zero, the most direct approach might seem to be to carry out the reaction at this temperature and measure the entropy change. Obviously this is not feasible, since, in the first place, reaction velocities decrease rapidly with temperature and become unobservably small well above the absolute zero. In the second place entropy changes may be directly measured only for reversible reactions, for which $\Delta S = q_r/T$, and aside from the difficulties with this mathematically indeterminate form, we could hardly expect to find a reaction which would be reversible at the absolute zero. The so-called "direct" verification of the third law involves the application of the second law, namely, the principle that the entropy of a system is a function of its state, and the net entropy change experienced by a system undergoing a cycle is therefore zero.

Let us consider the cycle shown schematically in Fig. 8-1. The starting point of this cycle is a system composed of reactants at the absolute zero. The first step of the cycle consists in heating the separated reactants to some temperature T where from previous experience it is known that the reactants are in equilibrium with the products. In the second step the reaction is allowed to proceed to completion at the temperature T. The separated products are cooled to the absolute zero in the third step, and in the fourth step, which is imaginary, the reverse reaction is allowed to

[1] G. O. Jones and F. E. Simon, *Endeavor*, **8**, 175 (1949).

proceed. Designating the entropy change during each step by the corresponding subscript, we have from the second law that

$$\Delta S_{\mathrm{I}} + \Delta S_{\mathrm{II}} + \Delta S_{\mathrm{III}} + \Delta S_{\mathrm{IV}} = 0$$

or $\Delta S_{\mathrm{IV}} = -(\Delta S_{\mathrm{I}} + \Delta S_{\mathrm{II}} + \Delta S_{\mathrm{III}})$. Since the entropy change during the first three steps can be measured experimentally, the entropy change at the absolute zero can be determined.

One of the simplest types of chemical reaction is the allotropic transformation of a chemical element. As discussed in an earlier chapter, tin is capable of existing in two crystalline modifications, white tin and gray tin. White tin, the common form, is stable above 19°C, and gray tin is stable below this temperature. The heat of transition at this equilibrium temperature is 541 cal/gram atom. Under ordinary circumstances white tin can readily be cooled below 19°C without the occurrence of the transition, and consequently the heat capacity of both forms can be measured from low temperature to the transition temperature. Writing the reaction

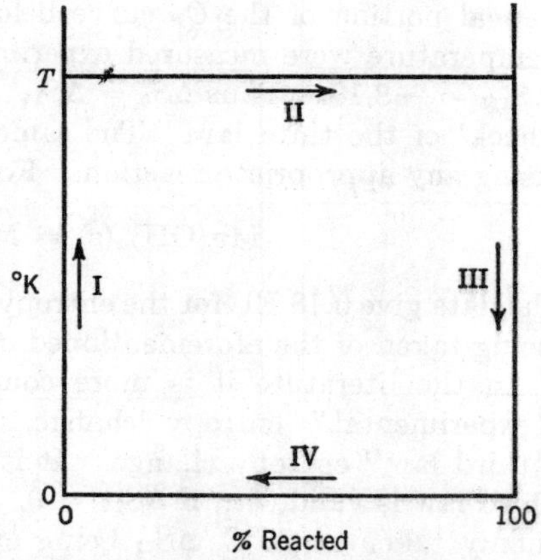

FIG. 8-1. Schematic diagram illustrating a method of evaluating the entropy change of a reaction at the absolute zero.

$$\mathrm{Sn(gray)} = \mathrm{Sn(white)}$$

the entropy changes ior the first three steps of the above cycle (Fig. 8-1) are

$$\Delta S_{\mathrm{I}} = \int_{0°}^{292°} C_{P(\mathrm{gray})} d \ln T = 9.11 \text{ EU [cal/(mole)(deg)]}$$

$$\Delta S_{\mathrm{II}} = \frac{q_r}{T} = \frac{541}{292} = 1.85$$

$$\Delta S_{\mathrm{III}} = \int_{292°}^{0°} C_{P(\mathrm{white})} d \ln T = -11.04$$

Thus the entropy change at the absolute zero, ΔS_0, is

$$\Delta S_{\mathrm{IV}} = -(\Delta S_{\mathrm{I}} + \Delta S_{\mathrm{II}} + \Delta S_{\mathrm{III}}) = -(9.11 + 1.85 - 11.04) = 0.08 \text{ EU}$$

This is probably smaller than the experimental error, and thus it has been shown that in this case ΔS_0 is zero within the experimental error. This case is one of the earliest confirmations of the third law. It should be added that the heat capacities below 80°K were obtained from

Debye's formula and those above this temperature were determined experimentally.

Phosphine (PH_3) is also capable of existing in two forms, α and β, which are in equilibrium at 49.43°K. Carrying out a similar calculation for the reaction

$$PH_3(\beta) = PH_3(\alpha)$$

it is found that $\Delta S_I = 4.38$ EU of which only 0.34 arises from the theoretical portion of the C_P curve below 15°K; heat capacities above this temperature were measured experimentally. Also $\Delta S_{II} = 3.76$ EU and $\Delta S_{III} = -8.13$. Thus $\Delta S_0 = \Delta S_{IV} = -0.01$ EU, providing an excellent check[1] of the third law. The same type of evidence may be obtained using any appropriate reaction. For the reaction

$$Mg(OH)_2(s) = MgO(s) + H_2O(g)$$

the data give 0.18 EU for the entropy change at the absolute zero (account being taken of the aforementioned difficulty with H_2O).

In the literature it is more common to compare ΔS_{II}, termed the "experimental" entropy change, with $-(\Delta S_I + \Delta S_{III})$, termed the "third law" entropy change. It is seen that these will be equal if the third law is valid, *i.e.*, if $\Delta S_{IV} = 0$. The temperature of step II is commonly taken as 25°C, ΔS_{II} being evaluated from equilibrium and heat measurements by use of the relation $\Delta S_{II} = (\Delta H_{II} - \Delta F_{II})/T$. In the immediately preceding example the results may be summarized as follows:

$$\Delta S \text{ (third law)} = 36.85 \pm 0.08 \text{ EU}$$
$$\Delta S \text{ (experimental)} = 36.67 \pm 0.10 \text{ EU}$$

Verification through Statistical Mechanics. By the methods of statistical mechanics the thermodynamic functions of chemical species in the ideal gaseous state may be computed from spectroscopic data. This computation involves a large, important, and growing field which is outside the scope of this book, and the metallurgist, even though unable to perform the calculation himself, should be familiar with the finished product and be able to use it. The final result of such a work is commonly a table containing, for the substance investigated, values of the following thermodynamic functions at a series of temperatures: $H° - H°_0$, $(F° - H°_0)/T$, and sometimes $S°$, $(H° - H°_0)/T$ or others. It will be observed that all the rest may be obtained from the first two. The superscript zero (°) is used to denote the standard state, *i.e.*, the ideal-gas state at 1 atm pressure. The subscript zero, as usual, refers to the

[1] C. C. Stevenson and W. F. Giauque, *J. Chem. Phys.*, **5**, 1949 (1937).

absolute temperature. If such tables are available for all the substances involved in a particular reaction, then $\Delta(H^\circ - H_0^\circ)$, $\Delta(F^\circ - H_0^\circ)/T$, and ΔS° can be computed; the first two are equal, of course, to $\Delta H^\circ - \Delta H_0^\circ$ and $(\Delta F^\circ - \Delta H_0^\circ)/T$.

A comparison of the spectroscopically evaluated S° for a substance (or ΔS° for a reaction) with that resulting from experimental heat capacities and the third law affords a check on the latter. Table 8-1 shows values

TABLE 8-1. ENTROPY OF A FEW GASES AT 298°K, ENTROPY UNITS, CAL/(MOL)(DEG)

Gas	Third law	Spectroscopic
O_2	49.1	49.003
N_2	45.9	45.767
HCl	44.5	44.617
HBr	47.6	47.437
HI	49.5	49.314
H_2S	49.1	49.15
SCO	55.3	55.34
CO_2	51.1	51.061
NH_3	45.9	46.01
CO*	46.2	47.32
H_2O*	44.3	45.13
H_2*	29.6	31.22

* See text for explanation of discrepancy in these cases.

of the entropy of several gases at 25°C as computed from spectroscopic data and from the third law. It will be observed that in several cases the value of S° obtained from spectroscopic data appears more reliable than that resulting from calorimetric data and the third law. It should be pointed out that the statistical mechanical derivation involves an assumption which we may call the statistical mechanical "equivalent" of the third law. Hence any comparison such as that just mentioned does not constitute an entirely independent check of the third law, but merely a partial check.

It is interesting to note the comparison for carbon monoxide. Here S_{298}° is found from calorimetric data (measured heat capacities) and the third law to be 46.2 EU, whereas the statistical mechanical treatment of spectroscopic data gives 47.32 EU. The "frozen-in" disorder of crystalline CO, discussed previously, undoubtedly leads to a low experimental C_P which in turn (through $\int C_P \, d \ln T$) leads to a low value for the entropy at room temperature. This conclusion is strengthened by the fact that ΔS as determined by equilibrium measurements (discussed in the next section) is in agreement with the spectroscopic value.

This frozen-in disorder or lack of equilibration at low temperature is frequently a source of error in the calorimetric determination of C_P, or more strictly a source of error in the application of the third law thereto. In spite of the superiority of the spectroscopic method in this respect it should be mentioned that this method involves the summation of a number of terms, and if one of these is neglected or inaccurately computed a rather large (and consistent) error may result. Although such errors and oversights have occurred in the past, it is generally believed that they are few in modern tabulations.

Verification through Equilibrium Measurements. The definitional relation for free energy [Eq. (7-28)] when applied to a reaction becomes $\Delta F = \Delta H - T\,\Delta S$. The entropy change if the reactants and products are in their standard states is, therefore,

$$\Delta S^\circ = \frac{\Delta H^\circ - \Delta F^\circ}{T}$$

and the measurement of ΔH° and ΔF° at any definite temperature serves to evaluate ΔS°. As will be seen in the next chapter, an equilibrium measurement of any one of several types serves to evaluate ΔF°. Consequently, the experimental determination of a single heat of reaction and of a single equilibrium gives ΔS° and affords another comparison with the entropy change obtained from the third law by integration of heat-capacity data.

Applying Eq. (7-36) to a reaction whose constituents are in their standard states,

$$\left(\frac{\partial \Delta F^\circ}{\partial T}\right)_P = -\Delta S^\circ$$

Thus equilibrium measurements alone over a range of temperature can be used to give ΔS° and provide a test of the third law.

Comparison of Methods. It is thus seen that three general methods may be used to evaluate the entropy of a substance at a given temperature: (1) the third law method, involving integration of experimental heat capacities; (2) the spectroscopic method (giving the entropy in the ideal gaseous state), involving statistical mechanical calculations based on spectroscopic data; and (3) the equilibrium method, involving equilibrium measurements over a range of temperature or equilibrium and heat measurements at the temperature. Although a comparison of the results of all three methods is seldom possible, one is available in the case of chlorine gas at 25°C.[1] The third law entropy of 53.31 ± 0.10 EU per mole Cl_2 results from summation of the following terms: entropy of

[1] K. K. Kelley, *U.S. Bur. Mines Bull.* 477, 1950.

solid by extrapolation of heat capacity below 15°K, 0.33; entropy of solid between 15°K and the melting point from measured heat capacity, 16.57; entropy of fusion by dividing heat of fusion 1531 cal by the melting point 172.12°, 8.89; entropy of liquid up to boiling point from measured heat capacity, 5.23; entropy of vaporization by dividing heat of vaporization 4878 cal by the boiling point 239.05°K, 20.41; entropy of heating gas to 298.1°K from measured heat capacity, 1.88. The equilibrium entropy of 53.0 ± 0.4 results from combination of the ΔS_{298} for the reaction $2Ag + Cl_2 = 2AgCl$, namely, -27.4, obtained by measurement of the electromotive force of cells, with the third law entropies of silver and silver chloride, namely 10.2 ± 0.1 and 23.0 ± 0.1, respectively, according to the relation $\Delta S = 2S_{AgCl} - 2S_{Ag} - S_{Cl_2}$. The spectroscopic value is calculated to be 53.31 ± 0.01. It will be seen that the three determinations agree well, the spectroscopic value seeming the most precise.

Coefficient of Thermal Expansion at Absolute Zero. So far in our statement that the entropy change of a reaction approaches zero at absolute zero we have placed no restriction on the pressure or volume prevailing during this theoretical reaction. Since the third law, like all other scientific laws, is based upon experiment, we must appeal to experiment to determine whether such a restriction is required. In the examples previously considered in the experimental verification of the third law the pressure was 1 atm throughout. The question naturally arises whether the third law would have been equally valid had the pressure been, for example, 10,000 atm. It is usually presupposed that this would be the case, *i.e.*, that the third law would be valid no matter at what pressure the reaction takes place and also that it is equally valid at constant volume. Let us now consider the implications of such a presupposition.

From one of Maxwell's relations [Eq. (7-33)] we have, for any substance, $(\partial S/\partial P)_T = -(\partial V/\partial T)_P \equiv -\alpha V$, the latter equality following from the definition of the coefficient of thermal expansion,

$$\alpha \equiv \frac{1}{V}\left(\frac{\partial V}{\partial T}\right)_P$$

Hence, by summing up for all substances involved in a reaction,

$$\left(\frac{\partial \Delta S}{\partial P}\right)_T = \alpha_1 V_1 + \alpha_2 V_2 + \cdots - \alpha_1' V_1' - \alpha_2' V_2' - \cdots$$

the unprimed quantities refer to the reactants, and the primed quantities to the products. If the third law is valid in the general form suggested above, then at the absolute zero the entropy change is not a function of pressure and $(\partial \Delta S/\partial P)_{T=0} = 0$. If such is the case, then, at the absolute zero,

$$\alpha_1 V_1 + \alpha_2 V_2 + \cdots - \alpha_1' V_1' - \alpha_2' V_2' - \cdots = 0$$

Since none of the volumes approaches zero at the absolute zero, it follows, in general, that either the coefficients of expansion are finite and interrelated in such a manner that the above expression reduces to zero at all pressures or that the coefficient of expansion of all substances is zero at the absolute zero regardless of the pressure. The latter possibility seems more credible and indeed seems to be borne out by the limited experimental evidence available. Hence we may conclude that it is likely that *the coefficient of expansion of all solid crystalline materials which are at internal equilibrium is zero at the absolute zero* and that *the third law is valid at all pressures.*

It can be shown also that, if the coefficient of thermal expansion of each substance involved in a reaction approaches zero at the absolute zero, then the entropy change is independent of the volume change and *the entropy change of a reaction taking place at the absolute zero at constant volume is zero no matter what that volume may be.* By another of Maxwell's relations [Eq. (7-32)],

$$\left(\frac{\partial S}{\partial V}\right)_T = \left(\frac{\partial P}{\partial T}\right)_V = -\left(\frac{\partial P}{\partial V}\right)_T \left(\frac{\partial V}{\partial T}\right)_P = -\alpha V \left(\frac{\partial P}{\partial V}\right)_T$$

the second equality following from the fundamental relation of partial differentiation and the third from the definition of α. If α approaches zero at absolute zero, then $(\partial S/\partial V)_{T=0} = 0$.* Hence the entropy of each product and reactant is not affected by the volume change accompanying the reaction, and the entropy change accompanying the reaction is the same (namely, zero) as it would have been if the reaction had occurred at constant pressure. It should be noted here that, although existing evidence indicates that α approaches zero at absolute zero, there is no evidence that β does likewise.

EQUILIBRIUM IN ORDER-DISORDER TRANSFORMATIONS AT LOW TEMPERATURE

The order-disorder type of transformation, sometimes referred to as a phase change of the second kind, was discussed in Chap. 4. There the experimental evidence, chiefly from X-ray diffraction, was briefly reviewed and the conclusion reached that in several substitutional alloys

* This follows provided that $(\partial P/\partial V)_T$ does not approach infinity at absolute zero. Since $(\partial P/\partial V)_T = -1/\beta V$, where β is the coefficient of compressibility, and since, as noted later, there is no evidence that β approaches zero at absolute zero, the above condition is satisfied.

the atomic arrangement tends to be random at higher temperature and ordered at lower temperature. For certain compositions there is a particular intermediate temperature at which, upon slow cooling, transition takes place isothermally from a relatively disordered structure to a relatively ordered one. As mentioned previously, there is much evidence, *e.g.*, in the relation between C_P and T shown in Fig. 4-16, that, as cooling is continued below the transition temperature, the degree of order increases; and it would not seem an unreasonable guess that, if the equilibrium state were approached at a temperature in the vicinity of the absolute zero, the state would be a perfectly ordered one. This guess is supported by the modern statistical mechanical interpretation of the third law, according to which every phase approaches (or would approach if homogeneous equilibrium prevailed, *i.e.*, if not "frozen" in some metastable form) a unique lowest energy state at the absolute zero.

The third law was previously stated thus: "the entropy of any phase whose quantum states and atomic arrangement correspond to a unique lowest energy state at the absolute zero is zero." The above-mentioned statistical mechanical addition to this is that every phase tends to approach such a unique lowest energy state at the absolute zero. If this statement is to be taken at face value, we are forced to the conclusion that every alloy phase, no matter what its composition, tends to approach the perfectly ordered state at very low temperature. For if there is a unique atomic arrangement corresponding to the lowest energy, then obviously the repetition of such an atomic arrangement, or superlattice cell, in a crystal constitutes an ordered arrangement. If the atom-fraction ratio is not the ratio of two small integers, then the unit superlattice cell is not small; and if this ratio is not a rational number, then the superlattice cell coincides with the crystal itself.

The tendency toward order at low temperature may be interpreted rather simply. In a binary alloy composed of A and B atoms the total energy, to a first approximation, may be regarded as the negative sum of the bond energies of nearest neighbor pairs. Since the most stable configuration is that with the lowest free energy, it follows from the definition of F that at the absolute zero the most stable configuration is that one with the lowest enthalpy H, which does not differ significantly from E. Of course, this conclusion might have been reached from mechanical considerations alone. If the energy of the $A—B$ bonds is greater[1] than that of the $A—A$ and $B—B$ bonds which are broken in its formation, then the $A—B$ bond is stable and will form to the maximum

[1] That is, the ΔE of formation is greater in absolute value. Since ΔE is negative, it is seen that the E of the $A—B$ bond is, in fact, low. This corresponds to a low free energy and a stable configuration.

extent possible under the limitations of the particular crystal structure and composition. In other words, every A atom will have the maximum possible number of B atoms as nearest neighbors, and vice versa. Thus for any particular number of A and B atoms on a given lattice there will be one particular arrangement which has the greatest number of A—B bonds; this is the stable ordered arrangement at the absolute zero. At higher temperature the thermal agitation will give rise to departure from this perfectly ordered arrangement. On the other hand, if the A—B bond is less stable than the A—A and B—B bonds from which it might be formed, then the stable arrangement will involve the lowest possible number of A—B bonds—a number which obviously may approach zero since the two types of atoms may segregate, each migrating to a different part of the crystal.

It should be noted that the segregation considered here is on the particular lattice under consideration. For the present purpose there is no need to consider that there may be other phases more stable than the one under consideration. Thus even a solution which is supersaturated with respect to some other phase tends at sufficiently low temperature to approach the completely ordered state.

The present indications are, therefore, that all phases tend at low temperature to approach the perfectly ordered state and that the entropy of this perfectly ordered state is zero at absolute zero. It is well known that reaction rates decrease rapidly with temperature and become unobservably small in the vicinity of absolute zero. We could hardly hope to observe an order-disorder transformation at very low temperature, and as noted in Chap. 4, the transformations that have been observed occur well above room temperature. The entropy of formation of a solid solution at ordinary temperature could not be calculated by the third law from measured heat capacities if the equilibrium temperature of the order-disorder transformation were much below room temperature. The disorder would be "frozen" in the crystal, and the heat capacity observed would not be that of a phase at internal equilibrium. However, there is hope that in the future such an entropy may be calculable by the methods of statistical mechanics using the third law in the extended form stated above.

LIMITING LAWS

Some of the most important laws of the physical sciences are limiting laws; *i.e.*, they become precise only in the limit that certain specified quantities become vanishingly small. In a sense even Newton's law for acceleration in a gravitational field is a limiting law; it becomes valid only as friction becomes negligibly small. The first law of thermodynamics is usually believed not to be of this type, but to be of universal validity. Similarly, the second law is commonly believed to be generally valid provided the system under consideration is large compared with atomic dimensions (doubt is sometimes expressed as to its application to living organisms).

However, limiting laws are frequently used in conjunction with the first two laws of classical thermodynamics. Two of the most common and useful limiting laws are the ideal-gas law and Raoult's law (of which Dalton's law for gases is a special case). The ideal-gas law is a limiting law in that it becomes precisely valid only at vanishingly small pressure, and Raoult's law for the solvent of a solution becomes precise only at a vanishingly small concentration of solute. The third law is also a limiting law, in that it strictly applies to a system only at vanishingly small temperature. Guggenheim,[1] with considerable justification, groups these three limiting laws together as a "third principle" of thermodynamics. It will be observed that all three are derivable from statistical mechanics.

[1] E. A. Guggenheim, "Thermodynamics," Interscience Publishers, Inc., New York, 1949.

CHAPTER 9

FUGACITY, ACTIVITY, AND THE EQUILIBRIUM CONSTANT

It is well to recall at this point our main purpose in developing in detail the thermodynamic consequences of the first, second, and third laws. This purpose is to derive the relations between the compositions of coexisting phases at equilibrium and such measurable quantities as the temperature, pressure, heat capacity, heat of reaction, etc. All the general equilibrium relations derived from only the first and second laws involve variables such as S, A, or F which can hardly be considered directly measurable and which cannot be regarded as independent, since they cannot be varied directly by the experimenter at will. The fruitfulness of these equilibrium relations is somewhat limited unless some additional relation pertaining to the substance under consideration is available. In the case of a gas this additional information is usually its equation of state, and for a solution it is a relation expressing its departure from ideality. In this chapter we shall utilize the ideal-gas law and the ideal-solution law in developing relations of the desired type, namely, those involving only measurable quantities.

FUGACITY

The combined statement of the first and second laws for a system doing work only against pressure is $dF = V\,dP - S\,dT$ [Eq. (7-29)]. At constant temperature $dF = V\,dP$; and for 1 mole of an ideal gas at constant temperature, since $V = RT/P$,

$$dF = RT\,d\ln P \tag{9-1}$$

In a gas mixture the partial pressure of the ith constituent, as noted in Chap. 2, is defined by the relation $p_i = N_iP$ (Dalton's law). Let us now consider again the semipermeable diaphragm introduced in Chap. 2 in the discussion of partial pressure. If two parts of a box containing a gas mixture are separated by a diaphragm which is permeable to component 1 of the mixture but not to the others, then there will be established eventually an equilibrium or, more strictly, a partial equilibrium, at which the partial pressure of component 1 is the same in the two parts. If initially one side is evacuated or contains only component 1, the total pressure at equilibrium on this side will be equal to the partial pressure of component

1 on the other side. Also, at equilibrium, the transfer of a small amount of component 1 across the diaphragm is accompanied by no change in the total free energy, this by virtue of the fact that the system is at equilibrium and the transfer of a small amount does not alter the temperature or pressure. Hence the free energy of pure component 1 at a particular pressure is equal to its free energy in the mixture.

The free energy of 1 mole of component 1 in the mixture, designated $\bar{F}_1$, is known as the partial molal free energy and will be discussed in Chaps. 10 and 13 on solutions. It represents simply the change in total free energy of the mixture when 1 mole of component 1 is added to a large amount of the mixture. From the above it is now apparent that the molal free energy F_1 of component 1 on the side of the box containing only component 1 is equal to the partial molal free energy $\bar{F}_1$ in the mixture on the other side. Also the total pressure P on the first side is equal to the partial pressure p_1 on the other. Since on the first side $dF_1 = RT\, d \ln P$, it follows, for the mixture, that $d\bar{F}_1 = RT\, d \ln p_1$. Obviously similar reasoning may be applied to each component of a mixture, and we may write in general for the ith component

$$d\bar{F}_i = RT\, d \ln p_i \tag{9-2}$$

Fugacity of an Ideal Gas. Equations (9-1) and (9-2) apply only to ideal gases or to mixtures thereof. Their form is so convenient that it is desirable to define a new function such that the form of these equations is preserved even though the gas is not ideal. This new function is called the *fugacity*, designated by the symbol f, and is partially defined for a pure species by the isothermal relation

$$RT\, d \ln f = dF \tag{9-3}$$

The fugacity of a species in a gas mixture is partially defined by the similar isothermal relation

$$RT\, d \ln f_i = d\bar{F}_i \tag{9-4}$$

Equations (9-3) and (9-4) do not furnish complete definitions of the fugacity, since, upon integration, an integration constant must be introduced. Thus for a pure species

$$RT \ln f_i = F_i + I \tag{9-5}$$

The integration constant I is always chosen so that the fugacity approaches the pressure as the pressure approaches zero. For an ideal gas the fugacity is numerically equal to the pressure at all pressures. Similarly, for a gas mixture, the fugacity of each component approaches its partial pressure as the total pressure, and hence the partial pressure, approaches zero. For an ideal-gas mixture the fugacity of each component is numerically equal to the partial pressure of that component.

Fugacity of a Nonideal Gas. Although the principal interest of the metallurgist in gases is limited to the conditions of relatively high temperature and substantially 1 atm pressure, where most gases behave ideally or nearly so, it is well to know the magnitude of the error involved in setting the fugacity equal to the pressure or partial pressure. Let us consider, for example, the following equation of state, which expresses moderately well the behavior of gases under conditions where slight departure from ideality is exhibited.

$$V = \frac{RT}{P} - \alpha \tag{9-6}$$

It will be noted that, if α is zero, this expression reduces to the ideal-gas law. For moderate departures therefrom α may be considered a function of temperature only and therefore a constant at constant temperature.

Recalling the definition of fugacity [Eq. (9-3)] and the expression for dF at constant temperature, from the combined statement of the first and second laws [Eq. (7-29)] we have at constant temperature

$$RT\, d \ln f = V\, dP \tag{9-7}$$

This equation is valid for any substance no matter what its equation of state. Applying it to a gas obeying the above equation of state [Eq. (9-6)], it is found that

$$RT\, d \ln f = RT\, d \ln P - \alpha\, dP$$

or

$$d \ln \frac{f}{P} = -\frac{\alpha}{RT}\, dP$$

Let us integrate from $P = 0$ to P, bearing in mind (from the definition of fugacity) that f/P approaches unity as P approaches zero.

$$\ln \frac{f}{P} = -\frac{\alpha P}{RT} \tag{9-8}$$

Alternatively this may be written

$$\frac{f}{P} = e^{-\alpha P/RT}$$

Since the equation of state is valid only for small values of P and hence of $\alpha P/RT$, we may utilize the mathematical approximation $e^{-x} \cong 1 - x$, valid for values of x small compared with 1, so that

$$\frac{f}{P} = 1 - \frac{\alpha P}{RT}$$

Substituting for α from the equation of state [Eq. (9-6)],

$$\frac{f}{P} = \frac{PV}{RT}$$

The expression RT/V is frequently called the "ideal pressure," meaning the pressure which would exist in the volume V at temperature T if the volume contained 1 mole of ideal gas. Designating the ideal pressure as P_i, the foregoing expression becomes

$$\frac{f}{P} = \frac{P}{P_i} \tag{9-9}$$

from which it will be noted that the actual pressure is the geometric mean of the fugacity and the ideal pressure. Also it is seen that the percentage error involved in assuming the fugacity equal to the pressure is the same as the percentage departure from the ideal-gas law. It should be borne in mind that the above considerations apply only to a gas whose behavior exhibits only moderate departure from that of an ideal gas at the temperature and pressure under consideration.

Table 9-1 shows the variation of fugacity with pressure and temperature for several common gases. The largest departures from ideality occur at low temperature, at high pressure, and for gases such as H_2O, NH_3, and CO_2 which have a high critical temperature and a high condensation temperature at 1 atm pressure. In fact when temperature and pressure are expressed in terms of the ratio to the critical value (called the reduced temperature or pressure), the relationship between f/P and the reduced temperature and pressure is essentially constant for all gases.[1]

The magnitude of the effect of nonideality even at 1 atm pressure is illustrated in the following example of the volumetric analysis for CO by combustion thereof. Combustion with oxygen of a gas mixture consisting of 90 per cent CO and 10 per cent CO_2 results in a slightly smaller volume than would be observed if the resulting CO_2 behaved ideally. If due account is not taken of departure from ideality (about $\frac{1}{2}$ per cent) the initial CO_2 content would be reported as 9.5 per cent instead of 10 per cent, and the ratio of CO to CO_2 as 8.5 instead of 9.0—a 5 per cent error occasioned by a $\frac{1}{2}$ per cent departure from ideality.[2] This is a rather exceptional case. In most cases of metallurgical interest the gases involved exhibit such slight departure from ideal behavior that we shall subsequently consider the ideal-gas law to be adequate.

[1] R. H. Newton, *Ind. Eng. Chem.*, **27**, 302 (1935).

[2] The alternative analysis of the gas mixture by absorption of the CO_2 in caustic does not involve a large error because of nonideality of CO_2. However, other errors are encountered here, namely, the decomposition of CO when strong potassium hydroxide solution is used and the large and somewhat questionable correction for water vapor when dilute caustic solution is used.

TABLE 9-1. FUGACITY OF SEVERAL GASES

Gas	P, atm			
	50	100	500	1000
0°C				
H_2	51.5	106	697	2000
O_2	48.5	92	440	1180
CH_4	45.2	82	315	920
100°C				
H_2	51.2	105	650	1700
O_2	50.0	100	540	1390
CH_4	48.5	94	475	1320
CO_2	44.2	79	230	520
NH_3	39.0	*	*	*
200°C				
H_2	51.0	104	617	1540
O_2	50.5	102	575	1455
CH_4	50.0	100	535	1400
CO_2	47.8	92	375	830
NH_3	45.0	81	215	330

* Condenses.

ACTIVITY

The *activity* of a substance is defined as the ratio of the fugacity of the substance in the state in which it happens to be to its fugacity in its standard state. Thus

$$a \equiv \frac{f}{f^\circ} \tag{9-10}$$

or

$$a_i \equiv \frac{f_i}{f_i^\circ} \tag{9-10a}$$

f° and f_i° being the fugacity in the standard state for a pure substance and for the ith component of a solution, respectively. The *standard state* of a substance is commonly chosen as the pure liquid or solid form of the substance at 1 atm pressure and at the temperature under consideration or as the gas at 1 atm pressure and at the temperature under consideration.[1] Obviously only one of the forms—solid, liquid, or gas—is stable

[1] More strictly, the standard state of a gas is usually chosen as that state in which the fugacity is unity. Other choices of standard state for components of liquid and solid solutions will be considered later.

under these conditions, except for a transition temperature, and it is this one which is usually chosen as the standard state. The activity of a substance in its standard state is seen [Eq. (9-10)] to be unity. It will be noted that the standard state defined above varies with experimental variation of temperature but not with experimental variation of pressure.

In view of the isothermal differential equation defining fugacity [Eq. (9-4)] and the fact that f° is constant at constant temperature, we may write the following (isothermal) relations:

$$RT\, d \ln a = dF \tag{9-11}$$

and

$$RT\, d \ln a_i = d\bar{F}_i \tag{9-11a}$$

The identity in form between Eqs. (9-11) and (9-3) gives further evidence of the close similarity between fugacity and activity, which was, of course, first apparent from the definition of activity [Eq. (9-10)]. From this equation it is immediately seen that for a gas (*i.e.*, a substance which is gaseous at 1 atm pressure at the temperature under consideration) the activity is numerically equal to the fugacity. The usefulness of the activity function as well as of the fugacity is found in the treatment of condensed phases, *i.e.*, liquid and solid substances, particularly those of low vapor pressure.

Integration of Eq. (9-11) at constant temperature from the standard state, designated by superscript zero, to any arbitrary state gives

$$RT \ln a = F - F^\circ \tag{9-11b}$$

since by definition a° is unity. Similar integration of Eq. (9-11*a*) gives, for the *i*th component of a solution,

$$RT \ln a_i = \bar{F}_i - F_i^\circ \tag{9-11c}$$

These four equations may be regarded as alternative definitions of the activity which will be used more frequently than Eq. (9-10) or (9-10*a*). Differentiation of Eq. (9-11*c*) at constant temperature gives

$$RT\, d \ln a_i = d(\bar{F}_i - F_i^\circ) \tag{9-12}$$

which is obviously identical with Eq. (9-11*a*) since F_i° is constant under this condition.

To understand more fully the meaning of the activity of a component of a liquid or solid solution, let us consider Raoult's law. *Raoult's law* (in its original form) states that the partial pressure of any component of an ideal solution is equal to the product of its mole fraction and the vapor pressure of the pure component at the temperature of the solution. This statement, however, implies the ideality of the vapor, and the definition of an *ideal solution* is now stated in terms of fugacity instead of partial pressure: A solution is said to be ideal if each component follows the relation

$$f_i = f_i^\circ N_i \tag{9-13}$$

This is seen to be identical with the similar relation involving pressures in the case that the vapor obeys the ideal-gas law (the only case we shall consider). For all solutions it is found, by proper choice of components, that Raoult's law is approached as a limiting law by each component as its mole fraction approaches unity. Hence Raoult's law serves as a very convenient norm or reference which expresses the behavior of an ideal solution and the limiting behavior of any solution.

Let us now consider a component i of an ideal solution of any composition or a component i of any solution in which N_i is nearly unity. In either case Eq. (9-13) applies. This may be rewritten $N_i = f_i/f_i^\circ$, and from the definition of activity [Eq. (9-10)] it is seen immediately that, whenever Raoult's law holds, the activity of a component is equal to its mole fraction. It should be borne in mind, however, that the activity *in all circumstances* is defined by Eq. (9-10) or, equivalently, by Eq. (9-11) and the additional information that the activity is unity in the standard state.

EQUILIBRIUM CONSTANT

Let us consider the general chemical reaction at constant temperature and pressure

$$l\text{L} + m\text{M} + \cdots = q\text{Q} + r\text{R} + \cdots \tag{9-14}$$

where the capital letters are used as general symbols for chemical elements or species and the small letters for the number of gram atoms or gram moles of each. Although we wish to consider the reaction under isothermal and isobaric conditions, it is obvious that any one species involved still may, according to the circumstances, exist in any one of an infinite variety of states. For example, it may exist as a pure substance, or it may be in solution in any of a variety of solvents, including those composed of the other species involved in the reaction.

The general free-energy change may be written

$$\Delta F = q\bar{F}_\text{Q} + r\bar{F}_\text{R} + \cdots - l\bar{F}_\text{L} - m\bar{F}_\text{M} - \cdots$$

and in the special case that all products and reactants are in their standard states,

$$\Delta F^\circ = qF_\text{Q}^\circ + rF_\text{R}^\circ + \cdots - lF_\text{L}^\circ - mF_\text{M}^\circ - \cdots$$

Subtracting these two equations,

$$\Delta F^\circ - \Delta F = q(F_\text{Q}^\circ - \bar{F}_\text{Q}) + r(F_\text{R}^\circ - \bar{F}_\text{R}) + \cdots - l(F_\text{L}^\circ - \bar{F}_\text{L}) - m(F_\text{M}^\circ - \bar{F}_\text{M}) - \cdots$$

Substituting from Eq. (9-11*c*),

$$\Delta F^\circ - \Delta F = -qRT \ln a_Q - rRT \ln a_R - \cdots + lRT \ln a_L + mRT \ln a_M + \cdots$$

In view of the mathematical identity $x \ln y = \ln y^x$, the foregoing equation becomes

$$\Delta F^\circ - \Delta F = -RT \ln \frac{a_Q^q a_R^r \cdots}{a_L^l a_M^m \cdots} \qquad (9\text{-}15)$$

Let us now consider the equilibrium state in which all reactants and products are in equilibrium with each other; in this case the activity product is defined as K, the *thermodynamic equilibrium constant.*

$$K \equiv \left(\frac{a_Q^q a_R^r \cdots}{a_L^l a_M^m \cdots}\right)^{eq} \qquad (9\text{-}16)$$

The superscript *eq* is used here to accentuate the fact that K is defined in terms of the activities at equilibrium; this designation will not generally be used. Introducing this identity in Eq. (9-15) along with the fact that at equilibrium $\Delta F = 0$,

$$\Delta F^\circ = -RT \ln K \qquad (9\text{-}17)$$

This is perhaps, the most frequently used thermodynamic relation in the consideration of chemical equilibria. It follows immediately from this equation that, since ΔF° is a function of temperature only, K is similarly a function of temperature only.

Numerical Calculations Involving the Equilibrium Constant. We shall now consider the isothermal application of the thermodynamic equilibrium constant to equilibria which do not involve liquid or solid solutions. As an example let us consider the equilibrium between iron, magnetite, CO, and CO_2 at a temperature below 560°C, where magnetite is the lowest stable oxide; this may be represented by the equation

$$3Fe(\alpha) + 4CO_2(g) = Fe_3O_4(\text{mag.}) + 4CO(g)$$

Let us suppose that a numerical value of ΔF° is known or can be computed by one of the methods to be discussed subsequently, also that the total pressure is 1 atm, and let it be required to find the partial pressure of CO and CO_2.

The numerical value of K may be found readily from Eq. (9-17), which may be written in more convenient form in terms of common logarithms thus:

$$\log K = -\frac{\Delta F^\circ}{2.303RT} \qquad (9\text{-}18)$$

$\Delta F°$ is conventionally expressed in calories per gram atomic weight of all atoms taking part in the reaction as written. The value of R in corresponding units is 1.987, and Eq. (9-18) becomes

$$\log K = -\frac{\Delta F°}{4.575T} \tag{9-19}$$

The numerical coefficient 4.575 is one of the most frequently used constants in thermodynamic computations and should be memorized by the reader intending to do any work in this field.

By virtue of Eq. (9-16) the thermodynamic equilibrium constant for this reaction takes the form

$$K = \frac{a_{Fe_3O_4}a_{CO}^4}{a_{Fe}^3 a_{CO_2}^4}$$

At all times in the application of the equilibrium constant to a specific case it is necessary to have knowledge or make some assumption as to the nature and extent of the solutions involved. In this case, as usual, we shall assume that the gaseous solution (CO and CO_2) behaves ideally; hence the activity of CO and of CO_2 may be replaced by the partial pressure. Also, at this temperature the solubility of carbon and oxygen in metallic iron and in magnetite may be ignored, and the composition of magnetite does not depart measurably from that implied by the formula Fe_3O_4. Hence the activity of iron and of magnetite may be taken as unity, since by virtue of the restricted solubility the composition of these phases remains substantially constant under all conditions where they are stable in the temperature and pressure range under consideration. Thus we have

$$K = \frac{p_{CO}^4}{p_{CO_2}^4}$$

Or, more conveniently,

$$K^{\frac{1}{4}} = \frac{p_{CO}}{p_{CO_2}}$$

It will be noted that so far we have but one equation in p_{CO} and p_{CO_2}, and in order to solve uniquely for these two unknowns, another relation is required. In such problems it is quite general, although not universal, to find that the numerical determination of the unknown quantities requires an equation (or equations) in addition to the expressions for the equilibrium constant. In this case the total pressure was specified as 1 atm, so that

$$p_{CO} + p_{CO_2} = 1$$

From these two relations, then, p_{CO} and p_{CO_2} can be determined as follows:

$$p_{CO_2} = \frac{1}{K^{\frac{1}{2}} + 1}$$

$$p_{CO} = \frac{K^{\frac{1}{2}}}{K^{\frac{1}{2}} + 1}$$

General Solution of Equilibrium Problems. Let us now consider a general method for the numerical solution of an equilibrium problem when the equilibrium constants are known at the temperature involved. In general there may be several independent equilibria, each corresponding to a chemical equation and each associated with an equilibrium constant. The equilibrium constants are related to the free energies by Eq. (9-17) and are evaluated as described above. If we are considering a closed system, corresponding to a batch process during which no material is added or subtracted, then for every component or element an expression for the conservation thereof may be written. Also for every independent equilibrium, an equation for the corresponding equilibrium constant may be written. For a system at complete equilibrium the number of such independent equilibria is equal to the number of species appearing in the chemical equations minus the number of components.[1] In addition an equation may be formulated representing the one possible arbitrary restriction on the system in addition to those of constancy of temperature and of masses;[2] this restriction is most commonly that of a fixed pressure or a fixed volume.

The equations which may be written may thus be classified in two types: (1) those expressing the state of equilibrium and (2) those expressing the arbitrary restrictions, including those of conservation of masses. If this system of independent simultaneous equations is to have a unique solution, the number of unknowns must be equal to the number of equations. The unknowns, in general, are the masses of the various components or species (the latter term applying to molecular substances) in each phase; the composition of each phase and the partial pressure of each gaseous constituent may be determined therefrom.

As an example of the general solution of the equilibrium problem let us consider a gaseous system composed of carbon, hydrogen, and oxygen.

[1] For metallurgical reactions, the number of components is generally equal to the number of chemical elements involved. Further discussion of the significance of the term *components* is deferred to Chap. 11 on heterogeneous equilibria.

[2] The state of a system at equilibrium is physically determined by fixing the masses of the several components and two other variables such as temperature and pressure, as discussed in Chap. 6. The equations expressing conservation of elements or components serve to fix the masses of the components; thus the assignment of a temperature leaves only one other possible arbitrary restriction.

In the gas mixtures generally used in the heat-treatment of steel the predominant molecular species at equilibrium (other than N_2 which is not considered here) are CO, CO_2, H_2, H_2O, and CH_4. The expressions for the conservation of masses are three in number, one for each component element. Using the symbol n with the appropriate subscript to denote the number of gram atoms or moles, these expressions are

$$n_C = n_{CO} + n_{CO_2} + n_{CH_4} \qquad (a)$$
$$n_H = 2n_{H_2} + 2n_{H_2O} + 4n_{CH_4} \qquad (b)$$
$$n_O = n_{CO} + 2n_{CO_2} + n_{H_2O} \qquad (c)$$

Since there are five species and three components in the system, there are two independent equilibria. The following two may be selected.[1]

$$CO_2 + H_2 = CO + H_2O \qquad (d)$$
$$CH_4 + H_2O = CO + 3H_2 \qquad (e)$$

The corresponding equilibrium constants (whose values at a given temperature are obtainable from tables such as those mentioned in Chap. 1 are

$$\frac{p_{CO}p_{H_2O}}{p_{CO_2}p_{H_2}} = K_1 \qquad (f)$$
$$\frac{p_{CO}p^3_{H_2}}{p_{CH_4}p_{H_2O}} = K_2 \qquad (g)$$

Of the five equations (*a*), (*b*), (*c*), (*f*), and (*g*), the variables in the first three are moles and those in the last two are partial pressures. For solution it is necessary to express all five in terms of either moles or partial pressures. Let us choose to use moles as the variables; therefore we convert Eqs. (*f*) and (*g*) to this basis by means of the relation

$$p_i = N_iP = \frac{n_i}{n_T}P$$

[1] There are numerous other equilibria which could be selected, such as the following:

$$CH_4 + CO_2 = 2CO + 2H_2$$
$$CH_4 + 2H_2O = CO_2 + 4H_2$$
$$CH_4 + 2CO_2 = 3CO + H_2 + H_2O$$
$$CH_4 + 3CO_2 = 4CO + 2H_2O$$

It will be noted, however, that all of these are obtainable from the two given in the text and hence are not independent, *i.e.*, contribute no new information. The best choice of the two independent equilibria is that one which leads to the easiest mathematical solution. Usually this means that simple equations involving fewer molecular species are better; furthermore, molecular species present in minor amount should appear no more than necessary in the equations, preferably only once. A favorable choice of equilibria may be made from experience or may be found by trial and error

The subscript i, as usual, refers to any particular species, and n_T is the total number of moles. Thus,

$$n_T = n_{CO} + n_{CO_2} + n_{H_2} + n_{H_2O} + n_{CH_4} \qquad (h)$$

Equations (f) and (g) then become

$$\frac{n_{CO}n_{H_2O}}{n_{CO_2}n_{H_2}} = K_1 \qquad (i)$$

and

$$\frac{n_{CO}n_{H_2}^3}{n_{CH_4}n_{H_2O}}\frac{P^2}{n_T^2} = K_2 \qquad (j)$$

Considering the number of gram atoms of carbon, oxygen, and hydrogen (n_C, n_O, n_H) and the total pressure (P) as known, it will now be noted that the six independent equations (a), (b), (c), (h), (i), and (j) contain six unknowns n_{CO}, n_{CO_2}, n_{H_2}, n_{H_2O}, n_{CH_4}, and n_T and are therefore soluble by the usual methods for solving simultaneous equations. It is relatively easy to eliminate four unknowns by solving the linear equations (a), (b), (c), and (h) for four unknowns in terms of the other two and substituting these in Eqs. (f) and (g). This leaves two simultaneous equations in two unknowns which can most readily be solved by the method of successive approximations (iteration). This procedure is rather tedious but is aided considerably by keeping the chemistry of the situation in mind.

The specific problem just discussed arises in the heat-treatment of steel.[1] It is usually further complicated by the presence of nitrogen, which is not considered here. If a gas sample is taken from a heat-treating furnace and analyzed for the five molecular species considered, this analysis pertains to the cold gas and does not necessarily, or indeed usually, pertain to the gas at furnace temperature. Let us suppose that the gas sample as collected has the following analysis:

Gas	*Volume %*
CO	30
CO_2	5
H_2	55
H_2O	3
CH_4	7

and that the problem is to find the equilibrium composition at some particular furnace temperature and at 1 atm pressure.

Since the analysis is given in volume percentage which is proportional to the number of moles, it is convenient to consider a total mass cor-

[1] This subject is discussed in detail by R. W. Gurry, *Trans. AIME*, **188**, 671 (1950).

responding to 1 mole of the sample as analyzed; the number of moles of each constituent is then equal to the volume percentage divided by 100. The number of gram atoms of each element carbon, hydrogen, and oxygen is independent of the molecular constitution, *i.e.*, is conserved during cooling of the sample for analysis, although the molecular constitution of the mixture changes. Thus

$$n_C = n'_{CO} + n'_{CO_2} + n'_{CH_4} = n_{CO} + n_{CO_2} + n_{CH_4}$$
$$n_H = 2n'_{H_2} + 2n'_{H_2O} + 4n'_{CH_4} = 2n_{H_2} + 2n_{H_2O} + 4n_{CH_4}$$
$$n_O = n'_{CO} + 2n'_{CO_2} + n'_{H_2O} = n_{CO} + 2n_{CO_2} + n_{H_2O}$$

where the primed quantities refer to the sample as analyzed and the unprimed quantities to the equilibrium composition of the mixture at furnace temperature. Evaluating,

$$n_C = 0.30 + 0.05 + 0.07 = 0.42$$
$$n_H = 2(0.55) + 2(0.03) + 4(0.07) = 1.44$$
$$n_O = 0.30 + 2(0.05) + 0.03 = 0.43$$

The six simultaneous equations for this case are

$$n_{CO} + n_{CO_2} + n_{CH_4} = 0.42$$
$$2n_{H_2} + 2n_{H_2O} + 4n_{CH_4} = 1.44$$
$$n_{CO} + 2n_{CO_2} + n_{H_2O} = 0.43$$
$$n_{CO} + n_{CO_2} + n_{H_2} + n_{H_2O} + n_{CH_4} = n_T$$
$$\frac{n_{CO}n_{H_2O}}{n_{CO_2}n_{H_2}} = K_1 = 0.508 \qquad \text{(at 1200°F or 649°C)}$$
$$\frac{n_{CO}n_{H_2}^3}{n_{CH_4}n_{H_2O}n_T^2} = K_2 = 2.66 \qquad \text{(at 1200°F or 649°C)}$$

From these, numerical values for the six n's may be obtained. In order to determine whether or not a furnace gas is oxidizing or carburizing to steel of known composition, it is desirable to convert to partial pressures. This is accomplished by using the relation $p_i = (n_i/n_T)P$, or, since P is 1 atm in this case, $p_i = n_i/n_T$.

General Method When Many Solutions Are Required. As mentioned in the preceding section, the simultaneous solution of the six simultaneous equations is tedious and time consuming. If, *e.g.*, in furnace control work, many numerical solutions of the six general equations are required it is much more convenient to use a modified method which will now be described. This consists of finding many sets of numerical values of the equilibrium partial pressures which are consistent with the numerical values of the equilibrium constants and the total pressure and determining n_C, n_H, and n_O therefrom. In effect this method consists of working the problem in reverse of the solution just cited for a single case. The results so found may then be tabulated or recorded graphically, and the solution of any particular problem found by interpolation. Thus, for example, values for p_{H_2} and p_{H_2O} may be selected arbitrarily; p_{CO}, p_{CO_2}, and p_{CH_4} may then be found by means of Eqs. (*f*) and (*g*) and the additional relation: $p_{CO} + p_{CO_2} + p_{H_2} + p_{H_2O} + p_{CH_4} = P$. The solution of these linear simultaneous equations is direct and easy. It is convenient to express these partial pressures, for tabular or graphical purposes, as functions of the ratios $n_C/(n_C + n_H + n_O)$, $n_H/(n_C + n_H + n_O)$, and

TABLE 9-2. VALUES OF SEVERAL EQUILIBRIUM CONSTANTS

°F	°C	$\frac{p_{CO}p_{H_2O}}{p_{CO_2}p_{H_2}}$	$\frac{a_C p_{CO_2}}{p^2_{CO}}$	$\frac{a_C p^2_{H_2}}{p_{CH_4}}$	Equilibrium with iron and its oxide* $\frac{p_{CO}}{p_{CO_2}}$	$\frac{p_{H_2}}{p_{H_2O}}$
800	427	0.111	3750	0.111	0.826*	7.46*
900	482	0.178	431	0.326	0.909*	5.10*
1000	538	0.267	64.5	0.847	0.990*	3.70*
1100	593	0.376	12.8	1.94	1.13	3.00
1200	649	0.508	3.00	4.08	1.33	2.62
1300	704	0.662	0.847	7.94	1.54	2.33
1400	760	0.833	0.273	14.3	1.74	2.09
1500	816	1.02	0.0990	24.6	1.93	1.89
1600	871	1.22	0.0402	39.8	2.12	1.74
1700	927	1.44	0.0175	62.1	2.30	1.60
1800	982	1.66	0.00833	93.5	2.47	1.49
1900	1038	1.89	0.00420	135.	2.63	1.39
2000	1093	2.12	0.00227	190.	2.80	1.32
2100	1149	2.36	0.00128	261.	2.96	1.25
2200	1204	2.61	0.000758	350.	3.11	1.19

Graphite is chosen as the standard state for carbon.

* Below 560°C the stable oxide in equilibrium with iron is magnetite (Fe_3O_4), data marked with an asterisk. At higher temperature it is wüstite (FeO_x where x has a value slightly greater than unity, depending upon temperature).

$n_O/(n_C + n_H + n_O)$, which may be determined immediately from the partial pressures. A graph whose coordinates are any two of these three ratios serves as a convenient means of expressing gas composition.

It will be noted that this method involves only the solution of linear equations; successive approximations and the solution of second- and higher order equations are completely avoided. It is most useful when many numerical solutions involving the same equilibria at the same temperature are required. However, it is so much easier than the previously described method that it sometimes involves less labor when only a few or even one solution is required. Although exemplified here in only one type of problem, this general method is applicable to a wide variety of problems. It is of particular advantage in the numerical solution of problems involving many or complex equilibria leading to nonlinear equations such as slag equilibria and metal equilibria as well as gas equilibria.

Of particular interest in the heat-treatment of steel is the gas composition in equilibrium with iron and iron oxide,[1] from a knowledge of which we can predict whether or not a given furnace atmosphere will be oxidizing. It has already been shown that the equilibrium constant for the

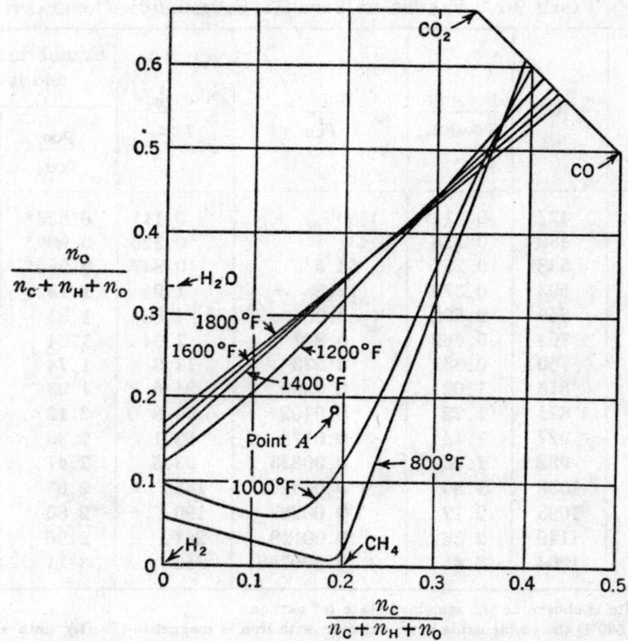

FIG. 9-1. Composition of nitrogen-free gas mixtures in equilibrium with iron and its oxide at several temperatures.

oxidation of iron to magnetite may be expressed as the ratio p_{CO}/p_{CO_2}. It is readily seen that the equilibrium constant for the iron wüstite equilibrium (for the range of temperature where wüstite is stable) also may be represented p_{CO}/p_{CO_2}. Similarly there is a definite ratio p_{H_2}/p_{H_2O} corresponding to the equilibrium of H_2 and H_2O with iron and its oxide.

[1] Below 560°C the lowest stable oxide of iron is magnetite, whereas above this temperature and up to the melting point of the oxide, 1371°C, it is wüstite. The wüstite in equilibrium with iron has a composition which is slightly temperature dependent; it contains appreciably more oxygen (or, more properly, less iron) than corresponds to the formula FeO.

Values of both ratios have been experimentally determined and are given for several temperatures along with some other gas constants in Table 9-2. Obviously the two ratios are related at any temperature through the water gas constant K_1 [Eq. (f)]. Actually, the experimental determination of these two ratios is one means of determining the water gas constant.

Returning now to the subject of the preceding paragraphs, it will be seen that, if the initial arbitrary choice of p_{H_2} and p_{H_2O} is made so that the

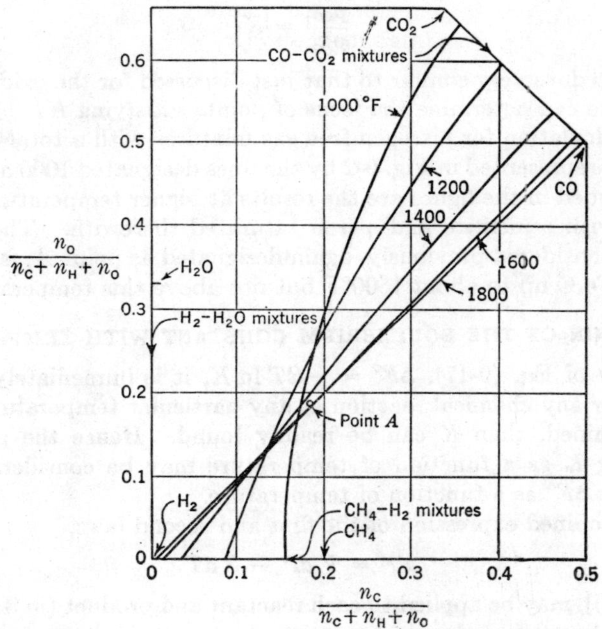

FIG. 9-2. Composition of nitrogen-free gas mixtures in equilibrium with cementite and ferrite (1000 and 1200°F), and with cementite and austenite saturated therewith (1400 to 1800°F).

ratio p_{H_2}/p_{H_2O} corresponds to equilibrium with iron and its oxide, then a series of such points forms the dividing line between atmospheres which are oxidizing and those which are reducing to plain carbon steel. The results of this calculation for nitrogen-free gas mixtures are shown graphically for several temperatures in Fig. 9-1.

As an example of the use of this figure, let us now consider whether the gas mixture whose analysis was given in the preceding section is oxidizing or reducing to steel at heat-treating temperatures. In this case n_C, n_H, and n_O have already been found to be 0.42, 1.44, and 0.43; hence

$n_C + n_H + n_O$ is 2.29, $n_C/(n_C + n_H + n_O)$ is 0.183, and $n_O/(n_C + n_H + n_O)$ is 0.188. This composition, designated as point A on the graph, is obviously reducing to iron oxide at 1200°F and higher but oxidizing to iron at 1000°F and lower.

In steel treating it is also of importance to know whether the furnace gas is carburizing or decarburizing. Below the eutectoid this is determined by whether the reaction $3Fe(\alpha) + 2CO(g) = Fe_3C(\text{cem.}) + CO_2(g)$ will proceed in a forward or reverse direction. The equilibrium constant for this reaction (taking the activity of iron and of cementite as unity) is

$$\frac{p_{CO_2}}{p_{CO}^2} = K_3$$

By a procedure very similar to that just discussed for the oxidation equilibrium we can determine the locus of points satisfying K_3. The results of this calculation for nitrogen-free gas mixtures with a total pressure of 1 atm are represented in Fig. 9-2 by the lines designated 1000 and 1200°F. Also included in the figure are the results at higher temperature for equilibrium with cementite and γ-iron saturated therewith. The gas composition considered previously, again designated as point A, is capable of forming Fe_3C up to about 1500°F but not above this temperature.

VARIATION OF THE EQUILIBRIUM CONSTANT WITH TEMPERATURE

In view of Eq. (9-17), $\Delta F° = -RT \ln K$, it is immediately apparent that, if for any chemical reaction at any particular temperature $\Delta F°$ can be determined, then K can be readily found. Hence the problem of expressing K as a function of temperature may be considered that of expressing $\Delta F°$ as a function of temperature.

The combined expression of the first and second laws,

$$dF = V\,dP - S\,dT$$

[Eq. (7-29)], may be applied to each reactant and product (in its standard state) of the general chemical reaction

$$l\text{L} + m\text{M} + \cdots = q\text{Q} + r\text{R} + \cdots$$

[Eq. (9-14)].

$$\begin{aligned} dF_Q° &= V_Q°\,dP - S_Q°\,dT \\ dF_R° &= V_R°\,dP - S_R°\,dT \\ &\cdots\cdots\cdots\cdots \\ dF_L° &= V_L°\,dP - S_L°\,dT \\ dF_M° &= V_M°\,dP - S_M°\,dT \\ &\cdots\cdots\cdots\cdots \end{aligned}$$

Multiplying each of these relations by the corresponding coefficient (q, r, etc.) in the chemical equation and subtracting the sum of the equations for the reactants from that for the products in the usual manner,

$$
\begin{aligned}
d(qF^\circ_Q + rF^\circ_R + \cdots - lF^\circ_L - mF^\circ_M - \cdots) \\
= (qV^\circ_Q + rV^\circ_R + \cdots - lV^\circ_L - mV^\circ_M - \cdots)\, dP \\
- (qS^\circ_Q + rS^\circ_R + \cdots - lS^\circ_L - mS^\circ_M - \cdots)\, dT
\end{aligned}
$$

The expressions in parentheses correspond to the free-energy change, volume change, and entropy change occurring when the reactants in their standard states are transformed to the products in their standard states according to the chemical reaction as written. These quantities are designated ΔF°, ΔV°, and ΔS°, respectively, so that

$$d\,\Delta F^\circ = \Delta V^\circ\, dP - \Delta S^\circ\, dT \qquad (9\text{-}20)$$

We wish to consider at present only a variation with temperature at constant pressure (in fact the standard state is that at 1 atm pressure); hence, setting dP equal to zero,

$$d\,\Delta F^\circ = -\Delta S^\circ\, dT \qquad (9\text{-}21)$$

From the definitional equation for F, $F = H - TS$, it follows by the same reasoning as above that

$$\Delta F^\circ = \Delta H^\circ - T\,\Delta S^\circ$$

or

$$\Delta S^\circ = \frac{\Delta H^\circ - \Delta F^\circ}{T}$$

Substituting from this equation for ΔS° in Eq. (9-21),

$$d\,\Delta F^\circ = -\frac{\Delta H^\circ - \Delta F^\circ}{T}\, dT$$

This and the following equation are forms of the *Gibbs-Helmholz equation*, and are valid at constant pressure without restriction to standard states.

$$T d\,\Delta F^\circ - \Delta F^\circ\, dT = -\Delta H^\circ\, dT$$

The integrating factor for this equation is $1/T^2$;* *i.e.*, if the equation is divided by T^2, the left side becomes equal to $d(\Delta F^\circ/T)$. Hence,

$$d\,\frac{\Delta F^\circ}{T} = -\frac{\Delta H^\circ\, dT}{T^2}$$

and

$$\frac{d(\Delta F^\circ/T)}{dT} = -\frac{\Delta H^\circ}{T^2}$$

or $\qquad (9\text{-}22)$

$$\frac{d(\Delta F^\circ/T)}{d(1/T)} = \Delta H^\circ$$

* As in the derivation of Eq. (7-37).

Also, by virtue of Eq. (9-17),

$$\frac{d \ln K}{dT} = \frac{\Delta H^\circ}{RT^2}$$

or (9-22a)

$$\frac{d \ln K}{d(1/T)} = -\frac{\Delta H^\circ}{R}$$

The former is commonly known as the *van't Hoff equation.* It will be seen that, if ΔH° can be expressed as a function of temperature, this equation can be integrated. From the first law we found [Eq. (6-20)] that $(\partial \Delta H/\partial T)_P = \Delta C_P$ or, at constant pressure, $d\,\Delta H = \Delta C_P\,dT$, which for the standard states becomes $d\,\Delta H^\circ = \Delta C_P^\circ\,dT$. The superscript zero is usually omitted from C_P, as data usually are available for substances only in their standard states. If the heat capacity of each product and reactant can be represented by a simple analytical expression in terms of temperature, then ΔC_P also can be so represented.

The Principle of Le Chatelier. Equation (9-22*a*) is a quantitative expression of van't Hoff's law of mobile equilibrium which states: When the temperature of a system in equilibrium is raised, a reaction takes place within the system which is accompanied by the absorption of heat; and conversely, when the temperature is lowered, a reaction occurs which is accompanied by the evolution of heat.

Le Chatelier (1884) showed that this law is a special case of a more general law, known as the theorem or *principle of Le Chatelier*, which may be stated as follows. *If a system in equilibrium is subjected to a constraint (addition of heat, increase of volume, etc.) which alters the equilibrium, the direction of the reaction taking place is such as to oppose the constraint, i.e., partially to nullify its effect.* For example, if the system under consideration consists of a liquid and its vapor, the addition of heat at constant volume will be accompanied by an endothermic reaction—vaporization. If, on the other hand, the process is adiabatic, an increase in the volume will give rise to vaporization and hence to a lowering of the temperature and a decrease in the vapor pressure. Le Chatelier's principle is of very broad applicability.

Empirical Representation of Heat Capacity. As noted in Chap. 6, the heat capacity of a solid cannot validly be represented by a simple analytical expression at temperatures where C_v is much below the Dulong and Petit value. At higher temperatures the departure from linearity is frequently not great and the heat capacity can adequately be represented by an analytical expression. For example, the heat-capacity curve for platinum (Fig. 6-8) departs only slightly from linearity in the range from room temperature up to the melting point. In such a region the heat capacity is often represented by a power series, thus

$$C_P = a + bT + cT^2 + \cdots \tag{9-23}$$

However, if the coefficients a, b, c are evaluated from measurements at relatively low temperature, the use of the equation at a temperature even slightly higher may result in serious error on account of the increasing predominance at higher temperature of the terms involving the higher powers of T. Although the use of such an empirical expression beyond the temperature range upon which it is based is dangerous, it is necessary on many occasions because of the unfortunate lack of experimental data. In order to minimize this error Kelley[1] uses an empirical equation of the form

$$C_P = a + bT - \frac{c}{T^2} \tag{9-24}$$

It will be noted that in this equation the final term approaches zero with rising temperature, so that the expression represents a curve which is concave downward at lower temperature and which approaches linearity at higher temperature. Again referring to Fig. 6-8, it is apparent that this function should be fairly satisfactory from somewhat below 0 to over 1000°C. The slight upward concavity, which is barely noticeable in the figure (but fairly general), is not reproduced by this function; this deficiency is not usually serious. Neither of these, nor any common type of equation, is adequate for α-iron or for any substance undergoing magnetic transformation or other so-called phase change of the second kind in the range of temperature under consideration. Equation (9-24) (sometimes without the final term) is commonly used for most gases in the region of metallurgical interest. On account of the lack of experimental data on the heat capacity of liquid metals, slags, and other liquids such as sulfides, halides, etc., the linear equation $C_P = a + bT$ is commonly adequate for these. Frequently, on account of the scarcity of data, the heat capacity, or more properly ΔC_P, must be taken as constant. In many cases the error thus involved is small compared to other errors in the numerical coefficients of the free-energy equation.

Integration of the Free-energy Equation. Adopting, then, Eq. (9-24) as most suitable for the temperature range of metallurgical interest, ΔC_P may be represented

$$\Delta C_P = \Delta a + \Delta bT - \Delta cT^{-2} \tag{9-25}$$

where $a = qa_Q + ra_R + \cdots - la_L - ma_M - \cdots$, etc., where q, r,

[1] K. K. Kelley, *U.S. Bur. Mines Bull.* 371, 1934; 406 and 407, 1937; 476, 1949.

. . . , l, m, . . . have the same significance as in Eq. (9-14). From Eq. (6-20),

$$d\,\Delta H^\circ = \Delta C_P\,dT = \Delta a\,dT + \Delta bT\,dT - \Delta cT^{-2}\,dT$$

Integrating,

$$\Delta H^\circ = \Delta H_0 + \Delta aT + \frac{\Delta b}{2}T^2 + \frac{\Delta c}{T} \tag{9-26}$$

In this expression ΔH_0 is merely the integration constant—no other significance is to be attached to it. It is mere convention to denote this constant ΔH_0 instead of I, a symbol which is reserved for the integration constant in the free-energy equation [Eq. (9-27)].

Having thus evaluated ΔH° as a function of temperature at constant pressure, the free-energy equation [Eq. (9-22)] may now be integrated. Substituting therein the expression for ΔH° from Eq. (9-26), and noting that $d(1/T) = -dT/T^2$,

$$d\left(\frac{\Delta F^\circ}{T}\right) = \Delta H_0\,d\left(\frac{1}{T}\right) - \Delta a\frac{dT}{T} - \frac{\Delta b}{2}\,dT - \Delta c\frac{dT}{T^3}$$

Integrating,

$$\frac{\Delta F^\circ}{T} = I + \frac{\Delta H_0}{T} - \Delta a \ln T - \frac{\Delta b}{2}T + \frac{\Delta c}{2T^2} \tag{9-27}$$

This can also be written

$$\Delta F^\circ = IT + \Delta H_0 - \Delta aT \ln T - \frac{\Delta b}{2}T^2 + \frac{\Delta c}{2T} \tag{9-28}$$

The Sigma Function. Substituting $-RT \ln K$ for ΔF° [Eq. (9-17)] in Eq. (9-27) and transposing terms,

$$-R \ln K + \Delta a \ln T + \frac{\Delta b}{2}T - \frac{\Delta c}{2T^2} = I + \frac{\Delta H_0}{T} \tag{9-29}$$

The left side of this equation is commonly designated Σ and is called the sigma function. Thus,

$$\Sigma \equiv -R \ln K + \Delta a \ln T + \frac{\Delta b}{2}T - \frac{\Delta c}{2T^2}$$

These terms are grouped together and designated by a special symbol because they are usually determined by experiment and hence may be regarded as known for the evaluation of I and ΔH_0. Equation (9-29) may be rewritten

$$\Sigma = \frac{\Delta H_0}{T} + I \tag{9-30}$$

If known values of Σ are plotted against $1/T$, a straight line should be obtained whose slope is ΔH_0 and whose intercept at $1/T = 0$ is I. However, the extrapolation to this intercept is commonly too great to give sufficient precision in the determination of I. It is more common to evaluate the slope graphically and to insert this back into Eq. (9-30), thus obtaining a value of I for each experimental value of K. The average of these values of I is then used to represent the constant in Eq. (9-30).

The student should realize that, although the terms have been represented in natural logarithms, it is not common to use a table of natural logarithms. Rather it is customary to rewrite these equations in terms of common logarithms; inserting the numerical value of R in calories, Eq. (9-29) becomes

$$-4.575 \log K + 2.3026\, \Delta a \log T + \frac{\Delta b}{2} T - \frac{\Delta c}{2T^2} = I + \frac{\Delta H_0}{T} \qquad (9\text{-}29a)$$

the left side being Σ.

Attention should be called to the fact that the second term in Eq. (9-29) or (9-29a) must be calculated to as many figures beyond the decimal point as is the first term, even though the precision of Δa does not warrant this number of significant figures. It frequently happens that, although this term varies but little over the temperature range under consideration, its absolute value is large. If the equation is used directly in the form given, the slide rule is usually sufficiently precise for the evaluation of all terms except that containing Δa. This term also may be evaluated by slide rule if some convenient temperature in or near the range under consideration is selected and Δa times the natural logarithm of this temperature is subtracted from both sides of the equation. Thus, for example, if 1000°K is selected as the reference temperature, Eq. (9-29a) becomes

$$-4.575 \log K + 2.3026\, \Delta a \log \frac{T}{1000} + \frac{\Delta b}{2} T - \frac{\Delta c}{2T^2} = (I - 2.3026\, \Delta a \log 1000) + \frac{\Delta H_0}{T} \qquad (9\text{-}29b)$$

The left side of this equation is no longer equal to Σ but may be designated Σ', which in this case equals $\Sigma - 2.3026\, \Delta a \log 1000$. The terms in parentheses may be designated I' ($I' = I - 2.3026\, \Delta a \log 1000$), and we thus obtain a relation similar in form and usage to Eq. (9-30).

$$\Sigma' = \frac{\Delta H_0}{T} + I' \qquad (9\text{-}30a)$$

Obviously any other convenient temperature could be used instead of 1000°K.

If no information is available as to the heat capacities, nor any basis for a reasonable guess, the heat capacity of the products is sometimes assumed to be equal to that of the reactants; *i.e.*, ΔC_P is assumed to be equal to zero. Equation (9-29) then reduces to the relation

$$-R \ln K = \frac{\Delta H_0}{T} + I \tag{9-31}$$

It follows from Eq. (9-22) that, if $\Delta F^\circ/T$ (or its equivalent, $-R \ln K$) is plotted against $1/T$, the slope of the resulting curve at any point is ΔH°, the enthalpy of reaction at that temperature. ΔH° is to be carefully distinguished from ΔH_0. Only in case the plot of $\ln K$ against $1/T$ is strictly linear is ΔH° constant and thus equal to ΔH_0. Very seldom is the precision of the data sufficient to testify to such linearity, and the crude assumption of zero ΔC_P should be made only when absolutely necessary. On the other hand, the plot of Σ against $1/T$ should always give a straight line; any departure from linearity in this plot indicates experimental error in the evaluation of K or the C_P's or the inadequacy of one or more of the equations used to represent C_P as a function of temperature.

The Free Energy of Vaporization, Fusion, and Sublimation of Zinc.[1] The method just developed will now be applied to the vaporization and sublimation of zinc. Gaseous zinc is monatomic in the temperature range under consideration and its heat capacity corresponds thereto, being constant at 5.0 cal/(gram atom)(deg) like that of the rare gases. Experimental data on the heat capacity of liquid zinc may be represented adequately by the equation

$$C_{P\ \mathrm{Zn(l)}} = 7.09 + 1.15 \times 10^{-3}T \qquad 419.5 \text{ to } 850^\circ\mathrm{C}$$

Thus, for the reaction

$$\mathrm{Zn(l)} = \mathrm{Zn(g)}$$

$\Delta C_P = -2.09 - 1.15 \times 10^{-3}T$; and since $\ln K = \ln p$, where p is the equilibrium vapor pressure of zinc,

$$\Sigma = -1.9885 \ln p - 2.09 \ln T - 0.575 \times 10^{-3}T$$

Numerous experimental data are available on the vapor pressure of liquid zinc. The values of Σ computed from these are plotted against $1/T$ in Fig. 9-3. The resulting straight line, designated L (for liquid), has a slope equivalent to a ΔH_0 of +31,087 cal. The calculation of the

[1] This example is taken from C. G. Maier's classic monograph, Zinc Smelting from a Chemical and Thermodynamic Viewpoint, *U.S. Bur. Mines Bull.* 324 (1930). This is one of the first thorough thermodynamic treatments of the chemical processes of a metallurgical industry.

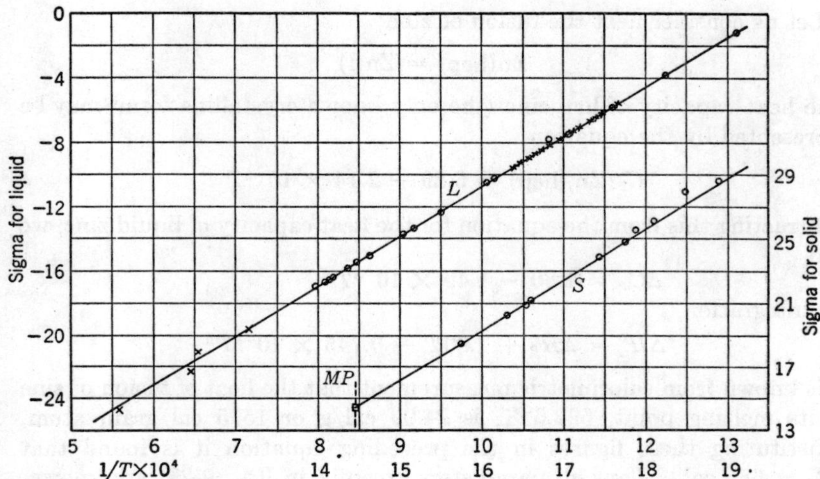

FIG. 9-3. Sigma function plot for vaporization and sublimation of zinc metal (lines L and S, respectively). [*From C. G. Maier, U.S. Bur. Mines Bull.* 324 (1930).]

constant I as the difference between Σ and $\Delta H_0/T$ [see Eq. (9-30)] is summarized as follows:

Investigator	No. of determinations	I
Braune	5	-41.876 ± 0.044
Braune	7	-41.813 ± 0.007
Greenwood	4	-42.001 ± 0.184
Jenkins	9	-41.832 ± 0.030
Rodebush and Dixon	30	-41.863 ± 0.013
Weighted mean		-41.837 ± 0.018

In computing the mean, the weight factor for each investigator was taken as the reciprocal of the average deviation of his I's from the mean value of his I's. By substitution of these values in Eq. (9-28), the equation for the free energy of vaporization of zinc becomes

$$\text{Zn(l)} = \text{Zn(g)} \qquad \Delta F^\circ = -RT \ln p = 31{,}087 + 2.09T \ln T + 0.575 \times 10^{-3}T^2 - 41.84T \quad (9\text{-}32)$$

or $\log p = 9.1373 - (6789.5/T) - 1.051 \log T - 1.255 \times 10^{-4}T$, p being expressed in atmospheres as usual.

Let us consider next the fusion of zinc:

$$\text{Zn(hcp)} = \text{Zn(l)}$$

The heat capacity of hcp zinc (the only known crystalline form) may be represented by the equation

$$C_P(\text{Zn, hcp}) = 5.29 + 2.14 \times 10^{-3}T$$

Subtracting this from the equation for the heat capacity of liquid zinc, we find

$$\Delta C_P = 1.80 - 1.49 \times 10^{-3}T$$

By integration,

$$\Delta H^\circ = \Delta H_0 + 1.80T - 0.745 \times 10^{-3}T^2$$

It is known from calorimetric measurement that the heat of fusion of zinc at its melting point, 692.6°K, is 24.09 cal/g or 1575 cal/gram atom. Substituting these figures in the preceding equation it is found that $\Delta H_0 = 685$ cal. Thus all parameters except I in Eq. (9-28) are known, and this equation becomes

$$\Delta F^\circ = 685 - 1.80T \ln T + 0.745 \times 10^{-3}T^2 + IT$$

The constant I may be evaluated from the fact that $\Delta F^\circ_{692.6} = 0$. This substitution gives I as 10.268, and hence, for the fusion of zinc,

$$\text{Zn(hcp)} = \text{Zn(l)} \qquad \Delta F^\circ = 685 - 1.80T \ln T + 0.745 \times 10^{-3}T^2 + 10.27T \qquad (9\text{-}33)$$

Equations (9-32) and (9-33) may now be added to obtain the free energy of sublimation of solid zinc.

$$\text{Zn(l)} = \text{Zn(g)} \qquad \Delta F^\circ = 31{,}087 + 2.09T \ln T + 0.575 \times 10^{-3}T^2 - 41.84T \qquad (9\text{-}32)$$

$$\text{Zn(hcp)} = \text{Zn(l)} \qquad \Delta F^\circ = 685 - 1.80T \ln T + 0.745 \times 10^{-3}T^2 + 10.27T \qquad (9\text{-}33)$$

$$\text{Zn(hcp)} = \text{Zn(g)} \qquad \Delta F^\circ = 31{,}772 + 0.29T \ln T + 1.32 \times 10^{-3}T^2 - 31.57T \qquad (9\text{-}34)$$

The vapor pressure of solid zinc may now be computed by inserting any arbitrary temperature in Eq. (9-34) to give ΔF° and substituting this in the relation $\Delta F^\circ = -RT \ln p$. One might reasonably inquire why the vapor pressure of solid zinc was not obtained from experimental data in the same way as was that of the liquid. The reason for this lies in the fact that it is highly desirable that Eqs. (9-32) to (9-34) be consistent, *i.e.*, that Eq. (9-34) be equal to the sum of Eqs. (9-32) and (9-33). In order to guarantee this, free-energy equations are computed for the two reactions for which the data are most reliable, and the free-energy equation of the third reaction is obtained by addition or subtraction. In this case the experimental determination of the vapor pressure of solid zinc is less precise than that of the vapor pressure of liquid zinc and the melting point.

It is, of course, important to compare the vapor pressure of solid zinc with direct experimental values, as any serious discrepancy would cast doubt upon the validity of the whole set of equations [Eqs. (9-32) to (9-34)], *i.e.*, upon the accuracy of the experimental data included therein. Maier chose to make the comparison in the following way. He computed a "theoretical" Σ for Eq. (9-34) from the relation $\Sigma = (31{,}772/T) - 31.57$ [from Eq. (9-30)]. This is represented by the straight line marked S (for solid) in Fig. 9-3. The value of Σ corresponding to the experimental vapor pressure was obtained in the usual way and is represented by the points. The computed line, as drawn, adequately expresses the experimental results at higher temperature, and the discrepancy at lower temperature is not more than might be expected in view of the experimental difficulties here. In making the visual comparison of the experimental points with the computed line it is well to bear in mind that a small error Δ in Σ corresponds to an error $\Delta/R \cong \Delta/2$ in $\ln p$, or 50Δ per cent error in p. Thus the greatest departure, about 0.4 unit in Σ (Fig. 9-3), corresponds to about 20 per cent error in the measured vapor pressure.

TABULAR METHOD OF RECORDING THERMODYNAMIC DATA

On account of the frequent inadequacy of the empirical representation of ΔC_P, as discussed in the beginning of the preceding section, it is becoming common practice to express thermodynamic data tabularly rather than analytically. Even for gases it is found, as more precise data become available, that the analytical representation of the heat capacity [as Eq. (9-25)] is not sufficiently precise. Since the equations for enthalpy and free energy as functions of temperature depend upon the heat capacity, an erroneous expression for the heat capacity is reflected in each of these.

At first thought it might seem that the situation would be improved by giving the heat capacity in tabular form rather than by an empirical equation. However, if this were done, the evaluation of $\Delta F°$ and K for a given reaction would require two graphical integrations.

In order to avoid this awkward situation, a different procedure has been developed. Let us add Eq. (9-22)

$$d\left(\frac{\Delta F^\circ}{T}\right) = \Delta H^\circ\, d\left(\frac{1}{T}\right) \tag{9-22}$$

pertaining to any given reaction, to the identity

$$-d\left(\frac{\Delta H_0^\circ}{T}\right) = -\Delta H_0^\circ\, d\left(\frac{1}{T}\right)$$

ΔH_0° is the enthalpy of the reaction at the absolute zero of temperature, the superscript zero denoting standard states for reactants and products and the subscript zero denoting 0°K. ΔH_0° is to be carefully distinguished from ΔH°, the enthalpy of the reaction at any arbitrary temperature, and ΔH_0, which was introduced in the preceding section as an integration constant. The sum indicated above is

$$d\left(\frac{\Delta F^\circ - \Delta H_0^\circ}{T}\right) = (\Delta H^\circ - \Delta H_0^\circ)\, d\left(\frac{1}{T}\right)$$

Integrating from 0°K to any arbitrary temperature T,

$$\frac{\Delta F^\circ - \Delta H_0^\circ}{T} - \lim_{T\to 0}\left(\frac{\Delta F^\circ - \Delta H_0^\circ}{T}\right) = \int_0^T (\Delta H^\circ - \Delta H_0^\circ)\, d\left(\frac{1}{T}\right)$$

Since $\lim_{T\to 0}\,[(\Delta F^\circ - \Delta H_0^\circ)/T] = -\Delta S_0^\circ$, by the definition $F = H - TS$, and since ΔS_0° (*i.e.*, the entropy change accompanying the reaction at 0°K) is zero by the third law, it follows that

$$\frac{\Delta F^\circ - \Delta H_0^\circ}{T} = \int_0^T (\Delta H^\circ - \Delta H_0^\circ)\, d\left(\frac{1}{T}\right) \tag{9-35}$$

As $\Delta H^\circ - \Delta H_0^\circ$ represents the sum of the heats required to heat the products from 0°K to T, minus the corresponding sum for the reactants, it follows that the integral can be evaluated from a knowledge of the heat capacities alone. After the integral is evaluated it is a simple matter to find ΔF° if ΔH_0° is known. If in the compilation of tables the integral had to be evaluated for all reactions involving certain substances, it is evident that the computation for all temperatures (or even at, say, 100° intervals) would be quite a formidable undertaking. However, if tables of $(F^\circ - H_0^\circ)/T$ were available for each substance involved, then

$$\frac{\Delta F^\circ - \Delta H_0^\circ}{T}$$

which is $\Delta(F^\circ - H_0^\circ)/T$, would be readily obtainable by simple addition and subtraction (just as ΔC_P was obtained from the C_P's involved) and the compilation task would be much simplified.

For any homogeneous substance of fixed composition we may, then, write a relation similar to Eq. (9-35):

$$\frac{F^\circ - H_0^\circ}{T} = \int_0^T (H^\circ - H_0^\circ)\, d\left(\frac{1}{T}\right) \tag{9-36}$$

from which the function $(F° - H°_0)/T$ can be evaluated if the heat capacity is known from 0°K to T. For gases this function can be evaluated by the methods of statistical mechanics from spectroscopic data. Tables of this function, as well as of $H° - H°_0$, are now available for many gases and for a few other substances.[1] When available for the reactants and products of the reaction under consideration, these tables afford a much easier method for evaluating K than the analytical method described in the preceding section. It is hoped and expected that this method of tabulating thermodynamic data will find even wider application, so that eventually tables will be available for all common substances.

The Use of Tabulated Thermodynamic Data. Suppose that we wish to find the equilibrium constant for the reaction

$$CH_4(g) + CO_2(g) = 2CO(g) + 2H_2(g)$$

at 1000°K. From tables published by the Bureau of Standards[2] we find the following information:

Gas	$\frac{F° - H°_0}{T}$ *at* 1000°K
CO	−48.860
H_2	−32.738
CH_4	−47.65
CO_2	−54.109

For the reaction,

$$\frac{\Delta F°_{1000} - \Delta H°_0}{1000} = 2(-48.860) + 2(-32.738) - (-47.65) - (-54.109) = -61.437$$

$\Delta H°_0$, evaluated as described below, is 55,552 cal; hence

$$R \ln K_{1000} = \frac{-\Delta F°_{1000}}{1000} = 61.437 - 55.552 = 5.885$$

Whence $\log K_{1000} = 1.286$, and $K_{1000} = 19.32$. The relative ease of this computation is obvious as compared with the task of setting up and evaluating the free-energy expression as described in the preceding section.

[1] Sources of such data are given in Chap. 1. For many substances C_P is known for elevated temperature but not for the vicinity of 0°K. In such a case the functions $H° - H°_0$ and $(F° - H°_0)/T$ cannot be evaluated. For this reason similar tables are sometimes compiled [for example, by K. K. Kelley in *U.S. Bur. Mines Bull.* 476 (1949)] wherein are tabulated the functions $H° - H°_{298}$ and $(F° - H°_{298})/T$. These functions which are used in a similar manner must, of course, be consistent in a single calculation.

[2] Wagman, Kilpatrick, Taylor, Pitzer, and Rossini, *Natl. Bur. Standards J. Research,* **34,** 143 (1945).

The evaluation of ΔH_0°, needed in the above computation, may be achieved through a knowledge of ΔH_{298}° which is found by adding and subtracting heats of formation of the substances involved in the usual way. Thus ΔH_0° is evaluated with the aid of the identity

$$\Delta H_0^\circ = \Delta H_{298}^\circ - \Delta(H_{298}^\circ - H_0^\circ)$$

The necessary tables of $H^\circ - H_0^\circ$ for various substances usually accompany those of $(F^\circ - H_0^\circ)/T$ in the literature. Occasionally ΔH_0° may be found directly in the literature. If neither ΔH_0° nor ΔH_{298}° is available from calorimetric measurements, ΔH_0° may be evaluated from a single measurement of the equilibrium constant, using tables of $(F^\circ - H_0^\circ)/T$ and performing a calculation essentially the inverse of the above example.

It will be noted that, although it is usual practice to tabulate both $(F^\circ - H_0^\circ)/T$ and $H^\circ - H_0^\circ$, these two functions are by no means independent but are related to each other by Eq. (9-36), which in differential form is

$$\frac{d[(F^\circ - H_0^\circ)/T]}{d(1/T)} = H^\circ - H_0^\circ$$

Thus either type of table can be prepared from the other. Furthermore, the entropy can be calculated, if desired, from these tables by means of the relation

$$S^\circ = \frac{H^\circ - F^\circ}{T} = \frac{H^\circ - H_0^\circ}{T} - \frac{F^\circ - H_0^\circ}{T}$$

The heat capacity can be obtained graphically as the slope of a plot of $H^\circ - H_0^\circ$ against temperature, or, frequently with sufficient precision, from the difference of two successive tabular values:

$$C_P = \frac{d(H^\circ - H_0^\circ)}{dT}$$

If desired, $E^\circ - E_0^\circ$ and $A^\circ - E_0^\circ$ can be obtained from such tables and the additional knowledge of the molal volume.

The tabular values and the thermodynamic functions derived therefrom all pertain to the standard state, *i.e.*, 1 atm pressure and fixed composition. Variation of the thermodynamic functions with pressure was discussed in Chap. 7; their variation with composition will be discussed in the chapters on solutions. Thus the tables of $(F^\circ - H_0^\circ)/T$ and $H^\circ - H_0^\circ$ are a convenient formulation which serves as the basis of calculations involving all thermodynamic substances and hence all reactions.

CHAPTER 10

SOLUTIONS

It is common knowledge that most substances are conglomerate in nature, *i.e.*, they consist of several phases, and that the phase regions are seldom limited to a single composition. In other words solutions are common—a pure substance exists only in our minds and represents a limiting state which we may strive to approach but never actually obtain. There is, in fact, a whole branch of chemistry concerned with the impurities in ordinary water, the common units of which are parts per million (ppm). The purest iron obtainable at the present time contains impurities expressible in thousandths of a per cent. It might be noted in passing that the principal difference between bessemer and open-hearth steel is about 0.01 per cent nitrogen, whose effect is by no means negligible from an industrial viewpoint. Nearly all substances of metallurgical interest are to be regarded as solutions—gaseous, liquid, or solid, dilute or otherwise.

COMPOSITION

For practical purposes the composition of a solution is expressed most often in terms of percentage by weight. However, we have already found it convenient to express thermodynamic functions on a molal or gram atomic basis. Another very decided advantage in retaining the molal or gram atomic formulation in dealing with solutions is due to the dominant position of Raoult's law in the application of thermodynamics to solutions. This law, originally found empirically and subsequently afforded a firm theoretical basis as a limiting law for practically all solutions, is expressed in terms of mole or atom fraction. In theoretical work it is usually of considerable advantage to adhere to this mode of expressing composition.

The absence of molecular structure in solid or liquid metals and in metallic solutions has already been discussed in considerable detail in Chaps. 4 and 5. Hence, for these solutions the atom fraction rather than the mole fraction is the appropriate unit.[1] As mentioned previously, the *atom fraction* of a component in a solution is defined as the ratio of the

[1] In the theoretical treatment of interstitial solutions the ratio of the number of atoms of interstitial component to the number of interstitial sites is sometimes convenient.

total number of gram atoms of that component to the total number of gram atoms of all components present in any given amount. Accordingly the atom fraction of component 1, N_1, is defined thus:

$$N_1 = \frac{n_1}{\sum_i n_i} \qquad (10\text{-}1)$$

i referring, as usual, to the general component. If the composition is given in weight per cent and we wish to convert to atom fraction, it is convenient to consider a total mass of 100 g so that the number of gram atoms of each component contained therein is

$$n_1 = \frac{\%_1}{M_1} \qquad n_2 = \frac{\%_2}{M_2} \qquad \text{etc.}$$

M_1, M_2, etc., representing atomic weights. Hence

$$N_1 = \frac{\%_1/M_1}{\sum_i(\%_i/M_i)} \qquad (10\text{-}2)$$

Atom Per Cent from Weight Per Cent—Graphical Method. The conversion from weight per cent to atom per cent is required so frequently that it seems worth while to call attention to a simple graphical method which saves time if a number of conversions must be made for a solution of two components. Let a vertical straight line OA (Fig. 10-1) be drawn on graph paper with a length corresponding to M_2, the atomic weight of component 2, and OB be laid off thereon with a length corresponding to M_1, the atomic weight of component 1. Let an isosceles triangle be drawn with base AB and altitude such that the vertex C is 100 units from the base. If now the point x is located on the line BC so that its distance wx from AB corresponds to the known weight per cent of component 1 and the line Ox be constructed and extended to determine the point y on AC, then the distance zy, divided by 100, gives the atom fraction of component 1. Clearly by similar procedure the weight per cent may be determined if the atom per cent is known. The proof of this construction is of simple geometric nature and may be supplied by the student. If greater precision is desired, it is frequently convenient to construct a plot of $N_1/\%_1$ vs. $\%_1$, requiring calculation for perhaps a half dozen points.

Stoichiometric and Species Mole Fraction. In the discussion of molten oxides and slags, it is customary to speak of the stoichiometric mole fraction of the oxidic constituents, without thereby necessarily implying a molecular structure. Thus in a synthetic slag composed of CaO and SiO_2 it is sometimes convenient to speak of a mole fraction of

CaO or of SiO_2, meaning the ratio of the total moles of CaO or SiO_2 to their sum. Care should be exercised in distinguishing this formal representation of composition from a representation which does imply a molecular constitution. For example, if it is thought that CaO and SiO_2 form molecules of $2CaO{\cdot}SiO_2$ in open-hearth slag, then it is appropriate to refer to a mole fraction thereof, $N_{2CaO{\cdot}SiO_2}$ which would equal the ratio of the postulated or calculated number of molecules of $2CaO{\cdot}SiO_2$ to the total number of gram molecules of all such postulated molecular species.

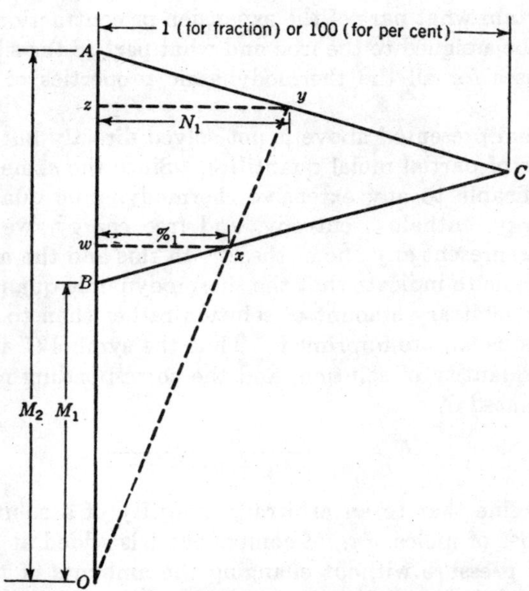

Fig. 10-1. Graphical method for converting weight per cent to atom per cent, or vice versa.

Other Composition Units. In aqueous solutions it is customary to designate composition by *molality*, defined as the number of moles[1] of solute per 1000 g of solvent, *i.e.*, of water or water plus alcohol, etc. *Normality*, gram equivalents per liter of solution, and *molarity*, moles per liter of solution, are also occasionally used, principally in analytical work.

PARTIAL MOLAL QUANTITIES

In dealing with solutions, one of the first questions which naturally arises is how to express molal quantities for a substance in solution. For example, although the molal volume of liquid iron or of liquid silicon is in principle uniquely and unambiguously determined by experiment, the

[1] The term *mole* is used to denote gram molecular weight or gram atomic weight.

method of determining the molal volume of either in a solution of both is by no means obvious and can, in fact, be carried out in no unique and unambiguous way, although the total volume is quite determinate. Some light may be shed on the interpretation of this difficulty by reflecting that, when two liquids are mixed, the total volume of the solution is, in general, not equal to the sum of the individual volumes before mixing; this reflects the difference of the interatomic forces in the pure substance and in the solution. If we wish to assign a fraction of the total volume of the solution to iron and one to silicon in the preceding example, there is no way to ascertain what part of the expansion or contraction[1] occurring on mixing is to be assigned to the iron and what part to the silicon. Similar difficulty arises for all the thermodynamic properties of substances in solution.

The problem presented above is not solved directly but is avoided by the invention of partial molal quantities. Since the same general treatment is applicable to any extensive thermodynamic quantity, such as volume, energy, enthalpy, entropy, and free energy, we shall use the symbol G to represent any one of these. In this and the next section we shall use a prime to indicate that the thermodynamic quantity so primed refers to any arbitrary amount of solution rather than to 1 mole; molal quantities, as usual, are unprimed. Thus the symbol G' will be applied to the total quantity of solution, and the corresponding molal quantity will be designated G.

$$G = \frac{G'}{n_1 + n_2 + n_3 + \cdots}$$

Let us imagine that to an arbitrary quantity of a solution an infinitesimal number of moles, dn_1, of component 1 is added at constant temperature and pressure without changing the amounts of the other constituents. The corresponding increment in the property G' is dG'; the ratio $(\partial G'/\partial n_1)_{P,T,n_2,n_3,\ldots}$ is known as a *partial molal quantity* and is designated $\bar{G}_1$. Thus

$$\bar{G}_1 \equiv \left(\frac{\partial G'}{\partial n_1}\right)_{P,T,n_2,n_3,\ldots} \tag{10-3}$$

Analogous relations are defined similarly for the other components. $\bar{G}_1$ may be interpreted equally well as the increment of G' when 1 mole of the first component is added to a very large quantity of the solution. For

[1] In general, it is to be expected that, if the interatomic forces between unlike species are greater than between like, a contraction and an evolution of heat will occur on homogenization. Similarly, if the attractive forces between unlike species are less than between like, either an expansion and heat absorption or a failure to homogenize is to be expected.

example, if the volume increase accompanying the addition of 1 gram atom of copper to a large amount of a liquid alloy is observed to be 8.5 cc, then the partial molal volume of copper in the alloy at the particular composition, temperature, and pressure is 8.5 cc. This is written $\bar{V}_{Cu} = 8.5$ cc. Or if copper is designated as component 1, we may write $\bar{V}_1 = 8.5$ cc. The symbol $\bar{V}_1$ is usually read "vee-one-bar," although sometimes "bar-vee-one."

From the fundamental theorem of partial differentiation, we have, at constant temperature and pressure,

$$dG' = \left(\frac{\partial G'}{\partial n_1}\right)_{n_2,n_3,\ldots} dn_1 + \left(\frac{\partial G'}{\partial n_2}\right)_{n_1,n_3,\ldots} dn_2 + \left(\frac{\partial G'}{\partial n_3}\right)_{n_1,n_2,\ldots} dn_3 + \cdots$$

Or

$$dG' = \bar{G}_1\, dn_1 + \bar{G}_2\, dn_2 + \bar{G}_3\, dn_3 + \cdots \tag{10-4}$$

If we add to a large quantity of solution n_1 moles of component 1, n_2 moles of component 2, etc., the increment in G' after mixing is $n_1\bar{G}_1 + n_2\bar{G}_2 + \cdots$. If now we mechanically remove a portion containing $n_1 + n_2 + \cdots$ moles, the extensive quantity G' for the main body of solution is now decreased by $(n_1 + n_2 + \cdots)G$. Since at the end of these processes the main body of solution is the same in composition and amount as it was initially, G' has the same value finally as initially, and the increment in G accompanying the individual additions is equal to the decrement accompanying their mass withdrawal.

$$(n_1 + n_2 + \cdots)G = n_1\bar{G}_1 + n_2\bar{G}_2 + \cdots$$

Dividing by $n_1 + n_2 + \cdots$, and noting that $n_i/(n_1 + n_2 + \cdots) = N_i$,

$$G = N_1\bar{G}_1 + N_2\bar{G}_2 + \cdots\ * \tag{10-5}$$

It will be noted that the only assumption used in deriving this relation is that G' is an extensive quantity. As usual, we are considering here massive phases, or at least phases of sufficient particle size that surface energy effects are negligibly small.

Equation (10-5) may be multiplied by $n_1 + n_2 + \cdots$ to give the relation

$$G' = n_1\bar{G}_1 + n_2\bar{G}_2 + \cdots \tag{10-6}$$

* This equation may be derived also by integrating Eq. (10-4) term by term from zero to one total mole under the condition of constant composition in order that the $\bar{G}$'s be constant. The physical process corresponding to this integration is the addition by infinitesimals of each of the constituents in the proportion in which they exist in the final solution. Equation (10-5), although obtained under the restriction of constant composition, is generally applicable, since all terms therein are functions of state. As properties of the solution they are independent of the mode of formation of the solution.

which upon complete differentiation (at constant temperature and pressure) gives

$$dG' = n_1\,d\bar{G}_1 + n_2\,d\bar{G}_2 + \cdots + \bar{G}_1\,dn_1 + \bar{G}_2\,dn_2 + \cdots \qquad (10\text{-}7)$$

Subtracting Eq. (10-4) from Eq. (10-7),

$$n_1\,d\bar{G}_1 + n_2\,d\bar{G}_2 + \cdots = 0 \qquad (10\text{-}8)$$

Dividing by $n_1 + n_2 + \cdots$, we have

$$N_1\,d\bar{G}_1 + N_2\,d\bar{G}_2 + \cdots = 0 \qquad (10\text{-}9)$$

If G is the free energy F, then, for a two-component system, Eq. (10-9) becomes

$$N_1\,d\bar{F}_1 + N_2\,d\bar{F}_2 = 0 \qquad (10\text{-}9a)$$

which is one form of the Gibbs-Duhem equation [see Eq. (10-53)]. This is perhaps the most important of all equations dealing with solutions, a large number of the subsequent relations being derived therefrom.

A General Method for Obtaining Partial Molal Quantities from Molal Quantities. Differentiating the expression for the molal property G of a two-component system at constant temperature and pressure [Eq. (10-5)],

$$dG = N_1\,d\bar{G}_1 + \bar{G}_1\,dN_1 + N_2\,d\bar{G}_2 + \bar{G}_2\,dN_2 \qquad (10\text{-}10)$$

Combining Eq. (10-9), $N_1\,d\bar{G}_1 + N_2\,d\bar{G}_2 = 0$, with Eq. (10-10) we have

$$dG = \bar{G}_1\,dN_1 + \bar{G}_2\,dN_2 \qquad (10\text{-}11)$$

Multiplying through by N_1/dN_2, and noting that $dN_1 = -dN_2$ (since $N_1 + N_2 = 1$),

$$N_1\frac{dG}{dN_2} = -N_1\bar{G}_1 + N_1\bar{G}_2$$

Adding this to Eq. (10-5) and transposing all terms,

$$\bar{G}_2 = G + N_1\frac{dG}{dN_2} = G + (1 - N_2)\frac{dG}{dN_2} \qquad (10\text{-}12)$$

Obviously an analogous equation holds for $\bar{G}_1$, namely,

$$\bar{G}_1 = G + (1 - N_1)\frac{dG}{dN_1} \qquad (10\text{-}12a)$$

Similarly it may be shown for a multicomponent system that

$$\bar{G}_1 = G + (1 - N_1)\left(\frac{\partial G}{\partial N_1}\right)_{N_2:N_3:N_4:\cdots} \qquad (10\text{-}13)$$

where the subscript $N_2{:}N_3{:}N_4{:}\ \cdots$ indicates that the relative amounts of all components other than component 1 are held constant. For each component an equation of like form may be written.

By means of Eqs. (10-12) $\bar{G}_1$ and $\bar{G}_2$ can be found from a knowledge of G either graphically or analytically. Analytically, it is evident that, if G, and hence dG/dN_1, may be represented by an empirical or other expression, then a corresponding expression for $\bar{G}_1$ results from Eq. (10-12*a*). A graphical method of particular interest, known as the *method of intercepts*, is illustrated in Fig. 10-2. Let us suppose that G has been determined experimentally over the entire composition range of a binary solution. G is represented as a function of N_2 by the curve in Fig. 10-2. Let us select any arbitrary point A on the curve and construct the tangent as shown. It will now be demonstrated that the ordinates of this tangent at $N_2 = 0$ and $N_2 = 1$ are equal to $\bar{G}_1$ and $\bar{G}_2$, respectively. At point A the value of G is equal to the length of line BC, and the slope of the tangent, as determined from the right triangle ACD, is $\overline{CD}/(1 - N_2)$. Equation (10-12), written in terms of these lengths, becomes

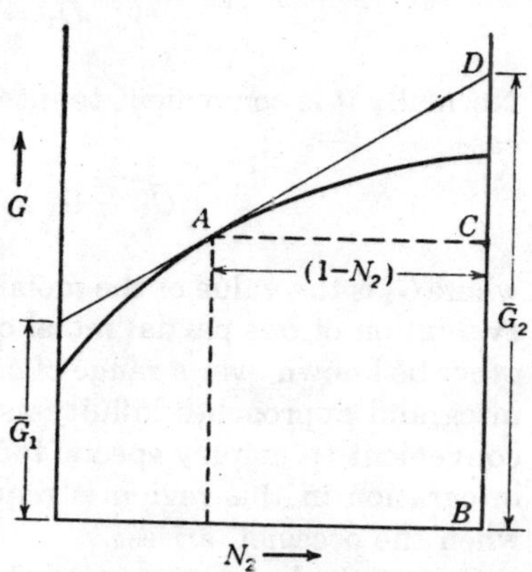

Fig. 10-2. Graphical method for obtaining partial molal quantities from the molal quantity.

$$\bar{G}_2 = \overline{BC} + (1 - N_2)\left(\frac{\overline{CD}}{1 - N_2}\right) = \overline{BC} + \overline{CD} = \overline{BD}$$

thus demonstrating that the ordinate of the tangent at $N_2 = 1$ is $\bar{G}_2$. In like manner it may be shown that the ordinate of the tangent at $N_2 = 0$ is $\bar{G}_1$.

For a three-component system compositions may be represented as points on an equilateral triangle. Corresponding values of any molal property G may be plotted upward from the triangular base, thus producing a surface bounded by the sides of the triangular prism. If at any point on this surface the tangent plane is constructed, then it may be shown that the intercepts of this plane on the edges of the prism, corresponding to $N_1 = 1$, $N_2 = 1$, and $N_3 = 1$, are equal, respectively, to $\bar{G}_1$, $\bar{G}_2$, and $\bar{G}_3$. This theorem is obviously an extension of that illustrated in Fig. 10-2.

Methods of Determining G, $\bar{G}_1$, and $\bar{G}_2$ from Each Other. In the preceding section a method was described for finding a partial molal quantity when the corresponding molal quantity is known over a range of composition. If a partial molal quantity for one component of a binary solution is known, that for the other may be found by graphical or analytical integration of Eq. (10-9); thus

$$\bar{G}_1 = -\int \frac{N_2}{N_1} d\bar{G}_2 \qquad (10\text{-}14)$$

Normally it is convenient to integrate from $N_1 = 1$ to $N_1 = N_1$, in which case

$$\bar{G}_1 - G_1 = -\int_{N_1=1}^{N_1=N_1} \frac{N_2}{N_1} d\bar{G}_2 \qquad (10\text{-}15)$$

where G_1 is the value of the molal property G for pure component 1. This evaluation of one partial molal quantity from the other requires that the other be known over a range of composition. It will be observed that the integrand approaches infinity as N_1 approaches 0; hence it is sometimes convenient to employ special functions to circumvent the difficulty when integration in this region is required. Such functions will be discussed when the occasion arises.

One method of determining G when $\bar{G}_2$ or $\bar{G}_1$ is known is first to find $\bar{G}_1$ from $\bar{G}_2$, or vice versa, by means of Eq. (10-15), then to evaluate G from Eq. (10-5). An alternative method is by integration of Eq. (10-12*a*), which can be rewritten in the form

$$d\left(\frac{G}{N_2}\right) = -\frac{\bar{G}_1}{N_2^2} dN_2 \qquad (10\text{-}16)$$

General Interrelations of Partial Molal Quantities for Solutions of Fixed Composition. The definition of the free energy of a solution is

$$F' = H' - TS'$$

By differentiation,

$$\left(\frac{\partial F'}{\partial n_1}\right)_{P,T,n_2,n_3,\ldots} = \left(\frac{\partial H'}{\partial n_1}\right)_{P,T,n_2,n_3,\ldots} - T\left(\frac{\partial S'}{\partial n_1}\right)_{P,T,n_2,n_3,\ldots}$$

Whence, by Eq. (10-3),

$$\bar{F}_1 = \bar{H}_1 - T\bar{S}_1 \qquad (10\text{-}17)$$

From the definition of H' and A' and similar differentiation,

$$\bar{H}_1 = \bar{E}_1 + P\bar{V}_1 \qquad (10\text{-}18)$$

and

$$\bar{A}_1 = \bar{E}_1 - T\bar{S}_1 \qquad (10\text{-}19)$$

Obviously similar equations may be written for each of the other components.

From the fundamental combined statement of the first and second laws in terms of the free energy,

$$dF' = V'\,dP - S'\,dT$$

a relation which is valid for all substances of constant composition, including solutions, it follows immediately that

$$\left(\frac{\partial F'}{\partial P}\right)_{T,n_1,n_2,\ldots} = V' \tag{10-20}$$

and

$$\left(\frac{\partial F'}{\partial T}\right)_{P,n_1,n_2,\ldots} = -S' \tag{10-21}$$

Differentiating the first of these equations partially with respect to n_1,

$$\left[\frac{\partial}{\partial n_1}\left(\frac{\partial F'}{\partial P}\right)_{T,n_1,n_2,\ldots}\right]_{T,P,n_2,n_3,\ldots} = \left(\frac{\partial V'}{\partial n_1}\right)_{T,P,n_2,n_3,\ldots} = \bar{V}_1$$

But since, in taking a second derivative, the order of differentiation is immaterial, the left side above may be written

$$\left[\frac{\partial}{\partial P}\left(\frac{\partial F'}{\partial n_1}\right)_{T,P,n_2,n_3,\ldots}\right]_{T,n_1,n_2,\ldots}$$

and inserting $\bar{F}_1$ for its definitional equivalent,

$$\left(\frac{\partial \bar{F}_1}{\partial P}\right)_{T,n_1,n_2,\ldots} = \bar{V}_1 \tag{10-22}$$

Similarly, from Eq. (10-21) it is found that

$$\left(\frac{\partial \bar{F}_1}{\partial T}\right)_{P,n_1,n_2,\ldots} = -\bar{S}_1 \tag{10-23}$$

Since $\bar{F}_1$ is a function of state, it follows that at constant composition it may be represented as a function of temperature and pressure; hence, from the fundamental theorem of partial differentiation,

$$d\bar{F}_1 = \left(\frac{\partial \bar{F}_1}{\partial P}\right)_{T,n_1,n_2,\ldots} dP + \left(\frac{\partial \bar{F}_1}{\partial T}\right)_{P,n_1,n_2,\ldots} dT$$

By substitution from Eqs. (10-22) and (10-23),

$$d\bar{F}_1 = \bar{V}_1\,dP - \bar{S}_1\,dT \tag{10-24}$$

which, it will be noted, is identical in form with the original combined statement of the first and second laws.

Adding Eqs. (10-17) and (10-18),

$$\bar{F}_1 = \bar{E}_1 + P\bar{V}_1 - T\bar{S}_1$$

Totally differentiating this,

$$d\bar{F}_1 = d\bar{E}_1 + P\,d\bar{V}_1 + \bar{V}_1\,dP - T\,d\bar{S}_1 - \bar{S}_1\,dT$$

and subtracting from Eq. (10-24), we have

$$d\bar{E}_1 = T\,d\bar{S}_1 - P\,d\bar{V}_1 \tag{10-25}$$

Now recalling that all the thermodynamic relations pertaining to systems of fixed composition were derived from the original combined expression of the first and second laws, $dE = T\,dS - P\,dV$, and the definitional relations for H, A, and F, and noting, as already pointed out, that Eqs. (10-25), (10-18), (10-19) and (10-17), respectively, are identical in form therewith, we see that every relation developed for a substance of fixed composition in terms of molal quantities is also valid for each component of a solution in terms of partial molal quantities. Thus, for example, having shown that $[\partial(F/T)/\partial(1/T)]_P = H$, we may immediately write

$$\left[\frac{\partial(\bar{F}_1/T)}{\partial(1/T)}\right]_P = \bar{H}_1 \qquad \left[\frac{\partial(\bar{F}_2/T)}{\partial(1/T)}\right]_P = \bar{H}_2 \qquad \text{etc.} \tag{10-26}$$

In this case, as in previous and subsequent chapters, the constancy of the n's is not indicated by subscripts when it is evident from the context.

IDEAL SOLUTIONS—RAOULT'S LAW

An *ideal solution* is one which obeys Raoult's law over all temperatures and pressures at which it is capable of existing. *Raoult's law* we shall define by the relation

$$a_i = N_i \tag{10-27}$$

Raoult[1] (1887) originally found empirically that for a limited number of binary solutions the partial pressure of each constituent is equal to the product of the mole fraction and the vapor pressure of the pure constituent:

$$p_i = p_i^\circ N_i \tag{10-28}$$

However, as already noted, if the vapor behaves as an ideal gas, the partial pressure of a constituent is equal to its fugacity, and Eq. (10-28) becomes

$$f_i = f_i^\circ N_i \tag{10-29}$$

Since the activity of a constituent of a liquid is defined as f_i/f_i°, it is seen immediately that Eq. (10-28) reduces to Eq. (10-27) in case the vapor

[1] F. M. Raoult, *Compt. rend.*, **104**, 1430 (1887).

behaves ideally. It does not seem reasonable, when inquiring as to the ideality of a solution, to have to inquire further as to the ideality of the vapors of the constituents; hence it is now common practice to rephrase Raoult's law as Eq. (10-29) or (10-27) rather than to use it in the form of Eq. (10-28). For many purposes the difference is insignificant.

Although Raoult's discovery was brilliant, most solutions are known to exhibit departures, in some cases quite drastic departures, from Eq. (10-27). This was recognized by Raoult, who observed, however, that

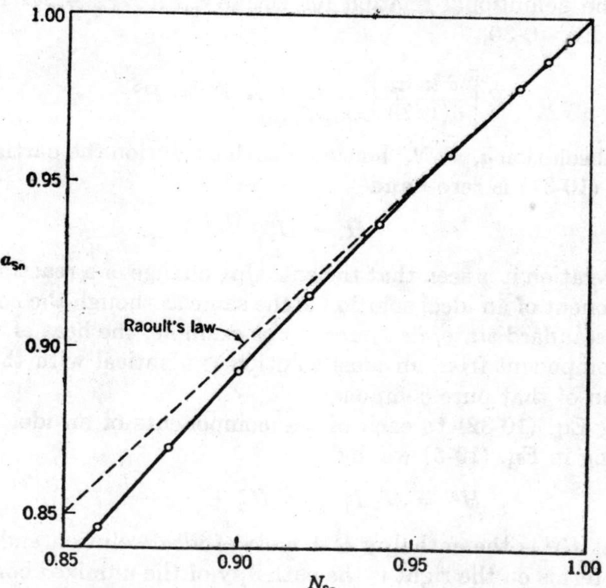

Fig. 10-3. Plot of the activity of tin against its atom fraction for the Sn-Sb system at 412°C. Demonstration of Raoult's law as a limiting law.

in substantially all solutions his law is obeyed by the solvent (major constituent), at low concentrations of solute (minor constituent). The approach to Raoult's law is illustrated by Fig. 10-3, where the activity of tin[1] in molten tin-antimony alloys is plotted against the atom fraction of tin in the high-tin concentration range. The departure from Raoult's law is seen to approach zero rapidly as the concentration of solute approaches zero. Not only does the curve representing the observed data approach the Raoult law line, but also its curvature (second derivative) approaches zero.

[1] J. A. Yanko, A. E. Drake, and F. Hovorka, *Trans. Electrochem. Soc.*, **89**, 357 (1946).

Heat of Formation of an Ideal Solution. Substracting from Eq. (10-26) the corresponding equation for the standard state, here taken to be the pure component,[1]

$$\left[\frac{\partial\left(\dfrac{\bar{F}_i - F_i^\circ}{T}\right)}{\partial\left(\dfrac{1}{T}\right)}\right]_{P,N_1,N_2,\ldots} = \bar{H}_i - H_i^\circ \tag{10-30}$$

Inserting the definitional relation for activity, $\bar{F}_i - F_i^\circ = RT \ln a_i$ [Eq. (9-11c)], in Eq. 10-30,

$$R\left[\frac{\partial \ln a_i}{\partial(1/T)}\right]_{P,N_1,N_2,\ldots} = \bar{H}_i - H_i^\circ \tag{10-31}$$

For an ideal solution $a_i = N_i$; hence for such a solution the partial derivative in Eq. (10-31) is zero,[2] and

$$\bar{H}_i = H_i^\circ \tag{10-32}$$

From this relation it is seen that the enthalpy change of a reaction involving a component of an ideal solution is the same as though the component were in its standard state, *i.e.*, pure. For example, the heat of vaporization of a component from an ideal solution is identical with the heat of vaporization of that pure component.

Applying Eq. (10-32) to each of the components of an ideal solution and inserting in Eq. (10-5) we find

$$H^{id} = N_1H_1^\circ + N_2H_2^\circ + \cdots \tag{10-33}$$

Noting that H^{id} is the enthalpy of 1 mole of ideal solution and that the sum of the terms on the right is the enthalpy of the unmixed components making up the solution, it follows immediately that the enthalpy, or heat, of mixing for an ideal solution, $\Delta H^{M,id}$, is zero.

$$\Delta H^{M,id} \equiv H^{id} - (N_1H_1^\circ + N_2H_2^\circ + \cdots) = 0 \tag{10-33a}$$

Volume Change in the Formation of an Ideal Solution. In a manner very similar to that above it will now be shown that the volume change in the formation of an ideal solution is zero. From Eq. (10-22) and the

[1] Sometimes it is convenient to choose as the standard state not the pure component but a solution of some definite composition; in this case the property G in the standard state would be the partial molal property and may be written $\bar{G}_i^\circ$.

[2] In Eq. (10-31) the subscripts N_1, N_2, . . . appear, contrary to our usual usage, in order to emphasize the constancy of composition. If the activity is a function of composition only, then the partial derivative at constant composition (being the derivative of a constant) is zero.

corresponding equation for the standard state in the form of the pure component we may write

$$\left[\frac{\partial(\bar{F}_i - F_i^\circ)}{\partial P}\right]_{T,N_1,N_2,\ldots} = \bar{V}_i - V_i^\circ \tag{10-34}$$

Inserting again the definitional relation for activity,

$$RT\left(\frac{\partial \ln a_i}{\partial P}\right)_{T,N_1,N_2,\ldots} = \bar{V}_i - V_i^\circ \tag{10-35}$$

Since for an ideal solution $a_i = N_i$, the partial derivative on the left is zero. Thus we see that

$$\bar{V}_i = V_i^\circ \tag{10-36}$$

and the partial molal volume of a component of an ideal solution is seen to be equal to the molal volume of the pure component.

Insertion of a value for each component from Eq. (10-36) into Eq. (10-5) yields, for an ideal solution,

$$V^{id} = N_1V_1^\circ + N_2V_2^\circ + \cdots \tag{10-37}$$

telling us that the volume of the unmixed components of an ideal solution is identical with that of the solution formed therefrom and thus that the volume change of mixing for an ideal solution, $\Delta V^{M,id}$, is zero.

$$\Delta V^{M,id} \equiv V^{id} - (N_1V_1^\circ + N_2V_2^\circ + \cdots) = 0 \tag{10-37a}$$

Free Energy and Entropy of Formation of an Ideal Solution. Multiplying Eq. (9-11c) by N_i,

$$N_i\bar{F}_i - N_iF_i^\circ = N_iRT \ln a_i$$

Summing such equations for all components of a solution,

$$(N_1\bar{F}_1 + N_2\bar{F}_2 + \cdots) - (N_1F_1^\circ + N_2F_2^\circ + \cdots) = RT(N_1 \ln a_1 + N_2 \ln a_2 + \cdots)$$

By Eq. (10-5) the terms in the far left parentheses are the free energy F of any solution, whereas those in the second parentheses are the free energy of the unmixed components. The free energy of formation ΔF^M of any solution from its components is therefore

$$\Delta F^M \equiv F - (N_1F_1^\circ + N_2F_2^\circ + \cdots) = RT(N_1 \ln a_1 + N_2 \ln a_2 + \cdots) \tag{10-38}$$

If the solution is ideal, $a_i = N_i$, and the free energy of formation is

$$\Delta F^{M,id} \equiv F^{id} - (N_1F_1^\circ + N_2F_2^\circ + \cdots) = RT(N_1 \ln N_1 + N_2 \ln N_2 + \cdots) \tag{10-39}$$

Differentiation of Eq. (10-39) with respect to temperature at constant pressure and composition gives, by virtue of Eq. (10-23) and its counterpart for a pure component, the entropy of mixing for an ideal solution $\Delta S^{M,id}$.

$$\Delta S^{M,id} = S^{id} - (N_1S_1^\circ + N_2S_2^\circ + \cdots)$$
$$= -R(N_1 \ln N_1 + N_2 \ln N_2 + \cdots) \quad (10\text{-}40)$$

Thus, the entropy of formation of an ideal solution from its components is independent of temperature and pressure. Figure 10-4 shows the entropy of formation of an ideal binary solution as a function of the mole

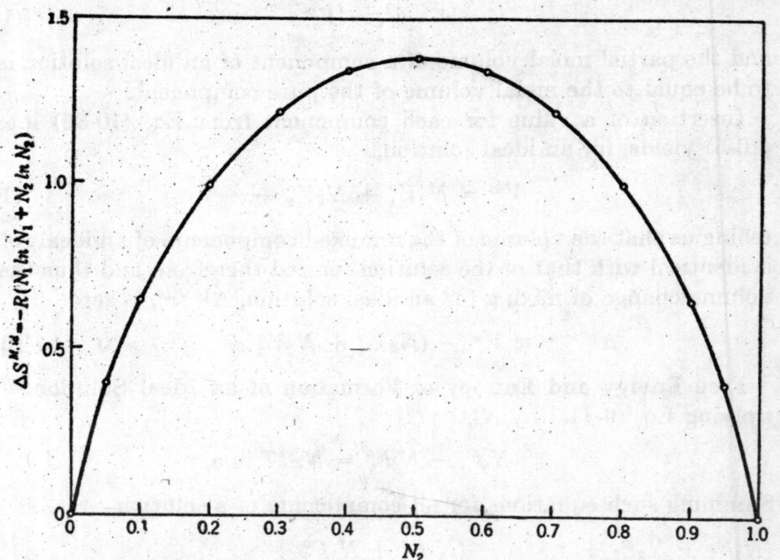

Fig. 10-4. ΔS of formation of 1 mole of an ideal solution from the pure components. The two limiting tangents are vertical.

fraction. It will be observed that the equations for free energy of formation [Eq. (10-39)], enthalpy of formation [Eq. (10-33*a*)], and entropy of formation [Eq. (10-40)] are consistent with the definitional equation $\Delta F = \Delta H - T\,\Delta S$.

ACTUAL (NONIDEAL) SOLUTIONS

It will be recalled that the problem of determining equilibria may be regarded primarily as one of determining free-energy change. The free-energy change of reactions involving pure substances or phases of fixed composition was discussed in Chap. 9. When dealing with a phase of variable composition it is necessary to know how the free energy changes with composition, *i.e.*, to know the free energy of formation of the solution or the partial molal free energies of the components (these are interrelated as discussed in preceding sections). The general problem of dealing thermodynamically with solutions may be considered one of properly formulating the partial molal free energies of the components thereof. The activity function was invented by G. N. Lewis to aid in this general problem. Its value lies in its close relation to composition; with appropriate choice of standard state the activity approaches the mole fraction as the mole fraction approaches unity.

The Activity Coefficient. It is found convenient to introduce a new function, the activity coefficient γ, which may be defined as the ratio of the activity to the mole fraction.[1]

$$\gamma_i = \frac{a_i}{N_i} \tag{10-41}$$

Commonly the pure component is chosen as the standard state, so that γ_i approaches unity as N_i approaches unity. The activity coefficient is seen to be unity for an ideal solution. In terms of the activity coefficient the activity may be written

$$a_i = \gamma_i N_i \tag{10-42}$$

and the partial molal free energy from Eq. (9-11*c*) becomes

$$\bar{F}_i - F_i^\circ = RT \ln a_i = RT \ln N_i + RT \ln \gamma_i \tag{10-43}$$

It will be noted that the final term, $RT \ln \gamma_i$, is zero for an ideal solution.

[1] If a concentration unit other than the mole fraction is used, it is convenient to use a standard state other than the pure component, and in this case, it is conventional to use an activity coefficient different from that defined above. For example, in dealing with aqueous solutions of electrolytes the concentration unit for the solute is usually the molality m; the activity coefficient γ is then defined as the ratio of activity to molality a/m, the standard state being chosen so that the activity approaches the molality as the molality approaches zero.

Furthermore, it is frequently found that the pure substance is not liquid at the temperature of the liquid solution under consideration or that it does not crystallize in the same system as the solid solution under consideration. For example, in the case of austenite, the solution of carbon in γ-iron, the concentration range of interest extends from zero to relatively small value of N_C; here it is convenient to choose a standard state for carbon such that a_C/N_C approaches unity as N_C approaches zero. Accordingly, γ_C is defined as in Eq. (10-41) but is equal to unity at $N_C = 0$ rather than at $N_C = 1$. The change from one standard state to another will be discussed later in this chapter.

For any actual (nonideal) solution the problem of evaluating the free energy may now be rephrased as one of evaluating the activity coefficients. In general no help in this problem can be expected from pure thermodynamics, *i.e.*, from the three laws. Appeal must be made to experiment or to statistical mechanics.

Let us investigate how the activity of cadmium in a liquid cadmium-lead solution varies with composition. This system is typical of many, but by no means all, metallic solutions exhibiting departures from

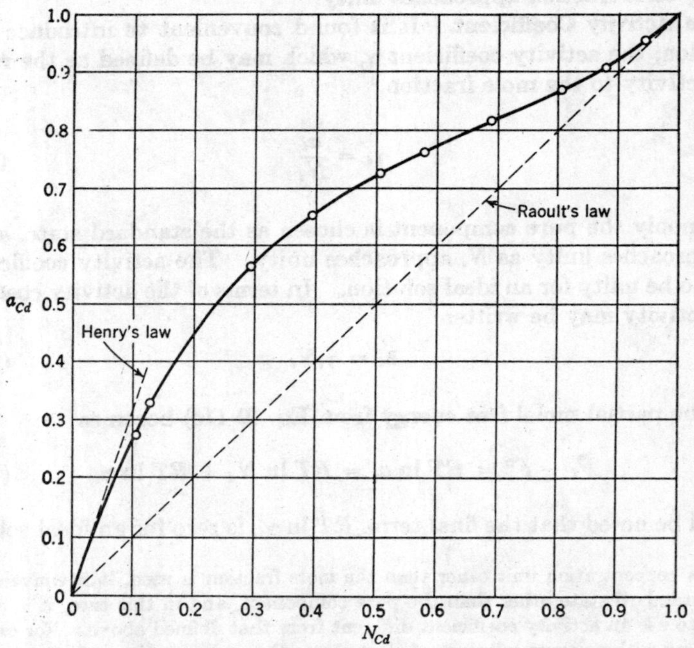

FIG. 10-5. Activity of cadmium in the Cd-Pb system at 500°C.

ideality. From the data of Elliott and Chipman[1] the activity of cadmium calculated from Eq. (10-41) is shown as a function of N_{Cd} in Fig. (10-5). The activity coefficient, $\gamma_{Cd} = a_{Cd}/N_{Cd}$, is shown as a function of N_{Cd} in Fig. (10-6), and log γ_{Cd} as a function of N_{Cd} in Fig. 10-7. The form of the latter curve suggests a parabola; to test this log γ_{Cd} is plotted against $(1 - N_{Cd})^2$ in Fig. 10-8. The resulting curve is not linear but does have finite slope throughout. The function $(\log \gamma_{Cd})/(1 - N_{Cd})^2$, which is plotted against N_{Cd} in Fig. 10-9, is always finite, and the plot always of finite slope.

[1] J. F. Elliott and J. Chipman, *Trans. Faraday Soc.*, **47**, 138 (1951).

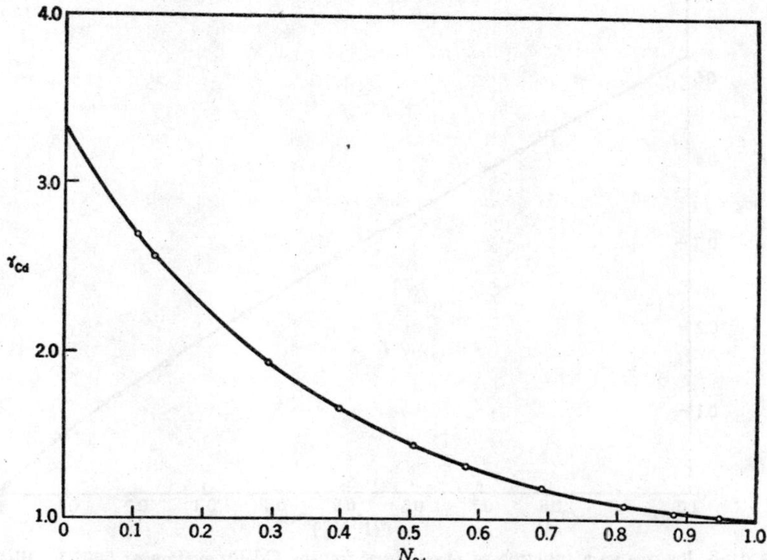

Fig. 10-6. Activity coefficient of cadmium in the Cd-Pb system at 500°C.

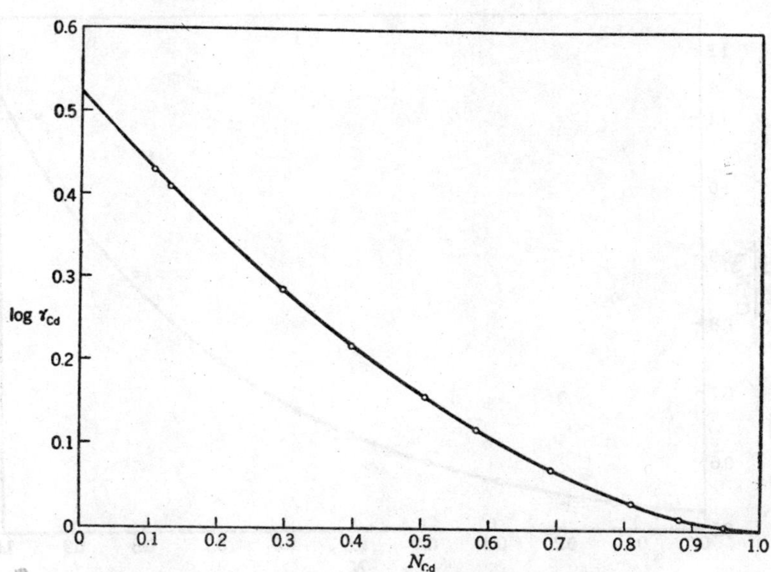

Fig. 10-7. log γ_{Cd} in the Cd-Pb system at 500°C.

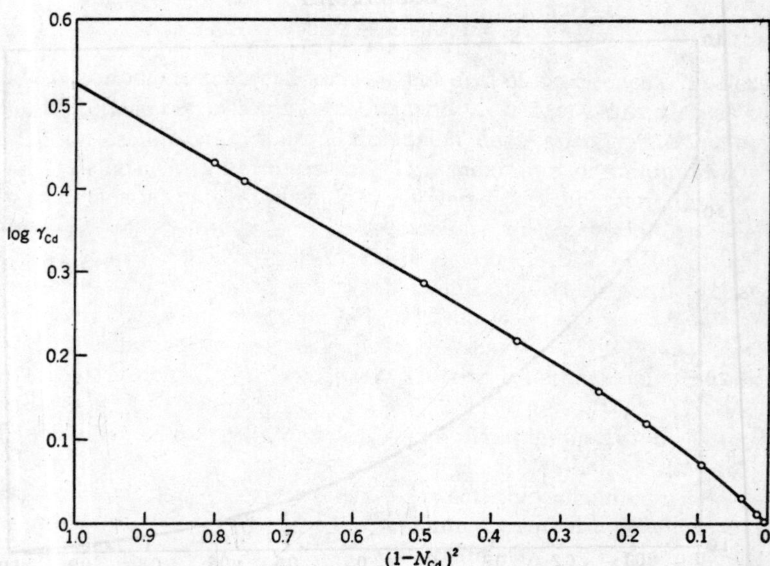

FIG. 10-8. log γ_{Cd} as a function of $(1 - N_{Cd})^2$ for the Cd-Pb system at 500°C. Slope is finite throughout.

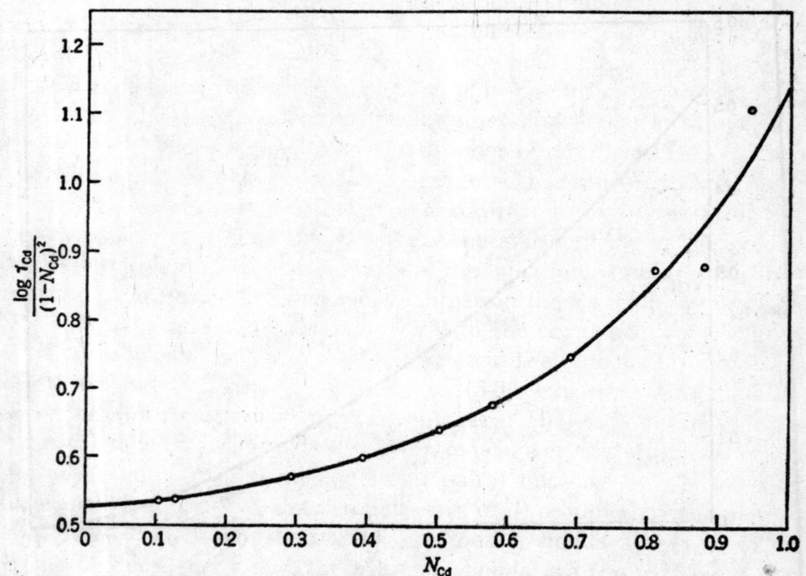

FIG. 10-9. Plot of log $\gamma_{Cd}/(1 - N_{Cd})^2$ against N_{Cd} for the Cd-Pb system at 500°C. Slope is finite throughout.

A similar series of plots for the tin-gold system at 600°C* is shown in Figs. 10-10 to 10-13. In addition to the measured activity of tin, Fig. 10-10 includes the activity of gold as derived from the activity of tin by integration of the Gibbs-Duhem equation. The activities are seen to be less than they would be if Raoult's law were obeyed; in such a case the departure from ideality is said to be negative, in distinction from the positive type shown in Fig. 10-5. The magnitude of the departure from

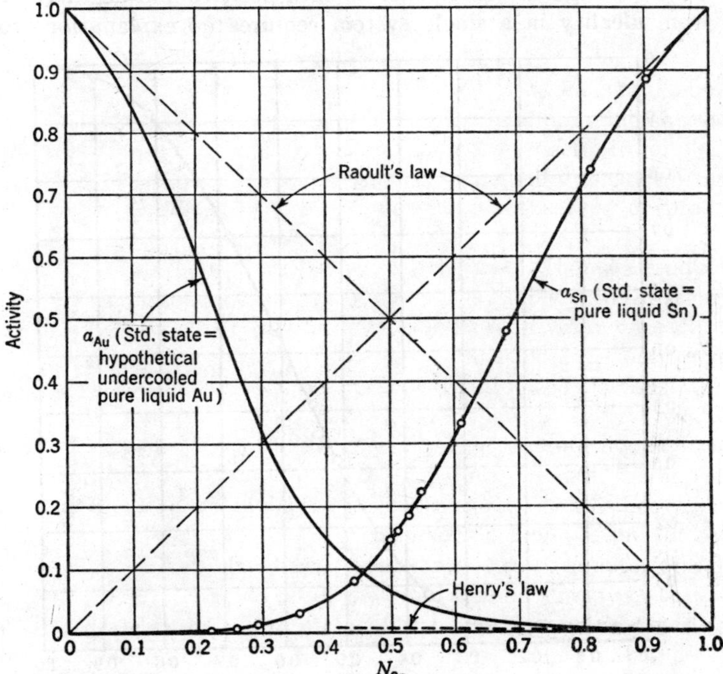

FIG. 10-10. Activity of tin and gold in the Sn-Au system at 600°C.

ideality is unusually great in this system; extrapolation of the line of Fig. 10-13 indicates an activity coefficient of tin of the order of 0.0001 at infinite dilution. The $(\log \gamma_{Sn})/(1 - N_{Sn})^2$ function is again finite, and moreover, the plot against N_{Sn} is in this case quite linear over the whole concentration range. This function will be discussed more extensively later; in a few systems it is nearly constant.

Occasionally both negative and positive departures from Raoult's law occur in the same binary system. An outstanding example of this behavior is found in the zinc-antimony system illustrated in Fig. 10-14.

* From the data of O. J. Kleppa, *J. Am. Chem. Soc.*, **72**, 3346 (1950).

The activity of zinc is less than its atom fraction at low concentration but greater than its atom fraction at high concentration. The activity coefficient, initially less than unity, rises continuously to values greater than unity and ultimately falls back to unity as Raoult's law is approached in the vicinity of pure zinc. A similar but less pronounced behavior is exhibited by the cadmium-bismuth and cadmium-antimony systems. Although the occurrence of both positive and negative departures from ideality in a single system requires no explanation from a

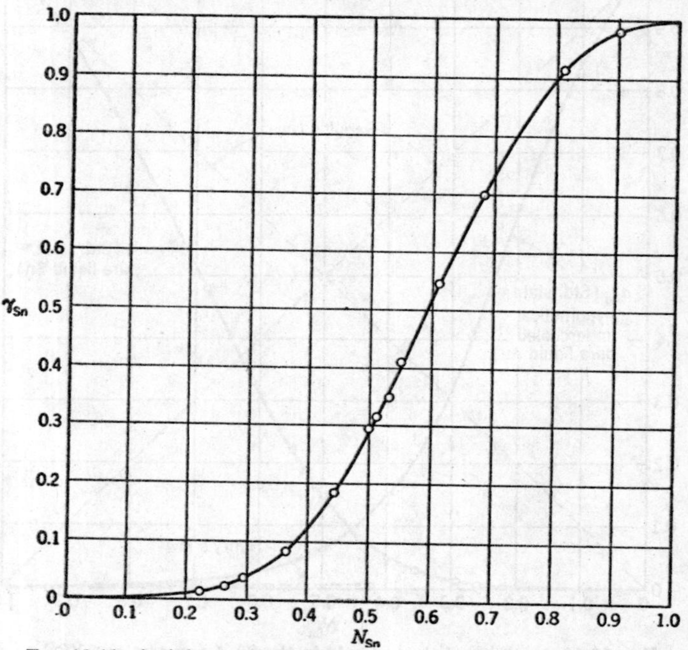

FIG. 10-11. Activity coefficient of tin in the Sn-Au system at 600°C.

thermodynamic viewpoint, it does from an atomistic viewpoint. We associate a negative departure with a tendency of one type of atom to group with the other and a positive departure with a tendency of each type of atom to group with similar atoms. If this reasoning is pursued in the present case, we are forced to the conclusion that such tendencies may change rather drastically with composition. If the grouping tendency is a function of the nature of the solution, then the change of this tendency in the examples cited is not too surprising, since each system ranges from metallic to semimetallic.

The sequences of Figs. 10-5 to 10-9 for the Cd-Pb system and 10-10 to 10-13 for the Sn-Au system reflect a very close conformity to Raoult's

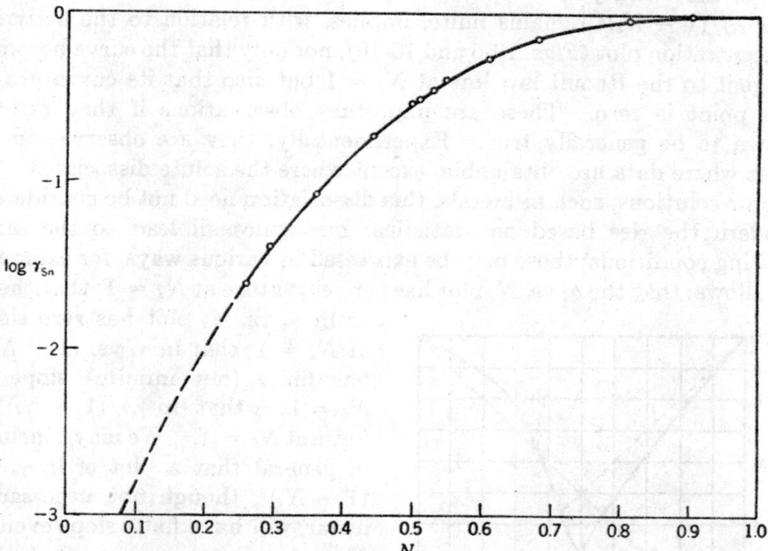

FIG. 10-12. Plot of log γ_{Sn} against N_{Sn} for the Sn-Au system at 600°C.

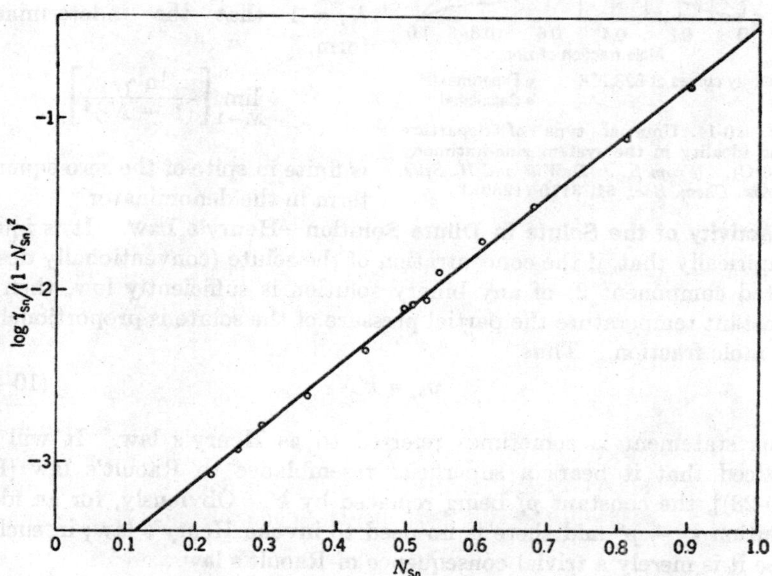

FIG. 10-13. Plot of log $\gamma_{Sn}/(1 - N_{Sn})^2$ against N_{Sn} for the Sn-Au system at 600°C. Slope is a constant.

law in the vicinity of the pure component. The fact that the function $(\log \gamma_i)/(1 - N_i)^2$ remains finite, implies, with relation to the activity-concentration plot (Figs. 10-5 and 10-10), not only that the curve becomes tangent to the Raoult law line at $N_i = 1$ but also that its curvature at this point is zero. These are important observations if they can be shown to be generally true. Experimentally, they are observed in all cases where data are obtainable, except where the solute dissociates. In atomic solutions, such as metals, this dissociation need not be considered. Modern theories based on statistical mechanics all lead to the same limiting conditions; these may be expressed in various ways, for example, as follows: that the a_i vs. N_i plot has zero curvature at $N_i = 1$; that the γ_i or $\ln \gamma_i$ vs. N_i plot has zero slope at $N_i = 1$; that $\ln \gamma_i$ vs. $(1 - N_i)^2$ has finite (not infinite) slope at $N_i = 1$; or that $(\ln \gamma_i)/(1 - N_i)^2$ is finite at $N_i = 1$. We may conclude in general that a plot of $\ln \gamma_i$ vs. $(1 - N_i)^2$, though not necessarily linear, will have finite slope even at $N_i = 1$. Another way of stating this is that compliance with Raoult's law is so good in the vicinity of $N_i = 1$ that the indeterminate form,

$$\lim_{N_i \to 1} \left[\frac{\ln \gamma_i}{(1 - N_i)^2} \right]$$

is finite in spite of the zero squared term in the denominator.

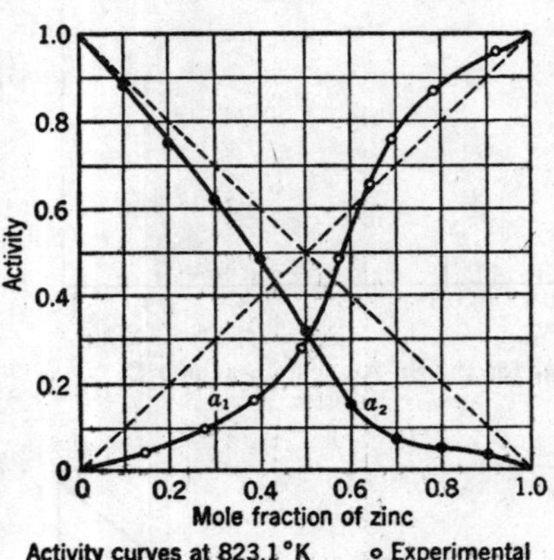

FIG. 10-14. Unusual type of departure from ideality in the system zinc-antimony (550°C). [*From B. J. DeWitt and H. Seltz, J. Am. Chem. Soc.*, **61**, 3170 (1939).]

Activity of the Solute in Dilute Solution—Henry's Law. It is found empirically that, if the concentration of the solute (conventionally designated component 2) of any binary solution is sufficiently low, then at constant temperature the partial pressure of the solute is proportional to its mole fraction. Thus,

$$p_2 = k'N_2 \tag{10-44}$$

This statement is sometimes referred to as Henry's law. It will be noticed that it bears a superficial resemblance to Raoult's law [Eq. (10-28)], the constant p_i° being replaced by k'. Obviously, for an ideal solution $k' = p_i^\circ$ and there is no need to invoke Henry's law; in such a case it is merely a trivial consequence of Raoult's law.

If the gas phase behaves ideally, as it usually does at low concentration and hence at low partial pressure of component 2, then $p_2 = f_2$ and Eq. (10-44) becomes

$$f_2 = k'N_2 \qquad (10\text{-}45)$$

Dividing through by f_2° and noting that $f_2/f_2^\circ = a_2$, we find that

$$a_2 = \frac{k'}{f_2^\circ} N_2$$

or

$$a_2 = kN_2 \qquad (10\text{-}46)$$

where $k = k'/f_2^\circ$. It is conventional at the present time to designate Eq. (10-46) rather than Eq. (10-44) as the primary statement of *Henry's law*. The advantage of this convention is that the validity of Henry's law does not depend upon ideality of the gas phase. Henry's law is now regarded as pertaining primarily to homogenous solutions rather than to the equilibrium between liquid or solid solution and gas.

It is important to realize that the validity of Henry's law depends upon the proper choice of solute species. In order to make this clear let us consider two cases: (1) a solution of nitrogen in water and (2) a solution of nitrogen in liquid iron. In the first case the nitrogen dissolves molecularly as N_2. The equilibrium may be written

$$N_2(g) = N_2(\text{dissolved in water})$$

The equilibrium constant is

$$K = \frac{a_{N_2(\text{in water})}}{a_{N_2(g)}}$$

Taking the activity of nitrogen in the gas phase as its partial pressure, and that in the aqueous phase from Eq. (10-46),

$$K = \frac{kN_{N_2}}{p_{N_2}}$$

or

$$N_{N_2} = \frac{K}{k} p_{N_2} = k'p_{N_2} \qquad (10\text{-}47)$$

a relation which is in accord with experiment.

In the second case under consideration nitrogen dissolves atomically (or ionically) in solid or liquid iron, as discussed previously. Here the equilibrium is

$$N_2(g) = 2N(\text{dissolved in iron})$$

and the equilibrium constant is

$$K = \frac{a^2_{N(\text{in iron})}}{a_{N_2(g)}}$$

Rewriting as before,

$$K = \frac{(kN_N)^2}{p_{N_2}}$$

and

$$N_N = \frac{\sqrt{K}}{k}\sqrt{p_{N_2}} = k'\sqrt{p_{N_2}} \tag{10-48}$$

This relation is also in accord with experiment; the marked difference from Eq. (10-47) will be noted. Since all the common diatomic gases N_2, O_2, H_2, etc., dissolve atomically in metals, Eq. (10-48) may be generalized to

$$N_2 = k'\sqrt{p_2} \tag{10-49}$$

where component 2 is diatomic in the gaseous state and has but slight solubility in the metal (component 1). This relation is known as *Sievert's law*, since it was established by his work on the solubility of gases in metals.

In case the gas is not entirely diatomic, Eq. (10-49) is still valid provided p_2 is interpreted as the partial pressure of the diatomic gaseous species. If for a solution of a normally diatomic gas in iron the temperature is so high or the pressure so low that the gas phase consists essentially of the monatomic species, then instead of Eq. (10-49) we find that $N_2 = k'p_2$, where p_2 now represents the partial pressure of the monatomic gas.

From these examples it is seen that Eqs. (10-44) and (10-46) are valid for dilute solutions provided the subscript 2 refers to the same molecular species. In dealing with metallic solutions there is little danger of confusion, since the solutions are atomic in all known cases. Thus it would be correct to say that the solubility of N_2 gas as N_2 molecules in liquid iron is proportional to p_{N_2} and also that the solubility of N gas is proportional to p_N. However, under normal circumstances nitrogen gas is predominantly N_2, and since nitrogen dissolves atomically in iron, it is more fruitful to say that the solubility of N_2 gas in liquid iron is proportional to $\sqrt{p_{N_2}}$.

DETERMINATION OF THE ACTIVITY OF ONE COMPONENT WHEN THAT OF THE OTHER IS KNOWN; THE GIBBS-DUHEM EQUATION

One of the most important relations involved in the thermodynamic treatment of solutions is Eq. (10-9). This relation, as noted previously, depends upon only one premise, namely, that the property G' under consideration is extensive; hence the molal property G depends only upon the temperature, pressure, and composition and not upon the amount of solution. Replacing G by the molal free energy F, Eq. (10-9) for a binary solution becomes

$$N_1 \, d\bar{F}_1 + N_2 \, d\bar{F}_2 = 0 \qquad (10\text{-}50)$$

Upon insertion of the definitional relation for activity, $d\bar{F}_i = RT \, d \ln a_i$, and division by RT this reduces to

$$N_1 \, d \ln a_1 + N_2 \, d \ln a_2 = 0 \qquad (10\text{-}51)$$

If the vapor in equilibrium with the solution is ideal, then $a_i = p_i$, and

$$N_1 \, d \ln p_1 + N_2 \, d \ln p_2 = 0 \qquad (10\text{-}52)$$

Or dividing by dN_1,

$$N_1 \left(\frac{\partial \ln p_1}{\partial N_1} \right)_{T,P} + N_2 \left(\frac{\partial \ln p_2}{\partial N_1} \right)_{T,P} = 0 \qquad (10\text{-}53)$$

the subscripts being written to call attention to the fact that this equation, like others in this section, is valid only at constant temperature and pressure. This equation is known as the *Gibbs-Duhem equation* or sometimes as the Duhem-Margules equation. As is customary, we shall refer to any equation immediately related thereto as the Gibbs-Duhem equation, whether it is written in terms of free energy [Eq. (10-9*a*)], fugacity, partial pressure [Eqs. (10-52) and (10-53)], activity [Eq. (10-51)], or activity coefficient [Eq. (10-57)].

Derivation of Raoult's Law for the Solvent When Henry's Law Is Obeyed by the Solute. By taking the logarithm of the Henry's law relation [Eq. (10-46)]

$$\ln a_2 = \ln k + \ln N_2$$

and differentiating,

$$d \ln a_2 = d \ln N_2$$

Insertion of this in Eq. (10-51), gives the relation

$$N_1 \, d \ln a_1 + N_2 \, d \ln N_2 = 0$$

valid for a solution so dilute that Henry's law holds for the solute. Rearranging,

$$d \ln a_1 = - \frac{N_2}{N_1} \, d \ln N_2$$

and utilizing the relations $d \ln N_i = dN_i/N_i$ and $dN_2 = -dN_1$,

$$d \ln a_1 = d \ln N_1$$

Integrating,

$$a_1 = IN_1$$

The integration constant I is seen to be unity, since, from the definition of activity, $a_1 = 1$ when $N_1 = 1$, whence

$$a_1 = N_1$$

a statement of Raoult's law [*cf.* Eq. (10-27)]. Hence we have demonstrated by means of the Gibbs-Duhem equation that, if Henry's law is valid for the solute, Raoult's law is valid for the solvent. Previously, we accepted Raoult's law for the solvent and Henry's law for the solute as separate empirical laws; however, we now see that they are related through the Gibbs-Duhem equation.

We shall find it convenient to regard Henry's law as empirical and Raoult's law as derived therefrom. In fact, if, by means of extra terms, the activity of the solute is expressed beyond the range where Henry's law is valid,

$$a_2 = kN_2(1 + bN_2 + \cdots)$$

then it may readily be shown by the above procedure that

$$a_1 = N_1(1 - bN_2^2 + \cdots)$$

Thus we see that, at reasonably low values of N_2, a small departure bN from Henry's law leads to a very small departure $-bN_2^2$ from Raoult's law. Attention has already been called to this high degree of validity of Raoult's law in the vicinity of the pure solvent.

Integration of the Gibbs-Duhem Equation. One of the problems arising frequently in dealing with binary solutions is the determination of the activity coefficient of one component when that of the other is known. This problem exists because it is often relatively easy to determine the activity, and hence the activity coefficient, of one component (by measurement of vapor pressure or by other methods to be discussed later) but the direct measurement of the activity of the other component is frequently difficult. The difficulty may arise, for example, from the low volatility of the other component, this preventing the experimental determination of its vapor pressure.

It is interesting to note, in anticipation of subsequent chapters, that all methods for determining the activity coefficient depend upon the experimental establishment of an equilibrium between phases. Aside from the emf method, this means that the activity coefficient of a component in one phase must be known in order that it may be found in the other phase. The phases in which the activity coefficient is known a priori are limited to ideal gases, phases of fixed composition, and dilute solutions. Hence in order to determine experimentally the activity coefficient of a component in solution it is necessary to equilibrate the solution either with a gas containing this component or with another phase either pure or nearly so. The measurement of an activity coefficient by the emf method requires an electrode which is reversible with respect to one component but irreversible with respect to the others. Hence it is seen that, in general, the determination of an activity coefficient may not be easy, and in some cases, we may be fortunate indeed to be able to measure the activity coefficient of even one component.

The fundamental thermodynamic relation involved in determining the activity coefficient of one component from that of the other is Eq. (10-9), G being the free energy:

$$N_1\,d\bar{F}_1 + N_2\,d\bar{F}_2 + \cdots = 0 \qquad (10\text{-}54)$$

In view of the definitional equation for activity, $d\bar{F}_i = RT\,d\ln a_i$, this becomes

$$N_1\,d\ln a_1 + N_2\,d\ln a_2 + \cdots = 0 \qquad (10\text{-}55)$$

or, for a two-component system,

$$N_1\,d\ln a_1 + N_2\,d\ln a_2 = 0$$

In principle this equation permits directly the evaluation of one activity from the other, for by transposing one term, dividing by N_1, and integrating, we find

$$\ln a_1 = \int_{N_1=1}^{N_1=N_1} -\frac{N_2}{N_1}\,d\ln a_2 \qquad (10\text{-}56)$$

This relation follows immediately from Eq. (10-14). As noted there, however, the integrand approaches infinity at $N_1 = 0$, and $\ln a_2$ approaches minus infinity at $N_1 = 1$; hence the graphical integration is difficult and lacks precision. The plot for this integration for the cadmium-lead system (Figs. 10-5 to 10-9) is shown in Fig. 10-15. Here $\log a_{Pb}$ at the arbitrary composition $N_{Cd} = N_{Pb} = \frac{1}{2}$ is represented by the shaded area to infinity; the difficulty of the evaluation is apparent.

Improved Integration of the Gibbs-Duhem Equation. We shall now obtain a relation similar to Eq. (10-55) in terms of activity coefficients. By differentiation of $N_1 + N_2 + \cdots = 1$, we obtain

$$dN_1 + dN_2 + \cdots = 0$$

Multiplying the first term by N_1/N_1, the second by N_2/N_2, etc. (each of which is unity),

$$N_1\frac{dN_1}{N_1} + N_2\frac{dN_2}{N_2} + \cdots = 0$$

or

$$N_1\,d\ln N_1 + N_2\,d\ln N_2 + \cdots = 0$$

Subtracting this from Eq. (10-55),

$$N_1\, d \ln \frac{a_1}{N_1} + N_2\, d \ln \frac{a_2}{N_2} + \cdots = 0$$

Now, by the definition of the activity coefficient, $\gamma_i = a_i/N_i$,

$$N_1\, d \ln \gamma_1 + N_2\, d \ln \gamma_2 + \cdots = 0 \tag{10-57}$$

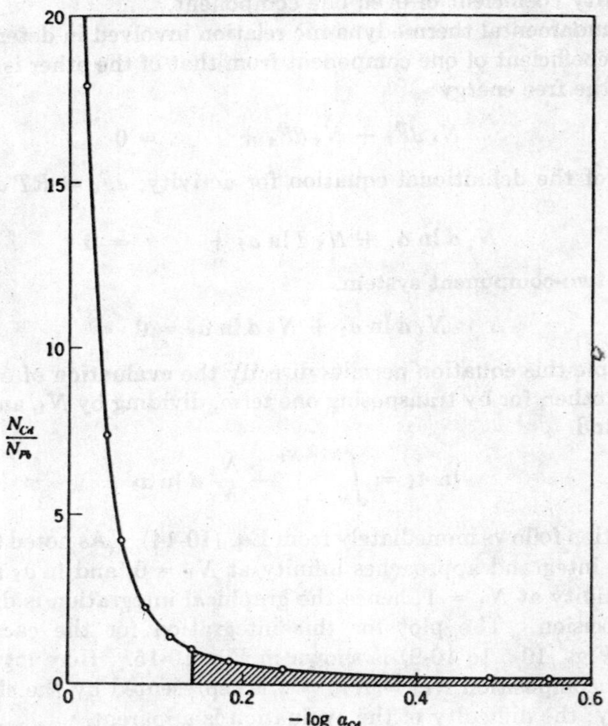

Fig. 10-15. Determination of a_{Pb} from a knowledge of a_{Cd} for the Cd-Pb system at 500°C by graphical integration of Eq. (10-56). Note the difficulty of evaluating the long tail to infinity.

From this relation, analogous to Eq. (10-55), the activity coefficient of one component of a two-component system can be obtained when that of the other is known over a range of composition. Again, by transposition of one term, division by N_1, and integration,

$$\ln \gamma_1 = - \int_{N_1=1}^{N_1=N_1} \frac{N_2}{N_1}\, d \ln \gamma_2 \tag{10-58}$$

Since γ_2 is always finite, the graphical evaluation of this integral presents no difficulty in the vicinity of $N_1 = 1$ as did the graphical integration of

Eq. (10-56). However, difficulty does arise if we wish to consider a solution low in concentration of component 1. Figure 10-16 illustrates the evaluation of log γ_{Pb} for the cadmium-lead system by this method.

From Eq. 10-58 it follows that, if γ_2 always increases as N_2 decreases, then γ_1 exceeds unity at all compositions. This follows, since, in such a case, $d \ln \gamma_2$ and the integral are always negative; hence $\ln \gamma_1$ is positive

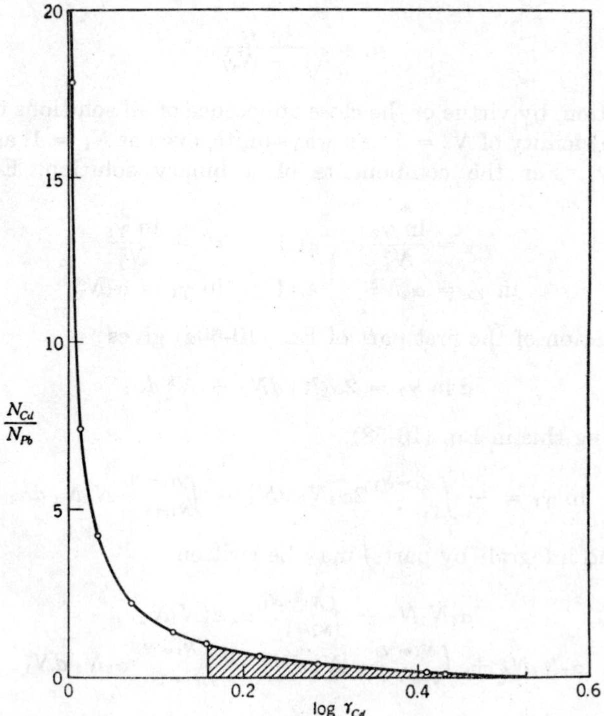

FIG. 10-16. Determination of γ_{Pb} from γ_{Cd} for the Cd-Pb system at 500°C by graphical integration of Eq. (10-58).

and γ_1 is greater than unity. Frequently the above condition accompanies a positive departure from ideality; however, this is not necessarily so. The statement sometimes made "if one component always exhibits positive departure from ideality then the other does also" is not universally correct—though frequently true. Conversely if γ_2 always decreases as N_2 decreases (the situation leading to negative departures), then γ_1 is less than unity at all compositions. Again, the fact that γ_2 is always less than unity is not sufficient to guarantee that γ_1 is also, as may be seen in Fig. 10-14. Unusual behavior of this type, involving both positive and negative departures from ideality in a single system, is also found in some silicate systems.

Further Aid in the Integration of the Gibbs-Duhem Equation; the Function α. Let us now consider a new function, here designated α, which considerably facilitates the integration of Eq. (10-58):

$$\alpha_i \equiv \frac{\ln \gamma_i}{(1 - N_i)^2} \tag{10-59}$$

This function, by virtue of the close adherence of all solutions to Raoult's law in the vicinity of $N_i = 1$, is always finite, even at $N_i = 1$, as discussed previously. For the components of a binary solution, Eq. (10-59) becomes

$$\alpha_2 = \frac{\ln \gamma_2}{N_1^2} \quad \text{and} \quad \alpha_1 = \frac{\ln \gamma_1}{N_2^2} \tag{10-60}$$

$$\ln \gamma_2 = \alpha_2 N_1^2 \quad \text{and} \quad \ln \gamma_1 = \alpha_1 N_2^2 \tag{10-60a}$$

Differentiation of the first part of Eq. (10-60*a*) gives

$$d \ln \gamma_2 = 2\alpha_2 N_1 \, dN_1 + N_1^2 \, d\alpha_2$$

Substituting this in Eq. (10-58),

$$\ln \gamma_1 = -\int_{N_1=1}^{N_1=N_1} 2\alpha_2 N_2 \, dN_1 - \int_{N_1=1}^{N_1=N_1} N_2 N_1 \, d\alpha_2$$

The second integral (by parts) may be written

$$\alpha_2 N_1 N_2 - \int_{N_1=1}^{N_1=N_1} \alpha_2 \, d(N_1 N_2)$$

or

$$\alpha_2 N_1 N_2 + \int_{N_1=1}^{N_1=N_1} \alpha_2 N_1 \, dN_1 - \int_{N_1=1}^{N_1=N_1} \alpha_2 N_2 \, dN_1$$

Whence

$$\ln \gamma_1 = -\alpha_2 N_1 N_2 - \int_{N_1=1}^{N_1=N_1} \alpha_2 \, dN_1 \tag{10-61}$$

In obtaining γ_1 from γ_2 by this method, the graphical integration does not involve the evaluation of an area with a tail to infinity. The graphical contribution for the evaluation of log γ_{Pb} from α_{Cd} for the cadmium-lead system is shown in Fig. 10-17. From Eq. (10-61) it will be seen that the activity coefficient of component 1 at infinite dilution, γ_1°, is given by the simple relation

$$\ln \gamma_1^\circ = \int_0^1 \alpha_2 \, dN_1 \tag{10-62}$$

thus showing that $\ln \gamma_1^\circ$ is the average value of α_2 over the entire composition range.

Although α is, in general, a function of composition, it is found occasionally that it is substantially constant. Eq. (10-60*a*) with α_2 constant, was originally suggested by Margules[1] (by an equation with an infinite series of terms, $\ln \gamma_2 = \alpha_2 N_1^2 + \beta N_2^3 + \cdots$); it was later reported on an empirical basis by Porter[2], and has more recently been derived by several others (including Scatchard, Rushbrooke, Kirkwood, Eyring, and

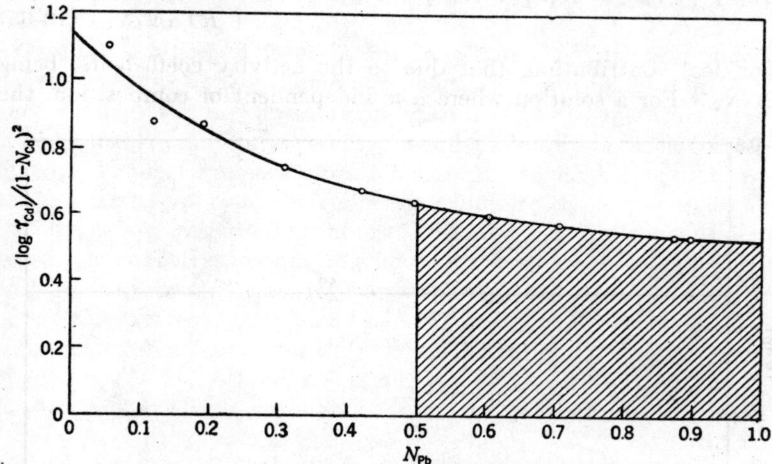

FIG. 10-17. Determination of γ_{Pb} from γ_{Cd} for the Cd-Pb system at 500°C by use of the α function. The shaded area represents the contribution of the final term of Eq. (10-61).

Hildebrand), using the methods of statistical mechanics. If α_2 is constant, Eq. (10-61) becomes

$$\ln \gamma_1 = -\alpha_2 N_1 N_2 - \alpha_2(N_1 - 1) = -\alpha_2 N_1 N_2 + \alpha_2 N_2 = \alpha_2 N_2(1 - N_1)$$

or

$$\ln \gamma_1 = \alpha_2 N_2^2 \tag{10-63}$$

which by comparison with Eq. (10-60*a*) shows these two equations to be symmetrical; *i.e.*, if α_2 is constant, γ_2 is the same function of N_1 as γ_1 is of N_2. Consequently, if α_2 is constant,

$$\alpha_1 = \alpha_2 = \alpha$$

so that

$$\ln \gamma_1 = \alpha N_2^2 \qquad \text{and} \qquad \ln \gamma_2 = \alpha N_1^2 \tag{10-64}$$

The thalium-tin system is one for which the α's are substantially constant and equal. Figure 10-18, from the data of Hildebrand and Sharma,[3]

[1] M. Margules, *Sitzber. Akad. Wiss. Wien.*, **104,** 1243 (1895).

[2] A. W. Porter, *Trans. Faraday Soc.*, **16,** 336 (1921).

[3] J. H. Hildebrand and J. N. Sharma, *J. Am. Chem. Soc.*, **51,** 462 (1929).

shows the constancy of $(\log \gamma_{Tl})/(1 - N_{Tl})^2$, which is $\alpha/2.303$; and Fig. 10-19 is a plot of the symmetrical relations [Eq. (10-64)] from the α so evaluated.

In such a case where it has been shown by experiment or adopted as an approximation that α_1 and α_2 are constant, we shall replace them by the single constant α. In this case the molal free energy takes the form

$$F = N_1\bar{F}_1 + N_2\bar{F}_2 = N_1F_1^\circ + N_2F_2^\circ + RT(N_1 \ln N_1 + N_2 \ln N_2) + RT\alpha N_1N_2 \quad (10\text{-}65)$$

the nonideal contribution, that due to the activity coefficients, being $RT\alpha N_1N_2$. For a solution where α is independent of composition, the

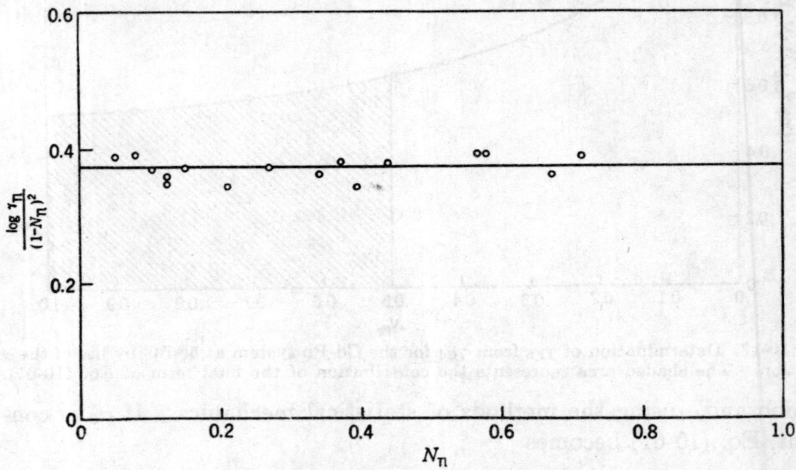

FIG. 10-18. Plot of α_{Tl} for the Tl-Sn system at 414°C, indicating it to be constant.

condition for imminent immiscibility may readily be found by setting the second and third derivatives of Eq. (10-65) equal to zero. The condition is found to correspond to $\alpha = 2$ and $N_1 = N_2 = \frac{1}{2}$. Thus any solution whose isothermal behavior can be described by Eq. (10-64) will exhibit a miscibility gap if α is greater than 2, or alternatively if F^{xs}, the nonideal contribution to the free energy, at $N_1 = N_2 = \frac{1}{2}$ is greater than $RT/2$. This is discussed in more detail in Chap. 13.

REGULAR SOLUTIONS

A concept which has proved to be of some use is that of the regular solution, originally introduced by Hildebrand.[1] A regular solution is defined as one for which the entropy of formation and hence the partial molal

[1] J. H. Hildebrand, *J. Am. Chem. Soc.*, **51**, 66 (1929).

entropies are the same as for an ideal solution. In order to see the significance of this, it is convenient to divide the thermodynamic properties of the solution into two parts—an "ideal" part and an "excess" part. Thus, in general, we may write

$$G = G^{id} + G^{xs} \tag{10-66}$$

where G is the value of the thermodynamic function for the actual solution, G^{id} is the value which the function G would have if the solution were ideal, and G^{xs}, defined by the equation, is the difference between the two.

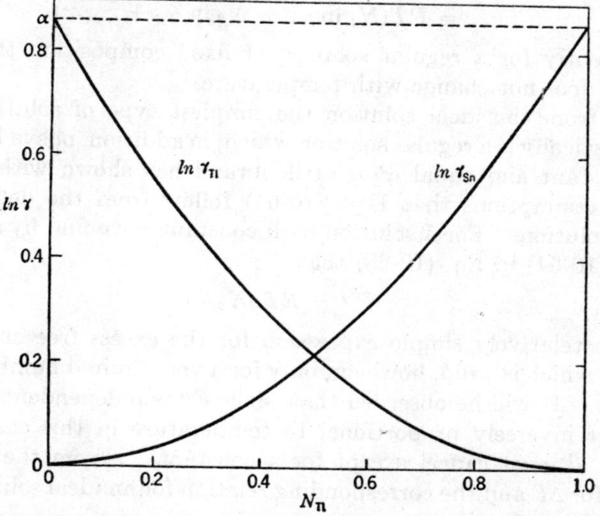

FIG. 10-19. Symmetrical plots of lnγ for the components of the Tl-Sn system at 414°C, as computed from the constant α evaluated in Fig. 10-18.

Subtracting the corresponding thermodynamic function for the pure components ($N_1G_1^\circ + N_2G_2^\circ + \cdots$) from each side, the change in the property during mixing becomes

$$\Delta G^M = \Delta G^{M,id} + G^{xs} \tag{10-67}$$

According to Eqs. (10-33*a*), (10-40), and (10-39), $\Delta H^{M,id} = 0$,

$$\Delta S^{M,id} = -R(N_1 \ln N_1 + N_2 \ln N_2 + \cdots)$$

and

$$\Delta F^{M,id} = RT(N_1 \ln N_1 + N_2 \ln N_2 + \cdots)$$

All thermodynamic relations valid at constant composition apply to G^{xs} as well as to G. For example, from the relation $(\partial\, \Delta F/\partial T)_P = -\Delta S$ it may readily be shown that $(\partial F^{xs}/\partial T)_P = -S^{xs}$, for by substitution in the former from Eq. (10-67), using the above values for $\Delta F^{M,id}$ and $\Delta S^{M,id}$,

$$R(N_1 \ln N_1 + N_2 \ln N_2 + \cdot\cdot\cdot) + \left(\frac{\partial F^{xs}}{\partial T}\right)_P = R(N_1 \ln N_1 + N_2 \ln N_2 + \cdot\cdot\cdot) - S^{xs}$$

which leads immediately to the relation sought. Hence it follows for a regular solution where S^{xs} is zero that $(\partial F^{xs}/\partial T)_P$ is also zero. Moreover, from the definition of the activity and the activity coefficient,

$$\begin{aligned} F^{xs} = \Delta F^M - \Delta F^{M,id} &= RT(N_1 \ln a_1 + N_2 \ln a_2 + \cdot\cdot\cdot) \\ &\quad - RT(N_1 \ln N_1 + N_2 \ln N_2 + \cdot\cdot\cdot) \\ &= RT(N_1 \ln \gamma_1 + N_2 \ln \gamma_2 + \cdot\cdot\cdot) \end{aligned} \tag{10-68}$$

Consequently for a regular solution of fixed composition the quantity $RT \ln \gamma_i$ does not change with temperature.

Aside from the ideal solution the simplest type of solution to treat mathematically is a regular solution which, in addition, obeys Eqs. (10-64) (has constant and equal α's). Hildebrand has shown with the aid of various assumptions that Eqs. (10-64) follow from the definition of a regular solution. For a solution with constant α we find by substitution of Eqs. (10-64) in Eq. (10-68) that

$$F^{xs} = RT\alpha N_1 N_2 \tag{10-69}$$

This is a relatively simple expression for the excess free energy of the solution, which is valid, however, only for a very limited number of actual solutions. It will be observed that, since F^{xs} is independent of temperature, α is inversely proportional to temperature in this case (since all terms in T must cancel except for a constant). From the definitional relation for ΔF and the corresponding relation for an ideal solution, noting that $\Delta H^{M,id} = 0$,

$$\Delta F^M = \Delta H^M - T\,\Delta S^M$$

and

$$\Delta F^{M,id} = \qquad - T\,\Delta S^{M,id}$$

Substituting these in Eq. (10-67),

$$F^{xs} = \Delta H^M - T(\Delta S^M - \Delta S^{M,id}) \tag{10-70}$$

Since for a regular solution $\Delta S^M = \Delta S^{M,id}$, it follows for such a solution that

$$F^{xs} = \Delta H^M \tag{10-71}$$

and from Eq. (10-69) we find for a regular solution obeying Eqs. (10-64) that

$$\Delta H^M = RT\alpha N_1 N_2 \tag{10-72}$$

Some metallic solutions approximately obey this parabolic equation.

A parabolic relationship for ΔH^M or F^{xs} should not, however, be taken as a demonstration that the solution is of simple regular type, for it frequently is found that ΔH^M and F^{xs} can adequately be expressed by the relations

$$\Delta H^M = bN_1N_2 \qquad \text{and} \qquad F^{xs} = b'N_1N_2 \tag{10-73}$$

where b and b' are unequal, and, by virtue of Eq. (10-70), ΔS^M is unequal to $\Delta S^{M,id}$. It is, in fact, unusual to find $\Delta S^M = \Delta S^{M,id}$ except for substantially ideal solutions. Figure 10-20 from the data of Elliott and

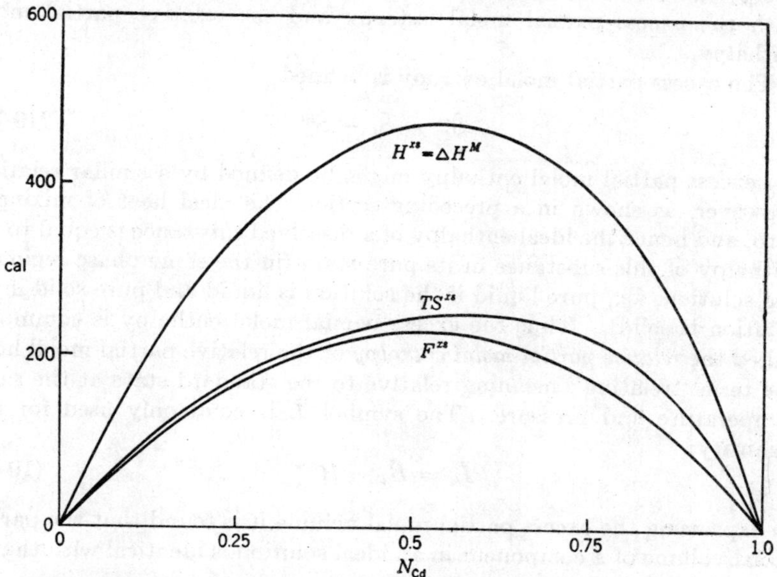

FIG. 10-20. Excess functions for the formation of 1 mole of molten cadmium-tin alloy at 500°C from the pure components.

Chipman[1] shows ΔH^M, TS^{xs}, and their difference F^{xs} for the cadmium-tin system at 500°C. ΔH^M and F^{xs} are nearly parabolic, but the system is far from regular.

EXCESS PARTIAL MOLAL QUANTITIES AND RELATIVE PARTIAL MOLAL ENTHALPY

It is frequently convenient to use excess partial molal quantities which are defined by a relation similar to that for the excess molal quantities [Eq. (10-66)]:

$$\bar{G}_i^{xs} = \bar{G}_i - \bar{G}_i^{id} \tag{10-74}$$

[1] J. F. Elliott and J. Chipman, *Trans. Faraday Soc.*, **47**, 138 (1951).

For the free energy,

$$\bar{F}_i^{xs} = \bar{F}_i - \bar{F}_i^{id} = RT \ln a_i - RT \ln N_i = RT \ln \gamma_i \qquad (10\text{-}75)$$

Thus it is seen that there is a close relationship between the excess partial molal free energy and the activity coefficient; in fact, the former may be avoided completely by use of the latter. The excess partial molal free-energy function is convenient, however, when comparison is to be made with the excess partial molal entropy and the relative partial molal enthalpy.

The excess partial molal entropy is defined

$$\bar{S}_i^{xs} = \bar{S}_i - \bar{S}_i^{id} \qquad (10\text{-}76)$$

An excess partial molal enthalpy might be defined by a similar relation. However, as shown in a preceding section, the ideal heat of mixing is zero, and hence the ideal enthalpy of a dissolved substance is equal to the enthalpy of this substance in its pure state (in the same phase region as the solution, *i.e.*, pure liquid if the solution is liquid and pure solid if the solution is solid). Thus the excess partial molal enthalpy is commonly called the *relative partial molal enthalpy* or the relative partial molal heat, the term "relative" meaning relative to the standard state at the same temperature and pressure. The symbol $\bar{L}$ is commonly used for this quantity:

$$\bar{L}_i = \bar{H}_i - H_i^{\circ}* \qquad (10\text{-}77)$$

In expressing the excess partial molal volume it is found that the partial molal volume of a component in an ideal solution is identical with that in the standard state.

All the general thermodynamic relations pertaining to partial molal quantities are valid also when expressed in excess partial molal quantities. This is immediately seen from the fact that any relation valid for a partial molal property of a component is valid for the ideal property and, hence, for their difference, the excess property. Thus, for example,

$$\left(\frac{\partial \bar{F}_i^{xs}}{\partial T}\right)_P = -\bar{S}_i^{xs} \qquad (10\text{-}78)$$

$$\left(\frac{\partial \bar{F}_i^{xs}}{\partial P}\right)_T = \bar{V}_i^{xs} \qquad (10\text{-}79)$$

$$\left[\frac{\partial(\bar{F}_i^{xs}/T)}{\partial(1/T)}\right]_P = \bar{L}_i \qquad (10\text{-}80)$$

* The final term of this equation might be written $\bar{H}_i^{\circ}$ to indicate that a standard state other than the pure component might be chosen.

Behavior of the α Function for Various Metallic Systems. The function $\ln \gamma_i/(1 - N_i)^2$, designated α_i, was introduced in a preceding section as an aid in the integration of the Gibbs-Duhem equation, to find the activity of one component of a binary system when that of the other is known over a range of composition. The α function is also found useful in obtaining the Henry's law constant by extrapolation to $N_i = 0$; Henry's law requires that α_i be finite at infinite dilution. This same requirement follows at $N_i = 1$ from Raoult's law. The behavior of the function at both these extremities has been found to be smooth.

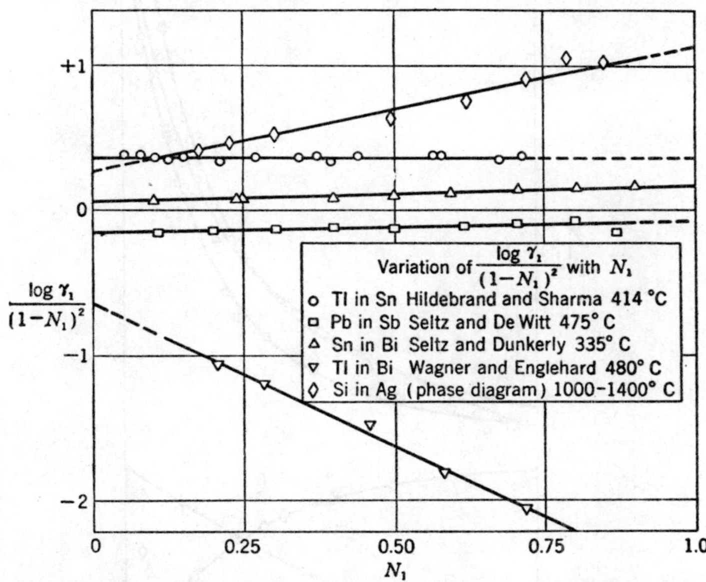

FIG. 10-21. Systems exhibiting a linear relation between α and composition.

In view of the importance of the α function, it seems worth while to consider further the general nature of its variation with composition for several systems. Before discussing the shapes of the various α function plots, it perhaps should be mentioned that any statement regarding the contour of the α function is also true of $\alpha/2.303 = \log \gamma/(1 - N)^2$. The simplest type of behavior is shown in Fig. 10-21, where the plots are linear, as is also the case for the tin-gold system (Fig. 10-13). In only one case, the thallium-tin system, is α constant, as shown previously in Figs. 10-18 and 10-19. Similar plots for several amalgams (Fig. 10-22) exhibit rather pronounced curvature; this may be regarded as a second type of behavior. Of course it could be argued that sufficiently precise data for the systems of Fig. 10-21 would probably indicate some slight curvature here, but it is rather surprising to find that relatively few of the systems investigated show intermediate behavior, *i.e.*, moderate curvature.

A third type is shown in Fig. 10-23, where the data can apparently be represented by two linear segments. It is not intended to imply that

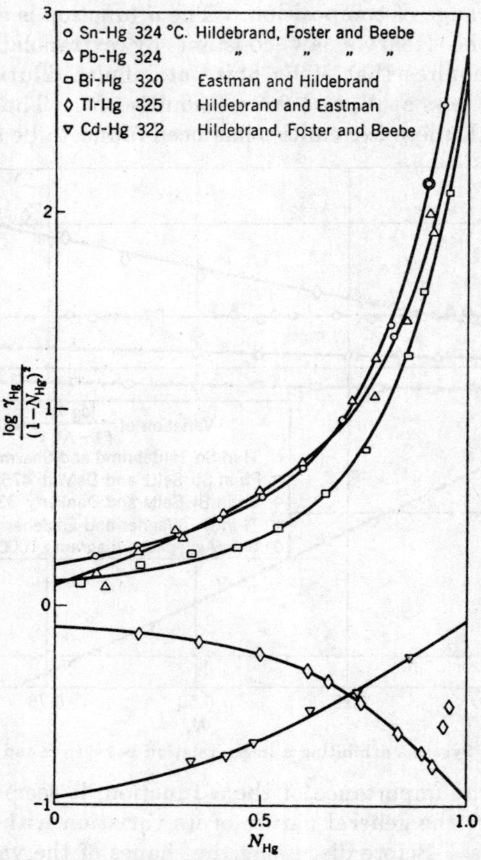

FIG. 10-22. Amalgams exhibiting a curved relation between α and composition.

there is an actual discontinuity in slope, as the data are far from adequate to rule out a gentler transition from one nearly linear segment to the other. In fact, the more recent data of Elliott and Chipman on the cadmium-lead system (Fig. 10-9) may be represented by a smooth curve more like those of the amalgams, in contrast to Taylor's data of Fig. 10-23, although the disparity in the two sets of data is small. Elliott and Chipman represented their data on the cadmium-bismuth system also by a smooth curve, whereas for the lead-bismuth system they show two linear segments, similar to those in Fig. 10-23. The most pronounced behavior of this type, as shown in Fig. 10-23, is that of the zinc-antimony system. It will be recalled that attention was previously brought to this system (Fig. 10-14) because it exhibited both positive and negative departures from ideality, again here indicated by the fact that α_{Sb} changes sign at about $N_{Sb} = 0.63$.

In considering the various contours of the α function plot it is of interest to consider the form of the α_2 plot implied via the Gibbs-Duhem equation

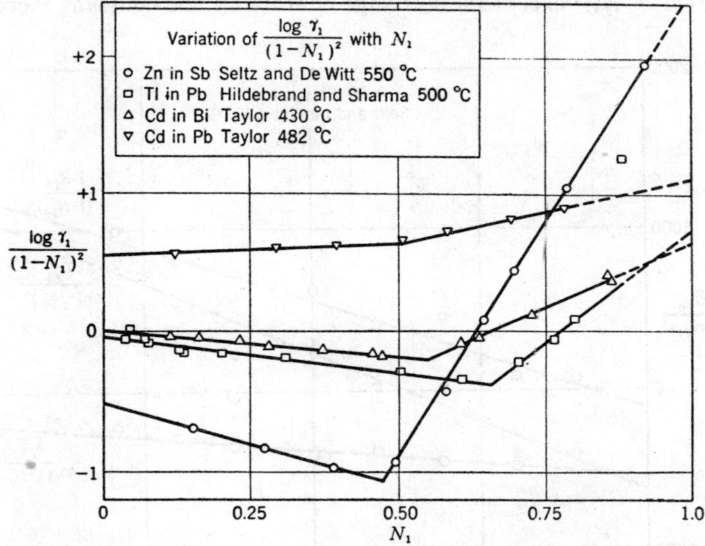

FIG. 10-23. Systems showing a third type of relation between α and composition.

by that of the α_1 plot. It was previously shown that, if α_1 is constant, as for thallium-tin (Fig. 10-21), then α_2 is also constant and moreover the α's are identical. It can also be shown that, if α_1 is a linear function of N, then α_2 is also a linear function with the same slope. Furthermore, the value of α_2 at $N_2 = 0$ is equal to the mean value of α_1, that is, the value of α_1 at $N_1 = N_2 = \frac{1}{2}$. In case α_1 is a smoothly curved function of composition (Fig. 10-22), no equally simple rule can be given for the determination of α_2; yet it can be stated, in general, that α_2 is in such case also a smoothly curved function of composition.

For the cases represented in Fig. 10-23 an entirely different situation prevails. If α_1 is represented as a function of N by two linear segments, then α_2 must be represented by one linear segment and one curved segment, the segments for the respective components joining at a single composition. In the case of the zinc-antimony system, α_{Sb} (derived from α_{Zn} indicated in Fig. 10-23) exhibits marked curvature in one segment, leading to a rather anomalous curve which suggests that either (1) more and better data would lead to a different curve for α_{Zn} than that shown in Fig. 10-23 or (2) a rather drastic change in the behavior of the antimony atom starts to occur at the composition where the segments join.

It has been noted previously that α_i defined as $\ln \gamma_i/(1 - N_i)^2$ is equal to $\bar{F}_i^{xs}/RT(1 - N_i)^2$; that is, an isothermal plot of α_i is equivalent to a plot of $\bar{F}_i^{xs}/(1 - N_i)^2$ except for a change of scale by the constant factor RT.

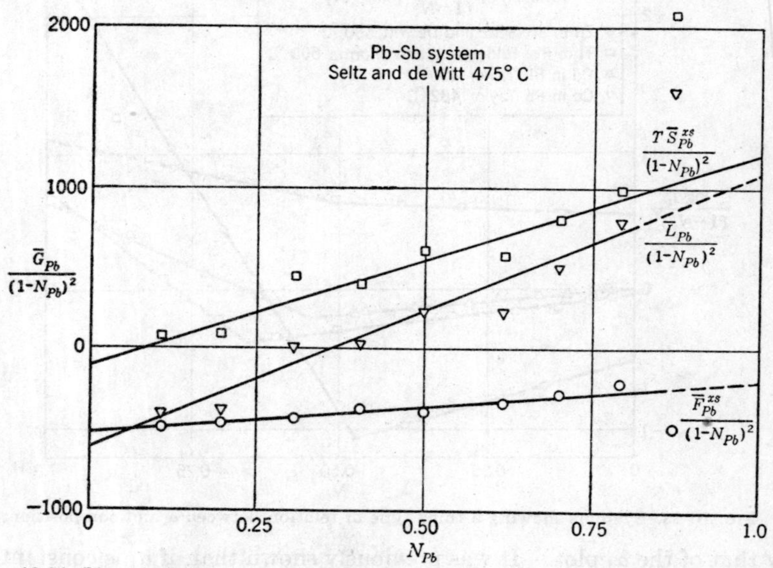

FIG. 10-24. Linear relation between composition and the ratio of various partial molal quantities to $(1 - N_i)^2$.

A similar statement is true for the plot $\log \gamma_i/(1 - N_i)^2$ (Figs. 10-21 to 10-23), the scale in this case being related to $\bar{F}_i^{xs}/(1 - N_i)^2$ by the factor $2.303RT$. The line for the lead-antimony system of Fig. 10-21 is thus replotted as the lowest line of Fig. 10-24. This linear relation for the free energy leads us to speculate as to whether a similar relation is true for the relative partial molal heat and the excess partial molal entropy. The upper curves of Fig. 10-24 indicate that such is the case for this system, although the experimental error is larger for these latter functions, and as usual is magnified as N_i approaches unity. In general it is to be expected that linearity of the heat and entropy curves as in Fig. 10-24 follows from linearity of the free-energy or α curve (Fig. 10-21). Similar generalizations from the types of curves in Figs. 10-22 and 10-23 must await further experimental data.

VOLUME FRACTION AND THE HILDEBRAND SOLUBILITY PARAMETER

Hildebrand and Scott[1] have found that departures from ideality in liquid metallic solutions can be represented more simply in terms of volume fraction than in terms of atom fraction. The volume fractions of the components of a binary solution are defined:

$$\varphi_1 = \frac{N_1V_1}{N_1V_1 + N_2V_2} \qquad \text{and} \qquad \varphi_2 = \frac{N_2V_2}{N_1V_1 + N_2V_2}$$

V being the molal volume of the pure component designated by the subscript. For several systems they found that

$$\bar{F}_1^{xs} = k_\varphi \varphi_2^2$$

where k_φ is substantially constant. For the case that k_φ is constant the excess free energy of formation of the solution ΔF^{xs} is equal to $k_\varphi\varphi_1\varphi_2$, an expression very similar to their theoretical expression for the energy of mixing ΔE^M of a regular solution. The latter is

$$\Delta E^M = V\left[\left(\frac{\Delta E_1^V}{V_1}\right)^{\frac{1}{2}} - \left(\frac{\Delta E_2^V}{V_2}\right)^{\frac{1}{2}}\right]^2 \varphi_1\varphi_2$$

where ΔE^V is the energy of vaporization of the component denoted by the subscript.

The attempts to correlate the respective coefficients of the product $\varphi_1\varphi_2$ are not particularly successful. Nevertheless, Hildebrand and Scott feel that the term $(\Delta E_1^V/V_1)^{\frac{1}{2}}$, which they call the solubility parameter, is of some value. For example, the condition for complete miscibility is given as

$$\frac{V_1 + V_2}{2}\left[\left(\frac{\Delta E_1^V}{V_1}\right)^{\frac{1}{2}} - \left(\frac{\Delta E_2^V}{V_2}\right)^{\frac{1}{2}}\right]^2 < 2RT$$

A check of this is difficult because of the scarcity of good data; for liquid-metal systems the results are not impressive. When the solubility parameter is plotted against atomic radius or electronegativity for a number of elements whose solubility in a given metal is known in the attempt to obtain a grouping of the soluble elements as in Fig. 4-8 or 4-9, the grouping is inferior to that obtained when electronegativity is plotted against atomic radius.

[1] J. H. Hildebrand and R. L. Scott, "The Solubility of Nonelectrolytes," Reinhold Publishing Corporation, New York, 1950.

CONVERSION FROM ONE STANDARD STATE TO ANOTHER

It frequently happens that $\Delta F°$ is known as a function of temperature for a particular choice of standard states of reactants and products, but it is desired to transform to a different set of standard states. For example, we may know the free energy of oxidation of a pure metal M according to the reaction

$$\text{M(pure l)} + O_2\text{(g)} = MO_2\text{(cryst.)} \qquad \Delta F_a^\circ = \Delta H_a^\circ - T\,\Delta S_a^\circ \tag{a}$$

but we wish to consider the oxidation of M in dilute solution in a solvent, say iron. Accordingly, we wish to express the activity of M relative to

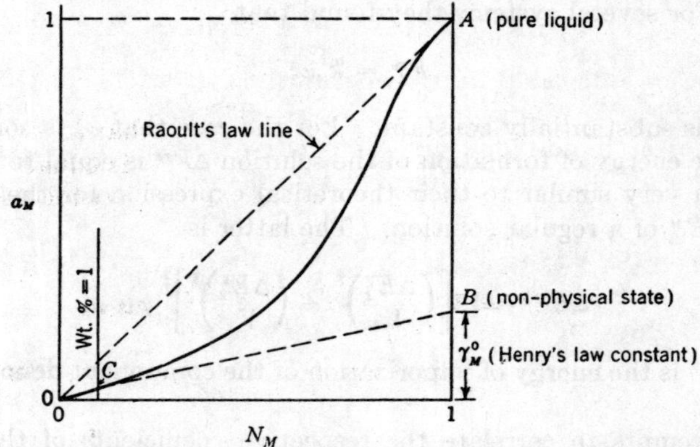

Fig. 10-25. Schematic diagram illustrating various standard states of a solution, to aid in understanding the conversion of activities from one standard state to another.

the nonphysical standard state such that a_M approaches N_M at infinite dilution, since then the activity can be expressed relatively simply in terms of concentration. The conversion, of course, involves a knowledge of the free energy of transfer of M from one standard state to the other and clearly requires some experimental knowledge of the activity-composition relation.

From Fig. 10-25 it is seen that we wish to find the free-energy change accompanying the isothermal transfer of M from state A to state B; this is $RT \ln (a_{M(B)}/a_{M(A)})$, which, since $N_M = 1$ in both cases and $\gamma_A = 1$, is $RT \ln \gamma_M^\circ$ where γ_M° is the activity coefficient of M at infinite dilution in iron. The transfer may be expressed

$$\text{M(pure l)} = \text{M(nonphysical state)} \qquad \Delta F_b = RT \ln \gamma_M^\circ \tag{b}$$

As γ_M° is in general a function of temperature, it should be so represented, if possible, and ΔF_b may be conveniently represented $\Delta F_b = \Delta H_b - T\Delta S_b$, where ΔH_b and ΔS_b are constants. Subtraction of Eq. (*b*) from Eq. (*a*) gives

$$M(\text{nonphys.}) + O_2(g) = MO_2(\text{cryst.}) \qquad \Delta F_c^\circ = \Delta F_a^\circ - RT \ln \gamma_M^\circ \quad (c)$$

Obviously $K_c = K_a \gamma_M^\circ$. The advantage of Eq. (*c*) over Eq. (*a*) is that at low concentration the activity of M (based on the nonphysical standard state) may be set equal to its atom fraction.

For further simplification at low concentration a still different choice of standard state is commonly adopted, namely, one such that the activity approaches the weight per cent at infinite dilution. This standard state is represented by point C of Fig. 10-25, situated at unit weight per cent on a straight line connecting the origin with B. Although, strictly, this state is also nonphysical, the departure from the curve is frequently negligibly small. The free-energy change on transferring a gram atom of M from state B to state C is $RT \ln (a_{M(C)}/a_{M(B)})$. From the figure it is seen that the activity ratio is equal to the corresponding atom fraction ratio. The atom fraction of M in a 1-weight-per-cent solution in iron is approximately 0.01(55.85)/(at. wt.)$_M$ and that in the nonphysical state B is unity, so that $a_{M(C)}/a_{M(B)} \cong 0.5585/(\text{at. wt.})_M$ and, at low concentration,

$$M(\text{nonphysical state}) = M(1\% \text{ in Fe}) \qquad \Delta F_d = RT \ln \frac{0.5585}{(\text{at. wt.})_M} \quad (d)$$

It will be noted that this is a purely formal type of equation and the evaluation of ΔF_d requires no experimental knowledge of the behavior of the solution. Subtraction of Eq. (*d*) from Eq. (*c*) gives

$$M(1\% \text{ in Fe}) + O_2(g) = MO_2(\text{cryst.})$$

$$\Delta F_e = \Delta F_a - RT \ln \gamma_M^\circ - RT \ln \frac{0.5585}{(\text{at. wt.})_M} \quad (e)$$

The expression for the equilibrium constant for Eq. (*e*) involves the concentration of M in iron simply as the weight per cent.

Similar procedure may be adopted for other types of equilibria and for solvents other than iron.

NONMETALLIC SOLUTIONS AT ELEVATED TEMPERATURE; MOLTEN SALT SOLUTIONS AND SLAGS

Thermodynamic data on molten nonmetallic solutions at elevated temperature are rather meager. Even the constitution of such solutions is still in the speculative stage. Perhaps the best guess as to their structure is provided by analogy with the corresponding solid solutions; the similarity of structure of liquids and solids has been pointed out previously. It will be recalled from Chap. 3 that solid chlorides, such as NaCl, are ionic rather than molecular in nature, and available evidence, such as that from the electrical conductivity, indicates that fused salts are usually ionic in nature. This raises the question as to whether we may legitimately expect these fused salt solutions to behave in substantially the same manner as metallic solutions and solutions of organic substances at room temperature. Should we expect them to obey Raoult's law under these conditions? It will be remembered that Raoult's law was developed empirically for molecular solutions (organic molecules) and subsequently was shown to hold, as a limiting law, for metallic (atomic) solutions.

The available evidence would seem to indicate that, for liquid binary solutions involving a common ion, Raoult's law is obeyed as a limiting law, that is, a_i approaches N_i as N_i approaches unity. Measurements by Greiner and Jellinek[1] of partial pressures in the three systems NaCl-KCl, NaI-KI, and KCl-KI at 1180°C indicate that all three systems are substantially ideal.

Nonideal Behavior of Some Salt Solutions. However, solid solutions of the halides of the alkali metals do not behave ideally. This is indicated by the measurable heat of formation of such solutions. Some of these heats, taken from a tabulation by W. E. Wallace,[2] are given below; the figures pertain to a solid solution containing $\frac{1}{2}$ mole of each of the indicated salts.

Salts	ΔH^M_{298}, *cal*
KCl, RbCl	203
KBr, KCl	232
NaCl, NaBr	335
KBr, KI	390
NaI, KI	625
NaBr, KBr	700
NaCl, KCl	1050

If we assume that the excess free energy is comparable to ΔH^M, that is, that the solutions are nearly regular; then it appears that some of the above solid solutions are metastable, since some of the tabulated values exceed $RT/2$ (see page 266).

By measurement of the electromotive force of suitable cells at 500 to 600°C Hildebrand and Salstrom[3] have investigated the activity of silver bromide in solution in the molten bromides of lithium, sodium, potassium, and rubidium. The activity of AgBr in each system is shown

[1] B. Greiner and K. Jellinek, *Z. physik. Chem.* **A165**, 97 (1933).

[2] W. E. Wallace, *J. Chem. Phys.*, **17**, 1095 (1949).

[3] E. J. Salstrom and J. H. Hildebrand, *J. Am. Chem. Soc.*, **52**, 4650 (1930). E. J. Salstrom, *ibid.*, **53**, 1794, 3385 (1931); **54**, 4252 (1932).

as a function of its mole fraction in Fig. 10-26. Hildebrand and Salstrom[1] point out that each system is regular in that the excess free energy of mixing is substantially independent of temperature and hence the entropy of mixing is equal to the ideal entropy of mixing. Furthermore, they show that the excess partial molal free energy of AgBr in each of

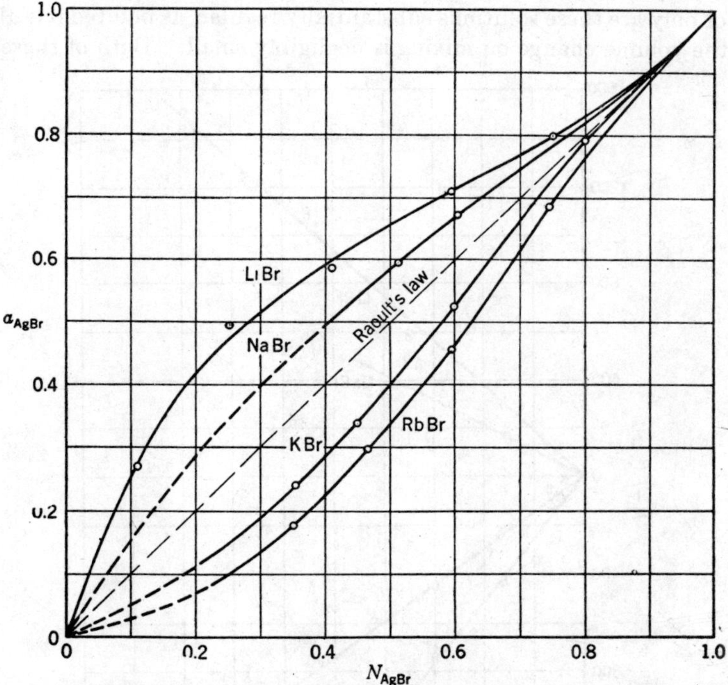

Fig. 10-26. Activity of silver bromide in liquid binary solutions of the alkali bromides at 550°C.

these systems may be represented by the relation

$$\bar{F}_1 - \bar{F}_1^{id} = bN_2^2$$

as illustrated in Fig. 10-27. The constant b, which is apparently independent of temperature, has the values 1880, 1050, −1480, and −2580 for solutions of AgBr in the bromides of lithium, sodium, potassium, and rubidium, respectively. From Eq. (10-75) it is seen that this relation is equivalent to Eq. (10-64), b being equal to $RT\alpha$. In so far as b is truly constant, then, the α's for the components of each system are constant and equal. Hildebrand and Salstrom attribute the departures from ideality to the fact that the alkali bromides are nearly completely ionic in

[1] J. H. Hildebrand and E. J. Salstrom, *J. Am. Chem. Soc.*, **54**, 4257 (1932).

nature whereas silver bromide is less ionic (partially covalent). They further suggest that the increase in size of the alkali ion in the series lithium, sodium, potassium, rubidium (see Figs. 3-11 and 3-12) is responsible for the change from positive to negative departure—a suggestion which may have some virtue but is not entirely satisfying.

Not only are these solutions substantially regular, as pointed out above, but the volume change on mixing is negligibly small. Both of these con-

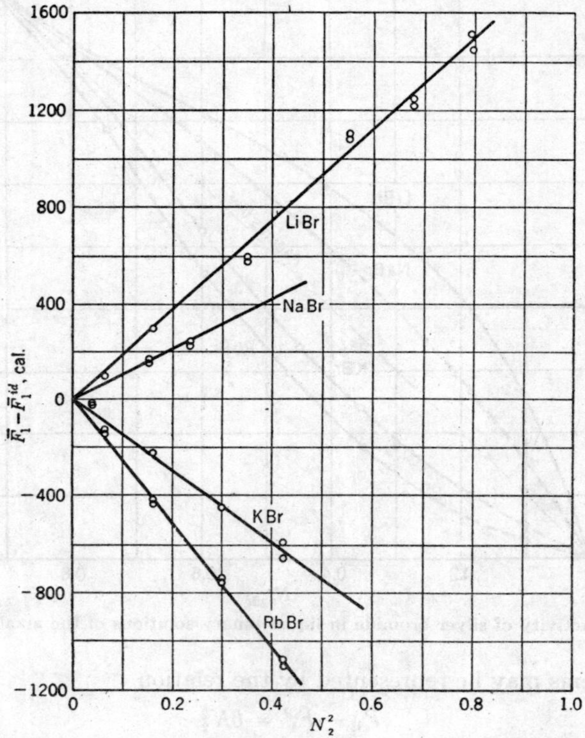

FIG. 10-27. Excess partial molal free energy of silver bromide, $\bar{F}_1 - \bar{F}_1{}^{id}$, in liquid binary solutions of the alkali bromides. [*From Hildebrand and Salstrom, J. Am. Chem. Soc.*, **54**, 4257 (1932).]

siderations militate against the postulation of the formation of a complex ion (such as $AgBr_2^-$) to explain the negative departures. According to Thurmond and Hildebrand[1] the molar refractivity of the solutions furnishes further evidence against this postulate.

All fused salts considered so far have been uni-univalent, and the pairs have had a common ion. Available data on uni-bivalent salts are typified

[1] C. D. Thurmond and J. H. Hildebrand, American Chemical Society Meeting, September, 1949.

by the work of Hildebrand and coworkers on the pairs $PbCl_2$-$PbBr_2$* and $PbBr_2$-$ZnBr_2$.† Hildebrand and Ruhle investigated the system $PbCl_2$-KCl‡ exemplifying a simple mixed type of electrolyte. Little work has been done on electrolytes of general mixed types, *i.e.*, those with anions different, cations different, and valences different. A new problem of a theoretical nature arises here in that it is no longer obvious, in view of the ionization, how an ideal solution is to be defined. A suggestion for the treatment of the general ionic solution is known as *Temken's rule*, which states that for an ideal solution of this type the activity of any particular ion species is equal to the ratio of the number of such ions to the total number of ions of like charge.

The above and similar evidence in the literature would seem to justify the conclusion that for a binary salt solution containing a single anion or a single cation Raoult's law is approached as a limiting law, just as in the case of a metallic solution. It seems natural to extend this conclusion to all solutions with a common ion, *e.g.*, the ternary solution of $PbCl_2$, AgCl, and NaCl; thus it would seem reasonable to suppose that in all such cases a_i approaches N_i as N_i approaches unity. The departure from ideality at finite concentration seems to be of the same order as that in metallic solutions. It also seems reasonable to extend these conclusions to systems of oxides or sulfides which do not involve a complex ion. For example, we might reasonably expect a solution of FeO and MnO to be nearly ideal; the meager available evidence substantiates this.

Complex Ions. Jander and Herrmann§ found definite evidence of the fluoaluminate ion, AlF_6^{3-}, in their investigation of the fused salt system AlF_3-NaF. This is hardly surprising, since this ion is well known elsewhere, including aqueous solutions; it might well be compared with the fluosilicate ion or even one of the silicate ions.

Silicate solutions are particularly complex. Silicon appears to have a strong tendency to exhibit tetrahedral coordination, and in oxidic solution it tends to form complex anions such that each silicon atom is surrounded by four oxygen atoms. Thus in a solution corresponding in composition to calcium metasilicate, $CaO{\cdot}SiO_2$, the simplest anion with tetrahedral silicon atoms is $(Si_2O_6)^{4-}$, which may be represented

```
 / O       O       O \ 4-
|    \   /   \   /    |
|     Si       Si     |
|    /   \   /   \    |
 \ O       O       O /
```

* E. J. Salstrom and J. H. Hildebrand, *J. Am. Chem. Soc.*, **52**, 4641 (1930).

† A. Wachter and J. H. Hildebrand, *J. Am. Chem. Soc.*, **52**, 4655 (1930).

‡ J. H. Hildebrand and G. C. Ruhle, *J. Am. Chem. Soc.*, **49**, 722 (1927).

§ W. Jander and H. Herrmann, *Z. anorg. allgem. Chem.*, **239**, 65 (1938).

The partial (about 50 per cent) covalent nature of the silicon-oxygen bond may be regarded as accounting for the stability of such a complex anion, the stability being regarded as the resistance to breaking down into simpler species such as Si^{4+} and O^{2-}. The subject has been discussed in some detail by Chipman.* In solutions with less silica, *i.e.*, higher CaO/SiO_2 ratio, which may be regarded as containing some orthosilicate, $2CaO{\cdot}SiO_2$, the simplest orthosilicate ion satisfying the tetrahedral coordination for silicon is $(SiO_4)^{4-}$. Strong evidence of the ionic nature of silicate solutions is furnished by their high electrical conductivity—of the order of that of aqueous solutions of strong electrolytes. It seems safe to conclude that they are highly if not completely ionized. The precise nature of the ions must as yet be regarded as not established.

Even the limited thermodynamic data available on silicate systems indicate anomalous behavior. The formation of intermediate phases or compounds is usually associated with negative departure from ideality on the part of the liquid solutions at higher temperature and this is borne out in the system FeO-SiO_2 by the thermodynamic investigation of Schuhmann and Enzio.† However, the miscibility gap on the high-silica side of the system at elevated temperature is incontrovertible evidence of positive departure from ideality. Thus the thermodynamic behavior of the system is somewhat similar to that of the zinc-antimony system considered previously. Binary melts of MgO-SiO_2, CaO-SiO_2,‡ or SrO-SiO_2 behave similarly. Atomistically, Richardson§ pictures this transition from positive to negative departure from ideality as the transition from a liquid composed predominantly of a silica network, similar to that of solid silica in which oxygen atoms are shared by silicon atoms, to a liquid somewhat similar to a randomized fayalite which may be viewed as composed of Fe^{++} and SiO_4^{4-} ions.

In view of the lack of knowledge of the fundamental nature of silicate solutions, it is difficult to formulate the thermodynamics thereof; an "ideal" behavior cannot be formulated, much less departures therefrom which we usually associate with solutions of strong electrolytes. Experimental determination of activities in such systems would seem to be a fruitful field for future investigation.

* J. Chipman, *Trans. ASM*, **30**, 817 (1942).

† R. Schuhmann and P. J. Enzio, *Trans. AIME*, **3**, 401 (1951).

‡ The drastic departure from ideality of the system CaO-SiO_2 is depicted in Fig. 13-15.

§ F. D. Richardson, Discussions of the Faraday Society, Vol. IV, p. 244, 1948.

CHAPTER 11

PHASE RELATIONS

So far our thermodynamic treatment has been centered upon homogeneous substances. Actually a single homogeneous phase cannot be isolated; it is patently impossible to contain or support it or to make any sort of physicochemical measurement upon it without introducing one or more other phases. Every system with which we deal has a boundary, and it is this boundary to which some other substance, *e.g.*, one serving as a measuring instrument, must be brought in order to carry out the desired measurement.

In determining the activity of a component of a phase, the presence of another phase is an inherent requirement. The activity, by definition, is a relative quantity, being taken as unity in some arbitrary phase. Either this arbitrary phase or some other phase of known activity must always be present in contact with the phase whose activity is to be determined. For example, if the activity of the volatile component of a liquid is to be measured by the vapor-pressure method it is obvious that the vapor phase must be present. It is similarly clear that liquid and vapor must be present in the determination of the boiling point and liquid and solid in the determination of the freezing point. Also in the measurement of activity by the emf method, to be described later, the introduction of appropriate electrodes is essential. Thus we see that, experimentally, heterogeneous systems are not only common—they are inescapable.

In dealing with equilibrium we are generally concerned with *heterogeneous equilibrium*—equilibrium involving more than one phase. The term *homogeneous equilibrium* applies to an equilibrium within a single phase and, as most frequently used, refers to molecular equilibria such as those involving gases (for example, $2NO_2 = N_2O_4$) or organic substances. Except in the case of gases we, as metallurgists, shall not deal much with homogeneous equilibrium as applied to specific reaction. We have already found that the solids and liquids with which we deal are not molecular in structure, and hence it would be meaningless to write chemical equations similar to the gas equilibrium above for metallic systems. For example, from our earlier considerations in Chap. 4, it is apparent that no meaning is assignable to the equation $Cu + Au = CuAu$ for a homogeneous copper-gold alloy, since the bonds resonate between any particular atom and all 12 of its nearest neighbors.

However, in this system ordering may prevail to a marked extent or, by quenching, a disordered state may be attained. Hence it would be quite proper to inquire whether homogeneous equilibrium prevails in an alloy of this system—the meaning of such inquiry being whether the equilibrium extent of order prevails. Similarly a cold-worked metal is not in the equilibrium state—in the sense that the atoms are not located at the sites of a regular lattice. Here also it would be quite proper to say that homogeneous equilibrium does not prevail. In such cases it is proper to speak of homogeneous (or internal) equilibrium or lack thereof, although a chemical equation in the usual sense cannot be written because of the absence of definite molecules. We shall now direct our attention to the thermodynamic treatment of heterogeneous equilibrium and phases of variable composition.

Combined Statement of First and Second Laws for a Phase of Variable Composition. Regarding the free energy of a single phase at internal equilibrium as a function of temperature, pressure, and composition, we may write, from the fundamental theorem of partial differentiation (in the absence of electric, magnetic and gravitational fields, surface energy effects, etc.),

$$dF' = \left(\frac{\partial F'}{\partial P}\right) dP + \left(\frac{\partial F'}{\partial T}\right) dT + \left(\frac{\partial F'}{\partial n_1}\right) dn_1 + \left(\frac{\partial F'}{\partial n_2}\right) dn_2 + \cdot \cdot \cdot \quad (11\text{-}1)$$

A prime is used in this chapter to denote a property of an indefinite amount of a phase; this is in distinction from the unprimed notation which, as usual, refers to a molal property. By comparison with the combined statement of the first and second laws for a phase of fixed composition, we see immediately that the first two partial differential coefficients are V' and $-S'$, respectively; the remaining coefficients we have already denoted as $\bar{F}_1$, $\bar{F}_2$, etc. Thus, we have

$$dF' = V'\,dP - S'\,dT + \bar{F}_1\,dn_1 + \bar{F}_2\,dn_2 + \cdot \cdot \cdot \quad (11\text{-}2)$$

We may also write the total derivative of the energy in terms of volume, entropy, and composition:

$$dE' = \left(\frac{\partial E'}{\partial S'}\right) dS' + \left(\frac{\partial E'}{\partial V'}\right) dV' + \left(\frac{\partial E'}{\partial n_1}\right) dn_1 + \left(\frac{\partial E'}{\partial n_2}\right) dn_2 + \cdot \cdot \cdot \quad (11\text{-}3)$$

The first two coefficients are immediately recognized from Eq. (7-13a) as T and $-P$. The coefficients $(\partial E'/\partial n_1)$, $(\partial E'/\partial n_2)$, etc., were designated *chemical potentials* by Gibbs and are represented by the symbols μ_1, μ_2, etc. Equation (11-3) becomes

$$dE' = T\,dS' - P\,dV' + \mu_1\,dn_1 + \mu_2\,dn_2 + \cdots \tag{11-4}$$

Combination of this equation with the derivative of the definitional equation for free energy, $F' = E' + PV' - TS'$, gives

$$dF' = V'\,dP - S'\,dT + \mu_1\,dn_1 + \mu_2\,dn_2 + \cdots \tag{11-5}$$

Comparison of this with Eq. (11-2) reveals that $\bar{F}_1 = \mu_1$, $\bar{F}_2 = \mu_2$, etc.; thus the partial molal free energy is identical with the Gibbs chemical potential. This identity may also be written [from Eqs. (11-4) and (11-5)]

$$\mu_1 = \left(\frac{\partial F'}{\partial n_1}\right)_{T,P,n_2,n_3,\ldots} = \left(\frac{\partial E'}{\partial n_1}\right)_{S',V',n_2,n_3,\ldots}$$

It may be shown in the same manner as above that

$$\mu_1 = \left(\frac{\partial H'}{\partial n_1}\right)_{T,S',n_2,n_3,\ldots} = \left(\frac{\partial A'}{\partial n_1}\right)_{T,V',n_2,n_3,\ldots}$$

Hence,

$$\mu_1 = \left(\frac{\partial E'}{\partial n_1}\right)_{S',V',n_2,n_3,\ldots} = \left(\frac{\partial H'}{\partial n_1}\right)_{T,S',n_2,n_3,\ldots} = \left(\frac{\partial A'}{\partial n_1}\right)_{T,V',n_2,n_3,\ldots} = \left(\frac{\partial F'}{\partial n_1}\right)_{T,P,n_2,n_3,\ldots} \tag{11-6}$$

It should be noted that this relation, except for $(\partial F'/\partial n_1)_{T,P,n_2,n_3,\ldots}$ does not involve partial molal quantities; *i.e.*,

$$\left(\frac{\partial E'}{\partial n_1}\right)_{S',V',n_2,n_3,\ldots} \neq \left(\frac{\partial E'}{\partial n_1}\right)_{T,P,n_2,n_3,\ldots} = \bar{E}_1$$

From Eq. (11-6) we see that μ, the chemical potential, may be written in terms of any one of the thermodynamic functions E, H, A, or F and is identical with the partial molal free energy $\bar{F}$. It is sometimes felt that the symbol μ rather than $\bar{F}$ stresses this generality of the chemical potential. We shall follow this procedure in the following section devoted to the development of the phase rule but shall subsequently return to the use of $\bar{F}$, since we are primarily interested in the free energy.

Equality of the Chemical Potential in Equilibrated Phases. We have already discussed in Chap. 6 the fact that at complete equilibrium a system is in mechanical, thermal, and chemical equilibrium. Although the concepts of mechanical equilibrium (uniform pressure) and thermal equilibrium (uniform temperature) usually present no particular difficulty, the concept of chemical equilibrium is more profound and can be fully understood only in terms of the chemical potential. It will now be demonstrated that for any heterogeneous system at equilibrium the chemical potential of any component is identical in all phases.

Let us write the differential equation [Eq. (11-2)] for the free energy of each phase of the equilibrated system under consideration.

$$\begin{aligned} dF'^{\text{I}} &= V'^{\text{I}}\,dP - S'^{\text{I}}\,dT + \mu_1^{\text{I}}\,dn_1^{\text{I}} + \mu_2^{\text{I}}\,dn_2^{\text{I}} + \cdots \\ dF'^{\text{II}} &= V'^{\text{II}}\,dP - S'^{\text{II}}\,dT + \mu_1^{\text{II}}\,dn_1^{\text{II}} + \mu_2^{\text{II}}\,dn_2^{\text{II}} + \cdots \\ &\cdots\cdots\cdots\cdots\cdots\cdots \end{aligned}$$

Each arabic subscript refers, as usual, to a component, and each roman superscript to a phase. Since at equilibrium the temperature and pressure are uniform throughout the system, no superscript is necessary on P and T. Also, with no loss of generality, we may consider the system to be held at constant temperature and pressure so that dT and dP are zero. Hence,

$$\begin{aligned} dF'^{\text{I}} &= \mu_1^{\text{I}}\,dn_1^{\text{I}} + \mu_2^{\text{I}}\,dn_2^{\text{I}} + \cdots \\ dF'^{\text{II}} &= \mu_1^{\text{II}}\,dn_1^{\text{II}} + \mu_2^{\text{II}}\,dn_2^{\text{II}} + \cdots \\ &\cdots\cdots\cdots\cdots\cdots \end{aligned}$$

Let us now consider the exchange, at equilibrium, of an infinitesimal amount of any component, *e.g.*, component 1, between any two phases, *e.g.*, from phase II to phase I. The resulting free-energy changes of these two phases are

$$\begin{aligned} dF'^{\text{I}} &= \mu_1^{\text{I}}\,dn_1^{\text{I}} \\ dF'^{\text{II}} &= \mu_1^{\text{II}}\,dn_1^{\text{II}} \end{aligned}$$

The total free-energy change of the system is the sum of these.

$$dF' = dF'^{\text{I}} + dF'^{\text{II}} = \mu_1^{\text{I}}\,dn_1^{\text{I}} + \mu_1^{\text{II}}\,dn_1^{\text{II}}$$

Since $dn_1^{\text{I}} = -dn_1^{\text{II}}$, that is, the amount of component 1 lost by one phase is equal to the amount gained by the other, we may write

$$dF' = (\mu_1^{\text{I}} - \mu_1^{\text{II}})\,dn_1^{\text{I}}$$

It has already been demonstrated in Chap. 7 that $dF' = 0$ for any change at equilibrium in a system of fixed mass and composition at constant temperature and pressure. It follows that

$$\mu_1^{\text{I}} = \mu_1^{\text{II}}$$

In like manner it may be shown that similar relations apply to all components in all phases; hence

$$\begin{aligned} \mu_1^{\text{I}} &= \mu_1^{\text{II}} = \mu_1^{\text{III}} = \cdots \\ \mu_2^{\text{I}} &= \mu_2^{\text{II}} = \mu_2^{\text{III}} = \cdots \\ &\cdots\cdots\cdots\cdots \end{aligned} \tag{11-7}$$

Thus we see that for any heterogeneous system at equilibrium (any heterogeneous thermodynamic substance), the chemical potential of each component has identically the same value in every phase. This equivalence is the very heart of our understanding of chemical equilibrium and is the basis of all equilibrium measurements. Since μ, like T and P, is characteristic of the system rather than of a particular phase, the superscript indicating the phase will henceforth be omitted.

As a special case let us consider a volatile component of a heterogeneous system. In view of the identity of the partial molal free energy and the chemical potential and the equality of the former in each phase to $RT \ln p + \mathrm{I}$, it follows that the statement of the equality of the chemical potentials is equivalent to a statement of the equality of the vapor pressure in all phases. Thus when water and ether are shaken together until equilibrated, it is found that, although the mutual solubility is not large, the two phases, upon separation, have the same partial pressure of ether—the aqueous phase has an odor of ether just as strong as that of the ether phase, in spite of the low concentration of ether therein. Similarly if an open vessel containing kerosene is exposed to the atmosphere for a long time, the partial pressure of the water in the kerosene will be equal to that of the atmosphere (roughly half that of pure water if the relative humidity is about 50 per cent), in spite of the relatively low solubility of water in kerosene. It is thus occasionally found that the slow corrosion of a metal (not sufficiently rapid to deplete the water in the kerosene) proceeds almost as rapidly in kerosene as in water.

THE PHASE RULE

Let us consider the formation of an arbitrary amount of a given phase at constant temperature and pressure, starting at zero quantity and adding infinitesimals at constant composition until the phase is built up to the preassigned amount. It is seen that this is a simple process consisting of the transfer of the components from some other phase to the phase under consideration. The energy of the phase formed by this process may be evaluated by integration of Eq. (11-4).

$$dE' = T\,dS' - P\,dV' + \mu_1\,dn_1 + \mu_2\,dn_2 + \cdots \qquad (11\text{-}4)$$

The integration is easy, since all the coefficients of the differentials are constants: T and P are arbitrarily fixed, and the chemical potentials, being functions of temperature, pressure, and composition only, are also constant. Hence we have for the energy

$$E' = TS' - PV' + \mu_1 n_1 + \mu_2 n_2 + \cdots + \text{a constant} \qquad (11\text{-}8)$$

This relation, which has lost no generality from the particular method of derivation, may now be differentiated completely.

$$dE' = T\,dS' + S'\,dT - P\,dV' - V'\,dP + \mu_1\,dn_1 + n_1\,d\mu_1 + \mu_2\,dn_2 + n_2\,d\mu_2 + \cdots \quad (11\text{-}9)$$

Subtracting Eq. (11-4) from Eq. (11-9).

$$0 = S'\,dT - V'\,dP + n_1\,d\mu_1 + n_2\,d\mu_2 + \cdots \quad (11\text{-}10)$$

and dividing by $n_1 + n_2 + \cdots$,

$$0 = S\,dT - V\,dP + N_1\,d\mu_1 + N_2\,d\mu_2 + \cdots \quad (11\text{-}11)$$

It will be noted that this equation is similar to Eq. (10-9), except that it is not restricted to isothermal and isobaric conditions; Eq. (10-9), however, is more general in that it pertains not only to the chemical potential but to any partial molal quantity.

In the absence of electric, magnetic, and gravitational fields, surface energy effects, etc., the state of a system consisting of a single phase may be regarded as a function of temperature, pressure, and composition. Defining the number of components as n, it is seen that there are altogether n composition variables, $N_1, N_2, \ldots, N_n$. However, in view of the relation $N_1 + N_2 + \cdots + N_n = 1$, only $n - 1$ of these are independent, any one being calculable from the other $n - 1$. Including the two variables, temperature and pressure, we see that the state of a single-phase system is a function of $n + 1$ independent variables which may be subjected to arbitrary infinitesimal or finite variation. Although the number of independent variables is thus determined, the particular choice of such variables is arbitrary. Instead of T, P, and any $n - 1$ of the mole fractions, we might have chosen T, V, and $n - 1$ mole fractions or T, P, and $n - 1$ of the chemical potentials, etc. The number of independent variables, or the number of variations which may independently be made, is called the *variance* or the number of *degrees of freedom*, and is designated v. The experimental counterpart of this definition is as follows: The variance is the number of variables to which values must be assigned, in order to specify completely the state of the system. Thus for a single-phase system, free from gradients in electric, magnetic, and gravitational fields, $v = n + 1$.

Let us now apply Eq. (11-11) to the system consisting of a single phase. This relation involves $n + 2$ differentials, dP, dT, $d\mu_1$, $d\mu_2$, . . . , $d\mu_n$. It will be noted that, unless there is another independent relation between these, all but one of these $n + 2$ differentials, or "variations" as they were termed by Gibbs, may be fixed arbitrarily, the remaining one then being calculable by Eq. (11-11). Since this is the same conclusion reached by elementary considerations in the preceding paragraph, we may conclude that for a single-phase system Eq. (11-11) is the only independent restriction. For this reason Gibbs referred to it as a fundamental equation. It is an alternate expression for the combined statement of the first and second laws [Eqs. (11-2) and (11-4)] and is sometimes known as "Gibbs' equation 97."*

Similarly for a system of n components and r coexistent phases at equilibrium there exist r relations or restrictions of the type imposed by Eq. (11-11).

$$
\begin{aligned}
0 &= S^{\mathrm{I}}\,dT - V^{\mathrm{I}}\,dP + N_1^{\mathrm{I}}\,d\mu_1 + N_2^{\mathrm{I}}\,d\mu_2 + \cdots + N_n^{\mathrm{I}}\,d\mu_n \\
0 &= S^{\mathrm{II}}\,dT - V^{\mathrm{II}}\,dP + N_1^{\mathrm{II}}\,d\mu_1 + N_2^{\mathrm{II}}\,d\mu_2 + \cdots + N_n^{\mathrm{II}}\,d\mu_n \\
&\cdots\cdots\cdots\cdots\cdots\cdots\cdots\cdots\cdots\cdots\cdots\cdots \qquad (11\text{-}12) \\
0 &= S^{r}\,dT - V^{r}\,dP + N_1^{r}\,d\mu_1 + N_2^{r}\,d\mu_2 + \cdots + N_n^{r}\,d\mu_n
\end{aligned}
$$

Since temperature, pressure, and all chemical potentials have the same values in all phases, the total number of derivatives or variations is again $n + 2$. The number of variations which may arbitrarily be made is equal to the number of possible variations $n + 2$, minus the number of restriction r, and hence is $n + 2 - r$,† so that the variance or number of degrees of freedom v is given by the relation

$$v = n + 2 - r \qquad (11\text{-}13)$$

This relation is known as the phase rule of Gibbs or simply the *phase rule.*

To illustrate the phase rule let us consider a system composed of a given mass of water and water vapor—a system of one component and two phases. The phase rule tells us that at equilibrium the number of degrees of freedom, or the variance, is 1—indicating that we may arbitrarily fix, within limits, either the temperature or the pressure, after which all the other intensive properties of each phase of the system become determinate. For example, if the pressure is arbitrarily fixed, the temperature is thereby uniquely determined; thus at 1 atm pressure water boils and is at equilibrium with its vapor only at 100°C.

* "The Collected Works of J. Willard Gibbs," Vol. I, p. 88, Longmans, Green & Co., Inc., New York, 1931.

† The reasoning here is exactly that applying to any system of linear simultaneous equations (which indeed the above are). If we have l equations involving m variables, then $m - l$ of the variables may arbitrarily be fixed and the remaining variables evaluated by solution of the equations. Thus in the system of two equations involving three variables,

$$
\begin{aligned}
x + y + z &= 0 \\
x + 2y - z &= 1
\end{aligned}
$$

only 3 − 2 or 1 variable, x, for example, may be arbitrarily fixed—say equated to zero. Then y and z may be evaluated numerically.

If our system consists of iron, wüstite, and a gaseous mixture of hydrogen and water vapor, the number of components is three (iron, hydrogen, and oxygen), the number of phases is three, and the phase rule tells us that at equilibrium the variance is 2. Thus upon arbitrarily fixing both the temperature and the total pressure, the chemical potentials (partial pressures) then assume particular determinate values. This is equivalent to saying that the composition of each phase, and hence its state, is determined when the temperature and pressure are arbitrarily established. It would, therefore, be a proper experiment (yield unique results) to determine at fixed temperature and pressure the composition of the gas or that of either of the two solids. It will be noted that, if only two phases are at equilibrium in a system of these three components, the variance is 3 and it would not be a proper experiment to measure at fixed temperature and pressure the composition of the gas alone, since this as well as temperature and pressure may be regarded as subject to arbitrary fixation. A fruitful experiment in this case would require a determination of the simultaneous composition of both phases.

The chemical equation for the burning of limestone may be written

$$CaCO_3(s) = CaO(s) + CO_2(g)$$

The number of components is two, which may be regarded as CaO and CO_2; the number of phases is three, two solids and a gas; hence the variance is 1, just as in the system liquid water–water vapor. Hence at any particular temperature the pressure is determinate, and this system has a "boiling point"—a temperature where the pressure equals 1 atm—just as a single-component system.

Phases, Phase Regions. Any homogeneous portion of a system is known as a *phase*. Different homogeneous portions at the same temperature, pressure, and composition—such as droplets—are regarded as the same phase. It will be recalled that a system or portion thereof is homogeneous if each volume element (small but containing many atoms) is identical with every other element in characteristics such as temperature, pressure, composition, and structure. Two phases are said to be in the same *phase region* if one may be obtained from the other by a *continuous* change of the variables under consideration (such as temperature, pressure, composition, structure); thus pure copper and an alloy of 5 per cent zinc in copper, at the same or slightly different temperature and pressure, lie in the same phase region, since the alloy may be obtained from the copper (*e.g.*, by diffusion) without a discontinuity in any of the properties.

Components. The number of components of a system is the smallest number of independently variable constituents by means of which the composition of each phase involved in the equilibrium may be expressed. Although the number of components is usually evident from chemical intuition, the above statement gives a precise formulation of the concept. In order to understand more fully the meaning of the term *component* as applied to the phase rule, let us tentatively identify the number of components with the number of elements. It is frequently true in pyrometallurgical systems that the number of components is equal to the number of elements. In the previous example of the burning of limestone, however, the number of components was two rather than three, since the composition of all phases could be expressed in terms of CaO and CO_2. It so happens from the chemistry of this system that in each phase involved in the equilibrium the number of atoms of oxygen is equal to the number of atoms of calcium plus twice the number of atoms of carbon. If such had not been the case, *e.g.*, if calcium oxide contained a variable amount of oxygen, as does wüstite, we should have to consider the system to be one of three components.

In the system involving the phases limestone, lime, and gaseous CO_2 the number of elements is three, but the number of components is two, since there exists for each phase the experimentally verifiable restriction that the number of atoms of oxygen is equal to the number of atoms of calcium plus twice the number of atoms of carbon. It frequently happens, even in nonmolecular systems, that this type of restriction is found. In the strict sense it cannot be deduced but can be determined only by experiment; however, it frequently can be inferred from a general chemical knowledge. For example, in the system CaO-MgO-SiO_2, a system involving four elements, it seems only natural to postulate under usual conditions that the number of oxygen atoms in the system of however many phases is equal to the number of calcium atoms plus the number of magnesium atoms plus twice the number of silicon atoms. If this is substantially the case, as indeed it is under all but the most drastic conditions, then the system is to be considered one of three components, since the four elements cannot be present in arbitrary amounts but only in amounts conforming to this restriction. If, on the other hand, we choose to investigate a system of these same four elements under highly reducing conditions so that a metallic phase is also present, the system is properly considered one of four components, since there is now no composition restriction similar to the above applicable to *all* phases.

The type of restriction mentioned above frequently applies to oxide and halide systems, *i.e.*, those with very stable compounds, and occasionally to sulfide systems. Depending upon experimental conditions, it might or might not be proper to assume such a restriction for a system involving an oxide of an element capable of two valences. For example, in the iron-oxygen-silicon system the oxide phases are not subject to a composition restriction such that the number of atoms of oxygen is equal to the number of atoms of iron plus twice the number of atoms of silicon, but rather it is commonly found experimentally that the number of atoms of oxygen is greater than this by virtue of the presence of some ferric iron. Hence the number of components here is equal to the number of elements, namely, three. Under special conditions, possibly those encountered in blast-furnace slags (highly reducing conditions with only a small amount of iron oxide), such a restriction may be considered to apply, and the number of components is one less than the number of elements.

Sometimes we wish arbitrarily to limit our consideration to certain restricted conditions in a multicomponent system. For instance, to investigate the fluxing action of fluorspar in open-hearth slags an experiment might be conducted to determine the solubility of dicalcium silicate in molten CaF_2. If it were found that the solid phase did not depart measurably from the composition corresponding to $2CaO \cdot SiO_2$, then both phases of interest could be regarded as admixtures of $2CaO \cdot SiO_2$ and CaF_2 (more strictly one phase as $2CaO \cdot SiO_2$ and the other as a mixture of $2CaO \cdot SiO_2$ and CaF_2). Thus, although there are four elements, there are two restrictions: one that the number of atoms of oxygen plus half the number of atoms of fluorine is equal to the number of atoms of calcium plus twice the number of atoms of silicon and the other, arbitrarily imposed by the manner of forming the system, that the number of atoms of oxygen is four times the number of atoms of silicon. To one familiar with such systems it is perhaps intuitively apparent that this is a system of two components, most conveniently considered as calcium fluoride and dicalcium silicate.

By this time it should be apparent that the number of components of a system depends upon the experimental conditions and the degree of precision which we wish to employ in dealing with the system. As one further example to illustrate the latter part of this statement, let us consider again the most reduced oxide phases in the iron-oxygen-silicon system. As an adequate approximation for many purposes, this may be considered a system of FeO and SiO_2 as shown in the lower part of Fig. 11-1. However, if we wish to be more precise, we must consider the fact that the lowest oxide, that in equilibrium with the solid or liquid metallic phase, contains excess oxygen (ferric iron). The amount of ferric iron is shown in the upper part of Fig. 11-1. It cannot be emphasized too strongly that the number of components of a system is to be determined empirically under the particular experimental conditions. For nonmolecular systems the number of components is equal to the number of elements minus the number of composition restrictions applicable to all phases under consideration.

In the case of molecular systems, such as those encountered in organic chemistry and in gases except at high temperature, the number of components not infrequently *exceeds* the number of elements. In many systems the fundamental particles which do not undergo decomposition in the processes concerned are molecules rather than atoms or ions. For

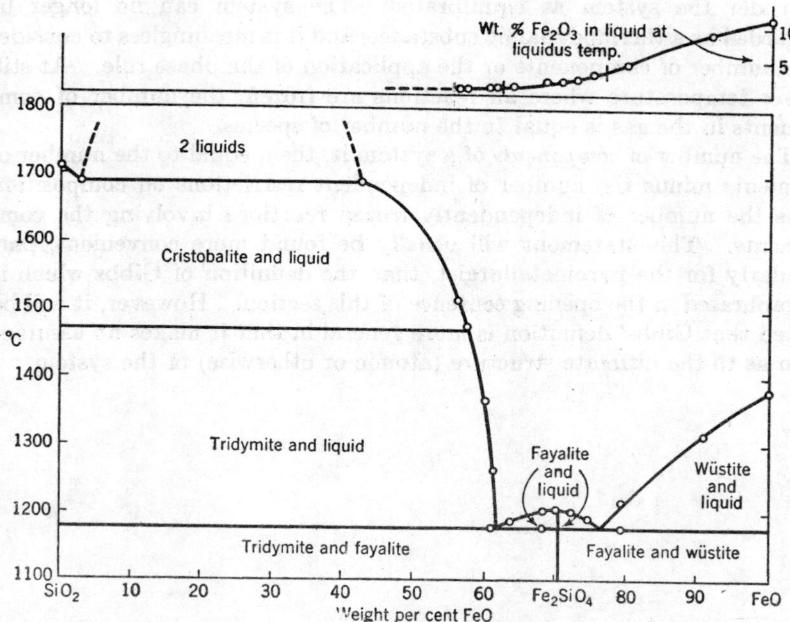

FIG. 11-1. Equilibrium diagram of the system FeO-SiO_2. The upper part of the figure shows the percentage of Fe_2O_3 in the melt; in the lower part all the iron present is calculated as FeO. [*From Bowen and Schairer, Am. J. Sci.*, **24**, 117 (1932).]

example, let us consider the equilibrium between an aqueous solution of hydrogen and oxygen and a vapor phase consisting of H_2, O_2, and H_2O. At a temperature below that at which reaction begins to take place, it is proper to consider this a system of three components although only two elements are involved. Thus the number of components exceeds the number of elements by the number of independent reactions which are "frozen," *i.e.*, do not proceed at all toward equilibrium. The system of H_2, O_2, and H_2O is therefore seen to be one of three components at low temperature and two components at high temperature. Such a system may not properly be treated by thermodynamics alone in the intermediate region where reaction occurs slowly but does not attain equilibrium in the time allotted for the experiment. Similarly in the gaseous atmospheres used in the treatment of steel at elevated temperature which contain carbon, oxygen, hydrogen, and nitrogen, there are four components at elevated temperature where equilibrium prevails between the molecular species N_2, CO, CO_2, H_2, H_2O, CH_4, C_2H_6, etc. At a lower temperature, however, the various possible reactions "freeze," each in a different temperature range, and it is, of course, improper to consider the system as equilibrated. The system can no longer be regarded as a thermodynamic substance, and it is meaningless to consider the number of components or the application of the phase rule. At still lower temperature where all reactions are frozen, the number of components in the gas is equal to the number of species.

The number of *components* of a system is, then, equal to the number of elements minus the number of independent restrictions on composition plus the number of independently frozen reactions involving the components. This statement will usually be found more convenient, particularly for the pyrometallurgist, than the definition of Gibbs which is paraphrased in the opening sentence of this section. However, it will be noted that Gibbs' definition is more general in that it makes no assumption as to the ultimate structure (atomic or otherwise) of the system.

CHAPTER 12

HETEROGENEOUS EQUILIBRIA

It may appear to the reader that a great deal of effort was wasted in deriving Eqs. (11-12), as the only use made of them so far was to count their number and thus arrive at the very simple relation [Eq. (11-13)] known as the phase rule. To dispense with the fundamental set of Eqs. (11-12) and preserve only Eq. (11-13) would, indeed, be analogous to glancing at a set of simultaneous algebraic equations and dismissing the subject with the bare statement that there is, or is not, a unique solution, without bothering to find the solution if one exists. As a matter of fact, the knowledge gained by application of Eq. (11-13) is frequently trivial, in the sense that it was already intuitively evident to one reasonably familiar with the case to be analyzed thereby. For example, anyone who did not know that water and ice are in equilibrium at a single temperature could hardly be expected to use the phase rule to ascertain this. Similarly, most of us learn to associate a unique temperature with the equilibrium allotropic transformation of an element before we learn of the phase rule.

Hence, our main interest in the set of Eqs. (11-12) lies, not in the relatively small amount of information to be deduced from Eq. (11-13), but in the actual solution of these equations for a particular set of circumstances, giving us complete quantitative information concerning the temperature, pressure, and the composition of each phase involved. Since each of the equations involved in Eqs. (11-12) is merely one form of the combined statement of the first and second laws, it is possible, and frequently convenient, to cast them in some other form, as, for instance, in terms of the equilibrium constant as was done in Chap. 9. This chapter will be devoted essentially to the application of Eqs. (11-12) or their equivalent to heterogeneous systems.

Atomistic Interpretation of Heterogeneous Equilibrium. Although, as mentioned previously, the thermodynamic treatment of heterogeneous equilibrium is based entirely on the first, second, and third laws of thermodynamics and is independent of any atomistic postulates or constructs, let us nevertheless consider the atomistic nature of heterogeneous equilibria and reactions simply to show the parallelism between the two viewpoints. The whole problem of chemical rate phenomena is incapable of solution by purely thermodynamic methods and depends upon the formulation of an atomic model. Although the subject of reaction rates can be touched upon only in a very superficial manner in this chapter, it seems worth while to discuss here a few elementary considerations pertaining thereto which form the foundation of the atomistic viewpoint of heterogeneous equilibrium.

Let us consider a liquid metal, *e.g.*, liquid sodium, at a temperature slightly above its melting point. From Chap. 5 it will be recalled that the structure of liquid sodium is rather closely related to that of the solid. The X-ray evidence shows this relation almost directly in that the nearer coordination shells of liquid and solid are very similar, the atomic positions being somewhat less well defined for the liquid than for the solid. The ordered atomic arrangement in the liquid, however, is local, and at sufficiently great distance from an atom the ordering effect fades out. In the case of several elements, *e.g.*, mercury, it was shown that the ordering effect is more extensive at low temperature (near the melting point) than at higher. It is to be recalled that the distribution curve of Chap. 5 represents an average and that the distribution about any particular atom in a liquid undergoes rapid changes with time. At some times the order is more pronounced, and at other times less, than that indicated by the curve; *i.e.*, the distribution about an atom in the liquid is sometimes more like that in the solid and sometimes more like that in a gas. In the vicinity of the melting point, then, it seems only natural to regard the larger ordered groupings, resembling the solid, as nuclei or potential nuclei for the start of crystallization. From this viewpoint the freezing point of a metal is the temperature below which nuclei of sufficient size will continue to grow and above which they will "dissolve." These nuclei were mentioned in Chap. 5 as "micro-icebergs" in the case of water.

It is now convenient to consider the free-energy change involved in the isothermal formation of one of these nuclei from the liquid. Adopting the simplification that the boundary of the nucleus is sharp, we may divide the free energy of formation into two parts—that of its interior and that of its surface. Adopting the further simplification that, as the nucleus grows (or shrinks), all linear dimensions grow (or shrink) in proportion, it follows that the free-energy change $\Delta F'$ accompanying the formation of one nucleus from the liquid can be represented as the sum of two terms, one proportional to the volume, *i.e.*, the cube of the "radius," and the other proportional to the surface, *i.e.*, the square of the "radius." Thus

$$\Delta F' = f(T)r^3 + g(T)r^2$$

or, at any particular temperature where the two functions of temperature may be regarded as constants,

$$\Delta F' = ar^3 + br^2$$

At a temperature above the melting point, $\Delta F'$ and its two component parts are represented schematically as functions of r in Fig. 12-1a. The

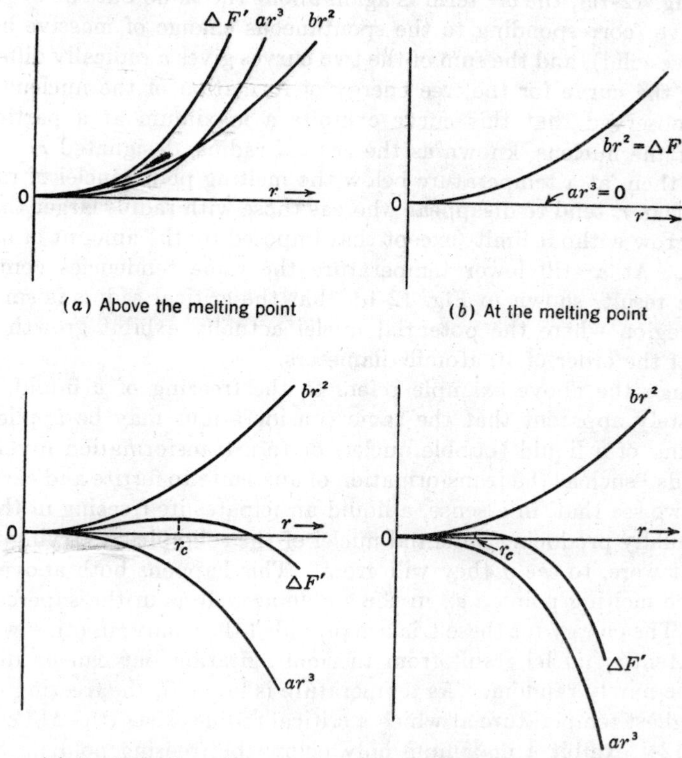

FIG. 12-1. The isothermal free energy of formation of a crystal nucleus in a melt, shown as the sum of surface (br^2) and bulk (ar^3) free energies. Spontaneous growth, which must be accompanied by a free-energy decrease, occurs only below the melting point and for a nucleus of radius greater than the critical radius r_c.

component of the free energy attributable to the surface is positive, as is always the case, and that attributable to the volume is positive, since above the melting point the formation of the massive solid from liquid does not occur. Hence the total free-energy change as shown in the figure increases indefinitely with increasing radius, illustrating that the nucleus is metastable and that the spontaneous decrease of free energy characteristic of every spontaneous process corresponds to the disappearance or solution of the nucleus. A similar plot corresponding to conditions at the melting point is shown in Fig. 12-1*b*. Here the contribution br^2 of the surface to the free energy change is about the same as before, but that of the volume ar^3 is zero, since, at the melting point, the free-energy change for a reaction involving massive phases (those with negligibly small surface to volume ratio) is zero. Below the melting point (Fig. 12-1*c*), the br^2 term is again about the same but the ar^3 term is negative (corresponding to the spontaneous change of massive liquid to massive solid), and the sum of the two curves gives a radically different form for the curve for the free energy of formation of the nucleus. It will be observed that this curve exhibits a maximum at a particular radius of the nucleus, known as the critical radius, designated r_c. It is evident, then, at a temperature below the melting point nuclei of radius smaller than r_c tend to disappear whereas those with radius larger than r_c tend to grow without limit (except that imposed by the amount of liquid present). At a still lower temperature the same tendencies continue with the result, shown in Fig. 12-1*d*, that the critical radius is smaller. In the region where the potential nuclei actually exhibit growth r_c is small—of the order of 10 atomic diameters.

Although the above example refers to the freezing of a liquid, it is immediately apparent that the same considerations may be applied to the boiling of a liquid (bubble nuclei) or to a transformation involving only solids,[1] such as the transformation of austenite to ferrite and carbide.

Thus we see that, in a sense, a liquid anticipates its freezing in that it is continually producing potential nuclei of the solid phase—trying them out, as it were, to see if they will grow. This happens both above and below the melting point, *i.e.*, in the stable as well as in the supercooled liquid. The energy for these trials is provided thermally; in other words, these potential nuclei result from thermal agitation but can hardly be said to be purely random. As temperature is lowered, the freezing point is the highest temperature at which a critical radius exists (the $\Delta F'$ curves of Fig. 12-1 exhibit a maximum only below the freezing point). Since the critical radius is very large (infinite) at the freezing point, we can hardly expect spontaneous formation of the crystalline solid. At a temperature slightly below the freezing point, the critical radius is sufficiently large that nuclei can seldom result from thermal agitation, and the liquid

[1] In the transition from one solid to another it is believed that usually a so-called transition state, or *coherent state*, is involved. The type of incipient nucleus here postulated is platelike and so thin that the treatment given above (consisting of breaking the free energy of formation of the nucleus into two parts, pertaining to the surface and the volume) is very crude, the forces at the boundary in this case producing an effect throughout the incipient nucleus.

must commonly be seeded or nucleation aided in some other way in order to start crystallization. At a lower temperature, the critical radius becomes small and the rate of nucleation rapid. At a still lower temperature, the critical radius does not become very much smaller; however, the thermal agitation is markedly less, so that the rate of nucleation is again small. Thus there is a temperature at which the rate of nucleation is a maximum.

Supercooling. Wang and Smith[1] have shown that many of the small droplets of tin in a 10 per cent Al-Sn alloy supercool as much as 100°C, thus supporting the contention of Turnbull and collaborators[2] that the nuclei normally effective at small degrees of supercooling are inhomogeneities or "dirt." When the metal is finely subdivided (as in the Al-Sn alloy) the droplets that happen to contain the most effective nucleating dirt solidify only slightly below the freezing point and most of the droplets undercool to lower temperature, some to much lower temperature. Wang and Smith consider that it remains questionable as to whether the nuclei effective at the greatest observed degree of supercooling are inhomogeneities or arise from the thermal fluctuations considered above.

Turnbull[3] has given a detailed mathematical analysis of the nucleation of a solid in a liquid by minute amounts of the solid retained in small conical or cylindrical cavities in the substrate or container. These small bits of solid by virtue of the capillary effect are not melted by prior heating of the solid for several degrees above the normal melting point. He finds that the degree of supercooling increases with the extent of prior superheating, asymptotically approaching a limiting degree of supercooling.

Fisher, Hollomon, and Turnbull[4] derived the following expression for the rate of homogeneous nucleation, *i.e.*, thermal nucleation in the absence of dirt, cavities, etc:

$$\ln \dot{n} = \ln \frac{RT}{h} - \frac{16\pi\sigma^3 V^2 T_0^2}{3kT\,\Delta H^2(T_0 - T)^2}$$

where $\dot{n}$ is the number of nuclei formed per unit time per mole of liquid at the temperature T, h is Planck's constant, k is Boltzmann's constant, σ is the interfacial tension between liquid and solid, V is the molal volume, ΔH is the molal heat of fusion, and T_0 is the normal melting point. As

[1] Wang and Smith, *Trans. AIME*, **188**, 136 (1950).

[2] Fisher, Hollomon, and Turnbull, *J. Applied Phys.*, **19**, 775 (1948). Turnbull and Fisher, *J. Chem. Phys.*, **17**, 71 (1949). D. Turnbull, *J. Applied Phys.*, **20**, 817 (1949).

[3] D. Turnbull, *J. Chem. Phys.*, **18**, 198 (1950).

[4] Fisher, Hollomon, and Turnbull, *Science*, **109**, 168 (1949).

pointed out in Chap. 5, this equation provides for a very rapid increase in the rate of nucleation with increased degree of supercooling. The effect of this is so marked that repeated measurements of the maximum degree of supercooling are usually reproducible to within a few degrees. However, it will be noted from the equation that at very low temperature the rate of nucleation again becomes small and hence that this rate must pass through a maximum, as observed above. In the case of metals this maximum rate is generally sufficiently great that the low-temperature branch of the curve is without significance. In certain cases, *e.g.*, glasses, it seems possible that the maximum rate may be slow enough to permit supercooling to the absolute zero.

At very great degrees of supercooling, supercooled liquids—particularly glasses—show at a characteristic temperature a rapid change in such properties as the heat capacity and coefficient of thermal expansion. The value of the viscosity at this temperature is usually about 10^{13} poises; common glasses have this viscosity at 400 to 600°C. Glycerin with a freezing point of 17°C may readily be supercooled to the vicinity of the absolute zero. The heat capacity exhibits no discontinuity as the liquid passes from the stable to the metastable liquid on cooling through the melting point. However, the heat capacity drops to about one-half its former value at about -100°C; this temperature becomes slightly lower the more slowly the experiment is performed. The interpretation placed on this by Jones and Simon[1] is that below this temperature internal equilibrium is not attained by virtue of the high viscosity and the disorder characteristic of a higher temperature is "frozen in."

Growth of Crystals. During the growth of a perfect crystal from a supersaturated solution or vapor, it is apparent that when the surface layer of a given face is completely filled with atoms or molecules, further growth of this face requires the nucleation of the next layer. This repetitive two-dimensional nucleation requires a supersaturation by a factor of about 1.5. Experimentally it is found that continued crystal growth requires only a much smaller degree of supersaturation, *e.g.*, 0.8 per cent for the growth of iodine crystals from the vapor.[2] To avoid this dilemma, Frank[3] postulated that crystal growth commonly occurs via screw-type dislocations. If just one such dislocation emerges near the center of a face, that crystal face can grow perpetually up a "spiral staircase"—without two-dimensional nucleation. In a most impressive experimental verification of this theoretical prediction Amelinckx[4]

[1] Jones and Simon, *Endeavor*, **8**, 175 (1949).
[2] Volmer and Schultze, *Z. physik. Chem.*, **A156**, 1 (1931).
[3] F. C. Frank, *Discussions Faraday Soc.*, **5**, 48 (1949).
[4] S. Amelinckx, *Nature*, **167**, 939 (1951).

observed the spiral growth of carborundum (SiC). Many other similar observations have since been made. This theory and these observations constitute one of the most important advances in our knowledge of crystal growth. An excellent review of this subject is presented by Shockley,[1] who illustrates the spiral staircases with drawings and photographs.

SYSTEMS OF ONE COMPONENT

Since in the case of a one-component system the composition is invariant, temperature and pressure may be considered the only independent variables. Let us imagine that a system consisting of a single component

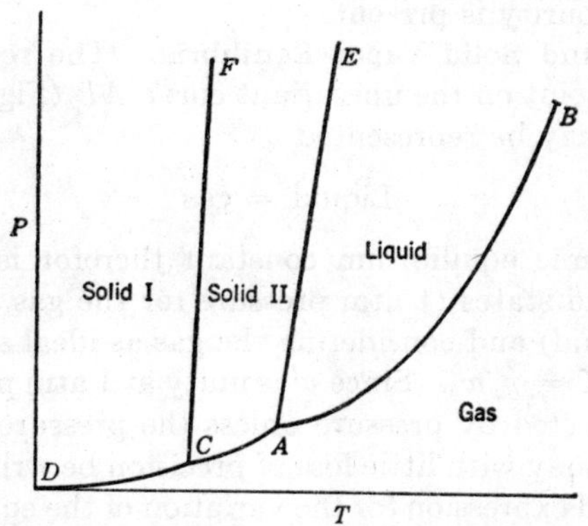

FIG. 12-2. Schematic phase diagram for a one-component system.

is subjected to a series of external conditions each characterized by a definite temperature and pressure. By noting the phases of which the system is composed under each of these sets of conditions, it is possible to construct a map, such as Fig. 12-2. At each temperature and pressure investigated where only a liquid phase is found, we place a characteristic mark, *e.g.*, a circle, to indicate this phase; and at each point where only a vapor phase is found, we place a cross, and so forth, for each solid modification. It is thus found empirically that our map or phase diagram consists of regions, in each of which only a single phase exists. These regions are said to be bivariant, since in any one of them the system has a variance of 2; at any point therein both temperature and pressure may, in general, be varied by a small amount without the appearance or disappearance of a phase. The curves separating the regions, called *univariant curves*, correspond to the simultaneous existence of two phases corresponding to the two regions they bound; the variance of the system along such a curve is 1. Three regions may have in common only a point

[1] W. Shockley, *J. Metals*, **4**, 829 (1952).

from which the univariant curves radiate; such a point, known as a *triple* or *invariant point*, represents an invariant system, *i.e.*, one with a variance of 0. This map aspect is the primary aspect of a phase diagram.

Certain rules of construction can be deduced from the phase rule which greatly facilitate the construction of a phase diagram from a limited number of experimental data or which aid in the interpretation of the data. For example, any two-phase system involving a single component is univariant and thus may be represented by a curve. If experimentally a zone is found rather than a line, *e.g.*, if a melting point is not sharp, we are forced to revise our opinion as to the number of components and conclude that an impurity is present.

Liquid-Vapor and Solid-Vapor Equilibria. The reaction to be associated with any point on the univariant curve AB (Fig. 12-2) (the vapor-pressure curve) may be represented

$$\text{Liquid} = \text{gas}$$

The thermodynamic equilibrium constant therefor is a^g/a^l. Adopting the usual standard states (1 atm pressure for the gas and pure liquid at 1 atm for the liquid) and considering the gas as ideal so that $a^g = p$, the vapor pressure, $K = p/a^l$. Since a^l is unity at 1 atm pressure and is usually but little affected by pressure unless the pressure is high, the equilibrium constant may with little loss of precision be written $K = p$. The previously derived expression for the variation of the equilibrium constant with temperature [Eq. (9-22a)] now becomes

$$\frac{d \ln p}{d(1/T)} = -\frac{\Delta H_v^\circ}{R} \tag{12-1}$$

where ΔH_v° is the molal heat of vaporization. Methods of handling this relation, including the use of the Σ function, were considered in detail in Chap. 9. It will be noted here merely that the simplified treatment, considering ΔH_v as constant, gives $\ln p = -(\Delta H_v/RT) + I$, or $p = e^I e^{-\Delta H_v/RT}$, thus accounting for the exponential nature of a vapor-pressure curve, depicted by AB in Fig. 12-2. The sublimation curve AC or CD may be treated in similar manner to give

$$\frac{d \ln p}{d(1/T)} = -\frac{\Delta H_s}{R} \tag{12-2}$$

p now being the vapor pressure of the solid and ΔH_s the heat of sublimation.

If, as when dealing with a liquid in the vicinity of the critical point (point B in the figure, which is the terminus of this univariant curve), it is no longer true that the molal volume of the liquid is negligible in comparison with that of the gas, a different treatment is required. Equations (11-12) become

$$0 = S_v\,dT - V_v\,dP + d\mu$$
$$0 = S_l\,dT - V_l\,dP + d\mu$$

Subtraction gives

$$(S_v - S_l)\,dT = (V_v - V_l)\,dP$$

or

$$\frac{dP}{dT} = \frac{\Delta S_v}{\Delta V_v} = \frac{\Delta H_v}{T\,\Delta V_v} \tag{12-3}$$

This relation, which could have been derived from Eq. (7-32), is one form of the *Clausius-Clapeyron equation.* Its integration requires an experimental knowledge of ΔH_v and ΔV_v. Its application to liquids in the vicinity of the critical point will not be pursued, since this is of little interest to metallurgists. If the volume of the liquid is small in comparison with that of the vapor, $\Delta V_v = V_v$; and if the vapor is ideal, $V_v = RT/P$. Substitution of this in Eq. (12-3) gives

$$\frac{d \ln P}{d(1/T)} = -\frac{\Delta H_v}{R}$$

a relation substantially the same as Eq. (12-1). The partial pressure p for a system of one component is identical with the total pressure P, and the enthalpy change accompanying the change from the liquid to the vapor state at equilibrium, ΔH_v, is nearly identical with that accompanying the change from one standard state to the other, ΔH_v°.

The temperature at which the vapor pressure of many elements has certain specified values is given in Table 12-1. It is customary to arrange the data in this fashion rather than to tabulate pressures at round temperatures, since compactness is gained in restricting the table to the pressure range of common interest for each element. In the left column is the designation (l) or (s) according as the element is liquid or solid at the lowest temperature recorded. The lowest tabulated temperature where a liquid is stable is indicated by the designation (l).

For the estimation of the vapor pressure of an element or compound when data are lacking, *Duhring's rule* is a valuable empirical aid. This rule states that the ratio of the absolute temperatures at which the vapor pressures of two similar substances are the same is a constant. Thus from the boiling points of Zn and Hg (1180 and 634°K, respectively) and the temperature (750°K) at which $p_{\mathrm{Zn}} = 10^{-3}$ atm, we may calculate the temperature at which $p_{\mathrm{Hg}} = 10^{-3}$ atm. This calculation gives

$$750^\circ \times \frac{634^\circ}{1180^\circ} = 403^\circ\mathrm{K},$$

as compared with the observed value 394°K (Table 12-1).

TABLE 12-1. VAPOR PRESSURE OF THE ELEMENTS*
(Temperature in degrees Kelvin for various partial pressures in atmospheres)

Species	10^{-6} atm	10^{-5} atm	10^{-4} atm	10^{-3} atm	10^{-2} atm	1 atm
A						87.3(l)
Ag	1200(s)	1305	1442	1607	1816	2485
Al	1290(l)	1405	1545	1725	1940	2600
As_4	477(s)	517	563	622	708	895
At_2	......	270(s)	320	350	390	500
Au	1570(l)	1720	1896	2112	2388	3239
B	1500(s)	1600	1750	1900	2150	2800(l)
Ba	810(s)	890	985(l)	1116	1293	1911
Be	1390(s)	1505	1655(l)	1830	2070	2780
Bi	873(l)	960	1060	1190	1360	1900
Br_2	173(s)	186	203	222	245	331(l)
C	2720(s)	2920	3170	3450	3800	4775
Ca	790(s)	867	961	1075	1231(l)	1755
Cb	2820(l)	3050	3340	3700	4120	5400
Cd	485(s)	530	585(l)	657	744	1038
Ce	1450(l)	1550	1700	1850	2100	2800
Cl_2	114(s)	123	139	153	169	239(l)
Co	1750(s)	1900(l)	2100	2300	2600	3370
Cr	1350(s)	1465	1600	1755	1960	2495(l)
Cs	383(l)	425	476	544	634	963
Cu	1400(l)	1530	1685	1875	2117	2868
F_2	......	......	......	......	58(l)	85
Fr(87)	......	......	490(l)	......	......	950
Fe	1550(s)	1680	1837(l)	2033	2277	3008
Ga	1225(l)	1350	1500	1690	1920	2700
Ge	1370(l)	1500	1670	1880	2150	2980
H_2	......	......	......	......	......	20.39(l)
He	......	......	......	......	......	4.22(l)
Hf	2850(l)	3100	3350	3750	4150	5500
Hg	287(l)	316	351	394	449	634
I_2	241(s)	260	282	308	341	456(l)
In	1100(l)	1210	1350	1510	1730	2440
Ir	2580(s)	2800(l)	3040	3350	3700	4800
K	429(l)	475	534	605	702	1052
Kr	......	......	......	......	......	119.9(l)
La	1500(l)	1650	1800	2000	2250	3000
Li	705(l)	775	865	980	1130	1640
Mg	653(s)	715	789	881	1000(l)	1399
Mn	1140(s)	1240	1360	1570(l)	1750	2370
Mo	2530(s)	2740	3000(l)	3330	3750	5077
N_2	......	......	......	......	......	77.4(l)
Na	510(l)	558	623	705	813	1187
Ne	......	......	......	......	......	27.3(l)
Ni	1630(s)	1765(l)	1930	2130	2380	3110

TABLE 12-1. VAPOR PRESSURE OF THE ELEMENTS.*—(*Continued*)

Species	10^{-6} atm	10^{-5} atm	10^{-4} atm	10^{-3} atm	10^{-2} atm	1 atm
O_2	......	......	......	......	......	90.2(l)
Os	2700(s)	2900	3160(l)	3470	3850	4900
P_4(yellow)	244(s)	268	296	334(l)	382	553
Pa	2250(l)	2450	2700	2950	3300	4500
Pb	887(l)	975	1088	1226	1408	2010
Pd	1660(s)	1800	2000(l)	2240	2530	3440
Po_2	......	660(s)	750	835(l)	945	1300
Pt	2160(l)	2340	2550	2820	3140	4100
Ra	650(s)	700	770	850	965	1410(l)
Rb	403(l)	445	496	561	650	952
Re	2900(s)	3150	3450(l)	3850	4300	5800
Rh	2200(s)	2400(l)	2600	2850	3200	4150
Rn	......	......	......	......	......	211(l)
Ru	2480(s)	2670	2900(l)	3180	3500	4500
S_2	......	......	500(l)			
Sb_2	860(s)	940(l)	1025	1160	1340	1890
Sc	1540(s)	1680(l)	1850	2050	2300	3000
Se_2	525(l)	568	620	679	755	1000
	(2 atm Se_6 and 10^{-4} atm Se at 1000°K)					
Si	1480(s)	1600	1740(l)	1920	2140	2750
Sn	1300(l)	1450	1600	1850	2150	3000
Sr	740(s)	810	900	1010	1150(l)	1657
Ta	3300(l)	3600	3900	4300	4800	6300
Tc(43)	2550(l)	2750	3000	3300	3700	5000
Te_2	655(s)	700	758(l)	825	907	1130
Th	2250(l)	2450	2700	2950	3300	4500
Ti	1640(s)	1800	1990	2210(l)	2500	3400
Tl	795(l)	870	965	1082	1235	1730
U	2000(l)	2150	2350	2580	2900	3800
V	1970(l)	2140	2340	2550	2900	3800
W	3230(s)	3490	3780(l)	4150	4625	5950
Xe	......	......	......	......	......	165.1(l)
Y	1750(s)	1900(l)	2100	2300	2700	3500
Zn	560(s)	610	672	750(l)	852	1180
Zr	2070(s)	2250	2450(l)	2700	3000	3850

* Taken from Leo Brewer, Report for the Manhattan Project, MDDC-438C, 1946.

Solid-Liquid Equilibria. A univariant curve, such as *AE* or *CF*, involving only condensed phases, *i.e.*, solids or liquids, normally has a much greater slope than one involving a gaseous phase. Any such equilibrium involving condensed phases may also be treated by the Clausius-Clapeyron equation. In a manner identical with that used in the derivation of Eq. (12-3), it may be shown for a fusion process, for example, that $dP/dT = \Delta H_f/T\,\Delta V_f$, where ΔH_f and ΔV_f are the molal enthalpy and volume changes on fusion. Thus, for a small change in the melting point ΔT_f, occasioned by a corresponding change in pressure ΔP, we have $\Delta T_f = (T_f \, \Delta V_f / \Delta H_f) \, \Delta P$. As mentioned in Chap. 5, the volume of the liquid is usually greater than that of the corresponding solid, water, bismuth, and gallium being notable exceptions, and the heat of fusion is always positive; hence the melting point usually increases with pressure. Ordinary ice is an exception, but the higher pressure forms of ice behave normally.

Solid-Solid Equilibria. An equilibrium between solid phases, *e.g.*, an allotropic transformation, may be treated in exactly the same way. A similar relation applies: $dP/dT = \Delta H_{tr}/T \, \Delta V_{tr}$, the subscript *tr* referring to the solid-solid transformation. In using this to find the effect of pressure on the temperature of the equilibrium α-γ transformation for iron, the following data are pertinent: The normal transformation temperature is 910°C, $\Delta H_{tr} = 215$ cal/gram atom, the densities of α and γ-iron are 7.571 and 7.633 g/cc, respectively, and 0.02421 cal equals 1 cc atm. The molal volume change ΔV_{tr} is

$$\frac{55.85}{7.633} - \frac{55.85}{7.571} = 0.060 \text{ cc}$$

Substitution in the reciprocal of the above expression gives

$$\frac{dT}{dP} = \frac{(910 + 273)(0.060)(0.02421)}{215} = 0.0082 \text{ deg/atm}$$

The rather rapid change of ΔH_{tr} with temperature for the above transformation as well as the variation of T, which was assumed constant in the above procedure, suggests that a more precise method be used if we wish to consider the effect of a large variation in pressure. With the aid of a table of ΔF as a function of T the following method is available: In Fig. 12-3 the univariant curve AB, to be determined quantitatively, represents the effect of pressure on the equilibrium temperature of the α-γ equilibrium. Since this is an equilibrium curve, the free-energy change for the transformation ΔF_{tr} is zero at all points thereon; it has a value other than zero at all points not on the curve AB, these points representing a transformation at nonequilibrium conditions. Hence we may write $-\Delta F_{tr(A)} + \Delta F_{tr(B)} = 0$ for the change in ΔF_{tr} in going from A to B. Adding and subtracting $\Delta F_{tr(C)}$,

$$[\Delta F_{tr(C)} - \Delta F_{tr(A)}] + [\Delta F_{tr(B)} - \Delta F_{tr(C)}] = 0$$

Referring to the diagram it is now seen that the first brackets represent the change in ΔF_{tr} in passing from one temperature to the other at constant pressure and the second brackets the change from one pressure to the other at constant temperature, equilibrium being then again attained. Taking the position of the arbitrary line $\overline{CB}$ as 1150°K, the quantity in the first brackets can be evaluated from Table 16-1 as 6.90 cal or 285 cc-atm/gram atom, and the quantity in the second brackets is seen to be equal to $\int \Delta V_{tr}\, dP$. For lack of data we again assume that ΔV_{tr} is constant and has a value of 0.060 cc/gram atom. Substituting in the above equation,

$$285 - 0.060P = 0$$

Hence P, the pressure required to maintain equilibrium between α and γ-iron at 1150°K, is 285/0.060 = 4750 atm. A similar calculation shows

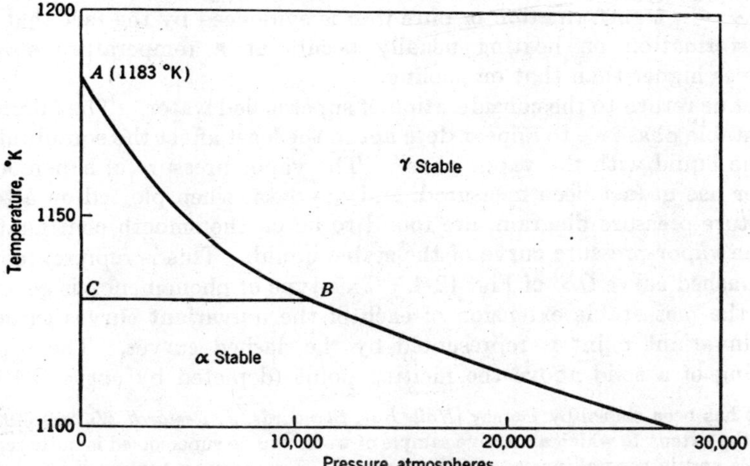

FIG. 12-3. The effect of pressure on the temperature at which α and γ-iron coexist in equilibrium.

that at 1100°K the pressure is 28,000 atm. At the latter temperature the aforementioned simpler solution gives a pressure of only 10,000 atm.

Metastable Phases. Thus far in our discussion of one-component systems we have considered only stable equilibria. We have considered those systems whose free energy has the lowest possible value at the temperature and pressure. However, as mentioned previously, thermodynamics is quite capable of dealing with a system which is in equilibrium in one respect even though it is out of equilibrium in another, providing that the reaction with respect to which the system is out of equilibrium does not proceed to any observable extent in the time under consideration. It is well known that water can be supercooled; the massive liquid can be cooled as much as 20°C below the ice point without the appearance of the solid phase.[1]

A metastable phase is also commonly encountered in the case of tin. The temperature at which white tin (tetragonal) is in equilibrium with gray tin (diamond type) is about 18°C. Thus, at ordinary winter temperatures white tin is metastable relative to gray tin, yet the transformation, known as "tin plague," occurs only occasionally. In fact, the reaction is so sluggish that the heat capacity of white tin has been measured down to the temperature of liquid nitrogen.[2]

The phenomenon of retained austenite is another example of a metastable phase, although this is not encountered at room temperature in a one-component system. However, a slight sluggishness in the $\alpha - \gamma$ and $\gamma - \alpha$ transformation of pure iron is evidenced by the fact that the transformation on heating usually occurs at a temperature several degrees higher than that on cooling.

Let us return to the consideration of supercooled water. The failure of the stable phase ice to appear does not in the least affect the equilibration of the liquid with the vapor phase. The vapor pressure of supercooled water has in fact been measured, and the data, when plotted on a temperature-pressure diagram, are found to lie on the smooth continuation of the vapor-pressure curve of the stable liquid. This is represented by the dashed curve OB' of Fig. 12-4. This type of phenomenon is general, and the metastable extension of each of the univariant curves through the invariant point is represented by the dashed curves. The superheating of a solid above the melting point (depicted by curve OA') is

[1] It has been shown by Dorsey [*Natl. Bur. Standards, J. Research*, **20**, 799 (1938)] that the extent to which a massive sample of water can be supercooled is quite reproducible and in general increases with its purity. For example, tap water may freeze at −5° and conductivity water at −20°C. The degree of supercooling obtainable does not seem particularly sensitive to minor disturbances or agitation, but crystallization of supercooled water can be induced by a sharp swirling motion. Thus, the spontaneous nucleation of crystals seems to involve the presence either of particles of impurity or of vortices, a conclusion attested by the fact that a sample from the top of a large settling chamber freezes without agitation at lower temperature than one from the bottom. Dorsey's work leaves one with the impression that continued purification of water would lead to even lower temperatures of spontaneous crystallization. This conclusion is supported by the fact that even greater degrees of supercooling are attained in the fine droplets of clouds. In such a fine dispersion each impurity particle nucleates only one droplet, so that the greater portion of all the water may be further supercooled. A similar technique may be used to supercool metals as noted in Chap. 5.

[2] E. Cohen and C. van Eyk [*Z. physik. Chem.*, **30**, 601 (1899)] found that the temperature of maximum velocity for the transformation of white tin to gray tin is about −50°C. The existence of such a maximum implies the existence also of a "*C* curve" similar to that used to express the transformation of austenite.

seldom, if ever, observed. However, the suppression of the gas phase (curve OC') is relatively easy by virtue of the difficulty of bubble formation; in fact, a liquid may be held under tension in a closed container which it wets. Thus it is seen that whether or not a metastable equilibrium prevails under a certain set of circumstances is a fact to be determined experimentally. If metastable equilibrium does prevail, we may be sure that the corresponding curve is a smooth extension of that for the stable equilibrium.

It is to be noted in the consideration of metastable equilibria that the fields on the pressure-temperature diagram are no longer to be considered

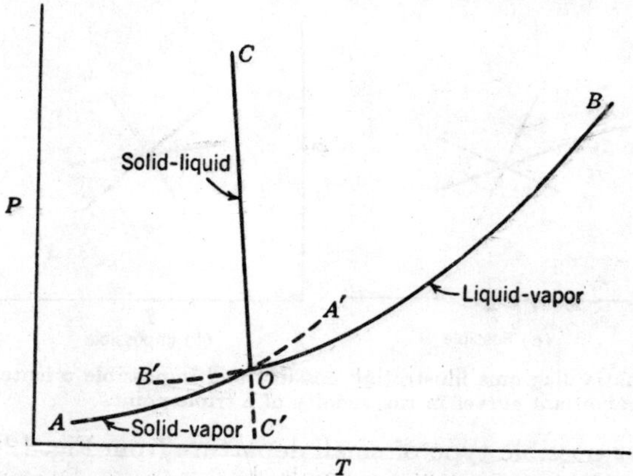

Fig. 12-4. Schematic phase diagram for water. Dashed lines represent metastable equilibria.

significant as they were in Fig. 12-2, where only stable equilibria were considered. Instead, it will be found fruitful to focus our attention upon the univariant curves, since, as will be evidenced later, little significance can be attached to the fields in other than one-component systems, even for stable equilibria.

Sequence of Univariant Curves about an Invariant Point. A fact of some interest is that the stable and metastable curves must alternate around a triple point. The relative orientation of the curves must be that shown in Fig. 12-5*a* rather than that shown in 12-5*b*. In order to demonstrate this, let us consider the free-energy relationship. At the temperature of the triple point the free energy of each of the phases present at the triple point is represented schematically as a function of pressure by a line in Fig. 12-6*a*; the point of crossing represents the triple point where the free energy is the same for each of the phases involved. At a higher pressure phase I has the lowest free energy of all three and hence is the stable phase, and at a lower pressure phase III has the lowest free energy and is the stable phase. Thus we see that, in the pressure-temperature diagram pertaining to stable phases only, we enter the region of phase I by proceeding upward from the triple point and the region of phase III by proceeding downward. Let us next consider (Fig. 12-6*b*) the free-energy diagram at a temperature slightly above (or below) that of the triple point. It is seen that at this temperature and at P_1 equilibrium involving phases I and II prevails and that these phases are metastable, since the stable configuration is phase III, this having a lower free energy. At P_2 there exists a stable equilibrium of phases I and III, and at P_3 another equilibrium metastable with respect to phase I.

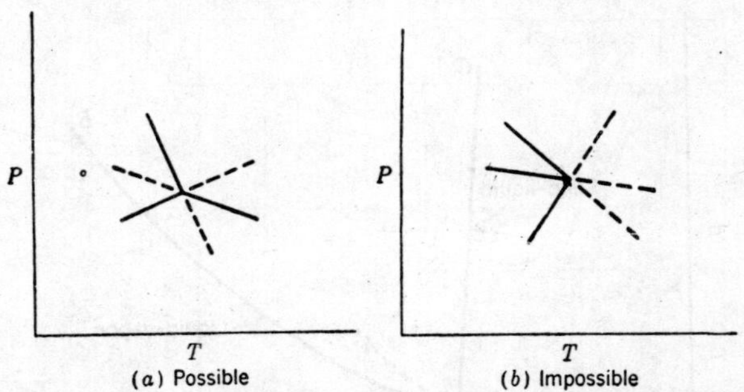

FIG. 12-5. Schematic diagrams illustrating possible and impossible orientations of stable and metastable univariant curves in the vicinity of a triple point.

The only other possible type of small departure from Fig. 12-6*a* is shown in Fig. 12-6*c*. This represents a small departure in temperature from the conditions of Fig. 12-6*a* in the opposite sense from that of Fig. 12-6*b*. Here there are two stable equilibria at pressures P_4 and P_6 and one metastable equilibrium at P_5. The pressure of the metastable equilibrium P_5 lies between those of the stable equilibria P_4 and P_6. Thus it has been shown that the metastable portion of each univariant curve lies between the stable portions of the other univariant curves, that the stable portion lies between metastable portions as shown in Fig. 12-5*a*, and hence that Fig. 12-5*b* represents an impossible construction.

It is similarly seen from free-energy considerations that a condensed phase with higher vapor pressure is metastable, or unstable, with respect to a condensed phase with lower vapor pressure. This follows immediately from the isothermal relation between free energy and partial pressure, $F = RT \ln p + I$. Hence if, at a particular temperature, a liquid has a higher vapor pressure than the solid, it has a higher free energy and is metastable with respect to the solid. An elementary view of this is that the phase with higher pressure will distill over to the phase of lower pressure until the former finally disappears. The distillation is, of course, unnecessary, the process usually taking place by nucleation and growth. Similar considerations apply to a solid-solid transformation. The concept of partial pressure is not particularly useful here, as, for example, in considering the transformation of γ to α-iron.

Ostwald's Rule. Discussion of metastable phases cannot end without mention of a principle commonly known as *Ostwald's rule*, called by him

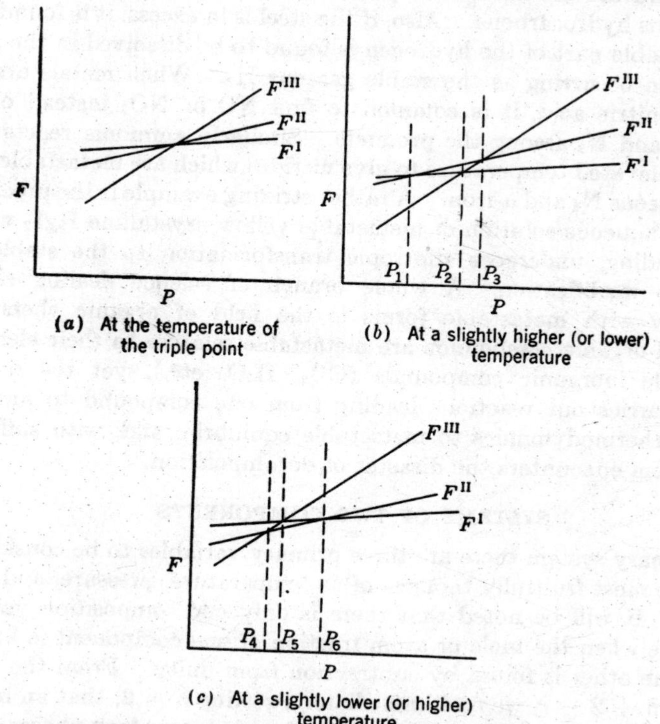

FIG. 12-6. Schematic diagrams illustrating the requirement that the stable and metastable curves alternate around a triple point, as in Fig. 12-5a.

the rule of successive reactions. This states that, when a system undergoing reaction proceeds from a less stable state, the most stable state is not formed directly but rather the next more stable state is formed, and so on, step by step until the most stable is formed. Ostwald's rule can hardly be said to have universal application, since many examples are known where the complete sequence of reactions through the metastable states has not been observed. For example, water vapor below the ice point does not usually first precipitate the metastable liquid water which subsequently freezes but transforms directly to ice, as frost or snowflakes.

However, it is very common to find metastable phases appearing before the stable one. Supercooled liquids, discussed previously, are common. The metastable phase cementite (Fe_3C) rather than the stable phase graphite is the usual constituent in steel. When steel is dissolved in acid, the products obtained are not the stable ones, ferrous salt, graphite, and hydrogen; rather the carbon and part of the hydrogen appear in the form of numerous hydrocarbons. Also, if the steel is in excess, it is found that an appreciable part of the hydrogen is found to be dissolved in the steel, rather than occurring as the stable gaseous H_2. When metals are dissolved in nitric acid, it is common to find NO or NO_2 instead of the stable O_2 and N_2 among the products. Similarly ammonia reacts with α-iron at elevated temperature to give nitrides which are metastable relative to gaseous N_2 and α-iron. A rather striking example is the precipitation from aqueous solution of metastable yellow crystalline HgI_2 which, upon standing, undergoes allotropic transformation to the stable red crystalline modification. A whole branch of science dealing almost exclusively with metastable forms is the field of organic chemistry. Nearly all organic compounds are metastable relative to their elements and simple inorganic compounds (CO_2, H_2O, etc.), yet the organic chemist carries out reactions leading from one compound to another, applying thermodynamics to metastable equilibria, and, with sufficient skill, seldom encounters the disaster of decomposition.

SYSTEMS OF TWO COMPONENTS

In a binary system there are three primary variables to be considered. These are most fruitfully thought of as temperature, pressure, and composition. It will be noted that there is only one composition variable here, since when the mole or atom fraction of one component is known, that of the other is found by subtraction from unity. From the phase rule, $v = n + 2 - r$, we find immediately, setting $n = 2$, that an invariant point ($v = 0$) corresponds to equilibrium between four phases and a univariant curve to equilibrium between three phases. Although the complete graphical representation of such a system involving three variables requires three dimensions, it is common practice to use sections or projections in order that the equilibrium behavior may be represented on paper. Of the various sections perhaps the most common is that at constant pressure, usually 1 atm, which is the temperature-composition diagram. It is convenient to think of this diagram as representing the range of possible equilibrium conditions of a system confined in a cylinder by a piston which exerts a pressure of 1 atm. If the system is substantially nonvolatile, the piston may be replaced, either in our minds or in the experiment, by an inert atmosphere exerting the same pressure. In so far as the system is volatile or the atmosphere used is not strictly inert, such substitution is an approximation.

Temperature-Composition Diagrams. For the present discussion a temperature-composition diagram is to be regarded primarily as a map determined by experimental investigation which shows that in certain regions certain phases are found to be present at equilibrium. Later it will be shown that certain restrictions are imposed by the laws of thermodynamics; *i.e.*, the map must conform to certain rules. From the metallurgist's viewpoint the equilibria of most interest in metallic systems are those involving only liquids and solids. Let us first consider an equi-

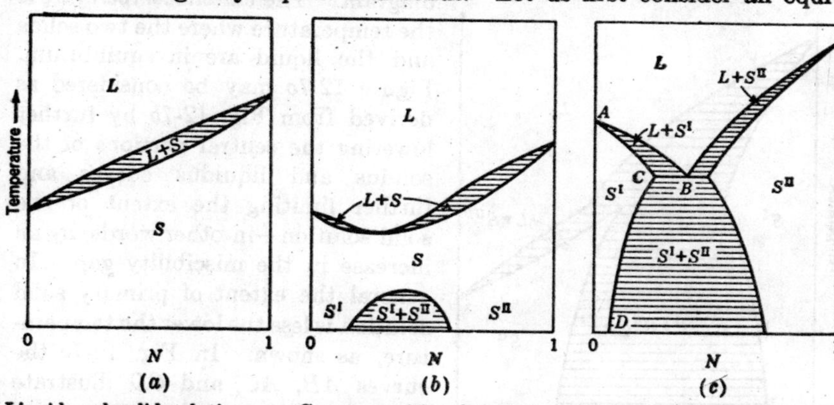

Liquid and solid solutions nearly ideal or having similar departures from ideality.

Greater positive departure from ideality for solid solution than for liquid.

Solid solution with still greater departure from ideality. Common eutectic-type diagram.

FIG. 12-7. Series of schematic binary temperature-composition diagrams showing the effect of increasing positive departure from ideality on the part of the solid solution.

librium between solid and liquid—perhaps the simplest case is that in which the binary system under consideration consists of a continuous series of solid solutions in the solid state and of liquid solutions in the liquid state. The temperature-composition diagram for this case appears as shown in Fig. 12-7*a*; the upper curve is known as the liquidus, and the lower as the solidus. This diagram is characteristic of a system in which both the solid and liquid solutions are ideal or in which the departures from ideality are similar in the two. From this it may be correctly inferred that the degree of nonideality of a system is often not apparent from the phase diagram.

A rather mild departure from ideality may give rise to a similar diagram in which the central portion of both curves is displaced slightly upward or downward; the more common is the downward type of departure corresponding to a greater positive departure from ideality on the part of the solid solution. An increased tendency in this direction leads to a diagram, shown in Fig. 12-7*b*, which is characterized by a minimum in both curves and which is usually accompanied by a miscibility gap in the series of solid solutions at lower temperature. In this series of diagrams the two-phase regions are represented by lined areas; these lines are known as *tie lines* or *conodes*. The termini of each tie line correspond to the phases which coexist at equilibrium; the tie lines are obviously horizontal, since the coexisting phases at equilibrium are at the same temperature.

A further increase in positive departure from ideality on the part of the solid solution leads to a diagram shown in Fig. 12-7*c*, the eutectic type of diagram. The *eutectic temperature* is the temperature where the two solids and the liquid are in equilibrium. Figure 12-7*c* may be considered as derived from Fig. 12-7*b* by further lowering the central portions of the solidus and liquidus curves and further limiting the extent of the solid solution—in other words, by an increase in the miscibility gap. In general the extent of primary solid solution is less the lower the temperature, as shown. In Fig. 12-7*c* the curves *AB*, *AC*, and *CD* illustrate *liquidus*, *solidus*, and *solvus* curves, respectively. Another type of diagram, which may be regarded as derived by the intrusion of a miscibility gap into the solidus curve of Fig. 12-7*a*, is shown in Fig. 12-8. This is a *peritectic* type of diagram, the temperature at which the two solids and the liquid are in equilibrium being the peritectic temperature.

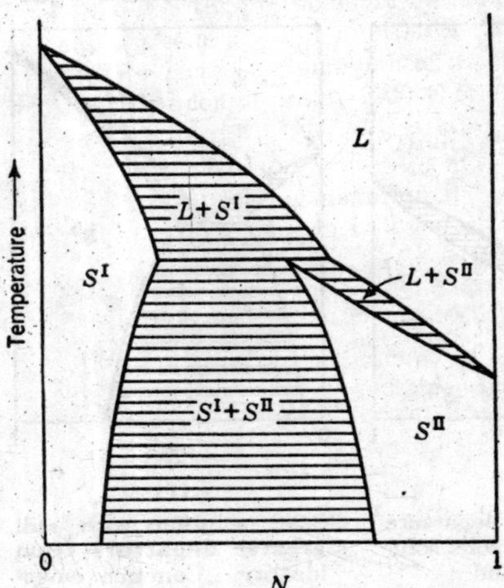

FIG. 12-8. Schematic illustration of the peritectic type of temperature-composition diagram.

Some systems exhibit a miscibility gap also in the liquid. It is more common for this gap to be wider at lower temperature and to disappear at higher temperature as shown in Fig. 12-9. The point *P* at which two liquids and a solid are in equilibrium is known as a *monotectic* point. Occasionally, however, it happens that the miscibility gap in the liquid is concave upward rather than downward, as in Fig. 12-10. In this case there is a strong presumption that the gap is also closed at the top as indicated by the dotted line in the figure.

Composition of Phases in a Two-phase Region Independent of Relative Amounts of the Phases. It was shown in the previous chapter that for any system at equilibrium the chemical potential of any component is the same in all the coexisting phases and independent of the amount of each phase. For a two-component system:

$$\mu_1^{\mathrm{I}} = \mu_1^{\mathrm{II}} \quad \text{and} \quad \mu_2^{\mathrm{I}} = \mu_2^{\mathrm{II}}$$

Thus, for the system, there exists only one chemical potential for each component, and μ_1 and μ_2 like T and P need no superscript pertaining to a particular phase. For each phase, however, the chemical potential may

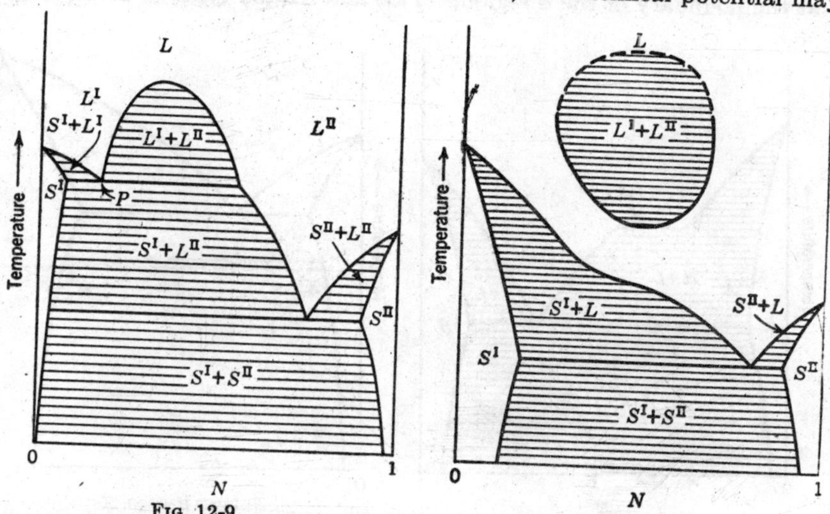

FIG. 12-9. FIG. 12-10.

FIG. 12-9. Schematic binary temperature composition diagram showing a region of liquid immiscibility ($L^{\mathrm{I}} + L^{\mathrm{II}}$) which closes at elevated temperature. P is known as a monotectic point.

FIG. 12-10. Schematic temperature-composition diagram showing a region of liquid immiscibility ($L^{\mathrm{I}} + L^{\mathrm{II}}$) which is bounded completely by liquid-phase regions. The decrease of liquid immiscibility with temperature, leading in the extreme to the lower closure of the two-liquid region, is relatively rare. The inflection of the liquidus curve is associated with this lower closure.

be regarded as a function of composition, temperature, and pressure. Inversely, at constant temperature and pressure, the composition of a stable phase is uniquely determined by the chemical potential. We conclude that, for two phases in equilibrium in a binary system, the chemical potential and hence the composition of each phase are uniquely determined and are independent of the relative amounts of the two phases.[1]

[1] The foregoing is not actually a proof. It was assumed in using the simplified combined statement of the first and second laws [Eq. (11-1)] that surface energies were so small as to be negligible. That this is so for massive phases may be regarded as derived from experience, *e.g.*, that the vapor pressure of a sample of water is independent of the volume provided this volume is large. It is this experience or assumption which leads to the omission of surface energy terms in Eq. (11-1) and the equations derived therefrom and hence to the above theroem.

The tie lines, as in Fig. 12-7, thus terminate at solubility curves and represent the unique compositions of the phases in equilibrium at the temperature and pressure under consideration. From this fact alone it is seen that a number of conceivable types of temperature-composition diagrams are incorrect. For example, the incorrect construction of Fig. 12-11 in the vicinity of point A shows a region presumably corresponding to equilibrium of α + liquid in which the tie lines do not have a terminus at the boundary of the α region. This and related types of construction

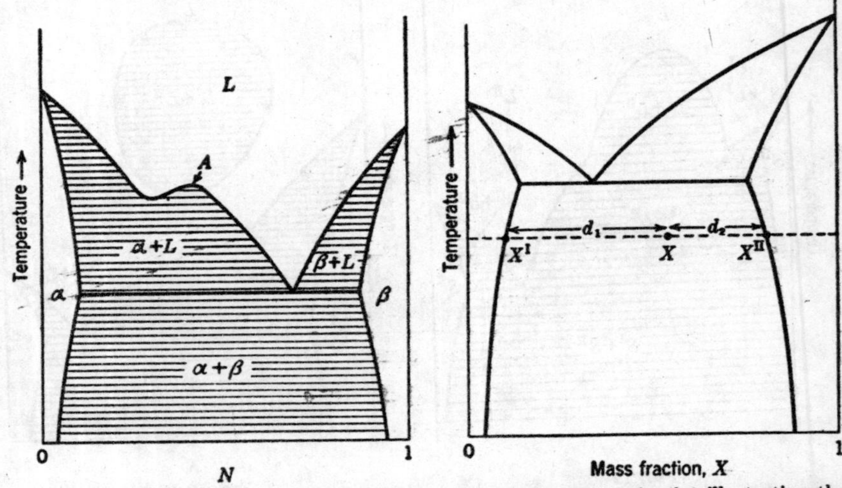

FIG. 12-11. Schematic temperature-composition diagram showing an improper construction in the vicinity of point A.

FIG. 12-12. Schematic plot illustrating the lever law, which relates the relative amounts of coexisting phases to distances on the temperature-composition diagram.

are obviously fallacious and may be regarded almost as violating common sense. Other less obvious fallacies will be pointed out later.

The Lever Law. Let us now consider a method of obtaining from the phase diagram the relative amounts of two coexisting phases when the gross composition, corresponding to a point lying within a two-phase region, is known. Suppose that the system is at some arbitrary temperature, indicated by the dotted line in Fig. 12-12 (similar to Fig. 12-7c) and that the gross composition corresponds to the mass fraction X of component 1. Under such conditions the equilibrium state of the system is represented by two phases I and II, whose compositions are denoted by mass fractions of component 1, X^{I} and X^{II}. It is convenient here to designate composition in terms of mass fraction rather than atom fraction. We now wish to find the individual masses m^{I} and m^{II} of phases I and II for unit total mass of the mixture. To do this let us express the mass fraction X of component 1 in terms of its fraction, X^{I} and X^{II}, in each phase and the fractional amount of each phase:

$$X = X^{I}m^{I} + X^{II}m^{II}$$

Since $m^{I} + m^{II} = 1$, we may substitute $X(m^{I} + m^{II})$ for X,

$$X(m^{I} + m^{II}) = X^{I}m^{I} + X^{II}m^{II}$$

Rearranging,

$$m^{I}(X - X^{I}) = m^{II}(X^{II} - X)$$

The expressions $X - X^{I}$ and $X^{II} - X$ are seen to be equal to the distances designated d_1 and d_2, and hence the foregoing expression may be written

$$m^{I}d_1 = m^{II}d_2 \tag{12-4}$$

Thus the mass of each coexisting phase is proved to be inversely proportional to the distance on the mass-fraction phase diagram from the point denoting the gross composition to that denoting the composition of the phase. It will be observed that this equation also expresses the condition for the balance of a lever with fulcrum at X and masses m^{I} and m^{II} at distances d_1 and d_2, respectively, and for this reason the law [Eq. (12-4)] is sometimes known as the lever law. It follows, for example, that, if X is midway between X^{I} and X^{II}, then the aggregate consists of 50 per cent by weight phase I and 50 per cent phase II; and if X is three-quarters of the way from X^{I} to X^{II}, then the aggregate consists of three-quarters phase II and one-quarter phase I, etc.

The whole reasoning of this section could have been carried through equally well if Fig. 12-12 had been in terms of atom fraction instead of weight fraction. In place of Eq. 12-4 the relation then found would be

$$n^{I}d_1' = n^{II}d_2' \tag{12-5}$$

The ratio n^{I}/n^{II} is the ratio of the total number of atoms in phase I to the total number of atoms in phase II, and d_1' and d_2' correspond on the atom-fraction diagram to d_1 and d_2 on the weight-fraction diagram.

Maximum or Minimum on the Temperature-Composition Diagram. It will now be demonstrated for a two-component system that, when the composition of two phases in equilibrium is identical, the temperature-composition curve for each phase passes through a maximum or minimum. Although this theorem is general and applies to any two phases, it may be visualized in terms of the solidus and liquidus curves, as in Fig. 12-7*b*.

For equilibrium at constant pressure between two phases in a two-component system we have, from Eq. (11-12),

$$0 = S^{I}\,dT + N_1^{I}\,d\mu_1 + N_2^{I}\,d\mu_2$$
$$0 = S^{II}\,dT + N_1^{II}\,d\mu_1 + N_2^{II}\,d\mu_2$$

Dividing through by dN_2 (where N_2 may be N_2^{I} or N_2^{II}) and subtracting,

$$-(S^{\mathrm{I}} - S^{\mathrm{II}})\frac{dT}{dN_2} = (N_1^{\mathrm{I}} - N_1^{\mathrm{II}})\frac{d\mu_1}{dN_2} + (N_2^{\mathrm{I}} - N_2^{\mathrm{II}})\frac{d\mu_2}{dN_2}$$

$-(S^{\mathrm{I}} - S^{\mathrm{II}})$ is ΔS_{tr}, the entropy change of the transformation, and is equal, for the equilibrium transformation, to $\Delta H_{tr}/T$.
Hence we may write

$$\frac{\Delta H_{tr}}{T}\frac{dT}{dN_2} = (N_1^{\mathrm{I}} - N_1^{\mathrm{II}})\frac{d\mu_1}{dN_2} + (N_2^{\mathrm{I}} - N_2^{\mathrm{II}})\frac{d\mu_2}{dN_2}$$

Under the condition that the composition of the two phases is identical at equilibrium, *i.e.*, $N_1^{\mathrm{I}} = N_1^{\mathrm{II}}$ and $N_2^{\mathrm{I}} = N_2^{\mathrm{II}}$, and since ΔH_{tr} is not zero, it is obvious that dT/dN_2 is zero unless $d\mu_1/dN_2$ or $d\mu_2/dN_2$ is infinite. For vanishingly small values of N_2, where by Henry's law $d\mu_2 = RT\,d\ln N_2$, it is seen that $d\mu_2/dN_2$ (and similarly $d\mu_1/dN_2$) does approach infinity, and hence this theorem is not valid in the immediate vicinity of the pure components. Also it occasionally happens, in the vicinity of a strong compound or intermediate phase, that $d\mu_2/dN_2$ is very large—or, for practical purposes, substantially infinite—and the theorem is not valid. Generally, the activity is an analytical function (*i.e.*, one that is continuous and all of whose derivatives are continuous) of the composition and is of such nature, as will be shown later, that it always increases with composition. The only instance where da_2/dN_2 or $d\mu_2/dN_2$ may be infinite is when an inflection is involved, and this is interpreted as ample evidence of a strong compound as just discussed. Except in the immediate vicinity of the pure components or of a strong compound, then, it has been demonstrated that, whenever the composition of two phases at equilibrium is identical, there is a maximum or minimum on the temperature-composition diagram.

Since in the above demonstration there is only one temperature and one composition for the two phases, there is only one value of dT/dN_2 for both phases at equilibrium; hence if one curve shows a maximum (or minimum), the other must also. Figure 12-7*b* represents a correct construction; Figs. 12-13*a* and 12-13*b* are examples of impossible construction.

The System SO_3-H_2O. In this system the extent of solid solution in the numerous solid phases is inobservably small, and these phases are represented as being of constant composition in the melting-point diagram (Fig. 12-14). Of the several peaks occurring at compositions corresponding to the many hydrates of SO_3, only the two on the right are pointed; the others are maxima (rounded). These features would be preserved if this figure were represented in terms of mole fraction instead of weight per cent.

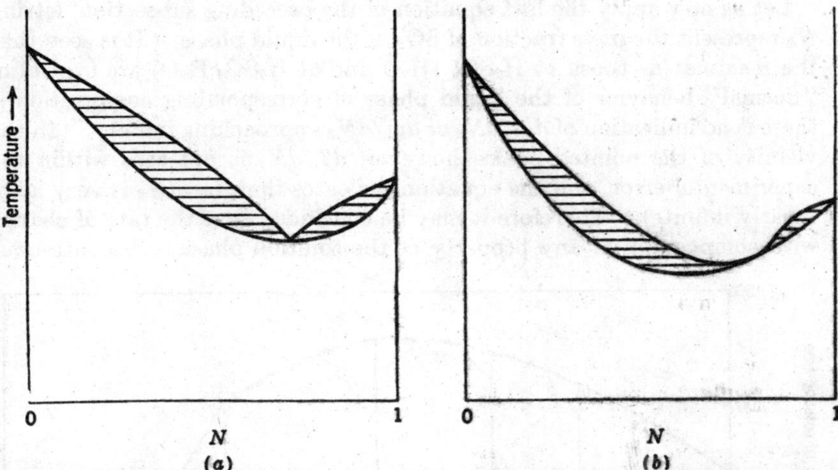

FIG. 12-13. Examples of erroneous construction. If two phases in equilibrium have the same composition at a given temperature, then both of the curves must pass through a minimum (or maximum) at the same point. In (*a*) the liquidus exhibits a discontinuity in slope rather than a minimum; in (*b*) the point of common composition does not correspond to the minimum of either curve.

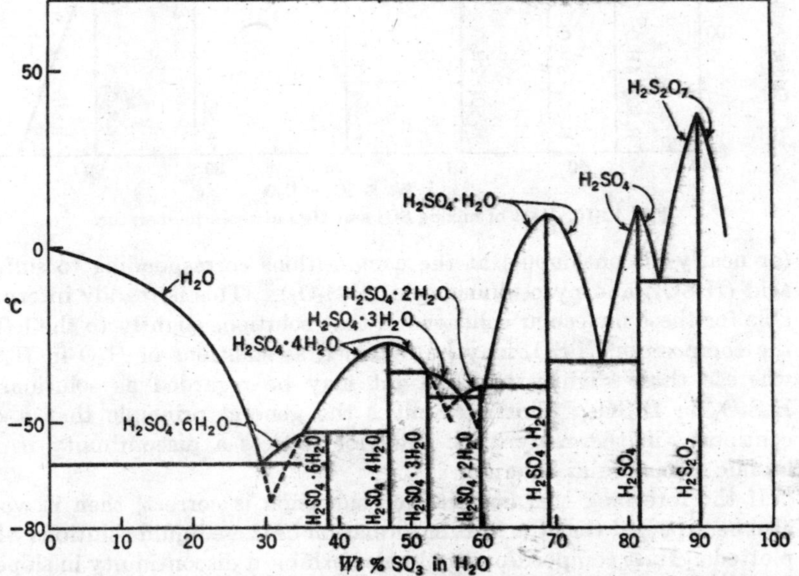

FIG. 12-14. Freezing diagram of the system SO_3-H_2O. Curves for observed metastable equilibria are shown dotted. The designation on each liquidus curve indicates the solid phase with which the liquid is in equilibrium. [*From data of C. M. Gable, H. F. Betz, and S. H. Maron, J. Am. Chem. Soc.*, **72**, 1445 (1950).]

Let us now apply the last equation of the preceding subsection, letting N_2 represent the mole fraction of SO_3 in the liquid phase. It is seen that the maxima (as those at $H_2SO_4 \cdot 4H_2O$ and at $H_2SO_4 \cdot H_2O$) are typical of "normal" behavior of the liquid phase of corresponding composition—there is no indication of $d\mu_1/dN_2$ or $d\mu_2/dN_2$ approaching infinity. In the vicinity of the pointed peaks, however, dT/dN_2 is not zero within the experimental error, and the equation indicates that $d\mu_2/dN_2$ is very large (nearly infinite). Therefore it may be concluded that the rate of change with composition of any property of the solution phase is discontinuous

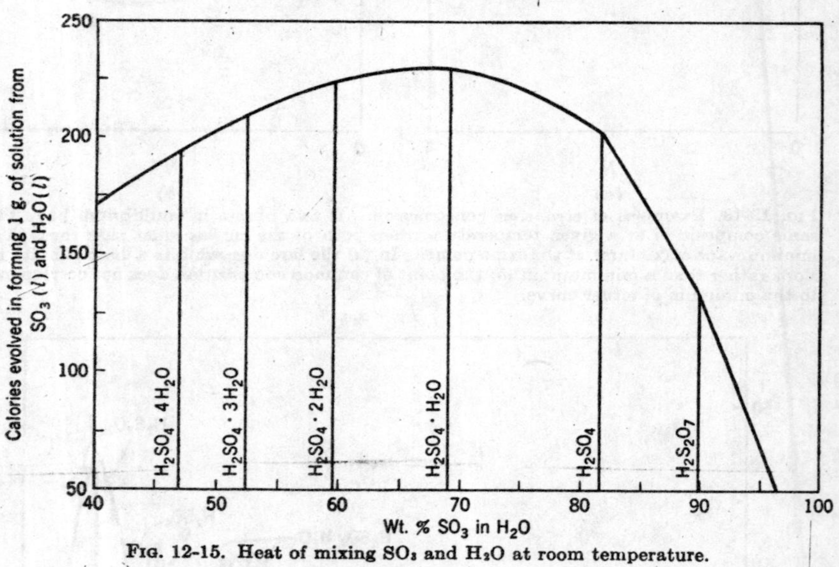

FIG. 12-15. Heat of mixing SO_3 and H_2O at room temperature.

(or nearly discontinuous) at the compositions corresponding to sulfuric acid (H_2SO_4) and pyrosulfuric acid ($H_2S_2O_7$). This is readily interpretable for these molecular solutions in that solutions slightly to the left of the composition H_2SO_4 may be regarded as solutions of H_2O in H_2SO_4 whereas those slightly to the right may be regarded as solutions of $H_2S_2O_7$ in H_2SO_4. This exemplifies the general principle that a discontinuity in thermodynamic behavior reflects a discontinuity in the atomic or molecular behavior.

If the foregoing thermodynamic conclusion is correct, then it would also be expected that the heat of formation of these liquid solutions when plotted against composition would also exhibit a discontinuity in slope at each of the compositions H_2SO_4 and $H_2S_2O_7$ but not at any other composition. That this is indeed so is illustrated in Fig. 12-15, which is taken from the data of Miles, Niblock, and Smith.[1] Similar behavior might be anticipated in a few systems of metallurgical interest, though few precise data are available.

Solubility of a Metastable Phase; Possible and Impossible Types of Curve Intersection. As will be shown in the next chapter, the solubility of a metastable phase in a given phase is always greater than the solubility of the corresponding stable phase. This is illustrated, for example, by the fact that the solubility of cementite in α or γ-iron is greater than that of graphite; similarly in the iron-nitrogen system, the metastable γ' (Fe_4N) phase has a greater solubility in α or γ-iron than does the stable nitrogen gas. An even simpler application of this rule may be made to the vapor pressure of a substance in the vicinity of its melting point. At a temperature below the melting point we may immediately conclude that the metastable liquid has a higher solubility in the vapor phase (or a higher vapor pressure) than the stable solid; the converse, of course, is true at a temperature above the melting point.

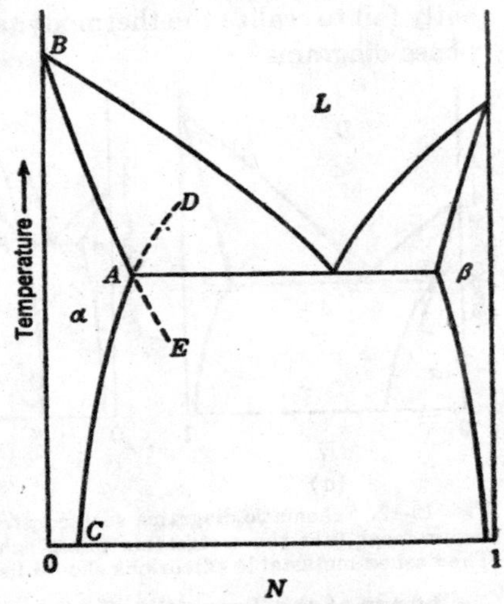

Fig. 12-16. Schematic diagram illustrating the principle that the solubility of a metastable phase is greater than that of a stable phase. Each dashed metastable extension must enter a two-phase region.

An interesting and informative application of this rule may be made to the intersection of solubility curves (boundaries of a single-phase region) on the temperature-composition diagram of a binary system. In Fig. 12-16 is shown a portion of common type of diagram with metastable extensions of two solubility curves for the α-phase region. The curve AC represents the composition of the α phase in stable equilibrium with β, or the solubility of β in α. The dashed curve AD, which is an extension of AC, represents the composition of α in metastable equilibrium with β. This dashed curve lies to the right of AB. As AB represents the composition of α in equilibrium with liquid (or the solubility of liquid in alpha), it is seen that, above the eutectic, the metastable phase β has a greater solubility in α than does the stable liquid. Similar considerations apply below the eutectic to the curves AC and AE. Here it is found that the metastable liquid has a greater solubility in α than does the stable β.

[1] Miles, Niblock, and Smith, *Trans. Faraday Soc.*, **40**, 281 (1944).

It is thus seen that the metastable extensions of the bounding curves of any single-phase region cannot lie within that single-phase region but must lie in the adjacent two-phase regions. Thus the constructions at points P, P', and P'' of Figs. 12-17*a*, *b*, and *c* are incompatible with thermodynamic requirements. Lipson and Wilson[1] point out that in Hansen's collection[2] there are at least 10 temperature-composition diagrams which violate this rule. Many investigators and authors apparently fail to realize the thermodynamic restrictions as to possible forms of phase diagrams.

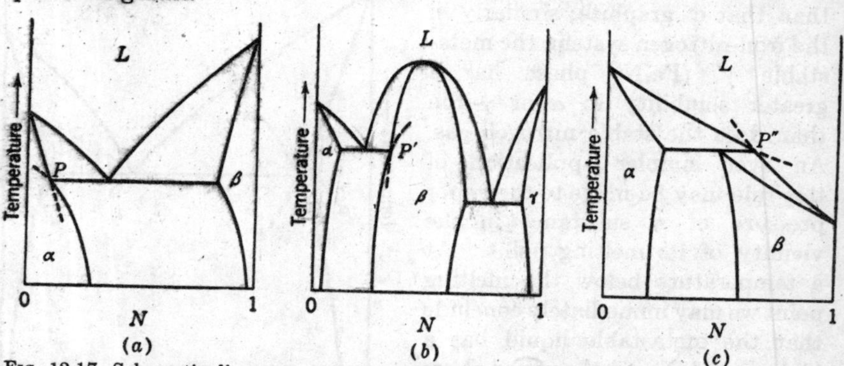

FIG. 12-17. Schematic diagrams with constructions incompatible with the thermodynamic requirement that the metastable phase exhibit greater solubility than the stable phase. The dashed metastable extensions should lie within the respective two-phase regions.

Change of the Transition Temperature. Although the thermodynamic treatment of the change in a transition temperature of a pure component produced by the presence of a second component is perfectly general, it is convenient to visualize it in terms of a specific transition, *e.g.*, the melting of a solid, designated phase I, to a liquid, phase II.

$$(\text{Component } 1)_{\text{phase I}} = (\text{component } 1)_{\text{phase II}}$$

Here again Eqs. (11-12) could be employed, but the treatment is simpler in terms of the equilibrium constant. We regard both components as being distributed between the two phases I and II. We may write for component 1

$$K_1 = \frac{a_1^{\text{II}}}{a_1^{\text{I}}} \tag{12-6}$$

As usual, we regard component one as the major component—the one whose transition temperature is under consideration. The standard

[1] H. Lipson and J. C. Wilson, *J. Iron Steel Inst.*, **142**, 107P (1940). See also J. C. Wilson, *J. Inst. Metals*, **70**, 543 (1944).

[2] M. Hansen, "Der Aufbau der Zweistofflegierungen," Springer-Verlag, Berlin, 1936.

state of component one in each phase is selected as the pure component. For component two we may write

$$(\text{Component 2})_{\text{phase II}} = (\text{component 2})_{\text{phase I}}$$

and

$$K_2 = \frac{a_2^{\text{I}}}{a_2^{\text{II}}} \tag{12-7}$$

It is convenient to choose a standard state for component 2 such that a_2 approaches N_2 as N_2 approaches zero in each phase. Expressing the activity in terms of the activity coefficient and the mole fraction,

$$K_1 = \frac{(1 - N_2^{\text{II}})\gamma_1^{\text{II}}}{(1 - N_2^{\text{I}})\gamma_1^{\text{I}}} \tag{12-8}$$

and

$$K_2 = \frac{N_2^{\text{I}}\gamma_2^{\text{I}}}{N_2^{\text{II}}\gamma_2^{\text{II}}} \tag{12-9}$$

It is now apparent that, if K_1 and K_2 are known (determined from free energies as discussed in Chap. 9) and the activity coefficients are known, then the above equations constitute a pair of simultaneous equations in the two unknowns N_2^{I} and N_2^{II}.

It is obvious that the determination of the composition of the two phases at a given equilibrium temperature is equivalent to a determination of the equilibrium temperature corresponding to a given amount of solute in one of the two phases. Thus if the standard free-energy change for the transfer of each component from one phase to the other is known, then K_1 and K_2 may be determined from the relation $\Delta F^\circ = -RT \ln K$. In the case of the solvent, ΔF_1° may usually be derived from a knowledge of the transition temperature, the heat of transition at that temperature and the heat capacities. ΔF_2° for the solute is usually not so readily attainable by this means, since the pure solute does not generally undergo the same type of transition as the solvent. Hence K_2 must usually be determined empirically at least at one and preferably at two temperatures; the value of K_2 at some other temperature may be obtained by assuming a linear relation between log K_2 and $1/T$. At the present state of our knowledge the activity coefficients for a solid or liquid solution must be determined by experiment. If at any particular temperature the equilibrium constants and activity coefficients are known, the composition of the two phases can be found by simultaneous solution of the two equations. Similarly if the composition of one phase is known, that of the other can be determined through the use of one of these equations, usually the relation in K_1.

Depression of the Freezing Point for a Dilute Solution. In applying the above relations to the freezing of a solution let us consider phase I to be the solid and phase II to be the liquid. If the solutions are sufficiently dilute that Raoult's law may be applied to the solvent, component 1, and Henry's law to the solute, component 2, the activity coefficients in Eqs. (12-8) and (12-9) may be set equal to unity, so that

$$K_2 = \frac{N_2^s}{N_2^l}$$

and

$$K_1 = \frac{(1 - N_2^l)}{(1 - N_2^s)} \cong 1 - N_2^l + N_2^s = 1 - N_2^l(1 - K_2)$$

whence

$$\ln K_1 \cong \ln[1 - N_2^l(1 - K_2)] \cong -N_2^l(1 - K_2)$$

the second approximate equality following from the fact that

$$\ln (1 - x) \cong -x$$

when $0 < x \ll 1$. Inserting now in Eq. (9-22*a*), $d \ln K/dT = \Delta H°/RT^2$, we have

$$\frac{d \ln K_1}{dT} = \frac{-d[N_2^l(1 - K_2)]}{dT} = \frac{\Delta H_m}{RT^2}$$

ΔH_m being the enthalpy of transfer of 1 mole of component 1 from the solid to the liquid (standard state in each case). Integrating from pure component 1 ($N_2 = 0$) with transition temperature T_m, assuming ΔH_m and K_2 to be substantially constant,

$$N_2^l(1 - K_2) = \frac{\Delta H_m}{R}\left(\frac{1}{T} - \frac{1}{T_m}\right) = \frac{\Delta H_m \, \Delta T}{RTT_m}$$

where $\Delta T = T_m - T$, that is, the depression of the transition temperature. ΔT is obviously negative if the transition temperature is raised by the presence of component 2. For a small depression of the freezing point T differs only slightly from T_m; hence we may set $T = T_m$, and

$$N_2^l = \frac{\Delta H_m \, \Delta T}{RT_m^2(1 - K_2)} \qquad (12\text{-}10)$$

Thus if the transition under consideration is a fusion, N_2^l is the atom fraction of the alloying constituent in the liquid phase, T_m is the melting point of the pure substance, ΔH_m is its heat of fusion, and ΔT is the melting-point depression. For example, in the case of iron, taking ΔH_m as 3670 cal/gram atom and the melting point as 1812°K, we find that $N_2^l = 0.00056\Delta T/(1 - K_2)$.

If, now, the depression of the freezing point ΔT for a certain low percentage of an alloying element is known, then K_2 (and hence the extent of solid solution) may be obtained from Eq. (12-10). Although K_2 was originally defined as a ratio of activities and simplified to a ratio of atom fractions at low concentration, it is now apparent by conversion to weight percentage that

$$K_2 = \frac{a_2^s}{a_2^l} = \frac{N_2^s}{N_2^l} = \frac{\dfrac{\%_2^s/M_2}{\%_2^s/M_2 + \%_1^s/M_1}}{\dfrac{\%_2^l/M_2}{\%_2^l/M_2 + \%_1^l/M_1}} \cong \frac{\%_2^s}{\%_2^l} \tag{12-11}$$

The final approximate equality follows by virtue of the fact that both of the two term denominators are substantially equal, the major term in each case being approximately the same since $\%_1^s \cong \%_1^l \cong 100$. From Eq. (12-11) it will be seen that K_2 is numerically equal to the percentage of component 2 in the solid phase which is in equilibrium with a liquid phase containing 1 per cent of component 2, within the error introduced by the approximations.

It will be noted from Eq. (12-10) that, if the solute is more soluble in the solid phase than in the liquid (*i.e.*, if the weight or atom fraction of the solute is greater in the solid than in the liquid), the freezing point is elevated rather than depressed.

Elevation of the Boiling Point for a Dilute Solution. Application of Eqs. (12-6) and (12-7) to the equilibrium between a dilute solution and its vapor leads, by the same steps used in the preceding section, to the following relation for a small elevation of the boiling point:

$$N_2^l = \frac{\Delta H_v \, \Delta T}{RT_b^2(1 - K_2)} \tag{12-12}$$

where N_2^l is the mole fraction of the solute in the liquid phase, ΔH_v is the heat of vaporization of pure component 1 (the solvent), ΔT is the boiling-point elevation, T_b is the boiling point in degrees Kelvin, and $K_2 = a_2^v/a_2^l$, the ratio of activity of solute in the vapor to that in the liquid. Equation (12-12) applies only when the species of molecules in the vapor phase are the same as those assumed in the liquid. This limitation on the validity follows from the use of the ideal-gas law; otherwise a more complicated equation would be obtained, as the student may verify by carrying through the derivation.

CHAPTER 13

FREE-ENERGY-COMPOSITION DIAGRAMS

As already discussed in Chap. 10 the free energy of 1 mole of a binary solution may be expressed by the relation

$$F = N_1\bar{F}_1 + N_2\bar{F}_2 \tag{13-1}$$

which for an isothermal system may also be written in terms of activities by employing Eq. (9-11c):

$$F = RT(N_1 \ln a_1 + N_2 \ln a_2) + N_1F_1^\circ + N_2F_2^\circ \tag{13-2}$$

The last two terms are the free energies of the components in the standard states and are constants for the isothermal, isobaric conditions we wish to

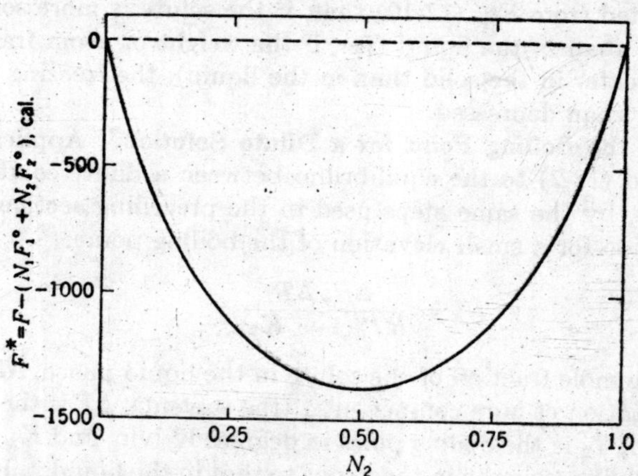

FIG. 13-1. Free-energy change for the formation of 1 mole of an ideal solution from the pure components. The scale for the ordinate applies at 1000°K only. *(From chapter by L. S. Darken in "Thermodynamics and Physical Metallurgy," American Society for Metals, Cleveland,* 1950.)

consider. For an ideal solution, where the activity of a component is equal to its mole fraction, the free energy in the above expression may be evaluated as a function of composition except for the two constant terms. Such a plot for an ideal solution at 1000°K, the constants being arbitrarily taken equal to zero, is shown in Fig. 13-1. A similar plot for several symmetric departures from ideality is shown in Fig. 13-2.

Vertical Tangents at Extremities of Free-energy-composition Curves. It is convenient to transfer the last two terms of Eq. (13-2) to the left side.

$$F - (N_1F_1^\circ + N_2F_2^\circ) = RT(N_1 \ln a_1 + N_2 \ln a_2)$$

This is the function actually plotted in Figs. 13-1 and 13-2. Differentiation with respect to N_2 gives

$$\frac{d[F - (N_1F_1^\circ + N_2F_2^\circ)]}{dN_2} = RT\left(\frac{N_1}{a_1}\frac{da_1}{dN_2} - \ln a_1 + \frac{N_2}{a_2}\frac{da_2}{dN_2} + \ln a_2\right)$$

At low concentration of component two Raoult's law ($a_1 = N_1$) is valid

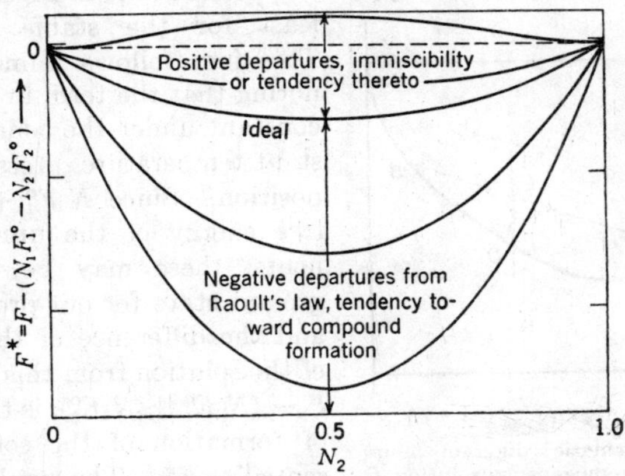

FIG. 13-2. Free energy of mixing for solutions exhibiting some types of departure from ideal-solution behavior. (*From chapter by L. S. Darken in "Thermodynamics and Physical Metallurgy," American Society for Metals, Cleveland,* 1950.)

for the solvent and Henry's law ($a_2 = kN_2$) is valid for the solute. Hence at small N_2

$$\frac{d[F - (N_1F_1^\circ + N_2F_2^\circ)]}{dN_2} = RT\left(\ln\frac{N_2}{N_1} + \ln k\right) \tag{13-3}$$

and as N_2 approaches zero, the derivative approaches $-\infty$. Hence all curves, as those of Figs. 13-1 and 13-2, whether or not the solution is ideal, have a vertical tangent at the left extremity; similar reasoning shows that they also have a vertical tangent at the right extremity. This fact may also be inferred from the general theorem [Eq. (10-12)] which as applied to the free energy becomes

$$\bar{F}_2 = F + N_1\frac{dF}{dN_2}$$

The graphical interpretation of this, as noted at the time, is that the intercepts of the tangent with the axes are equal to $\bar{F}_1$ and $\bar{F}_2$. From this equation it is seen that, as N_2 approaches zero and $\bar{F}_2$ (which equals $RT \ln a_2 + F_2^\circ$) approaches $-\infty$, the slope dF/dN_2 must also approach $-\infty$, F being always finite. Obviously $d[F - (N_1F_1^\circ + N_2F_2^\circ)]/dN_2$ also approaches $-\infty$.

It has already been shown from the second law that for a system of definite composition at constant temperature and pressure the most stable configuration[1] is the one which possesses the lowest free energy. This is the fundamental principle in the use of free-energy-composition diagrams. From the fact that the free energy is least for the stable configuration, it follows that, under the same conditions, the function $F - (N_1F_1^\circ + N_2F_2^\circ)$, which appears as the ordinate in Figs. 13-1*ff*, is also least for the stable configuration. This fact follows immediately upon noting that the term in parentheses is constant under the conditions of constant temperature, pressure, and composition. Since $N_1F_1^\circ + N_2F_2^\circ$ is the free energy of the unmixed components, these may be taken as the ground state for our present purpose, and the difference of the free energy of the solution from this ground state, $F - (N_1F_1^\circ + N_2F_2^\circ)$, is the free energy of formation of the solution, represented as F^*. The symbol F^* is used here instead of ΔF^M, used in Chap. 10, since we wish to focus attention not on the formation of a single solution but on F^* as a property of that solution, to be compared with F^* for other solutions or compounds of the same composition.

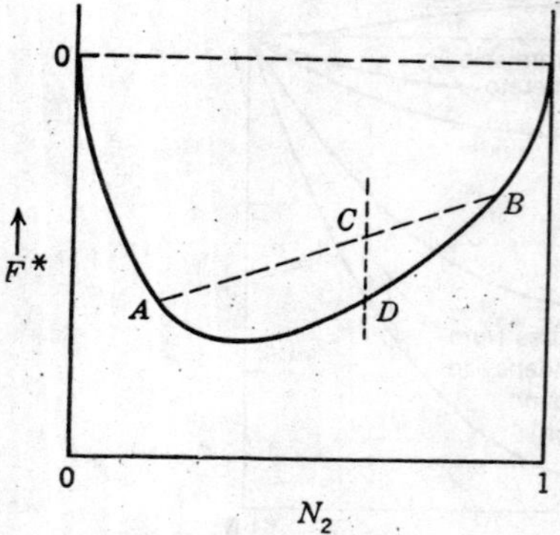

FIG. 13-3. Schematic diagram illustrating that a homogeneous solution D is stable relative to any mechanical mixture C of the same gross composition providing the F^*-composition curve is concave upward.

Free-energy Change on Mixing; Miscibility Gap. Let us compare the free energy of the solution with that of the unmixed components. It is obvious that F^* for a system consisting of two separated components is zero no matter what the relative amounts; this is represented by the horizontal dotted line in the upper part of Fig. 13-1. F^* for an ideal solution is represented by the curve and is seen to be lower than that for the unmixed components at all relative proportions. We next consider the mixing of the two solutions represented by points A and B in Fig. 13-3. The molal free energy of the two solutions considered as a single system before mixing corresponds to a point C on the dotted straight line AB, the

[1] By configuration is meant a phase or mixture of phases.

precise position depending upon the relative amounts.[1] The molal free energy F^* of the resulting solution is represented by point D which is below point C, and we see that the final solution is stable relative to the two unmixed portions. It is now evident that the single-phase solution is stable relative to any two unmixed portions for any phase region in which the curve representing F^* as a function of composition is concave upward.

Let us now consider a hypothetical series of solutions for which F^* may be represented by a curve of the type shown in the upper portion of Fig. 13-4. In this case it will be noted that in the central region two separated solutions represented by points B and C have an F^* corresponding to point D, which is lower than that for the single homogeneous solution represented by point A. The configuration with the lowest free energy for the gross composition $N_2 = x$ is obviously the point G corresponding to this composition on the double tangent EF. The stable configuration corresponding to this composition is thus two solutions of compositions y and z. It is seen, therefore, that stable systems of compositions from $N_2 = 0$ to $N_2 = y$ are composed of a single solution, whereas stable systems from $N_2 = y$ to $N_2 = z$ consist of two solutions of compositions y and z; from $N_2 = z$ to $N_2 = 1$ the stable system is again composed of a single solution. Single-phase solutions from $N_2 = y$ to $N_2 = z$ are metastable with respect to the two-phase system. This

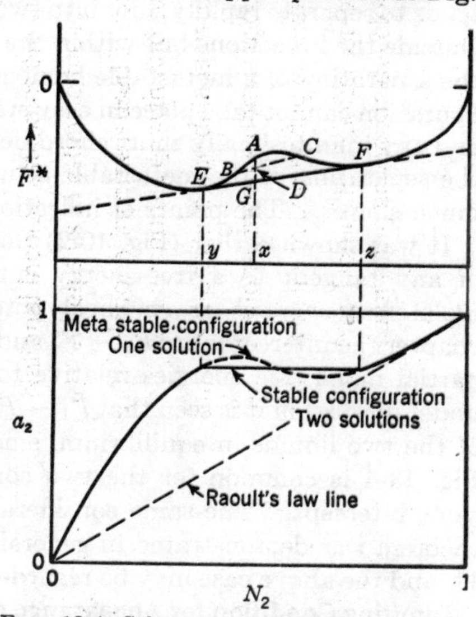

FIG. 13-4. Schematic free-energy and activity diagrams showing the conditions for immiscibility.

[1] To prove this, consider n moles of solution A, of molal free energy F^*_A, and $1 - n$ moles of solution B, of molal free energy F^*_B, as a composite but unmixed system. For this system $F^* = nF^*_A + (1 - n)F^*_B$. In terms of the mole fraction N_2 of the composite system and the mole fractions N_2^A and N_2^B of the original two solutions, $n = (N_2^B - N_2)/(N_2^B - N_2^A)$; and by substitution,

$$F^* = \frac{(N_2^B - N_2)}{(N_2^B - N_2^A)} F^*_A + \frac{(N_2 - N_2^A)}{(N_2^B - N_2^A)} F^*_B.$$

From this it is obvious that F^* for the composite but unmixed system is a linear function of N_2 as shown by line AB in Fig. 13-3.

situation is typical of systems exhibiting a miscibility gap and is associated with sufficiently great positive departure from Raoult's law. The lower part of Fig. 13-4 shows schematically the corresponding activity diagram.

Significance relative to reaction-rate phenomena is frequently attached to the inflections near points B and C on the F^* curve of Fig. 13-4. Between these inflections the curve is concave downward and even an infinitesimally short chord lies under the curve; any solution in this region tends to separate rapidly, first into two solutions of adjacent compositions. Outside the inflections but within the region where two phases are stable, the separation of a metastable homogeneous solution into the stable configuration cannot take place in easy stages, as the curve is concave upward and any infinitesimally short chord lies above the curve. In such regions the separation of the metastable solution into two solutions tends to be much slower. The points of inflection are sometimes called *spinodes.*

It was shown earlier (Fig. 10-2) that the intercepts (at $N_2 = 0$ and 1) of any tangent to a free-energy curve give the corresponding partial molal free energies or chemical potentials. For the F^* plots of this chapter the intercepts are $\bar{F}_1 - F_1^\circ$ and $\bar{F}_2 - F_2^\circ$, that is, the corresponding partial molal free energies relative to the standard states. In the case under discussion it is seen that $\bar{F}_1 - F_1^\circ$ and hence μ_1 are the same for each of the two liquids in equilibrium, since the tangent in the upper part of Fig. 13-4 is common for the two compositions and necessarily has the same intercept. The same consideration obviously applies to $\bar{F}_2$. This theorem was demonstrated in general for phases in equilibrium in Chap. 11, and the above case may be regarded as a graphical illustration thereof.

Limiting Condition for Appearance of a Miscibility Gap. It is apparent that there is a special intermediate case between those represented by Figs. 13-3 and 13-4 in which the system is just on the verge of exhibiting a miscibility gap. The conditions for this special case are that $d^2F^*/dN_2^2 = 0$ and $d^3F^*/dN_2^3 = 0$ at the composition where the miscibility gap is imminent. These derivatives are, of course, calculable only if F^* is known as a function of composition. In case the activity coefficients can be represented by Eqs. (10-64) namely, $\ln \gamma_1 = \alpha N_2^2$ and $\ln \gamma_2 = \alpha N_1^2$, where α is a constant, it is found that

$$F^* = RT(N_1 \ln N_1 + N_2 \ln N_2 + \alpha N_1 N_2) \tag{13-4}$$

so that

$$\frac{dF^*}{dN_2} = RT\left[\ln\frac{N_2}{N_1} + \alpha(N_1 - N_2)\right] \tag{13-5}$$

$$\frac{d^2F^*}{dN_2^2} = RT\left(\frac{1}{N_1} + \frac{1}{N_2} - 2\alpha\right) \tag{13-6}$$

and

$$\frac{d^3F^*}{dN_2^3} = RT\left(\frac{1}{N_1^2} - \frac{1}{N_2^2}\right) \tag{13-7}$$

Equating the third derivative to zero it is found that $N_1 = N_2 = \frac{1}{2}$. Equating the second derivative to zero and inserting this value, α is found to be 2.† Thus for a solution whose activity coefficients may be represented by Eqs. (10-64) an immiscibility will be found if $\alpha > 2$ and will not be found if $\alpha < 2$. The corresponding critical value of the activity coefficient at $N_1 = N_2 = \frac{1}{2}$ is 1.648, and the corresponding Henry's law coefficient, *i.e.*, the activity coefficient at infinite dilution of either component in the other, is 7.39.

Rewriting Eq. (10-12*a*),

$$\bar{F}_1 = F + N_2\frac{dF}{dN_1}$$

and differentiating with respect to N_1,

$$\frac{d\bar{F}_1}{dN_1} = N_2\frac{d^2F}{dN_1^2}$$

A second differentiation gives

$$\frac{d^2\bar{F}_1}{dN_1^2} = N_2\frac{d^3F}{N_1^3} - \frac{d^2F}{dN_1^2}$$

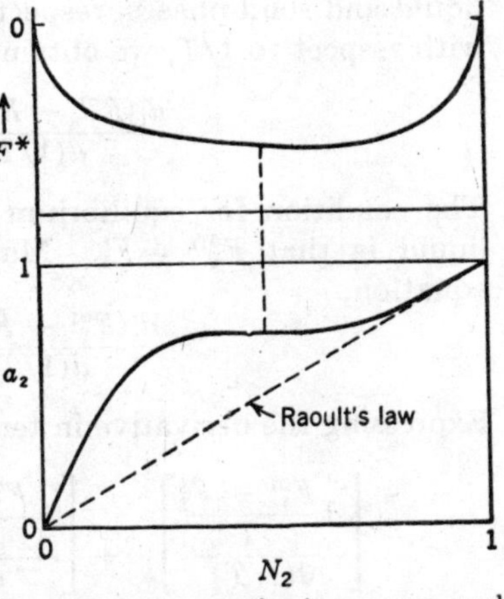

FIG. 13-5. Schematic free-energy and activity diagrams illustrating the conditions for imminent immiscibility, namely, that the second and third derivatives of the free energy and the first and second derivatives of the activity with respect to composition be zero. These conditions are satisfied at the composition denoted by the dotted line.

Obviously, similar relations may be derived for component 2. Hence under the condition of imminent unmixing of a solution, it is seen not only that d^2F/dN_1^2 and d^3F/dN_1^3 are both zero but also that the first and second derivatives of both of the partial molal free energies are zero. By virtue of the isothermal relation between the activity and the free energy it also follows that the first and second derivatives of the activities are zero, thus that the activity-composition plot shows a horizontal inflection. The functions for this case are shown schematically in Fig. 13-5.

It is now of interest to inquire as to the nature of the liquidus curve on the temperature-composition diagram under the condition investigated

† This value, as N_2 approaches zero, corresponds to $\ln \gamma_2 = 2$ and to an excess partial molal free energy $RT \ln \gamma_2 = 2RT$. The latter relation was used in Chap. 4 to derive Hume-Rothery's rule that 15 per cent or greater disparity in atomic radius leads to a severe limitation on the extent of solid solubility.

above, *i.e.*, that the liquid is on the verge of separating into two liquids. We shall designate as component 1 the component on whose liquidus the miscibility gap is imminent. The relation derived above for this case, namely, that at constant pressure and temperature $d\bar{F}_1/dN_1 = 0$, we now write as $(\partial \bar{F}_1/\partial N_1) = 0$, since we wish to consider a variation of temperature. We shall limit our consideration to a system in which the solid solubility of component 2 in component 1 is negligibly small.[1] Since the process here is one involving the crystallization of component 1, let us consider the free-energy change accompanying the melting of this pure component, $F_1^{\circ\text{I}} - F_1^{\circ\text{II}} = \Delta F_m$, where the superscripts I and II indicate the liquid and solid phases, respectively. Dividing by T and differentiating with respect to $1/T$, we obtain by virtue of Eq. (9-22)

$$\frac{d[(F_1^{\circ\text{I}} - F_1^{\circ\text{II}})/T]}{d(1/T)} = H_1^{\circ\text{I}} - H_1^{\circ\text{II}}$$

The condition for equilibrium between pure solid component one and liquid is that $F_1^{\circ\text{II}} = \bar{F}_1^{\text{I}}$. Making this substitution in the foregoing equation,

$$\frac{d[(F_1^{\circ\text{I}} - \bar{F}_1^{\text{I}})/T]}{d(1/T)} = H_1^{\circ\text{I}} - H_1^{\circ\text{II}}$$

Expressing the derivative in terms of its partials,

$$\left[\frac{\partial \dfrac{F_1^{\circ\text{I}} - \bar{F}_1^{\text{I}}}{T}}{\partial(1/T)}\right]_{N_1} + \left[\frac{\partial \dfrac{F_1^{\circ\text{I}} - \bar{F}_1^{\text{I}}}{T}}{\partial N_1}\right]_T \frac{dN_1}{d(1/T)} = H_1^{\circ\text{I}} - H_1^{\circ\text{II}}$$

Noting now that the first term may be replaced by $H_1^{\circ\text{I}} - \bar{H}_1^{\text{I}}$,

$$\left[\frac{\partial[(F_1^{\circ\text{I}} - \bar{F}_1^{\text{I}})/T]}{\partial N_1}\right]_T \frac{dN_1}{d(1/T)} = \bar{H}_1^{\text{I}} - H_1^{\circ\text{II}}$$

The enthalpy terms on the right represent the heat absorbed when 1 gram mole of pure solid component 1 is dissolved in a large amount of solution of the particular composition under consideration. Noting that

$$d\left(\frac{1}{T}\right) = -\left(\frac{1}{T^2}\right) dT$$

and that $1/T$ may be removed from the partial, we find upon rearranging that

$$\frac{dT}{dN_1} = \frac{T}{\bar{H}_1^{\text{I}} - H_1^{\circ\text{II}}}\left(\frac{\partial \bar{F}_1^{\text{I}}}{\partial N_1}\right)_T \tag{13-8}$$

[1] Since we are considering a system in which the liquid is on the verge of immiscibility it is reasonable to expect a rather limited solid solubility.

Under the condition for imminent unmixing of the solution, which is that the partial derivative is zero, we see that the slope dT/dN of the liquidus curve is also zero,[1] as illustrated in Fig. 13-6. It is readily seen that the second derivative d^2T/dN_1^2 is also zero.

Free-energy-composition Diagram for a System Involving a Solid. The type of diagram discussed in the previous sections may be used to interpret solubility relations. Let us consider a binary system (exhibiting complete miscibility in the liquid state) at a temperature below the melting point of one of the components, say component 1; let us suppose for the moment that the solid solubility in component 1 is vanishingly small. Since we are below the melting point of component 1, the pure solid is stable relative to the pure liquid and hence F^* for the solid is less than F^* for the liquid, as represented by point A in Fig. 13-7. It is important to note that in any diagram such as this the same standard state must be adopted for a component in all phases. The diagram is drawn as though F_1° and F_2° are the values characteristic of the pure liquids; any other standard states could, of course, be adopted. The composition of the liquid in equilibrium with the solid corresponds to the point of tangency of a line through the point A to the curve. Since the intercept of the tangent also corresponds to $\bar{F}_1 - F_1^\circ$ for the solution of composition corresponding to the point of tangency, it is seen that the chemical potential of solid component 1 is equal to the chemical potential of component 1 in the solution in equilibrium with the solid.

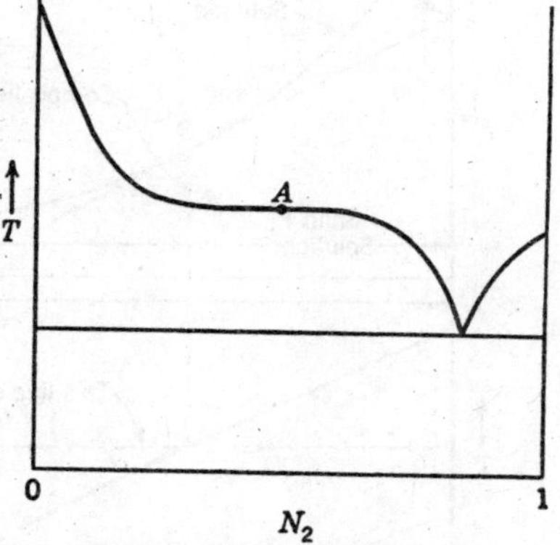

FIG. 13-6. Schematic diagram showing the contour of the liquidus for the condition of imminent unmixing of the liquid. A slightly greater positive departure from ideality would lead to the appearance of a two-liquid region in vicinity of point A, where there now exists a horizontal inflection.

Suppose we wish to consider the equilibrium between a solution and an intermediate compound of fixed composition. If the temperature under consideration is below the melting point of the compound, so that the compound is stable relative to the liquid of the same composition, then F^* for the compound must be below the corresponding F^* for the liquid as represented by point A in Fig. 13-8. The compositions of the

[1] The above reasoning is valid providing $\bar{H}_1^{\mathrm{I}} - H_1^{\circ\mathrm{II}}$ also is not zero, in which case the right side of the equation is indeterminate. However, $\bar{H}_1^{\mathrm{I}} - H_1^{\circ\mathrm{II}}$ is in effect the heat of fusion and can hardly be expected to be zero except as a special case.

liquid in equilibrium with this intermediate phase are found by constructing the tangents through point A as shown. At the melting point of the compound, the point corresponding to F^* for the compound falls on the curve for the liquid (point B), and at a temperature above the melting point, it lies above the curve for the liquid (point C).

If we wish to consider the equilibrium of a liquid with a primary solid solution, then F^* for the primary solid solution as well as for the liquid

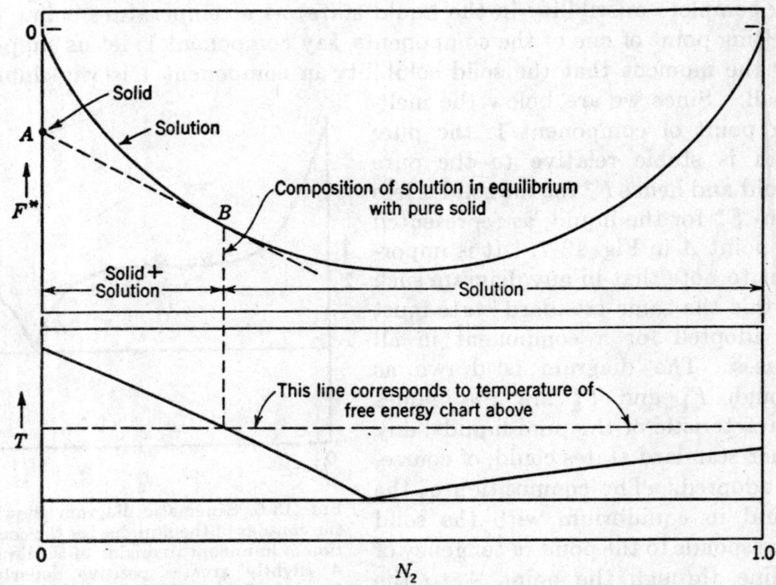

FIG. 13-7. Relation between free-energy chart and temperature-composition diagram for the equilibrium between a solution and a pure solid component. (*From chapter by L. S. Darken in "Thermodynamics and Physical Metallurgy," American Society for Metals, Cleveland,* 1950.)

solution must be represented by a curve, as shown in Fig. 13-9. The equilibrium compositions of the solid and liquid solution are found by constructing the common tangent as in the figure. A similar construction for an intermediate phase of variable composition is represented in Fig. 13-10. From tangent-intercept considerations it is seen that both of these figures again illustrate that the chemical potential of each component is the same for the two phases in equilibrium.

Relative Solubility of Allotropic Modifications. If one of the components is capable of existing in two allotropic forms (here considered pure), the situation at a particular temperature may be represented as in Fig. 13-11, the stable allotropic modification being assigned a lower free energy than the metastable form. The metastable form, in spite of its metastability, may be experimentally realizable; *e.g.*, the solubility of each of the forms of solid phosphorus in a given solvent may be determined. It will be noted that the *metastable form has the higher solubility*—an important theorem, easily demonstrated by use of the free-energy diagram. It is obvious that the only requirement is that the F^* curve for the liquid be concave upward, a requirement previously demonstrated to exist for

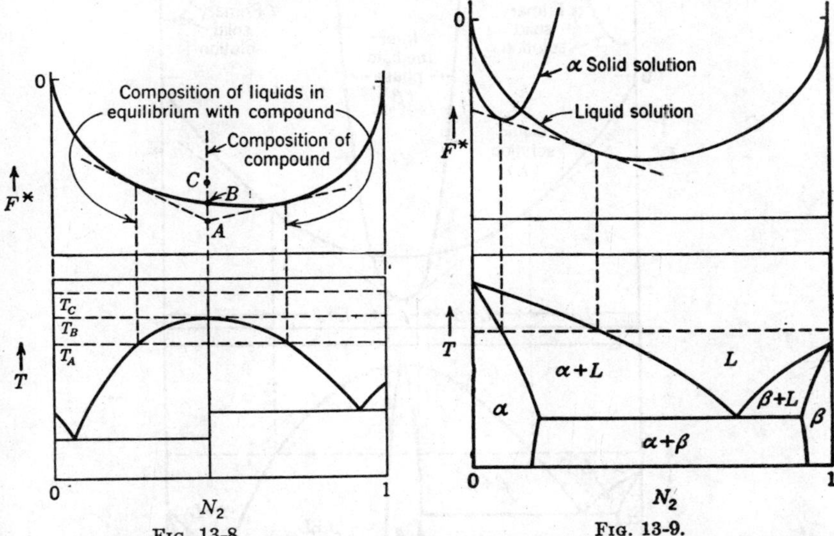

Fig. 13-8.

Fig. 13-9.

Fig. 13-8. Free-energy and temperature-composition diagrams illustrating equilibrium between solution and an intermediate phase of fixed composition. The relative position of the point denoting the free energy of the intermediate phase in the upper part of the figure depends upon the temperature as shown in the lower part.

Fig. 13-9. Free-energy and temperature-composition diagrams illustrating equilibrium between liquid and primary solid solution.

all stable liquids. This same theorem may be applied to solid solutions as well as liquid solutions.

We shall next consider the extension of the above theorem to a case where the solute phases concerned are not necessarily of the same composition. For example, we may wish to compare the solubility in water of anhydrous sodium sulfate with that of one of the hydrates, or we may wish to compare the solubility of graphite and cementite (Fe_3C) in austenite (fcc iron-carbon alloy). Before proceeding, it is desirable to call attention to the fact that in discussing solubility it is customary to speak of the solubility of one *phase* in another, *e.g.*, the solubility of graphite in austenite. This is a sort of metaphor; what is really meant is the composition of one phase when it is in equilibrium with another.

It is incorrect (or at least indeterminate) to speak of the solubility of a component in a phase; *e.g.*, "the solubility of carbon in austenite" is vague terminology for "the solubility of graphite in austenite" or "the solubility of cementite in austenite." The numerical value of the solubility may be given in any concentration units (weight per cent, atom fraction, etc.). Thus, although solubility refers to the composition of

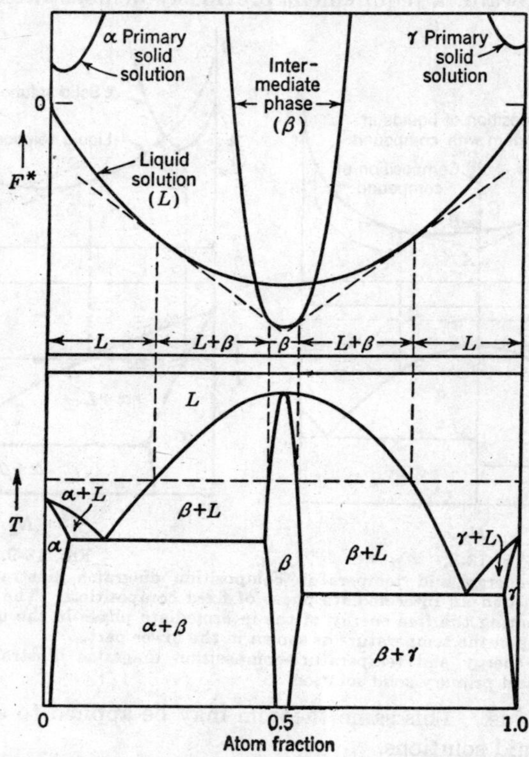

Fig. 13-10. Relation between free-energy chart and temperature-composition diagram for the case of a high-melting intermediate phase. (*From chapter by L. S. Darken in "Thermodynamics and Physical Metallurgy," American Society for Metals, Cleveland,* 1950.)

one *phase* when it is in equilibrium with another, the numerical value, like that of any concentration, is expressed in terms of the *components.* For example, the solubility of cementite in austenite (at some particular temperature) might be expressed as 0.9 per cent carbon.[1]

[1] The choice of components, as stated previously, is entirely arbitrary. Hence if we choose to regard the components as Fe and Fe_3C, instead of Fe and C (but not γ-iron and cementite, since these are phases), the solubility of either cementite or graphite in austenite may be expressed as per cent Fe_3C. Such a choice seems rather pointless in this case.

Let us now consider the relative solubility of graphite and cementite in austenite. A schematic diagram is shown in Fig. 13-12. This diagram is purposely distorted to make visible the separation of the points of tangency. It is immediately obvious from the diagram that, if the point corresponding to F^* for cementite lies above the tangent to the austenite curve drawn through the graphite point (as shown)—*i.e.*, if cementite is metastable relative to graphite and austenite saturated therewith—then the solubility of cementite in austenite is greater than that of graphite.

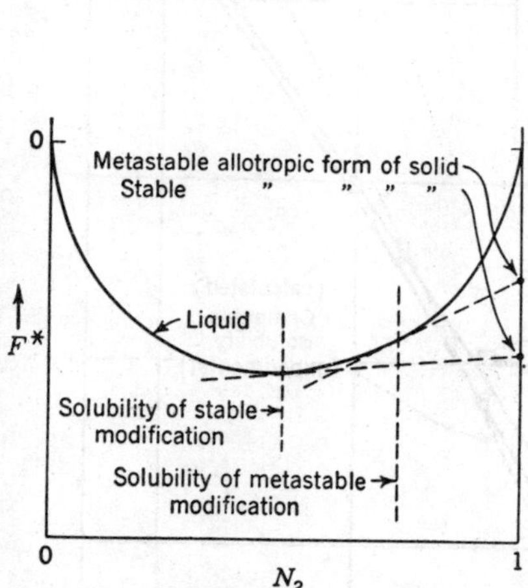

FIG. 13-11. Schematic diagram illustrating the general principle that, for two solids of the same composition, the metastable form is more soluble than the stable form.

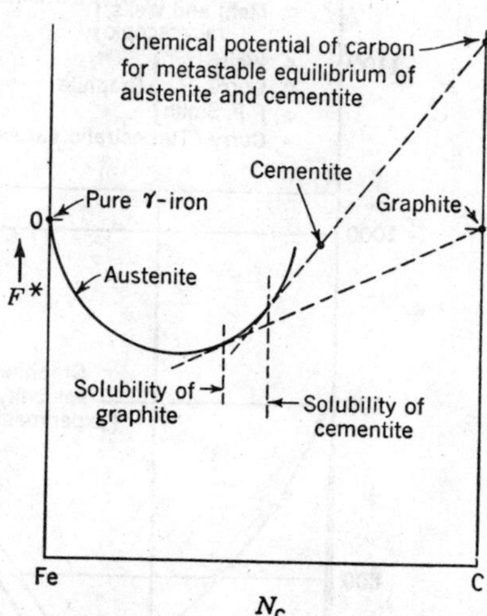

FIG. 13-12. Schematic diagram illustrating that the solubility in austenite of cementite (metastable) is greater than that of graphite (stable).

It has been observed at temperatures within the austenite range that alloys of iron and carbon composed of cementite and austenite transform spontaneously in sufficient time to austenite and graphite.[1] Hence, if this observation is accepted as correct, we may conclude that, at all temperatures between the eutectoid and the eutectic of the iron-carbon system, cementite is metastable with respect to graphite and austenite saturated therewith. In view of the theorem just demonstrated we may conclude also that the solubility of cementite in austenite is greater than that of graphite. Both of these solubilities have been measured. Among the more reliable measurements are those of Mehl and Wells[2] on the

[1] C. Wells, *Trans. ASM*, **26**, 289 (1938).
[2] R. F. Mehl and C. Wells, *Trans. AIME*, **125**, 429 (1937).

solubility of cementite, and those of Gurry,[1] Wells,[2] and Smith[3] on the solubility of graphite. The experimental results are shown in Fig. 13-13. Although the data on the solubility of cementite do not extend to the higher temperatures, it is obvious that these data indicate a crossing of the two solubility curves in the vicinity of 950°C. In view of the

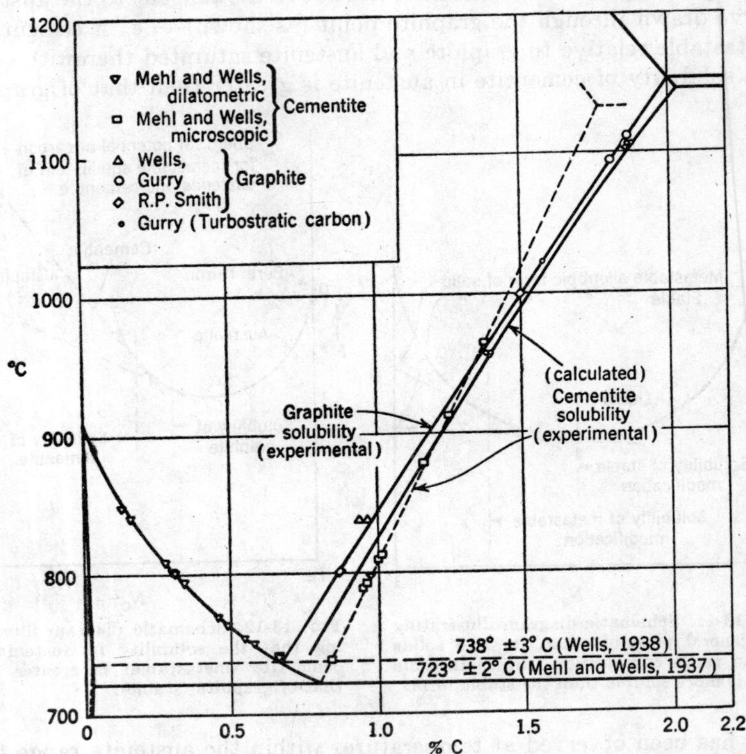

FIG. 13-13. The solubility of graphite and of cementite in austenite. The calculated cementite solubility and the measured graphite solubility are in accord with the observation that cementite is metastable throughout this temperature range. [*From Darken and Gurry, Trans. AIME*, **191**, 1015 (1951).]

foregoing theorem we should conclude that cementite is stable relative to graphite and austenite saturated therewith above this temperature. However, this conclusion is at variance with the previously mentioned direct observation of Wells. It is almost unbelievable that such a discrepancy should still exist in the iron-carbon system. The conclusion

[1] R. W. Gurry, *Trans. AIME*, **150**, 147 (1942).

[2] C. Wells, *Trans. ASM*, **26**, 289 (1938).

[3] R. P. Smith, *J. Am. Chem. Soc.*, **68**, 1163 (1946).

seems inescapable that one of the three observations (the graphite solubility, the cementite solubility, or the relative stability) is incorrect. Indications that the principal error lies in the observations of the cementite solubility led the authors[1] to calculate this solubility from the observed graphite solubility and other data. This calculated solubility of cementite is also shown in Fig. 13-13.

Application of the Foregoing Principle to the Construction of Phase Diagrams. The principle, derived above, that the solubility of a stable

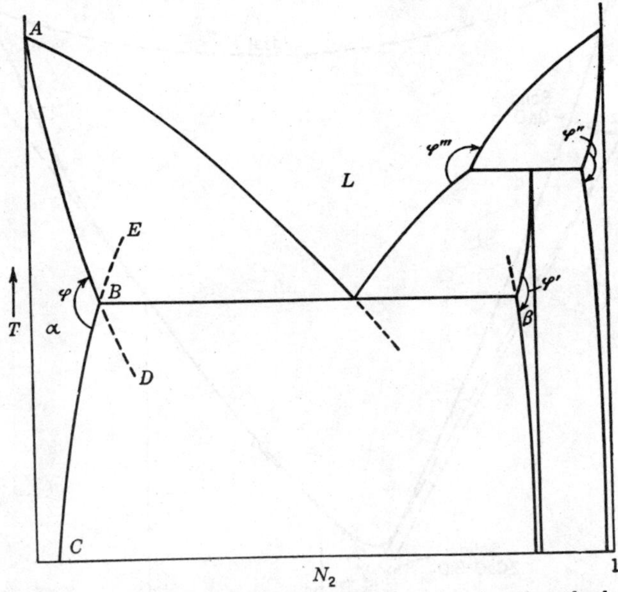

Fig. 13-14. Schematic temperature-composition diagram showing that the boundaries of a single-phase region meet in such a way that the angle φ within this region is less than 180°.

phase in a given phase is always less than that of a metastable phase,[2] also finds frequent application in the construction of phase diagrams. Let us consider the primary solid solution region at the left of the schematic diagram of Fig. 13-14. The boundary AB represents the composition of the primary solid solution α in stable equilibrium with the liquid L; the curve BC represents the composition of α in stable equilibrium

[1] L. S. Darken and R. W. Gurry, *Trans. AIME*, **191**, 1015 (1951).

[2] As mentioned previously the term stable (or metastable) as applied to a phase has meaning only when it is expressed or understood that the phase is stable (or metastable) with respect to some definite transformation. Thus, for example, we might say that the phases A and B are stable relative to the phase C—meaning that they will not react spontaneously to form C but rather that C will (or may) react to produce more A or B or both until completely consumed.

with the intermediate phase β. Let us construct the metastable extensions of both of these curves, shown broken. Thus a point on the extension BD represents the composition of the primary solid solution in metastable equilibrium with the supercooled liquid whose composition is also shown by a broken curve. Similarly a point on the extension BE represents the composition of the primary solid solution in metastable equilibrium with the intermediate phase β. In accord with the foregoing theorem all points on the metastable curve BD must lie to the right of the corresponding points on the stable curve BC, and thus the entire curve BD must lie to the right of BC. Similarly, BE must lie to the right of BA. Hence the angle φ must be less than 180°. Similar reasoning shows that the angles φ', φ'', and φ''' must also be less than 180°. The student unfamiliar with this application should carry through the reason-

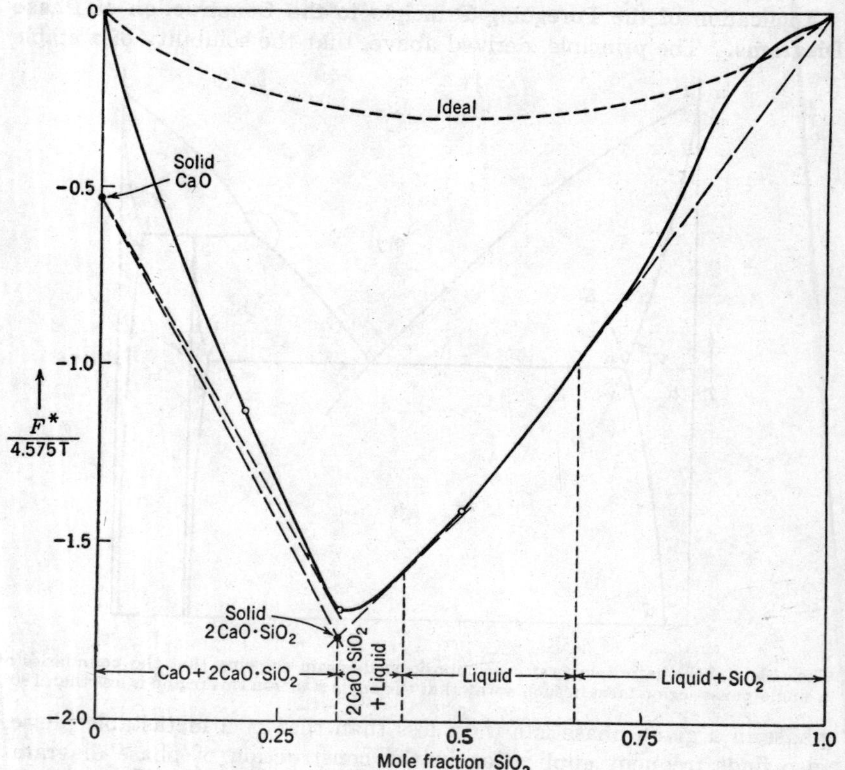

FIG. 13-15. Free-energy chart for the system CaO-SiO$_2$ at 1600°C. (*From chapter by L. S. Darken in "Thermodynamics and Physical Metallurgy," American Society for Metals, Cleveland,* 1950.)

ing for the other angles, noting the direction associated with "greater" solubility in each case. It will be observed that use is made of the properties of the metastable phases to gain information about the stable phases. This subject was considered previously in Figs 12-16 and 12-17 and the discussion thereof.

Free-energy Diagram for the System CaO-SiO₂. If our interest is focused on activity, it is convenient to plot $F^*/4.575T$, rather than F^*, against composition. If this is done, as in Fig. 13-15, the intercepts of the tangents are $(\bar{F}_1 - F_1^\circ)/4.575T = \log a_1$, and $(\bar{F}_2 - F_2^\circ)/4.575T = \log a_2$, and the logarithm of the activities may be directly determined by drawing the tangent or estimated by imagining one.

The standard states used in the construction of Fig. 13-15 are the metastable pure liquids CaO and SiO_2. The points on which the curve for the liquid solution was constructed are shown as open circles; these were estimated from data on the ternary system CaO-SiO_2-FeO†. Since the curve must have vertical tangents at its termini, and in view of the existence of miscibility gap at only slightly higher temperature, a fairly accurate sketching of the entire curve for the liquid is possible. The solid circles represent F^* for solid phases. Tangents from these touch the curve at points corresponding to the solubilities; these compositions are in agreement with the observed solubilities or extrapolation thereof. The stable configurations, easily recognized here from the principle of lowest free energy, are designated at the bottom of the figure.

The rather large negative departure from ideality, as evidenced by comparison with the ideal solution curve at the top of the figure, is conspicuous; at the minimum, $\log a$ for each component is seen (by imagining the horizontal tangent) to be about -1.7 or $a_{CaO} = a_{SiO_2} = 0.02$, less than one-tenth that for an ideal solution. The sharpness of the minimum indicates a very rapid change of both activities in this vicinity, so that this composition (two CaO to one SiO_2) seems a natural division between acid and basic compositions just as a pH of 7 is a natural division point for aqueous solutions at room temperature. In the acid region to the right of the minimum the curvature is small and the intercepts of the moving tangent change only slowly with composition; here the activity of CaO is very low (about $\frac{1}{1000}$) while that of SiO_2 is nearly one.

† C. R. Taylor and J. Chipman, *Trans. AIME*, **154**, 228 (1943).

CHAPTER 14

TEMPERATURE-PRESSURE DIAGRAMS

It follows immediately from the phase rule that a binary system is univariant when three phases are in equilibrium. In such a case one variable, *e.g.*, the temperature, may arbitrarily be fixed; the pressure and all the chemical potentials are then functions of the temperature and the properties of the particular system under consideration (providing, as we consider to be the case, that no external work is done other than that against pressure and that the energies associated with surfaces are negligibly small).

If the phases other than the gas are of fixed composition, *i.e.*, their composition does not change with temperature, and if the pressure is sufficiently low (not exceeding perhaps 10 atm) as to have negligibly small effect upon the free energy of the condensed phases,[1] then the problem is most conveniently handled in terms of the equilibrium constant. In fact, even if the composition of the condensed phases is not constant, it is frequently more convenient to handle the data in terms of the equilibrium constant, particularly when the temperature range is limited and the pressure not excessive. In the following sections the general treatment is given first and is followed by special cases including those most readily treated in terms of the standard free-energy change or equilibrium constant.

RELATION BETWEEN TEMPERATURE AND PRESSURE FOR A UNIVARIANT EQUILIBRIUM IN A BINARY SYSTEM

Let us investigate the general form of the relationship between pressure and temperature for a univariant equilibrium involving phases of variable composition. We first write the total differential of $\bar{F}_1/T$ in terms of its partials for a phase of variable composition (conveniently considered to be a liquid phase).

$$d\frac{\bar{F}_1}{T} = \left[\frac{\partial(\bar{F}_1/T)}{\partial(1/T)}\right]_{N_2,P} d\frac{1}{T} + \left[\frac{\partial(\bar{F}_1/T)}{\partial N_2}\right]_{T,P} dN_2 + \left[\frac{\partial(\bar{F}_1/T)}{\partial P}\right]_{T,N_2} dP \quad (14\text{-}1)$$

[1] The effect of pressure on the free energy at constant temperature is given by the relation $dF = V\,dP$. The gram-atomic volume of a solid metal is usually of the order of 10 cc; hence the free-energy change per atmosphere pressure is 10 cc-atm or about $\frac{1}{4}$ cal. Since the precision of the standard free-energy change of an oxidation reaction such as the above is seldom better than 10 cal, the effect of a few atmospheres pressure on the activity of a condensed phase is negligible.

It will be noted that in a binary system there are three independent variables: temperature, pressure, and one composition variable—here chosen as N_2. From Eqs. (10-26), (9-11*a*), and (10-22) the three partial derivatives can be evaluated and the above equation becomes

$$d\frac{\bar{F}_1}{T} = \bar{H}_1\, d\frac{1}{T} + R\left(\frac{\partial \ln a_1}{\partial N_2}\right)_{T,P} dN_2 + \frac{\bar{V}_1}{T}\, dP \tag{14-2}$$

Similarly for $\bar{F}_2/T$,

$$d\frac{\bar{F}_2}{T} = \bar{H}_2\, d\frac{1}{T} + R\left(\frac{\partial \ln a_2}{\partial N_2}\right)_{T,P} dN_2 + \frac{\bar{V}_2}{T}\, dP \tag{14-3}$$

In order to eliminate a_1 and a_2 by virtue of the Gibbs-Duhem relation [Eq. (10-51)], Eq. (14-2) is multiplied by N_1 and Eq. (14-3) by N_2. On adding the resulting equations,

$$N_1\, d\frac{\bar{F}_1}{T} + N_2\, d\frac{\bar{F}_2}{T} = (N_1\bar{H}_1 + N_2\bar{H}_2)\, d\frac{1}{T} + (N_1\bar{V}_1 + N_2\bar{V}_2)\frac{1}{T}\, dP \tag{14-4}$$

Since $N_1\bar{H}_1 + N_2\bar{H}_2 = H$ and $N_1\bar{V}_1 + N_2\bar{V}_2 = V$, we may write, dividing by $d(1/T)$,

$$\frac{N_1\, d(\bar{F}_1/T) + N_2\, d(\bar{F}_2/T)}{d(1/T)} = H + \frac{V}{T}\frac{dP}{d(1/T)} \tag{14-4a}$$

If the other two phases are also of variable composition, similar equations may be written for them.

$$\frac{N_1'\, d(\bar{F}_1/T) + N_2'\, d(\bar{F}_2/T)}{d(1/T)} = H' + \frac{V'}{T}\frac{dP}{d(1/T)} \tag{14-5}$$

$$\frac{N_1''\, d(\bar{F}_1/T) + N_2''\, d(\bar{F}_2/T)}{d(1/T)} = H'' + \frac{V''}{T}\frac{dP}{d(1/T)} \tag{14-6}$$

It will be observed that temperature, pressure, and partial molal free energy (chemical potential) are not primed, as each of these is the same for all phases in equilibrium. Simultaneous solution of this set of three equations for $dP/d(1/T)$ by elimination of the terms involving partial molal free energies, noting also that $N_1 + N_2 = 1$, $N_1' + N_2' = 1$, and $N_1'' + N_2'' = 1$, gives

$$\frac{dP}{d(1/T)} = -\frac{(N_1' - N_1'')H + (N_1'' - N_1)H' + (N - N_1')H''}{(N_1' - N_1'')V + (N_1'' - N_1)V' + (N_1 - N_1')V''}\, T \tag{14-7}$$

This is the general solution for the relation between temperature and pressure for three phases at equilibrium in a binary system. It will be observed that the coefficient of T on the right side is the ratio of the enthalpy change to the volume change when the equilibrium reaction occurs; neither the enthalpy terms (numerator) nor the volume terms (denominator) apply to the standard-state reaction but to the actual equilibrium reaction.

Case Where Only One Phase Is of Variable Composition. Let us now restrict our attention to a somewhat simpler case where only one phase (unprimed symbols), conveniently considered as a liquid or solid solution, is of variable composition and the other two phases, conveniently considered as solid (primes) and gas (double primes) are of fixed composition. In this case $N_2' = 0$ and $N_1' = 1$; $N_2'' = 1$ and $N_1'' = 0$. Equation (14-7) then reduces to

$$\frac{dP}{d(1/T)} = -\frac{H - N_1H' + (N_1 - 1)H''}{V - N_1V' + (N_1 - 1)V''}T \tag{14-8}$$

Substituting $H = N_1\bar{H}_1 + N_2\bar{H}_2$ and $V = N_1\bar{V}_1 + N_2\bar{V}_2$,

$$\frac{dP}{d(1/T)} = -\frac{N_1(\bar{H}_1 - H') + N_2(\bar{H}_2 - H'')}{N_1(\bar{V}_1 - V') + N_2(\bar{V}_2 - V'')}T \tag{14-9}$$

For the gas phase, from the ideal-gas law, $P = RT/V_2''$. Dividing Eq. (14-9) by this expression, and noting that, since the primed phase (solid) is pure component 1 and the double-primed phase (gas) is pure component 2, the heat content and volume may be so subscripted,

$$\frac{d\ln P}{d(1/T)} = -\frac{N_1(\bar{H}_1 - H_1') + N_2(\bar{H}_2 - H_2'')}{N_1(V_1 - V_1') + N_2(V_2 - V_2'')}\frac{V_2''}{R} \tag{14-10}$$

This, then, is the final equation for the univariant equilibrium in a two-component system between three phases, of which one is a pure solid (component 1), another a gas (component 2), and the third a solution. The only assumptions involved are that the solid phase is substantially pure component 1 and the gaseous phase behaves ideally and is composed substantially of component 2. Similar derivation for other special cases of interest may be performed by the student, starting with the general relation Eq. (14-7). It will be seen that it is frequently convenient to choose a compound of fixed composition rather than an element as a component, for in this case Eq. (14-10) may be applied to the equilibrium between compound, solution, and vapor, the compound being selected as component 1.

Equilibrium Involving a Salt, Its Aqueous Solution and Water Vapor. Let us now consider the equilibrium involving (1) a crystalline nonvolatile salt, (2) its saturated aqueous solution, and (3) the vapor phase. The temperature-composition diagram for the system potassium nitrate–water is shown in Fig. 14-1, and the temperature-pressure diagram for the above-mentioned equilibrium in this system in Fig. 14-2. Let us com-

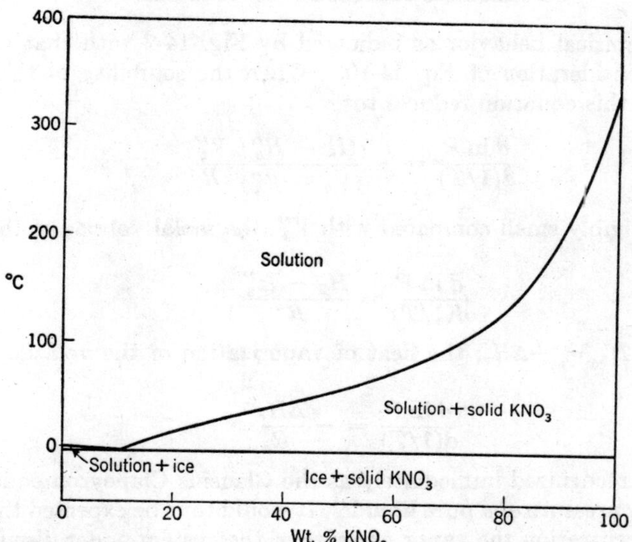

FIG. 14-1. Temperature-composition diagram for the system potassium nitrate–water. (*Replotted from G. W. Morey in "Commentary on the Scientific Writings of J. Willard Gibbs," Vol. I, Yale University Press, New Haven,* 1936.)

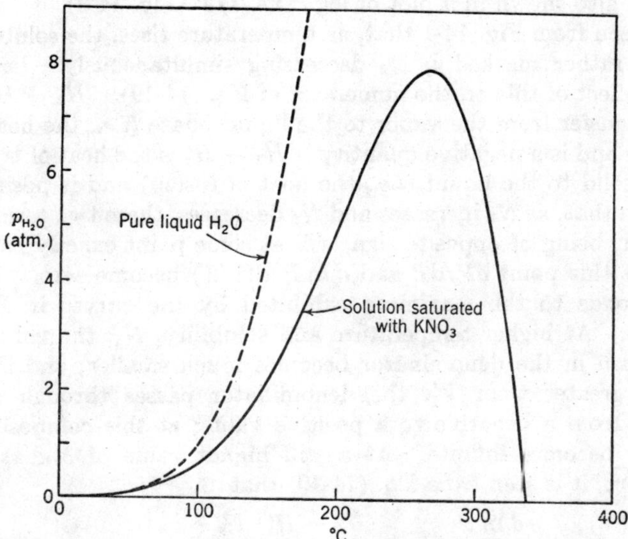

FIG. 14-2. Vapor pressure of water in equilibrium with saturated KNO_3 solution as a function of temperature. The dotted curve shows the vapor pressure of pure water. The composition of the saturated solution is given in Fig. 14-1. (*Replotted from G. W. Morey in "Commentary on the Scientific Writings of J. Willard Gibbs," Vol. I, Yale University Press, New Haven,* 1936.)

pare the empirical behavior as indicated by Fig. 14-2 with that anticipated by consideration of Eq. 14-10. When the solubility of the salt, N_1, is small this equation reduces to

$$\frac{d \ln P}{d(1/T)} = -\frac{(\bar{H} - H_2'')}{(\bar{V}_2 - V_2'')} \frac{V_2''}{R}$$

If $\bar{V}_2$ is negligibly small compared with V_2'', the molal volume of the gas phase,

$$\frac{d \ln P}{d(1/T)} = \frac{\bar{H}_2 - H_2''}{R}$$

Since $\bar{H}_2 - H_2'' = -\Delta H_v$, the heat of vaporization of the volatile component 2,

$$\frac{d \ln P}{d(1/T)} = -\frac{\Delta H_v}{R}$$

This will be recognized immediately as the Clausius Clapeyron equation for the vapor pressure of a pure liquid. It would thus be expected that at low salt concentration the vapor pressure of the system under discussion would approach the vapor pressure of pure water. The latter is shown dotted in the figure, and it is seen that this expectation is realized. The comparison is also shown in a plot of $\log P$ vs. $1/T$ (Fig. 14-3).

It will be seen from Fig. 14-1 that, as temperature rises, the solubility N_1 increases rather markedly, N_2 decreasing simultaneously. Let us consider the effect of this on the numerator of Eq. (14-10). $\bar{H}_2 - H_2''$ is the heat of transfer from the vapor to the liquid phase (*i.e.*, the heat of condensation) and is a negative quantity. $\bar{H}_1 - H_1'$ is the heat of transfer from the solid to the liquid (*i.e.*, the heat of fusion) and is positive. Thus it is seen that, as N_1 increases and N_2 decreases, these two terms in the numerator, being of opposite sign, will at some point exactly cancel each other; at this point dP/dT and $d \ln P/d(1/T)$ become zero. This point corresponds to the maximum exhibited by the curves in Figs. 14-2 and 14-3. At higher temperature and solubility, N_1, the value of the second term in the denominator becomes much smaller; and if, as usual, $\bar{V}_1$ is greater than V_1', the denominator passes through zero as it changes from a negative to a positive value; at this composition $d \ln P/d(1/T)$ becomes infinite. At a still higher value of N_1, as N_2 approaches zero, it is seen from Eq. (14-10) that

$$\frac{d \ln P}{d(1/T)} = -\frac{(\bar{H}_1 - H_1')}{(\bar{V}_1 - V_1')} \frac{V_2''}{R}$$

or

$$\frac{dT}{dP} = \frac{T(\bar{V}_1 - V_1')}{(\bar{H}_1 - H_1')}$$

This equation is identical with that developed previously to express the change of the melting point with pressure. Since this effect is very small for the pressures under consideration, the temperature of infinite slope practically coincides with the melting point of the salt.

The above example and particularly Fig. 14-3 have been discussed in some detail in order to make it clear that the relation between log P and

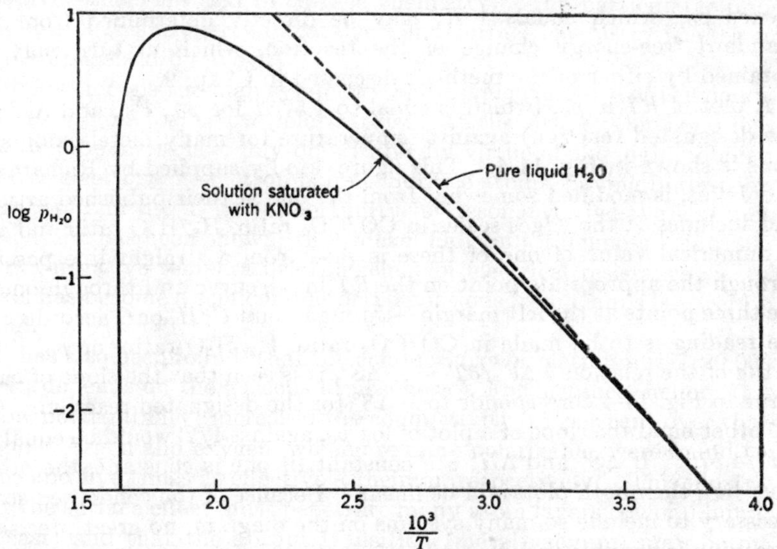

FIG. 14-3. Plot of log p_{H_2O} vs. $1/T$ for saturated KNO_3 solution and for pure water.

$1/T$ may depart very far from linearity when the phases involved are not of fixed composition.

METAL-OXIDE SYSTEMS

One of the simplest examples of an equilibrium between two condensed phases and a vapor phase is that between a metal, its oxide, and the vapor phase. If the condensed phases are of fixed composition, this problem is most readily handled in terms of the equilibrium constant. Let us consider the generalized reaction

$$x\text{M(cryst.)} + \text{O}_2\text{(g)} = \text{M}_x\text{O}_2\text{(cryst.)}$$

for which

$$K = \frac{a_{\text{M}_x\text{O}_2}}{a_{\text{M}}^x p_{\text{O}_2}} \tag{14-11}$$

Under the condition of a negligibly small range of solid solution (x is constant, possibly an integer), and further under the condition that the pressure is sufficiently near 1 atm that the metal and its oxide may be considered to be substantially in their standard states, the activity of these is unity and

$$K = \frac{1}{p_{O_2}} \tag{14-12}$$

Hence p_{O_2}, which equals $1/K$, may be directly determined from the standard free-energy change of the reaction, which in turn may be obtained by either of the methods discussed in Chap. 9.

A plot of $RT \ln p_{O_2}$ (which is equal to $4.575T \log p_{O_2}$, $\bar{F}_{O_2}$, and $\Delta F°$ for the designated reaction) against temperature for many metal-oxide systems is shown in Fig. 14-4. This figure, kindly supplied by Richardson and Jeffes, is modified somewhat from the one in their published article[1] and includes at the edges scales in CO/CO_2 ratio, H_2/H_2O ratio and p_{O_2}. A numerical value of one of these is read from a straight line passing through the appropriate point on the $RT \ln p_{O_2}$ curve and through one of the three points at the left margin—through point C, H, or O according as the reading is to be made in CO/CO_2 ratio, H_2/H_2O ratio, or p_{O_2}. By virtue of the relation $\partial\, \Delta F°/\partial T = -\Delta S°$, it is seen that the slope of each curve in Fig. 14-4 corresponds to $-\Delta S°$ for the designated reaction. On the other hand the slope of a plot of $\log p_{O_2}$ against $1/T$ would be equal to $\Delta H°/4.575$. If $\Delta S°$ and $\Delta H°$ are constant (if one is constant, the other must be), then both plots will be linear. Because of the condensed scale necessary to include so many systems on the diagram, no great precision in reading is possible and departures from linearity are hardly noticeable.

All the considerations of the preceding two paragraphs apply equally well if one or both phases are liquid or even if the metal or oxide is gaseous, provided the composition of the liquid or partial pressure of metal or oxide in the gas remains constant as the temperature changes. Condensed phases of variable composition are considered in the next section. Equilibria involving gaseous metal or oxide phases are most readily treated in terms of the equilibrium constant, Eq. (14-11); in this case the ordinate of Fig. 14-4 is to be interpreted as $\Delta F° = -RT \ln K$. As the activity of metal or oxide, or both, may no longer be set equal to 1, $\Delta F°$ is no longer equal to $\bar{F}_{O_2}$ or to $RT \ln p_{O_2}$.

Breaks in the Plot of $\Delta F°$ against Temperature. At the melting point of metal or oxide, since the liquid is in equilibrium with the solid, both are in equilibrium with the same phases; that is, p_{O_2} is the same at the melting point whether the oxide equilibrium involves solid or liquid. Thus $\Delta F°$ or $RT \ln p_{O_2}$ in Fig. 14-4 does not change discontinuously at the melting point of one of the phases involved. However, $\Delta S°$ does change at a

[1] Richardson and Jeffes, *J. Iron Steel Inst.*, **160**, 261 (1948).

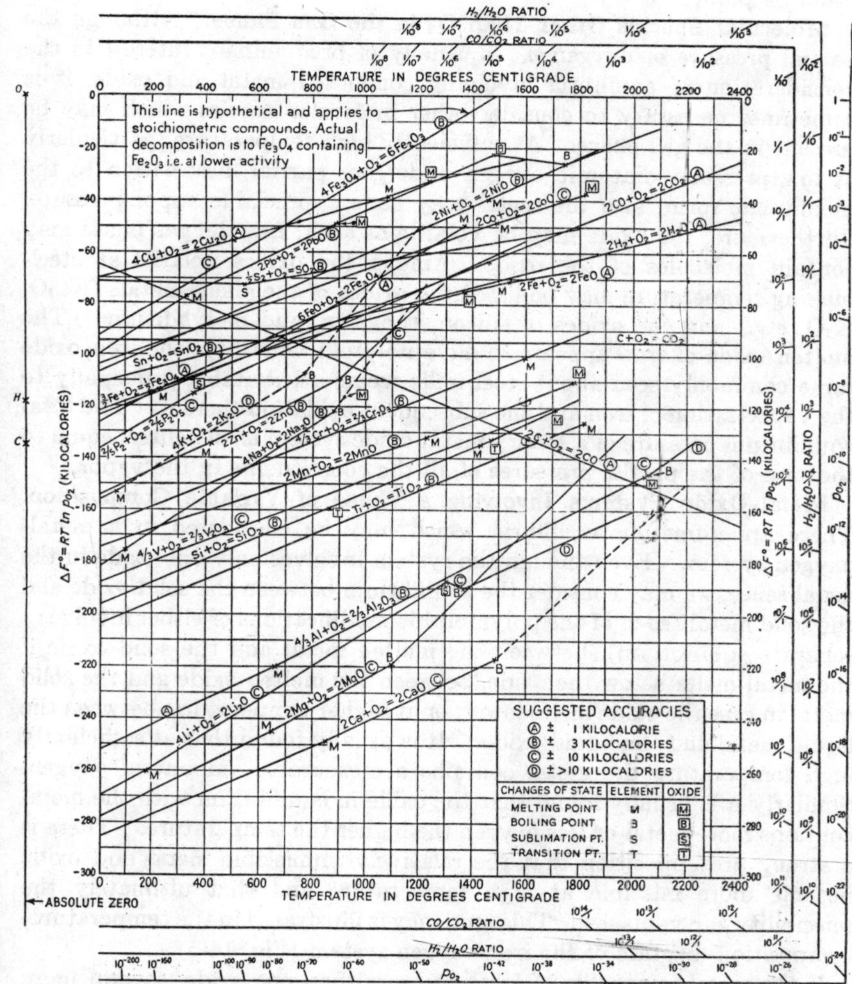

FIG. 14-4. The standard free energy of formation of many metal oxides as a function of temperature. [*From F. D. Richardson and J. H. E. Jeffes, substantially as in J. Iron Steel Inst.* **160**, 261 (1948).]

melting point, and hence $d\,\Delta F°/dT = -\Delta S°$, the slope of the curve, changes discontinuously at a melting point. Since the entropy change of melting is always positive, it follows that, if a product, the oxide, melts, $\Delta S°$ increases and the slope of the curve decreases. On the other hand, at the melting point of a reactant, the metal, $\Delta S°$ decreases and the slope of the curve increases. Many examples of this may be seen in the figure.

Similar considerations apply at a transition point, boiling point, or sublimation point.

Molecular Species Other Than O_2 in the Gas Phase. Although the partial pressure of oxygen p_{O_2} is usually of predominant interest in the consideration of equilibria involving condensed metal and oxide, it is sometimes necessary to consider other molecular species which may be present in the gas phase. At sufficiently high temperature, particularly at low pressure, diatomic oxygen undergoes partial dissociation to the monatomic form; also the metal may have appreciable vapor pressure. Furthermore, the oxide may be volatile as such; *i.e.*, the gas phase may contain molecules of the oxide. Among the oxides volatile at steelmaking temperature may be listed the oxides of the alkali metals (Na_2O, K_2O, etc.) and the oxides of silicon, tungsten, and molybdenum. The molten oxide of iron appears to have no great volatility; the iron oxide fume commonly seen about steel mills seems attributable principally to the vaporization of iron and the subsequent oxidation thereof. The total equilibrium pressure of a binary metal-oxide system is, of course, equal to the sum of the partial pressures of all the constituents in the vapor.

Metal-Oxide Systems Involving a Phase of Variable Composition. There are numerous equilibria which may be considered in a metal-oxygen system. Even though the system involves only one oxide in the usual sense, we may consider the equilibrium between the solid oxide and the solid metal (each of the polymorphic modifications of either involves a separate equilibrium), between the molten metal and the solid oxide if the metal melts below the oxide, between the molten oxide and the solid metal in case the oxide melts lower, or at higher temperature, between the liquid metal and the liquid oxide. It is usually found that at sufficiently high temperature the metal contains a measurable amount of oxygen. Similarly it is usually found that the oxide in equilibrium with the metal contains more metal or less oxygen the higher the temperature. There is a strong presupposition that the relatively immiscible metal and oxide become more miscible at high temperature and that ultimately the miscibility gap vanishes. This tendency is illustrated in the temperature-composition diagram of the iron-oxygen system (Fig. 14-5).

If the metal is capable of forming more than one oxide, several more equilibria may be considered. In the iron-oxygen system, for example, these equilibria correspond to the various regions of two condensed phases indicated in Fig. 14-5. It will be seen from the phase rule that an invariant point in a two-component system corresponds to an equilibrium between four phases (three condensed phases and the vapor phase) and that a univariant curve such as those of Fig. 14-4 corresponds to an equilibrium between three phases (two condensed phases and a gaseous

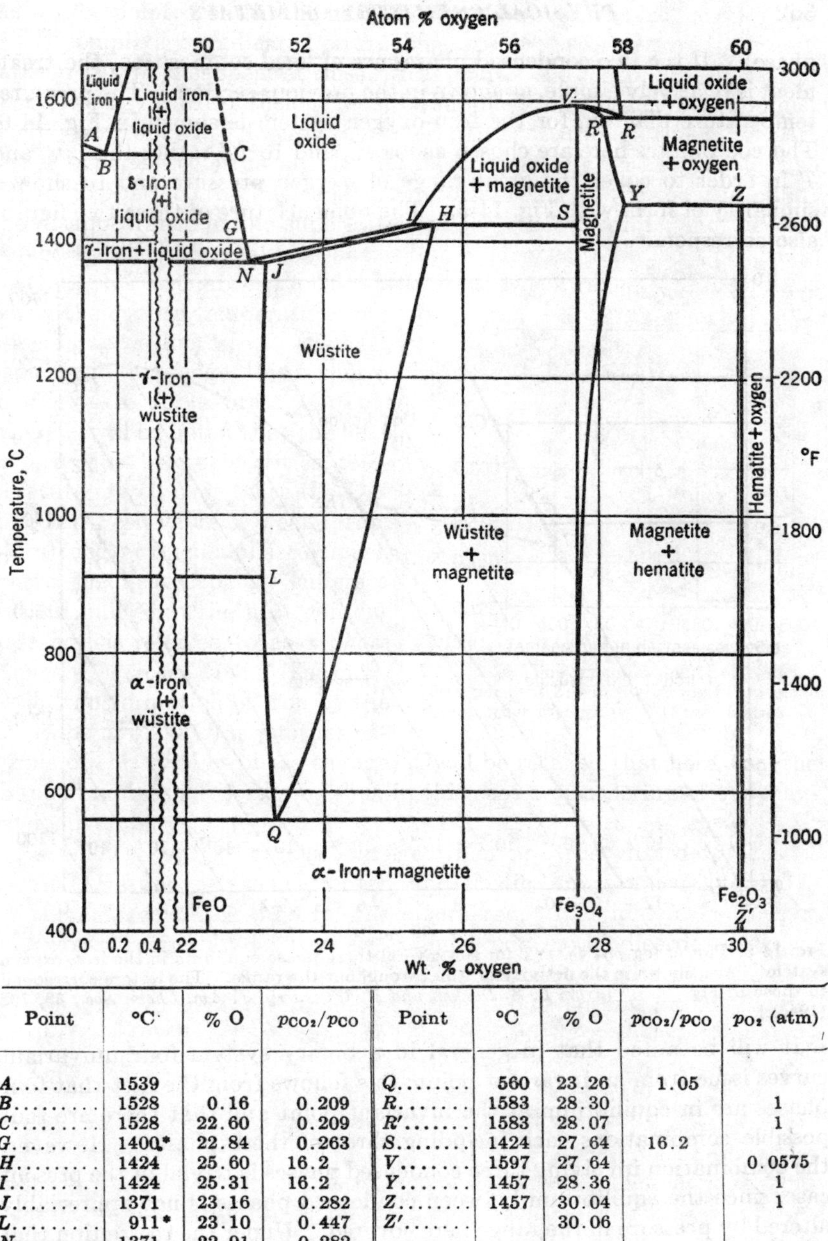

Point	°C	% O	p_{CO_2}/p_{CO}	Point	°C	% O	p_{CO_2}/p_{CO}	p_{O_2} (atm)
A........	1539			*Q*........	560	23.26	1.05	
B........	1528	0.16	0.209	*R*........	1583	28.30		1
C........	1528	22.60	0.209	*R'*.......	1583	28.07		1
G........	1400*	22.84	0.263	*S*........	1424	27.64	16.2	
H........	1424	25.60	16.2	*V*........	1597	27.64		0.0575
I........	1424	25.31	16.2	*Y*........	1457	28.36		1
J........	1371	23.16	0.282	*Z*........	1457	30.04		1
L........	911*	23.10	0.447	*Z'*.......		30.06		
N........	1371	22.91	0.282					

* Values for pure iron.

FIG. 14-5. Temperature-composition diagram for the iron-oxygen system. [*From L. S. Darken and R. W. Gurry, J. Am. Chem. Soc.* **68**, 798 (1946).]

phase). If the two condensed phases are of fixed composition, the treatment is relatively simple, as shown in the previous section. The pressure-temperature diagram for the iron-oxygen system is shown in Fig. 14-6. The coordinates here are chosen as $\log p_{O_2}$ and $10^4/T$ rather than p_{O_2} and T in order to cover the great range of oxygen pressure and to show a similarity of form with Fig. 14-5. The nomenclature of these two figures also corresponds.

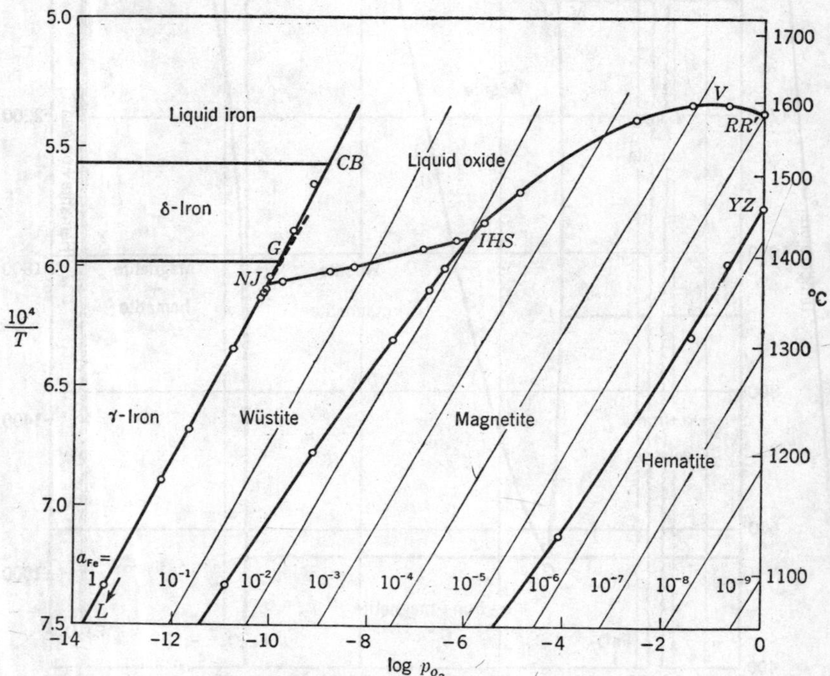

FIG. 14-6. Plot of $\log p_{O_2}$ vs. $1/T$ for the several three-phase equilibria in the iron-oxygen system. Also shown is the activity of iron throughout the range. The letters correspond to those in Fig. 14-5. [*From L. S. Darken and R. W. Gurry, J. Am. Chem. Soc.*, **68**, 798 (1946).]

It will be noted that in general in a binary system four univariant curves issue from an invariant point; this follows from the fact that four phases are in equilibrium at the invariant point and that there are four possible combinations each including three of these phases. However, the combination involving three condensed phases is trivial in the present case, since the equilibrium between condensed phases is not appreciably altered by pressure in the range here covered. Under the restriction that a gas phase always be present, significance may be attached to the regions of the temperature-pressure diagram, as in Fig. 14-6.

The equilibrium constant may readily be evaluated even though the phases may vary in composition if sufficient data are available to establish the standard free-energy change; however, the activity of the metal and of the oxide may not be equated to unity if the condensed phases are not of fixed composition. Thus p_{O_2} may not be evaluated even though log K is known unless information is available not only as to the equilibrium compositions but also as to the activity of the metal and of the oxide as functions of composition. Hence one method of determining the relation between the partial pressure and the temperature is from a knowledge of the standard free-energy change and the activities of the metal and of the oxide.

In the iron-oxygen system the univariant equilibrium between metallic iron, wüstite, and gas involves a substantially pure metallic phase, a gas phase, and an oxide phase which contains considerably more oxygen (or less iron) than corresponds to the composition FeO. Moreover, the wüstite phase involved here is not strictly of constant composition; the relative proportions of iron and oxygen vary slightly with temperature as shown in Fig. 14-5. A more extreme example of a phase of variable composition is commonly found, as discussed previously, in the univariant equilibrium involving a solid, a liquid, and a gaseous phase, *e.g.*, the equilibrium between magnetite and molten oxide. It is seen that the equilibrium between magnetite, liquid oxide, and gas involves a solid of substantially fixed composition (Fe_3O_4, on the low-oxygen side), a gaseous phase, and a liquid phase whose composition varies considerably as shown by the curve IV of Fig. 14-5. It is apparent in Fig. 14-6 that there is a large departure from linearity in the plot of log p_{O_2} against $1/T$ for this equilibrium, represented by the curve between points *IHS* and *V*. As already noted, this type of behavior is to be expected in such an equilibrium where the composition of at least one phase varies widely.

It might further be pointed out that the flatness of the curve at point *V* (the congruent melting point of magnetite, where magnetite of the composition Fe_3O_4 melts to a liquid of the same composition) is predictable from Eq. (14-7). Regarding the double-primed quantities as applying to the gas, the single-primed quantities to the liquid, and the unprimed quantities to the solid, it is seen that the principal term in the denominator, $(N_1 - N_1')V''$, approaches zero as N_1 approaches N_1', that is, as the composition of the liquid approaches that of the solid. Hence $dP/d(1/T)$ approaches infinity or $d(1/T)/d \ln P$ approaches zero in the very near vicinity of the composition Fe_3O_4; this conclusion is in accord with the observed behavior (Fig. 14-6).

Equilibrium between a Metal and Its Oxide of Unknown Variable Composition. Experimentally it is not uncommon to investigate the partial pressure of oxygen of the three-phase equilibrium, metal-oxide-gas, without knowledge as to whether the oxide phase is of fixed composition or not. In such a case the experimenter, when reporting in the literature, often plots $\log p_{O_2}{}^{\frac{1}{2}}$* vs. $1/T$, determines the slope of the linear or nearly linear curve, multiplies by $2.303R$, and calls the result the heat of formation of the oxide. Let us investigate the meaning that can be attached to such a procedure. It will be supposed that the metal phase is substantially pure—departures from this idealization can be handled in the same manner if necessary—and also that the gaseous phase consists exclusively of O_2. In this case Eq. (14-10) may be applied. The partial pressure of O_2 is usually so low, and therefore the partial molal volume of oxygen so large, that all other partial molal volumes may be ignored in comparison thereto. Equation (14-10) then takes the form

$$\frac{d \ln p_{O_2}}{d(1/T)} = \frac{N_1(\bar{H}_1 - H_1') + N_2(\bar{H}_2 - H_2'')}{N_2 R} \tag{14-13}$$

The numerator on the right is the enthalpy of formation of 1 gram mole of solid or liquid oxide from pure metal and gaseous oxygen, the number of moles of oxide phase being interpreted as the number of gram atoms of metal plus the number of gram moles of O_2 required to form a given quantity of oxide. When this heat, the numerator, is divided by N_2 in the denominator, the resulting expression is clearly the enthalpy change involved when 1 gram mole of O_2 reacts with sufficient metal to form oxide of whatever composition may be involved in the equilibrium. Hence it is seen that it is correct to write the foregoing expression in the form

$$\frac{d \ln p_{O_2}}{d(1/T)} = \frac{\Delta H}{R} \tag{14-14}$$

provided ΔH is interpreted as the enthalpy change *per mole of* O_2† for the

* Or whatever power of p_{O_2} is required to correspond to the subscript of oxygen in the (arbitrary) chemical formula for the oxide.

† It might seem advantageous to express the composition of the oxide phase in terms of atom fractions; this is readily accomplished by substituting $N_1 = n_1/(n_1 + n_2)$ and $N_2 = n_2/(n_1 + n_2)$ in Eq. (14-13) and multiplying numerator and denominator by $n_1 + n_2$, where n_1 is the number of gram atoms of metal and n_2 the number of gram moles of O_2. Equation (14-13) then becomes

$$\frac{d \ln p_{O_2}}{d(1/T)} = \frac{n_1(\bar{H}_1 - H_1') + n_2(\bar{H}_2 - H_2'')}{n_2 R}$$

Let us define the number of gram atoms of oxygen as n_2^a. In any given quantity of oxide phase $n_2^a = 2n_2$. The enthalpy change, $\bar{H}_2^a - H_2''^a$, accompanying the transfer of 1 gram atom of oxygen from the gas phase is just one-half that, $\bar{H}_2 - H_2''$, accom-

formation of the oxide. Thus, even though the plot of log p vs. $1/T$ may not be linear and the composition of the oxide not constant, the slope of the plot at any particular temperature gives the enthalpy of reaction per mole of O_2 at that particular temperature and for the oxide of whatever composition may be in equilibrium with the metal at this temperature. In the more general case that a substantial amount of oxygen is dissolved in the metal, it may be seen from Eq. (14-7) that the same relation holds provided ΔH is interpreted as the enthalpy change occurring when an appropriate amount of the equilibrium metallic phase reacts with 1 gram mole of O_2 to produce the corresponding amount of oxide phase.

Equilibrium between Iron and Wüstite. The partial pressure of oxygen in equilibrium with iron and its lowest oxide, wüstite, is seen from Fig. 14-6 to be too low for direct experimental determination. Experimentally a mixture of other gases, commonly CO-CO_2 or H_2-H_2O, is used to investigate such an equilibrium. Although in a strict sense the system thus investigated is a ternary one, it is found that the amount of the third component, carbon or hydrogen, in the condensed phase is negligibly small. Thus the equilibrium reaction actually investigated is

$$CO_2 + x\text{Fe(cryst. metal)} = \text{Fe}_x\text{O(wüstite)} + CO$$

It is common in ferrous metallurgy to underscore in a chemical equation any constituent of the metallic ferrous phase, particularly of the molten metallic phase. The above equation might therefore be written

$$CO_2 + x\underline{\text{Fe}}\text{(cryst. metal)} = \text{Fe}_x\text{O(wüstite)} + CO \qquad (A)$$

By virtue of the homogeneous gas equilibrium

$$CO + \tfrac{1}{2}O_2 = CO_2 \qquad (B)$$

---

panying the transfer of 1 gram mole. Hence $n_2(\bar{H}_2 - H_2'')$ equals $n_2^a(\bar{H}_2^a - H_2''^a)$ and, as $n_1 = n_1^a$, the foregoing equation becomes

$$\frac{d \ln p_{O_2}}{d(1/T)} = \frac{n_1^a(\bar{H}_1 - H_1') + n_2^a(\bar{H}_2^a - H_2''^a)}{\tfrac{1}{2}n_2^a R}$$

Since the atom fractions are $N_2^a = n_2^a/(n_1^a + n_2^a)$ and $N_1^a = n_1^a/(n_1^a + n_2^a)$,

$$\frac{d \ln p_{O_2}^{\frac{1}{2}}}{d(1/T)} = \frac{N_1^a(\bar{H}_1 - H_1') + N_2^a(\bar{H}_2^a - H_2''^a)}{N_2^a R}$$

From this point the reasoning follows the reasoning in the text, and the preceding equation may be written

$$\frac{d \ln p_{O_2}^{\frac{1}{2}}}{d(1/T)} = \frac{\Delta H}{R}.$$

providing ΔH in this case is interpreted as the enthalpy change accompanying the reaction of *one-half mole of* O_2 with sufficient metal to form the oxide in equilibrium therewith.

whose equilibrium constant is

$$K_B = \frac{p_{CO_2}}{p_{CO}p_{O_2}^{1/2}}$$

the partial pressure of O_2 for the iron oxide equilibrium is calculable from the pressures of CO and CO_2 and the equilibrium constant. The partial pressure of O_2 so determined may then be used in Eq. 14-14, just as though this pressure had been determined by direct experiment. The only approximations involved in so doing in addition to those used in deriving this equation are (1) that, as already stated, the third component is substantially insoluble in the condensed phases and (2) that ΔH is not affected by the excess of the experimental pressure (about 1 atm) over that of the truly binary system (about 10^{-10} atm). It has previously been shown that ΔH for an ideal gas is not affected by pressure and moreover that for condensed phases the effect of pressure is negligibly small.

Instead of converting from the experimental pressures p_{CO} and p_{CO_2} to p_{O_2} at each temperature and plotting as in Fig. 14-6, it is not uncommon to plot $\log p_{CO_2}/p_{CO}$ directly against $1/T$. Superficially it might appear that p_{CO_2}/p_{CO} is the reciprocal of the equilibrium constant for reaction (A); however, this is not so. The equilibrium constant for this reaction is $K_A = (a_{Fe_xO}/a^x_{Fe})(p_{CO}/p_{CO_2})$; and although a_{Fe} may be taken as unity, since the metallic phase is substantially pure iron at all temperatures under consideration, a_{Fe_xO} may not be taken as unity, since we wish to consider the variation in composition of wüstite. The activity of Fe_xO may arbitrarily be assigned a particular value at some particular composition, but this composition must be the same for all temperatures. Since the composition of wüstite at equilibrium with iron is not identical at all temperatures, p_{CO_2}/p_{CO} cannot strictly be taken as the reciprocal of the equilibrium constant for reaction (A).

Let us take the logarithm of the equilibrium constant for reaction (B) and differentiate with respect to $1/T$.

$$\frac{d \ln (p_{CO_2}/p_{CO}p_{O_2}^{1/2})}{d(1/T)} = -\frac{\Delta H_B}{R}$$

Adding this to one-half of Eq. (14-14),

$$\frac{d \ln (p_{CO_2}/p_{CO})}{d(1/T)} = \frac{\frac{1}{2}\Delta H - \Delta H_B}{R}$$

It will be seen that the difference of the heat terms, $\frac{1}{2}\Delta H - \Delta H_B$, is the enthalpy change accompanying reaction (A), conveniently designated ΔH_A; so that

$$\frac{d \ln (p_{CO_2}/p_{CO})}{d(1/T)} = \frac{\Delta H_A}{R}$$

Thus, if $\ln (p_{CO_2}/p_{CO})$ is plotted against $1/T$ (a straight line is not necessarily expected), the slope of the curve at any temperature gives the heat of the reaction at that temperature corresponding to the particular composition of wüstite that is in equilibrium with iron at that temperature. Since ΔH_A does not refer to the formation of the same wüstite composition over the range of temperature (it is not a $\Delta H°$), it cannot easily be represented in terms of ΔC_p and hence the Σ function may not be used to rectify the curve.

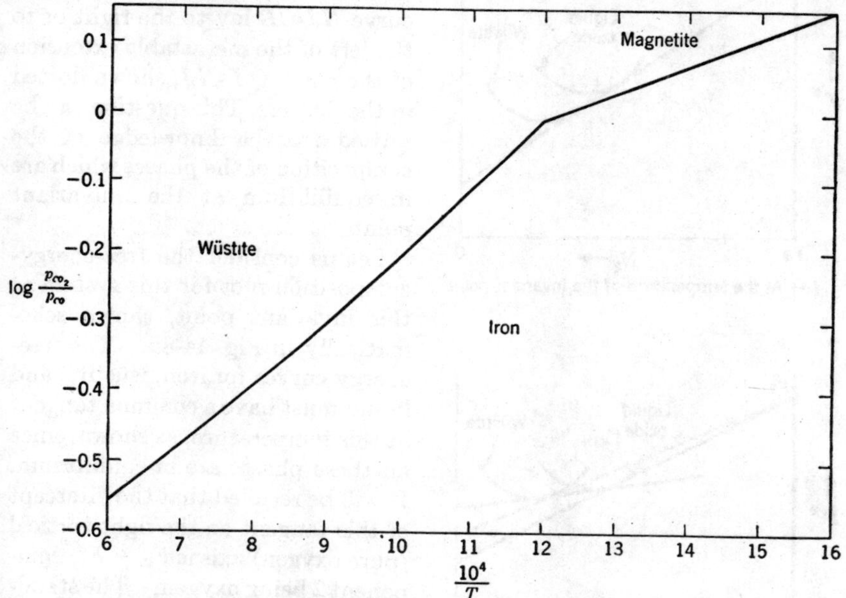

Fig. 14-7. Gas composition in equilibrium with iron and its lowest stable oxide: wüstite above 560°C and magnetite below this temperature.

Experimental values of $\log (p_{CO_2}/p_{CO})$ for the iron-wüstite equilibrium and also, at lower temperature, for the iron-magnetite equilibrium are plotted against $1/T$ in Fig. 14-7. Although the departure from linearity for the iron-wüstite equilibrium is small it seems somewhat greater than the experimental error; in fact the experimental curve shows evidence of an inflection in the vicinity of 1000°C. This reflects the minimum oxygen content of wüstite in equilibrium with iron (corresponding to the composition $Fe_{0.954}O$) at about this same temperature, as may be seen from curve QLJ in Fig. 14-5. The correspondence of these two effects may be seen by differentiating Eq. (14-13) with respect to $1/T$. Under the approximation that the partial molal enthalpies are constant, it is seen that $d^2 \ln p_{O_2}/d(1/T)^2$ is zero when dN_2/dT is zero.

Sequence of Univariant Curves about an Invariant Point. It is frequently desirable to know the sequence of the univariant curves about an invariant point. For example, if there were some question as to the precision of the experimental points determining the univariant curves radiating from the invariant point *NJ* in Fig. 14-6, it would be of value to know, from another source, if the curve *NJ-CB* lay to the right or to the left of the metastable extension of the curve *Q-L-NJ*, shown dotted in the figure. This question can be settled from a knowledge of the composition of the phases which are in equilibrium at the univariant point.

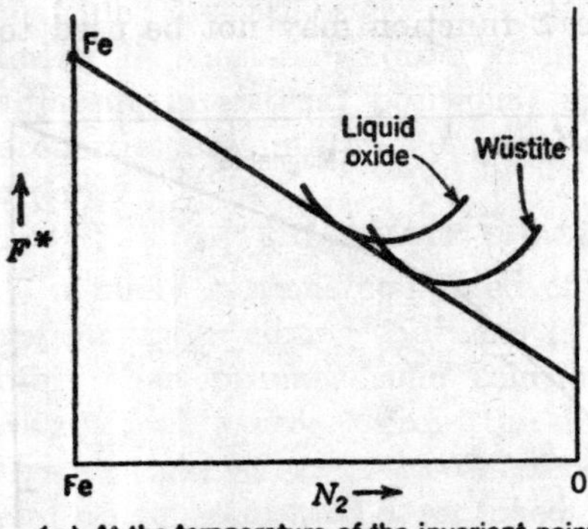

(*a*) At the temperature of the invariant point

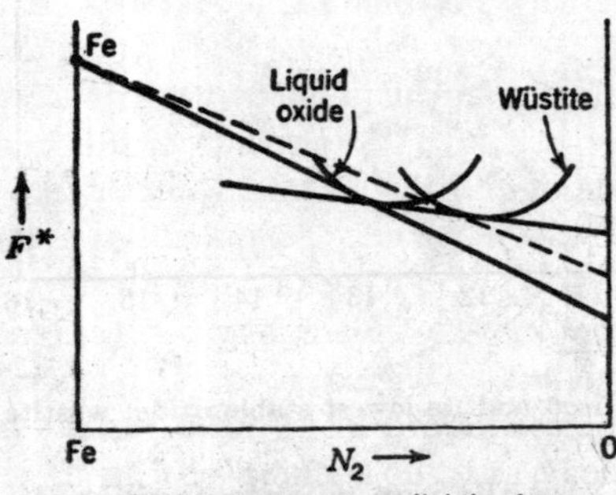

(*b*) At a temperature slightly above that of the invariant point

Fig. 14-8. Schematic free-energy plots for the iron-oxygen system in the vicinity of the invariant point *NJ* of Fig. 14-6, illustrating the general principle that the metastable extension of a univariant curve lies between the stable portions of the other two univariant curves.

Let us consider the free-energy-composition plot for this system at this invariant point, shown schematically in Fig. 14-8*a*. The free-energy curves for iron, wüstite, and liquid must have a common tangent at this temperature, as shown, since all these phases are in equilibrium. It will be recalled that the intercept of this tangent on the right vertical (pure oxygen) axis is $\bar{F}_2 - F_2^\circ$, component 2 being oxygen. The standard state of oxygen is taken as pure gaseous oxygen at 1 atm, and since N_2 in the figure represents the atom fraction of oxygen in the phase or mixture, this intercept is equal to $RT \ln p_{O_2}^{\frac{1}{2}}$. Let us next consider a similar plot (Fig. 14-8*b*) at a slightly higher temperature. It is apparent from Fig. 14-5 that the iron-wüstite equilibrium is here metastable, as this temperature is above the melting point of wüstite in equilibrium with iron; hence the curve for wüstite in Fig. 14-8*b* is shown schematically as being above the tangent to the curve for liquid oxide through the point for iron. Let us construct the tangent to the curve for wüstite through the point for iron, which represents the metastable equilibrium of iron with wüstite, and also the common tangent to the curves for wüstite and liquid oxide at this temperature. It is now seen immediately that the intercept of the line for the metastable iron-wüstite equilibrium lies between the intercepts for the two stable equilibria. Thus it has been demonstrated that the oxygen pressure of the metastable equilibrium lies between the oxygen pressures of the stable equilibria, in other words that the metastable extension of the curve *Q-L-NJ* lies between the curves *NJ-CB* and *NJ-IHS* in Fig. 14-6.

It is usual in a diagram such as Fig. 14-6 for the metastable extension of each univariant curve through an invariant point to fall between the stable portions of the other two univariant curves. The reader will notice that another example of this is to be found in the vicinity of the point *IHS*.

METAL-SULFIDE SYSTEMS

The equilibrium between a metal and the sulfide in stable equilibrium therewith is strictly analogous to the equilibrium between a metal and its oxide as discussed in the previous sections. The general equilibrium may be represented

$$x\text{M(cryst.)} + \text{S}_2\text{(g)} = \text{M}_x\text{S}_2\text{(cryst.)}$$

If the metal and sulfide phases are of substantially fixed composition, the equilibrium constant is

$$K = \frac{1}{p_{\text{S}_2}} \tag{14-15}$$

This constant may be obtained from $\Delta F°$ for the reaction above by the usual relation $\Delta F° = -RT \ln K$.

In order to compare the relative stability of the sulfides it is desirable that all equations be reduced to a common form such as the above, *i.e.*, that each equation involve the same number of atoms of sulfur. If, for example, we wish to know whether metallic iron will reduce silver sulfide to silver, we consider the equations for silver sulfide and iron sulfide.

$$4\text{Ag(cryst.)} + \text{S}_2\text{(g)} = \text{Ag}_4\text{S}_2\text{(or 2Ag}_2\text{S)(cryst.)} \qquad \Delta F°_A$$
$$2\text{Fe(cryst.)} + \text{S}_2\text{(g)} = 2\text{FeS(cryst.)} \qquad \Delta F°_B$$

It will be noted that the same coefficient of S_2 must appear in each equation in order that this term may cancel by direct subtraction. In the present case we obtain

$$2\text{Fe} + 2\text{Ag}_2\text{S} = 2\text{FeS} + 4\text{Ag} \qquad \Delta F° = \Delta F°_B - \Delta F°_A$$

Expressing $\Delta F°_A$ and $\Delta F°_B$ in terms of $p_{\text{S}_2(A)}$ and $p_{\text{S}_2(B)}$, the pressures of sulfur corresponding to the two equilibria,

$$\Delta F° = RT[\ln p_{\text{S}_2(B)} - \ln p_{\text{S}_2(A)}]$$

It is seen that the reaction will proceed spontaneously (and to completion in the absence of solid solution) in the forward direction if ΔF° is negative, *i.e.*, if $p_{S_2(A)} > p_{S_2(B)}$, thus illustrating that, in comparing the relative stability of sulfides, reactions containing the same number of molecules of S_2 are most conveniently considered.

Figure 14-9, taken from Richardson and Jeffes,[1] shows $\Delta F^\circ = RT \ln p_{S_2}$ as a function of temperature for a number of metal-sulfide equilibria. It is seen from the foregoing paragraph that the relative stability of the sulfides may be found by comparnig values of ΔF° at any temperature, the sulfide with the lower ΔF° being the stable one in the presence of both pure metallic phases.

It will be noted that at low temperature the partial pressure of S_2 will be very low; this does not at all detract from the use of the above method of comparing the thermodynamic stability. However, it may well happen that at low temperature the rate of reaction is so slow that the stable sulfide will not be formed at any appreciable rate from the unstable. The experimental investigation of equilibria involving sulfides is frequently conducted with the aid of H_2-H_2S mixtures, the reaction investigated being of the type $Fe + H_2S = FeS + H_2$. Data from this type of reaction may readily be converted to data for the type of reaction considered above by means of available data on the homogeneous gas reaction $H_2S = H_2 + \frac{1}{2}S_2$, also shown in the figure.

METAL-CARBIDE SYSTEMS

The equilibrium between a metal and its carbide is usually investigated experimentally by means of an equilibrium involving gases, usually CO and CO_2 or H_2 and CH_4. The general reaction may be written

$$x\text{M(cryst.)} + 2\text{CO(g)} = \text{M}_x\text{C(cryst.)} + \text{CO}_2\text{(g)} \qquad (A)$$

or

$$x\text{M(cryst.)} + \text{CH}_4\text{(g)} = \text{M}_x\text{C(cryst.)} + 2\text{H}_2\text{(g)} \qquad (B)$$

If the metal and carbide phases are of fixed composition, the equilibrium constants are

$$K_A = \frac{p_{CO_2}}{p_{CO}^2} \qquad \Delta F_A^\circ = -RT \ln \frac{p_{CO_2}}{p_{CO}^2} \qquad (14\text{-}16)$$

and

$$K_B = \frac{p_{H_2}^2}{p_{CH_4}} \qquad \Delta F_B^\circ = -RT \ln \frac{p_{H_2}^2}{p_{CH_4}} \qquad (14\text{-}17)$$

[1] F. D. Richardson and J. H. E. Jeffes, *J. Iron Steel Inst. (London)*, **171**, 167 (1952).

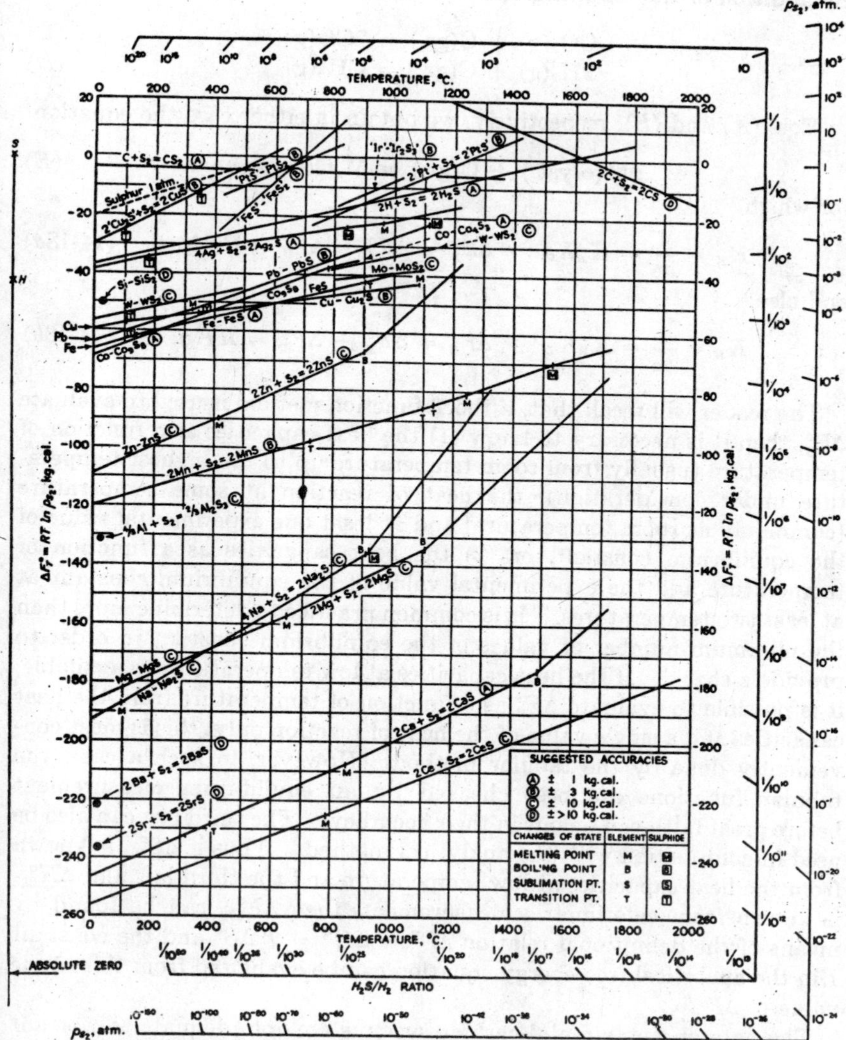

FIG. 14-9. The standard free energy of formation of a number of metal sulfides as a function of temperature. [*From F. D. Richardson and J. H. E. Jeffes, J. Iron Steel Inst.*, **171**, 167 (1952).]

By addition of the following two equations

$$CO_2(g) + C(gr.) = 2CO(g) \quad (C)$$
$$2H_2(g) + C(gr.) = CH_4(g) \quad (D)$$

to Eqs. (*A*) and (*B*), respectively, we obtain in either case the equation[1]

$$x\text{M(cryst.)} + \text{C(gr.)} = \text{M}_x\text{C(cryst.)} \quad (E)$$

for which

$$K_E = \frac{1}{a_C} = K_A K_C \qquad \Delta F^\circ_E = \Delta F^\circ_A + \Delta F^\circ_C = RT \ln a_C \quad (14\text{-}18a)$$

and also

$$K_E = \frac{1}{a_C} = K_B K_D \qquad \Delta F^\circ_E = \Delta F^\circ_B + \Delta F^\circ_D = RT \ln a_C \quad (14\text{-}18b)$$

The reader will recall that, if the Σ function method is used to evaluate ΔF°_E, then it is necessary to know (1) the heat capacities as a function of temperature (usually from room temperature up to the highest temperature under consideration), the heat of reaction at some temperature (commonly at room temperature) and at least one experimental value of the equilibrium constant, or (2) the heat capacities as a function of temperature and the experimental value of the equilibrium constant at at least two temperatures. It is common practice to determine more than the minimum number of values of the equilibrium constant in order to provide a check. If the heat capacities at low temperature are available, it is possible to evaluate ΔF°_E as a function of temperature from the heat capacities and a single value of the heat of reaction only; this is most conveniently done by the tabular method. However, in such a case, the tabular functions are best checked by an equilibrium measurement before great faith can be put in their accuracy. The third law can also be used in conjunction with the analytical method. Thus if ΔS°_{298} is known from the heat capacities at low temperature and the third law and ΔH°_{298} is known from calorimetric measurement, then ΔF°_{298} can be found by means of the definitional relation $\Delta F^\circ = \Delta H^\circ - T\,\Delta S^\circ$ and the constant I in the analytical free-energy equation can be evaluated from this single value of ΔF°.

The data on many metal-carbide systems are not adequate to warrant the extended treatment outlined above. ΔF°_{298}, ΔH°_{298}, and ΔS°_{298} for the formation of carbides are given in Table 14-1. ΔF° at other temperatures not too far from 298°K may be approximated by assuming ΔH° and ΔS°

[1] As is customary, the equation is written in terms of the standard states of the constituents. ΔF° for the reaction is readily expressed in terms of activities other than unity.

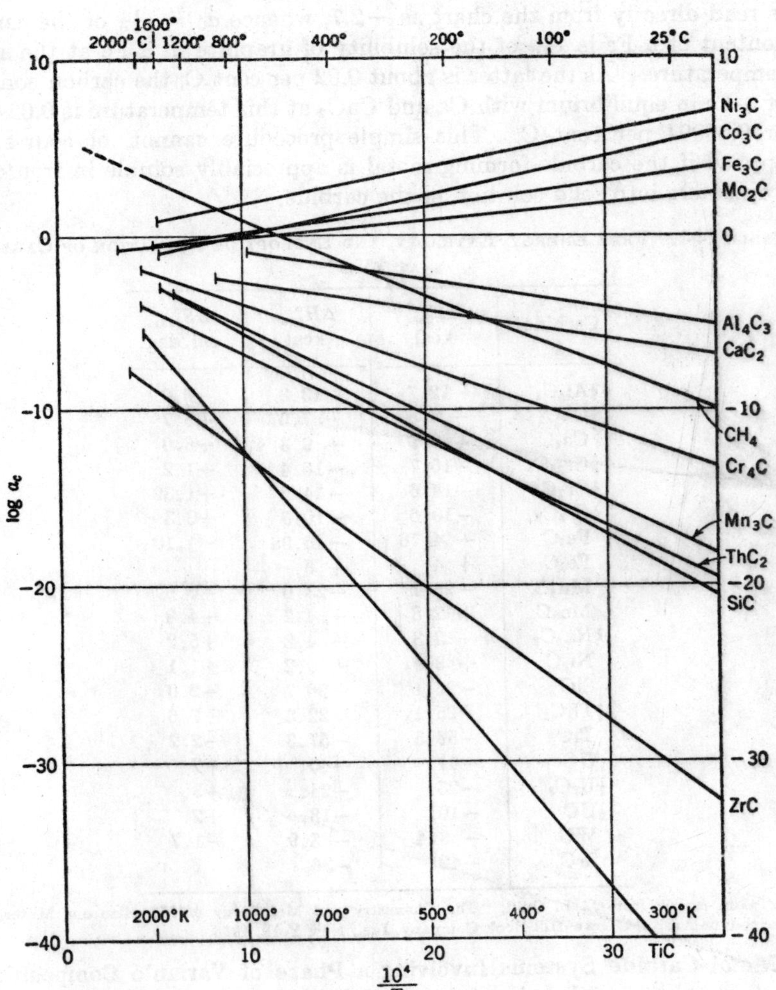

Fig. 14-10. The activity of carbon (standard state = graphite) for several metal-carbide systems as a function of temperature.

to be independent of temperature. The data collated by Kelley[1] were used to prepare Fig. 14-10.[2] This form of plot, using log a as ordinate, is convenient for many purposes. For example, if one wishes to know the carbon content of α-Fe in equilibrium with Ca and CaC_2 at 700°C, log a_C

[1] K. K. Kelley, *U.S. Bur. Mines Bull.* 407, 1937.

[2] We are indebted to Dr. Robert T. Howard for the calculations leading to this figure.

is read directly from the chart as −2.7, whence $a_C = \frac{1}{500}$ or the carbon content of α-Fe is $\frac{1}{500}$ of the solubility of graphite in α-Fe at the same temperature. As the latter is about 0.02 per cent C, the carbon content of α-Fe in equilibrium with Ca and CaC_2 at this temperature is 0.02/500 or 0.00004 per cent C. This simple procedure cannot, of course, be applied if the carbide-forming metal is appreciably soluble in iron or if iron enters into solid solution in the carbide.

TABLE 14-1. FREE ENERGY, ENTHALPY, AND ENTROPY OF FORMATION OF CARBIDES AT 25°C*

Carbide	ΔF°_{298}, kcal	ΔH°_{298}, kcal	ΔS°_{298}, cal/deg
$\frac{1}{3}Al_4C_3$	−12.7	−13.3	−2.0
$\frac{1}{2}CaC_2$	− 8.8	− 7.0	+5.7
Co_3C	+ 6.9	+ 9.3	+8.0
$\frac{1}{6}Cr_{23}C_6$	−16.7	−16.4	+1.2
$\frac{1}{3}Cr_7C_3$	−14.6	−14.2	+1.39
$\frac{1}{2}Cr_3C_2$	−10.6	−10.5	+0.3
Fe_3C	+ 4.76	+ 5.98	−4.10
Fe_2C	+ 4.	+ 6.	
Mn_3C	−23.1	−23.0	−0.6
Mo_2C	+ 2.8	+ 4.2	+4.8
$\frac{1}{2}Na_2C_2$	− 3.3	− 4.8	+5.2
Ni_3C	+ 8.9	+ 9.2	+1.1
SiC	−26.1	−26.7	−2.0
$\frac{1}{2}ThC_2$	−25.1	−22.8	+7.6
TiC	−56.5	−57.3	−2.2
UC	−41.	−40.	+2.
$\frac{1}{3}U_2C_3$	−25.	−24.	+3.
$\frac{1}{2}UC$	−19.	−18.	+2.
WC	− 3.4	− 3.9	−1.7
ZrC	−42.	−36.	?

* Taken mostly from L. L. Quill, "The Chemistry and Metallurgy of Miscellaneous Materials: Thermodynamics," McGraw-Hill Book Company, Inc., New York, 1950.

Metal-Carbide Systems Involving a Phase of Variable Composition. Of the carbide equilibria involving a condensed phase of variable composition, the most important is the equilibrium in the iron-carbide system between γ-iron (austenite) and cementite (Fe_3C).* The cementite participating in the equilibrium is of substantially fixed composition corresponding to the formula Fe_3C. The composition of the austenite participating in the equilibrium, however, varies from about 0.8 per cent carbon at the eutectoid to about 2.0 per cent at the temperature of the

* This equilibrium, along with others of the iron-carbon system, is discussed in detail in Chap. 16.

eutectic. The data of Mehl and Wells on the composition of austenite in equilibrium with cementite—conveniently referred to as the solubility of cementite in austenite—are shown in Fig. 13-13. As noted previously, cementite is metastable relative to graphite and austenite saturated with graphite, at least in the range of temperature that has been investigated. However, at the moment, we are not concerned with the graphite equilibrium, and, restricting our attention to temperatures considerably below the eutectic where it is found experimentally that graphite formation does not occur in the simple iron-carbon system except in a relatively long period of time, it is quite proper to consider the equilibrium between cementite and austenite.

This equilibrium could be investigated by means of one of the gas equilibria discussed in the preceding section, for example,

$$3\text{Fe}(\gamma) + 2\text{CO(g)} = \text{Fe}_3\text{C(cem.)} + \text{CO}_2\text{(g)} \qquad (14\text{-}19)$$

By combining with the equilibrium involving CO, CO_2, and carbon (graphite),

$$3\text{Fe}(\gamma) + \text{C(graph.)} = \text{Fe}_3\text{C(cem.)} \qquad (14\text{-}20)$$

Upon the assumption that oxygen enters only to a negligible extent into the condensed phases, the system can thus be considered binary. By choosing cementite and carbon as components 1 and 2, respectively, Eq. (14-10) may be applied to the present case. Since the vapor pressure of carbon is very low indeed, its pressure is set equal to its activity, and the volumes of the other phase are ignored in comparison with that of gaseous carbon. Eq. (14-10) then takes the form

$$\frac{d \ln a_C}{d(1/T)} = \frac{N_1(\bar{H}_1 - H_1') + N_2(\bar{H}_2 - H_2'')}{N_2 R} \qquad (14\text{-}21)$$

where the unprimed quantities on the right refer to austenite, the primes to cementite, and the double primes to graphite. The right side of the equation except for R is the enthalpy change involved when 1 gram atom of graphite reacts completely with austenite of the composition in equilibrium with cementite to produce cementite—which obviously contains more than 1 gram atom of carbon by virtue of the carbon dissolved in the austenite. Equation (14-21) is not fruitful for this particular case because of the absence of reliable data.

As frequently happens, more is to be gained by expressing the equilibrium [Eq. (14-20)] in terms of the equilibrium constant

$$K = \frac{1}{a_{Fe}^3 a_C} \qquad (14\text{-}22)$$

a relation valid for an equilibrium involving cementite of fixed composition. The activity of iron relative to pure γ-iron and that of carbon relative to graphite in γ-iron–carbon alloys have been determined by R. P. Smith,[1] and these data, in conjunction with the experimentally determined composition of austenite in equilibrium with cementite, serve to evaluate K. The enthalpy of formation of cementite from pure iron and graphite, $\Delta H°$ of Eq. (14-20), may be evaluated from a plot of ln K vs. $1/T$, whose slope is $-\Delta H°/R$.

There is no experimental evidence to indicate that the cementite in equilibrium with either austenite or ferrite in a system containing only iron and carbon is anything other than a phase of fixed composition corresponding to the formula Fe_3C. In a ternary system, such as the iron-manganese-carbon system, there is considerable evidence that part of the iron may be replaced by the other metal; in such a case the composition is conventionally represented $(Fe, Mn)_3C$.

METAL-NITRIDE SYSTEMS

The equilibrium between a metal and the corresponding nitride[2] is substantially identical in principle with that between the metal and the carbide. It may be represented

$$x\text{M(cryst.)} + \text{N}_2\text{(g)} = \text{M}_x\text{N}_2\text{(cryst.)}$$

In the absence of appreciable solid solution the equilibrium constant may be written

$$K = \frac{1}{p_{N_2}} \tag{14-23}$$

Available data on the free energy, enthalpy, and entropy of formation of metal nitrides are given in Table 14-2, which is similar in form to that for the carbides, Table 14-1. p_{N_2} for the univariant metal-nitride equilibrium may be evaluated from the relation $\Delta F° = RT \ln p_{N_2}$. The data collated by Kelley[3] are represented graphically in Fig. 14-11,[4] which is similar in form to Fig. 14-10 and may conveniently be used to obtain readily the partial pressure of nitrogen of the systems represented.

METAL-CHLORIDE SYSTEMS

The free energies of formation of many chlorides have been collected by Kellogg.[5] His plot is reproduced as Fig. 14-12. The general reaction is

[1] R. P. Smith, *J. Am. Chem. Soc.*, **68**, 1163 (1946).

[2] The various equilibria of the iron-nitrogen system are discussed in Chap. 15.

[3] K. K. Kelley, *U.S. Bur. Mines Bull.* 407, 1937.

[4] We are indebted to Dr. Robert T. Howard for the calculations leading to this figure.

[5] H. H. Kellogg, *Trans. AIME*, **188**, 862 (1950).

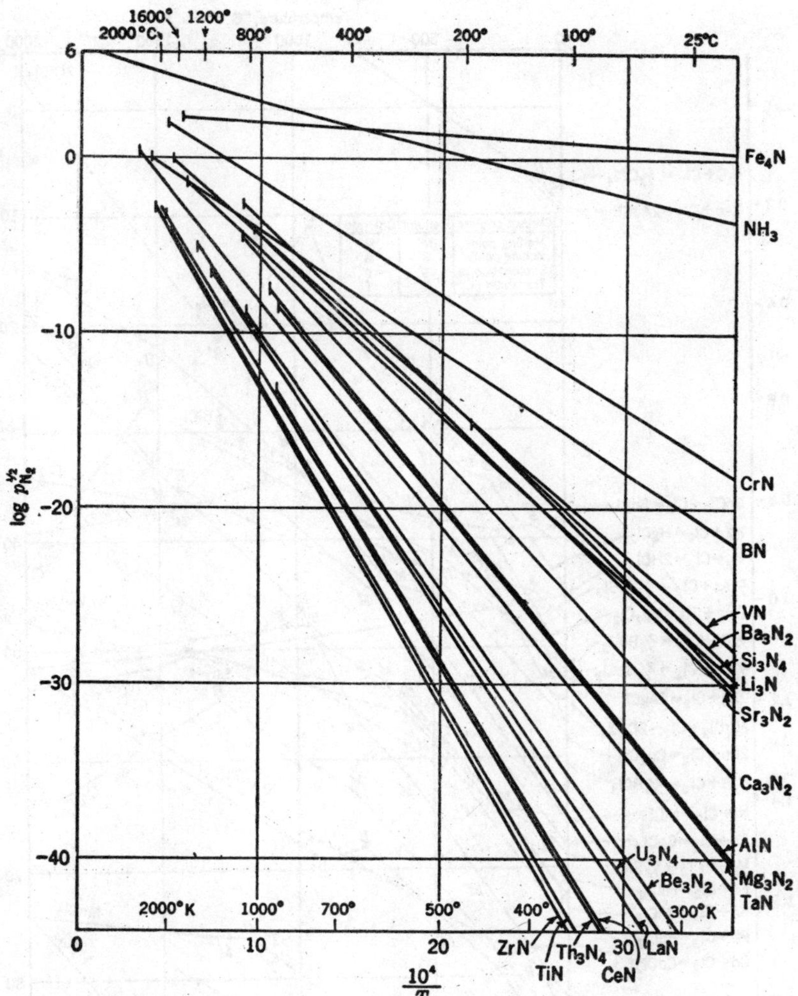

Fig. 14-11. The partial pressure of nitrogen for several metal-nitride systems as a function of temperature.

$$xM + Cl_2(g) = M_xCl_2 \qquad (A)$$

the state of the metal and the chloride being unspecified in order to permit application to solid, liquid, or gas. The irregular nature of the curves in the figure is occasioned by the phase changes of the metals and the chlorides. The melting, boiling, or sublimation point of the chloride is designated *M*, *B*, or *S*, and that of the metal *M′*, *B′*, or *S′*.

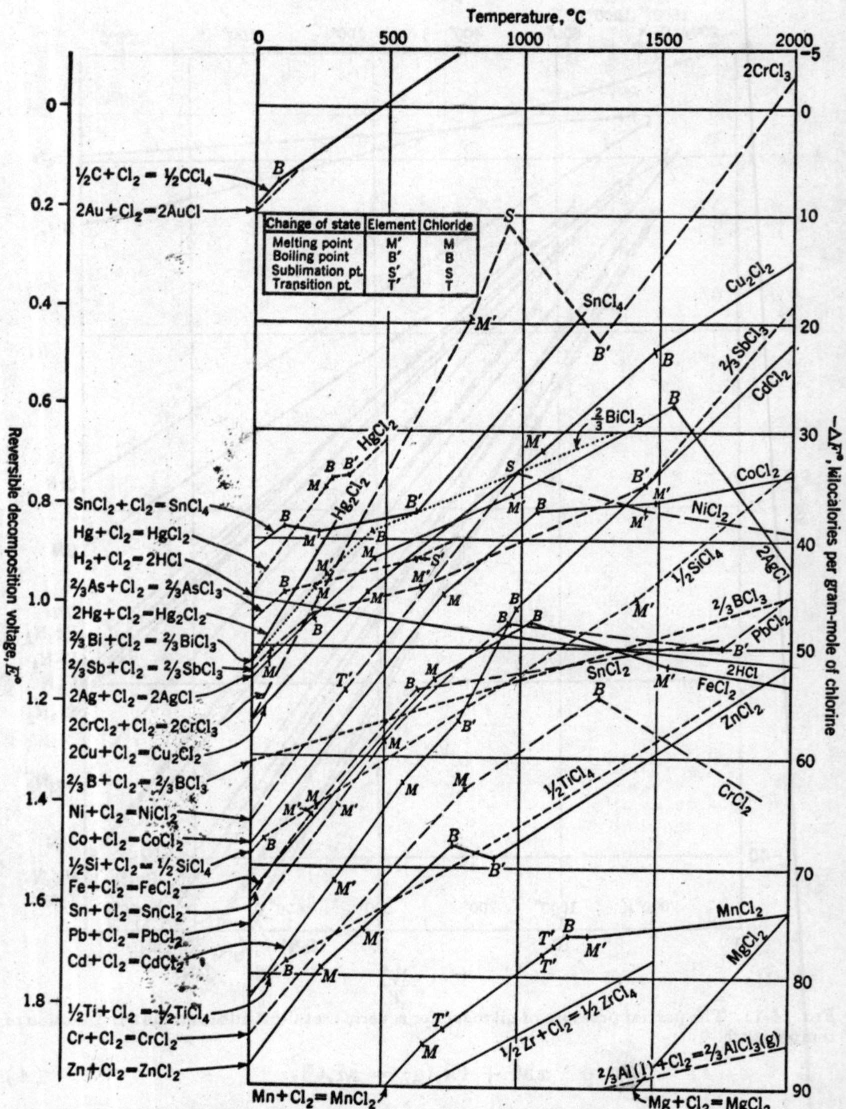

Fig. 14-12. The standard free energy of formation and the reversible decomposition voltage of several metal chlorides as a function of temperature. [*From H. H. Kellogg, Trans. AIME*, **188**, 862 (1950).]

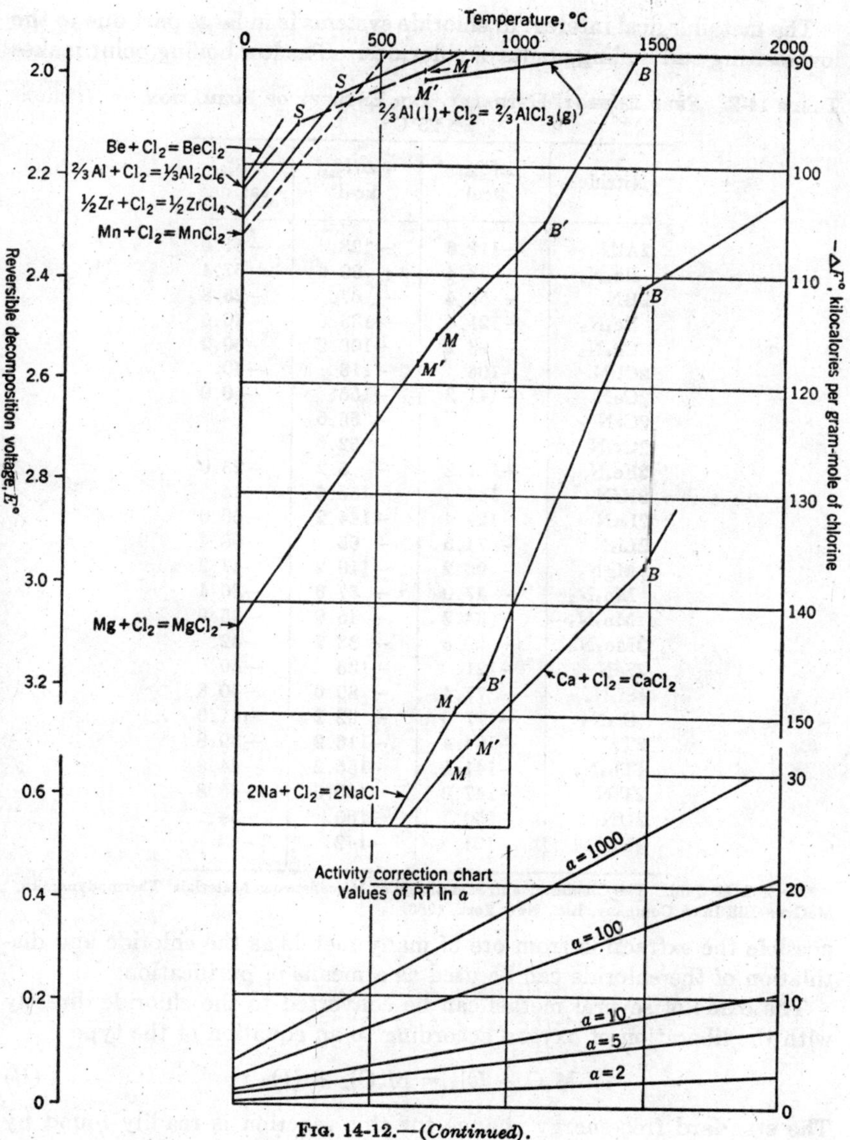

FIG. 14-12.—(*Continued*).

The metallurgical interest in chloride systems is in large part due to the low melting and boiling points of chlorides. The low boiling point makes possible the extraction from ore of many metals as the chloride and distillation of the chloride can be used as a means of purification.

TABLE 14-2. FREE ENERGY, ENTHALPY, AND ENTROPY OF FORMATION OF NITRIDES AT 25°C*

Nitride	ΔF°_{298}, kcal	ΔH°_{298}, kcal	ΔS°_{298}, cal/deg
2AlN	−112.6	−128.	−51.6
Ba_3N_2	− 73.4	− 90.6	−57.4
2BN	− 59.4	− 67.	−25.8
Be_3N_2	−121.4	−133.6	−40.6
Ca_3N_2	− 93.2	−108.2	−50.2
2CbN	−106.	−118.	−40.
2CeN	−141.2	−156.	−50.0
2CrN		− 56.6	
$2Cr_2N$		− 52.6	
$2Fe_4N$	+ 1.8	− 5.2	−23.0
2HfN	−144.	−156.6	−44.
2LaN	−129.4	−144.2	−50.0
$2Li_3N$	− 74.6	− 95.	−68.4
MgN_2	− 96.2	−110.2	−47.2
Mn_5N_2	− 47.0	− 57.8	−36.4
Mn_3N_2	− 34.2	− 45.8	−35.8
$2Mo_2N$	− 20.6	− 33.2	−42.
2ScN	−121.	−136.	−50.
$\frac{1}{2}Si_3N_4$	− 77.4	− 89.6	−40.8
Sr_3N_2	− 77.0	− 92.2	−51.0
2TaN	−104.4	−116.2	−39.8
$\frac{1}{2}Th_3N_4$	−141.8	−155.2	−44.8
2TiN	−147.0	−160.6	−45.8
2UN	−150.	−160.	−34.
$\frac{2}{3}U_2N_3$	−128.	−142.	−42.

* From L. L. Quill, "The Chemistry and Metallurgy of Miscellaneous Materials: Thermodynamics," McGraw-Hill Book Company, Inc., New York, 1950.

The oxide of several metals can be converted to the chloride directly with the liberation of oxygen according to an equation of the type

$$M_xO + Cl_2 = M_xCl_2 + \tfrac{1}{2}O_2 \qquad (B)$$

The standard free-energy change for this reaction is readily found by subtracting the free energy of formation of the oxide from that of the chloride (Figs. 14-4 and 14-10). Hydrogen chloride gas may also be used to convert the oxide to the chloride:

$$M_xO + 2HCl = M_xCl_2 + H_2O \qquad (C)$$

The standard free-energy change of this reaction is obtainable from these same figures. The formation of chloride from oxide is obviously favored by use along with the chlorine of a reducing agent which, in effect, combines with the oxygen on the right side of Eq. (*B*) and thus displaces the equilibrium to the right. The free-energy change of any such reaction may be computed in a manner similar to that above. If carbon is used as the reducing agent, the reaction may be written:

$$M_xO + C + Cl_2 = M_xCl_2 + CO \qquad (D)$$

CHAPTER 15

NITROGEN IN IRON AND STEEL

The iron-nitrogen system is of particular interest, since it furnishes several examples of an equilibrium of considerable industrial importance which is, nevertheless, highly metastable. To visualize the magnitude of the metastability let us consider first the stable equilibrium involving

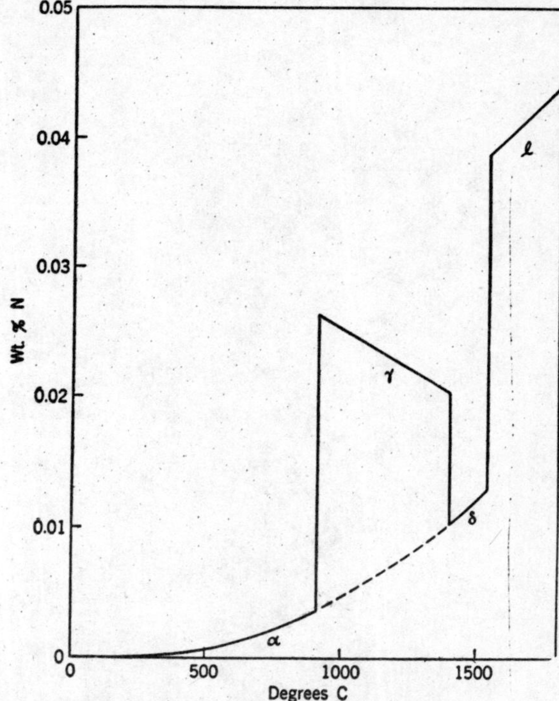

FIG. 15-1. Solubility of nitrogen gas at 1 atm in iron; stable equilibrium diagram at 1 atm.

nitrogen gas and the primary solid solution of nitrogen in iron, shown in Fig. 15-1. For comparison with this, the temperature-composition diagram for the iron-nitrogen system, taken from the work of Paranjpe, Cohen, Bever, and Floe,[1] is shown in Fig. 15-2. This entire diagram,

[1] V. G. Paranjpe, M. Cohen, M. B. Bever, and C. F. Floe, *Trans. AIME*, **188**, 261 (1950).

except for a very small region at the left side, corresponds to conditions which are metastable relative to nitrogen gas at 1 atm. In the iron-carbon system, cementite has a slightly greater solubility in austenite than has graphite, but in the iron-nitrogen system the solubility of the γ' phase (Fe_4N) in γ-iron (over 2 per cent at the eutectoid) is many fold

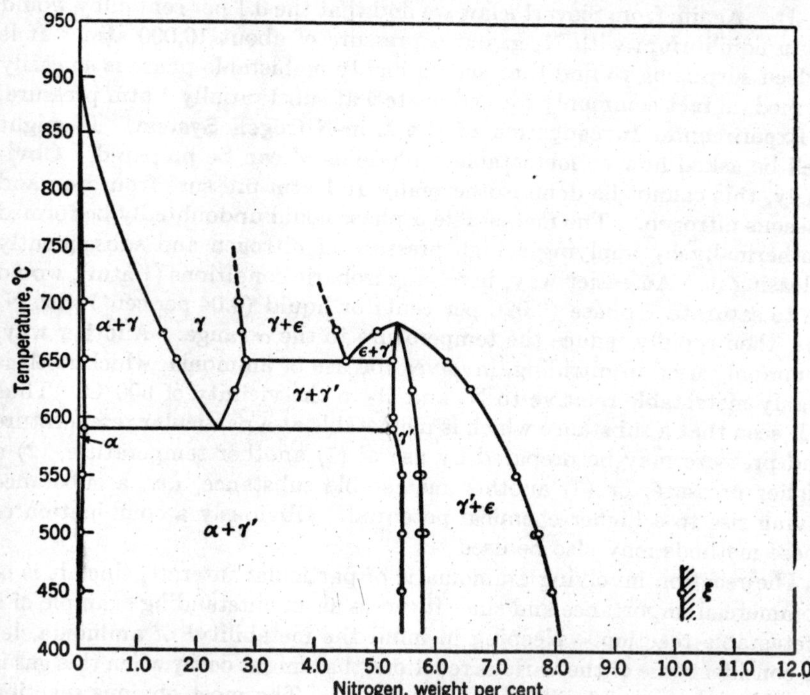

Fig. 15-2. Temperature-composition diagram of the iron-nitrogen system, metastable with respect to nitrogen gas at 1 atm except for the small region shown in Fig. 15-1. [*From Paranjpe, Cohen, Bever, and Floe, Trans. AIME*, **188**, 261 (1950).]

that (0.026 per cent) of gaseous N_2. This high degree of metastability prevails also at lower temperature, in the α region, and even higher degrees of metastability are encountered in the equilibria involving the higher nitrides of iron.

It is readily seen from Sievert's law that, if the solubility of N_2 gas in γ-iron at 910°C is 0.026 per cent and the solubility of an iron nitride at the same temperature is about 2 per cent, then the pressure of gaseous N_2 in equilibrium with this nitride and γ phase is given, at least approximately,[1] by the relation $P = (2/0.026)^2 \cong 6000$ atm. Obviously then,

[1] Sievert's law, stating that the solubility of a diatomic gaseous element in a metal is

an iron-nitrogen alloy (perforce containing over 2 per cent nitrogen) consisting of nitride and γ phase cannot be in equilibrium with nitrogen gas at 1 atm pressure, and any equilibrium that does prevail at atmospheric pressure must, therefore, be metastable. At 600°C it is possible to obtain an α phase containing 0.1 per cent nitrogen whereas the solubility of gaseous N_2 at this temperature in α-iron is about 0.001 per cent (Fig. 15-3). Again, from Sievert's law we find that the 0.1 per cent alloy would be in equilibrium with N_2 gas at a pressure of about 10,000 atm. It is indeed surprising to find that such a highly metastable phase is so easily formed, in fact commonly formed in steel at substantially 1 atm pressure.

Experimental Investigation of the Iron-Nitrogen System. It might well be asked how so metastable a phase as γ' can be prepared. Obviously, this cannot be done isothermally at 1 atm pressure from iron and gaseous nitrogen. The metastable α phase could undoubtedly be formed isothermally by applying a high pressure of nitrogen and subsequently releasing it. An easier way, involving isobaric conditions (1 atm), would be to saturate γ phase (0.026 per cent) or liquid (0.04 per cent) with N_2 gas, then rapidly reduce the temperature to the α range. Another way, commonly used in nitriding, involves the use of ammonia, which itself is highly metastable relative to N_2 and H_2 in the vicinity of 600°C. Thus it is seen that a substance which is metastable at a particular temperature and pressure may be prepared by use of (1) another temperature, (2) a higher pressure, or (3) another metastable substance, *i.e.*, a substance giving rise to a higher chemical potential. Obviously a combination of these methods may also be used.

The reaction involving ammonia is of particular interest, since it is of commercial importance and since it serves as an outstanding example of a metastable reaction. Keeping in mind the instability[1] of ammonia, let us consider some of the various reactions that might occur when this gas is brought into contact with α-iron at 600°C. The most obvious reaction involves the decomposition to gaseous N_2 and H_2, which happens to be catalyzed by iron.

$$NH_3(g) = \tfrac{1}{2}N_2(g) + \tfrac{3}{2}H_2(g) \qquad (A)$$

proportional to the square root of the pressure, is, of course, limited in applicability to dilute solutions and to the range of pressure where the gas behaves ideally. More strictly the above comparison should be made on the basis of fugacity.

[1] Every experimental measurement is, of course, characterized by a certain degree of precision. If the extent of a certain thermodynamically possible reaction is less than this precision, then the reactants may be regarded as metastable with respect to the possible products. If the extent is greater than this precision, the reactants are properly regarded as unstable relative to the products which form. In the above case ammonia is termed unstable because it usually decomposes to a measurable extent in the reactions considered.

Other possible reactions involve the solid solution of nitrogen, according to the equation

$$NH_3(g) = N(\text{in } \alpha\text{-Fe}) + \tfrac{3}{2}H_2(g) \qquad (B)$$

or of hydrogen, according to the equation

$$NH_3(g) = \tfrac{1}{2}N_2(g) + 3H(\text{in } \alpha\text{-Fe}) \qquad (C)$$

Nitride formation is not, for the moment, considered—*i.e.*, we now consider chemical potentials high compared with N_2 gas but less than those of nitride equilibria. Without experimental knowledge it would be impossible to decide which reaction or set of reactions prevailed. It is found experimentally that reaction (*C*) takes place to a negligibly small extent and that with proper precautions[1] the extent of reaction (*A*) is also small. Under these conditions the metastable equilibrium (*B*) and related equilibria may be investigated—in fact they are the subject of several investigations to be discussed later.

This metastable equilibrium between a gaseous ammonia-hydrogen mixture and the nitrogen dissolved in α-iron is treated thermodynamically in just the same way as a stable equilibrium. The equilibrium constant for reaction *B* is

$$K'_B = a_N \frac{p_{H_2}^{3/2}}{p_{NH_3}}$$

The concentration of nitrogen in α-iron is sufficiently small (the maximum being about 0.1 per cent N) that no appreciable error is introduced by the assumption of Henry's law. Furthermore, at such a low concentration the weight percentage may be considered proportional to the atom fraction, so that

$$K_B = [\%\text{N}] \frac{p_{H_2}^{3/2}}{p_{NH_3}} \qquad (15\text{-}1)$$

The brackets, as usual, are used to express concentration in the metallic phase. It will be noted that the convention, here adopted, of substituting percentage for activity implies a standard state (for nitrogen, at $a_N = 1$) of 1 per cent of nitrogen in α-iron, under the imposed condition that the assumptions made would be valid up to this composition if it were attainable. This composition of α-iron is not attainable under ordinary circumstances, since a nitride would separate from it very rapidly; moreover, if it were attained, it is quite likely that departures from Henry's law would be very noticeable. Thus the standard state in such a case is a

[1] A smooth, *e.g.*, electropolished, surface suppresses reaction (*A*) to a great extent at least up to a temperature of about 600°C. A small metal surface and a large gas flow also minimize the effect of this reaction.

nonphysical one and may be said to be the state that would exist if the solution were obtainable and if Henry's law were valid up to 1 per cent N. This choice of standard state is quite common in dealing with alloying constituents at low concentration.

From the foregoing expression for K_B it is apparent that we should expect [%N] to be proportional to $p_{NH_3}/p_{H_2}^{3/2}$. This is indeed found experimentally to be the case. As usual, the relation

$$\frac{d \ln K_B}{d(1/T)} = -\frac{\Delta H^\circ}{R}$$

applies to the change of the equilibrium constant with temperature. ΔH° is the enthalpy change of the reaction when all reactants and products are in their standard states. Because of the particular choice of the standard state for nitrogen (1 per cent N), it might seem that the value of the heat thus found is meaningless. However, since this nonphysical standard state is that which would exist if Henry's law were valid, it is apparent that ΔH° is the heat of reaction when 1 gram atom of nitrogen is dissolved in α-iron at any concentration sufficiently low that Henry's law is actually valid.[1]

It is more convenient to combine equations (A) and (B), obtaining

$$\tfrac{1}{2}N_2(g) = N(\text{in } \alpha\text{-Fe}) \qquad (D)$$

for which the equilibrium constant is $K_D = K_B/K_A$. In the low-temperature range K_D can be evaluated from data on the two reactions (A) and (B), whereas at higher temperature K_D can be evaluated from direct measurements of the solubility of N_2 gas at 1 atm in α or δ-Fe; it will be noted that K_D is the solubility of N_2 at 1 atm pressure in $\alpha(\delta)$-iron. A plot against $1/T$ of this, as well as of the solubility of γ' in $\alpha(\delta)$-iron, is shown in Fig. 15-3. The departure from linearity seems too great to attribute to experimental error and is seemingly to be attributed to ΔC_P. The value of ΔC_P required to fit the curve is about 6 cal/deg, which corresponds to about 10 cal/deg for the partial molal heat capacity $\bar{C}_P$ of nitrogen dissolved in α-iron. There are very few data on $\bar{C}_P$ in the metallurgical literature with which to compare this; comparing with C_P it is noted that this value is, of course, considerably above the Dulong and Petit value of 6 but is not out of the range of C_P for metallic elements at elevated temperature.

[1] This is made self-evident by the consideration that, if the standard state had been chosen as 0.1 or 0.01 per cent N in iron (where Henry's law is valid and the standard state has unambiguous physical meaning), K_B would merely be 10 or 100 times greater than the K_B used in the above text. ln K_B would be displaced one or two units, and the slope $d \ln K_B/d(1/T)$ would be unaltered.

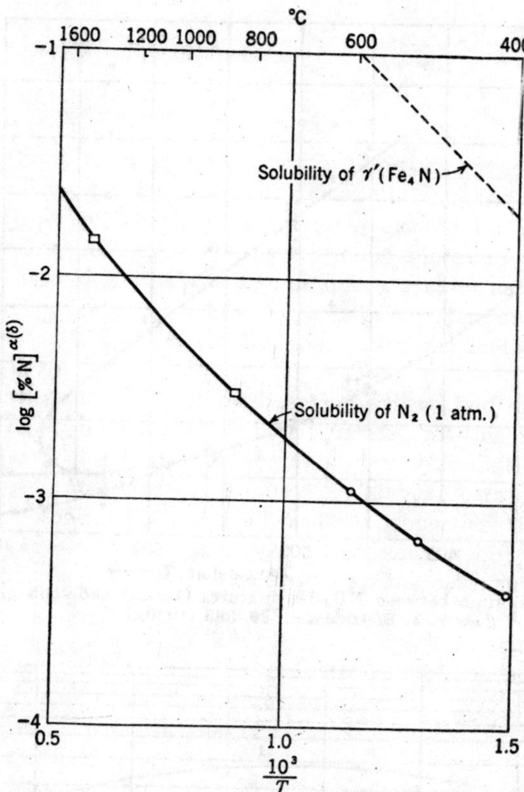

FIG. 15-3. The solubility of N_2 gas at 1 atm in $\alpha(\delta)$-iron as determined directly (squares), and by equilibration with NH_3-H_2 mixtures (circles); also the solubility of γ' (Fe_4N) in α-iron.

UNIVARIANT AND INVARIANT EQUILIBRIA

The metastable equilibria involving iron nitrides have been extensively investigated.[1] The experimental procedure involves equilibration of the solid phase or phases with the gas and subsequent chemical analysis or examination by metallographic or X-ray technique; a magnetometric method has also been used. The univariant equilibrium curves as found by Lehrer are shown in Fig. 15-4. By combination of these data with those of the homogeneous gas reaction (A) above, it is possible to calculate the fugacity of N_2 for each equilibrium at each temperature. log p_N , or

[1] E. Lehrer, *Z. Elektrochem*, **36**, 383 (1930). Eisenhut and Kaup, *ibid.*, **36**, 392 (1930). K. H. Jack, *Proc. Roy. Soc.*, **195A**, 34, 41, 56 (1948). Darken, Smith, and Filer, *Trans. AIME*, **191**, 1174 (1951). L. J. Dijkstra, *Trans. AIME*, **185**, 252 (1949). Paranjpe, Cohen, Bever, and Floe, *ibid.*, **188**, 261 (1950).

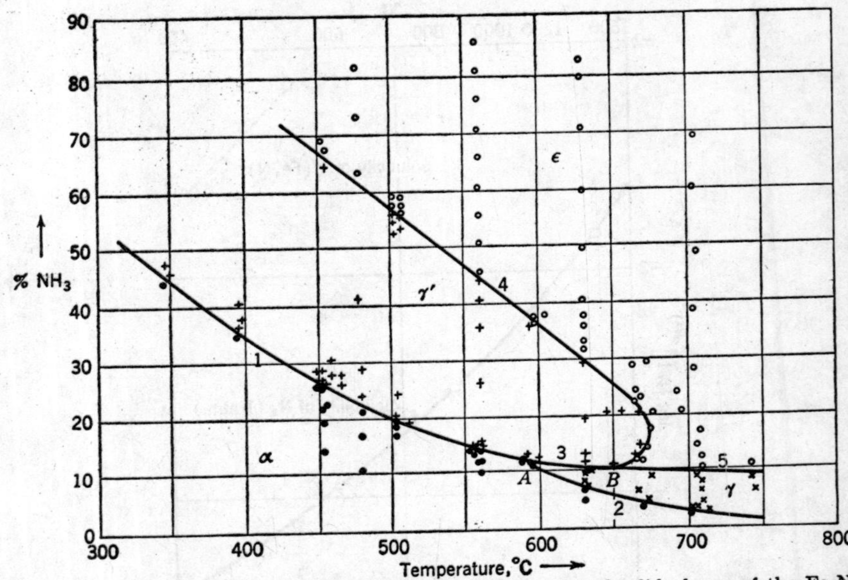

FIG. 15-4. Equilibrium between NH_3-H_2 mixtures (1 atm) and solid phases of the Fe-N system. [*From E. Lehrer, Z. Elektrochem.*, **36**, 383 (1930).]

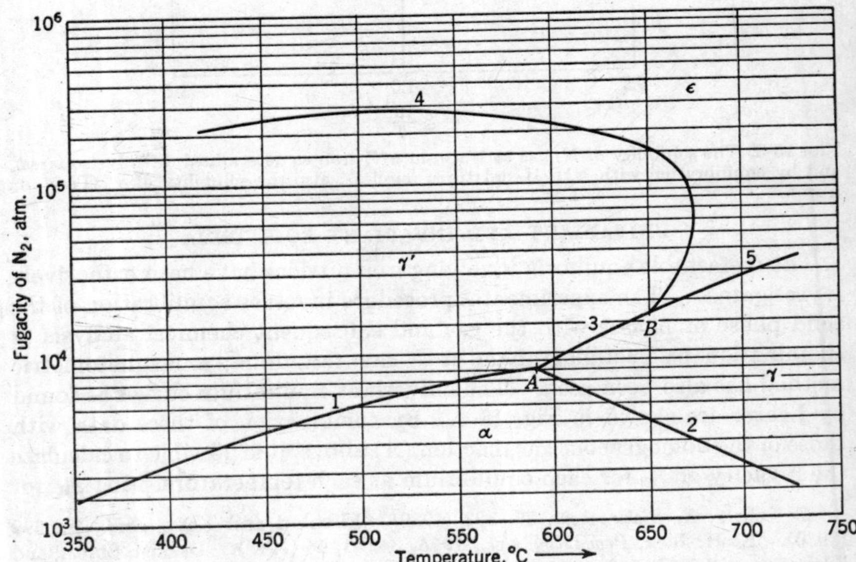

FIG. 15-5. Fugacity-temperature diagram for the Fe-N system, from the data of Fig. 15-4. [*From E. Lehrer, Z. Elektrochem.*, **36**, 383 (1930).]

strictly log f_{N_2}, thus calculated, is plotted against T in Fig. 15-5, taken from Lehrer, which shows the same univariant equilibria in slightly different form. The same data are again replotted in Fig. 15-6 where log $p_{NH_3}/p_{H_2}^{3/2}$ is used as the ordinate. In this plot curves representing equilibria between phases of fixed composition are linear, except for the slight curvature resulting from ΔC_P. The five univariant curves and the two invariant points of these figures will now be discussed.

1. **The α-γ' curve** of Fig. 15-6 exhibits negligible curvature, suggesting that the equilibrium involves only condensed phases of substantially

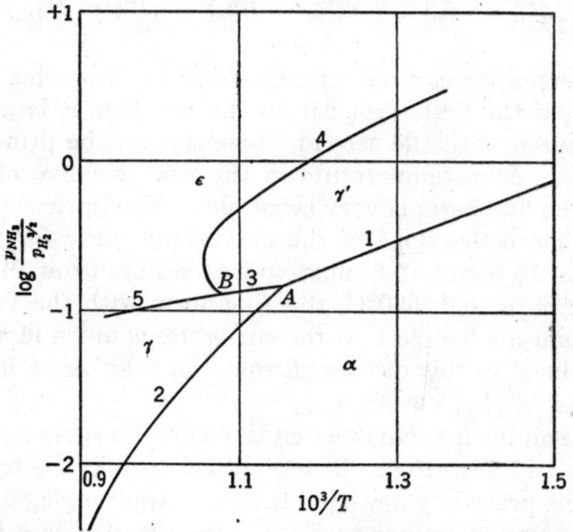

FIG. 15-6. Replot of the data of Fig. 15-4 for the Fe-N system.

constant composition. This suggestion is borne out by the temperature-composition diagram (Fig. 15-2).

2. **The α-γ equilibrium** curve also appears nearly straight in the range of the experimental measurements. This is because of the short experimental range, since, as may be seen from Fig. 15-2 or inferred from the general behavior of α-γ equilibria in ferrous systems, the α-γ equilibrium does not by any means involve phases of fixed composition.

A convenient way to write this equilibrium is

$$\alpha + \text{gas} = \gamma$$

It will be noted that this is a distinctly different type of equation from that commonly written; it involves phases whereas the usual chemical equation involves components or species. The coefficients are not written, since they obviously vary with temperature, and for this reason this equilibrium cannot be completely represented by a single chemical equation.[1] The complication of a phase of variable composition was present to a minor extent in the preceding α-γ' equilibrium. In that case, however, the solubility of nitrogen in α-iron was sufficiently small (less than 0.1 per cent) that the activity of iron in the α phase was not altered thereby, within the experimental error, and hence the presence of nitrogen in that phase could be ignored.

Again ignoring the presence of nitrogen in the α phase, we may apply Eq. (14-10), which now becomes

$$\frac{d \ln f_{N_2}^{\frac{1}{2}}}{d(1/T)} = \frac{1}{R}\left[\frac{N_{Fe}^{\gamma}}{N_{N}^{\gamma}}(\bar{H}_{Fe}^{\gamma} - H_{Fe}^{\alpha}) + (\bar{H}_{N}^{\gamma} - H_{\frac{1}{2}N_2}^{g})\right] \qquad (15\text{-}2)$$

where the superscripts as usual refer to phases. From this equation it is apparent that, if the first term within the brackets is large in absolute value as compared with the second, the slope will be principally determined thereby. At a temperature in the near vicinity of 910°C, it is obvious that the first term is very large, since N_N^{γ} approaches zero at this temperature; hence the slope of the univariant curve for the α-γ equilibrium in Figs. 15-5 and 15-6 must approach infinity at 910°C. In the vicinity of the eutectoid, 590°C, this first term with the variable coefficient has become smaller, so that the curvature is much less pronounced. It is to be realized in this discussion that the heat terms in parentheses are regarded as roughly constant.

3. **The γ-γ' equilibrium** curve extends over only a short range and is substantially linear in Fig. 15-6. The equilibrium may be handled in the same way as the preceding α-γ equilibrium. Applying again Eq. (14-10) we obtain an equation similar to Eq. (15-2). In this case the unprimed quantities refer to the γ phase, the primed quantity to the γ' phase, and the double-primed quantity to the gas phase. From Fig. 15-2 it is seen that the variation in composition of the γ' phase is rather small; hence N_{Fe}/N_N in the present counterpart of Eq. (15-2) undergoes no large variation. Ignoring the variation of the enthalpies with composition and temperature, we should expect the bracketed expression to change but little in value in the short range under consideration. This expectation appears verified by the near linearity of this univariant curve in Fig. 15-6.

The Invariant Equilibrium α, γ, γ'. Since all three univariant curves discussed above meet at a point A, it is apparent that there is an inter-

[1] In this case three independent chemical equations are required to describe the equilibrium completely. The following constitute one set of such equilibrium equations:

$$Fe(\alpha) = Fe(\gamma)$$
$$N(\alpha) = N(\gamma)$$
$$\tfrac{1}{2}N_2(g) = N(\alpha)$$

relation between the three univariant equilibria. Although it is possible to discuss this interrelation in terms of relations of the type of Eq. (14-7) or (14-10), it seems somewhat easier to discuss it first in terms of equilibrium constants. For the α-γ', α-γ, and γ-γ' equilibria considered above, let us write the chemical equations and equilibrium constants. The approximation is made here that the composition of γ' corresponds to the formula Fe_4N.

$$4\mathrm{Fe}(\alpha) + \frac{1}{2}\mathrm{N_2(g)} = \mathrm{Fe_4N}(\gamma') \qquad K_A = \frac{1}{f_{\mathrm{N_2}}^{\frac{1}{2}(\alpha,\gamma')}} \tag{A}$$

$$4\mathrm{Fe}(\alpha) = 4\mathrm{Fe}(\gamma) \qquad K_B = a_{\mathrm{Fe}}^{4(\gamma,\ \mathrm{equil.}\ \alpha)} \tag{B}$$

$$4\mathrm{Fe}(\gamma) + \frac{1}{2}\mathrm{N_2(g)} = \mathrm{Fe_4N}(\gamma') \qquad K_C = \frac{1}{a_{\mathrm{Fe}}^{4(\gamma,\ \mathrm{equil.}\ \gamma')} f_{\mathrm{N_2}}^{\frac{1}{2}(\gamma,\gamma')}} \tag{C}$$

In view of the substantially constant composition of the α and γ' phases, the activity of iron in α and of Fe_4N in γ' are considered unity and hence do not appear in the expression for the equilibrium constant. It will be noted that Eq. (*C*) results from subtraction of Eq. (*B*) from Eq. (*A*), hence that $K_C = K_A/K_B$. Substituting the above expressions in this relation,

$$f_{\mathrm{N_2}}^{\frac{1}{2}(\gamma,\gamma')} = \left[\frac{a_{\mathrm{Fe}}^{\gamma(\mathrm{equil.}\ \alpha)}}{a_{\mathrm{Fe}}^{\gamma(\mathrm{equil.}\ \gamma')}}\right]^4 f_{\mathrm{N_2}}^{\frac{1}{2}(\alpha,\gamma')} \tag{15-3}$$

Thus it is seen that the curve for the γ-γ' equilibrium is calculable from the α-γ' equilibrium curve (and its metastable extension) and a knowledge of the activity of iron in γ. By similar methods it may be shown in general that, when all but one of the univariant curves about an invariant point are known, this one is calculable from the others.

In this case it may be shown, by application of Eq. (14-10) to each of the univariant equilibria and consideration of the relative quantities of the reacting phases at the invariant point, that the following relation exists between the slopes of the three univariant curves at the invariant point:[1]

[1] This relation may also be derived in another way by considering the enthalpy changes of the three univariant reactions at the invariant point.

$$\text{Ferrite} + \mathrm{N_2(g)} = \text{austenite} \tag{A}$$

$$\text{Austenite} + \mathrm{N_2(g)} = \gamma' \tag{B}$$

$$\text{Ferrite} + \mathrm{N_2(g)} = \gamma' \tag{C}$$

Since enthalpy is a function of state, it is apparent that, if we convert 4 gram atoms of α-iron and $\frac{1}{2}$ mole of nitrogen gas to 1 gram formula weight of $Fe_4N(\gamma')$, the enthalpy change will be the same whether we proceed by steps (*A*) and (*B*) together or by

$$4\frac{N_N}{N_{Fe}}\frac{d\ln f_{N_2}^{\frac{1}{2}(\alpha,\gamma)}}{d(1/T)} + \left(1 - 4\frac{N_N}{N_{Fe}}\right)\frac{d\ln f_{N_2}^{\frac{1}{2}(\gamma,\gamma')}}{d(1/T)} = \frac{d\ln f_{N_2}^{\frac{1}{2}(\alpha,\gamma')}}{d(1/T)} \qquad (15\text{-}4)$$

This relation may be generalized to give the relation between the slopes of the three univariant equilibrium curves at an invariant point for any binary system by noting that the coefficient of each derivative is proportional to the number of gram atoms of the volatile component, or component whose fugacity is considered, which participates in the phase transformation with a fixed amount of the other component. The sum of the coefficients on one side of the equation is equal to that on the other.

Substituting in Eq. (15-4) the value of $f_{N_2}^{\frac{1}{2}}$ from the equilibrium constant for the dissociation of ammonia, $f_{N_2}^{\frac{1}{2}} = K_{NH_3}(f_{NH_3}/f_{H_2}^{\frac{3}{2}}) \cong K_{NH_3}(p_{NH_3}/p_{H_2}^{\frac{3}{2}})$,

$$4\frac{N_N}{N_{Fe}}\frac{d\log(p_{NH_3}/p_{H_2}^{\frac{3}{2}})^{\alpha,\gamma}}{d(1/T)} + \left(1 - 4\frac{N_N}{N_{Fe}}\right)\frac{d\log(p_{NH_3}/p_{H_2}^{\frac{3}{2}})^{\gamma,\gamma'}}{d(1/T)} = \frac{d\log(p_{NH_3}/p_{H_2}^{\frac{3}{2}})^{\alpha,\gamma'}}{d(1/T)} \qquad (15\text{-}5)$$

Around the invariant point A of Fig. 15-6 the measured slopes are

$$\frac{d\log(p_{NH_3}/p_{H_2}^{\frac{3}{2}})^{\alpha,\gamma}}{d(1/T)} = 5100$$

$$\frac{d\log(p_{NH_3}/p_{H_2}^{\frac{3}{2}})^{\gamma,\gamma'}}{d(1/T)} = 850$$

---

step (*C*) alone. Let us now write chemical reactions corresponding to the three equations above.

$$4\text{Fe}(\alpha) + \frac{x}{2}\text{N}_2(\text{g}) = x\text{N}(\gamma) + 4\text{Fe}(\gamma) \qquad (A')$$

$$4\text{Fe}(\gamma) + x\text{N}(\gamma) + \frac{(1-x)}{2}\text{N}_2(\text{g}) = \text{Fe}_4\text{N}(\gamma') \qquad (B')$$

$$4\text{Fe}(\alpha) + \tfrac{1}{2}\text{N}_2(\text{g}) = \text{Fe}_4\text{N}(\gamma') \qquad (C')$$

where x is a constant equal to number of atoms of nitrogen associated with four of iron in γ of eutectoid composition. Writing the equilibrium constants in terms of activities, taking the derivatives of the logarithms thereof (utilizing the fundamental theorem of partial differentiation and also the Gibbs-Duhem equation), and making use of the relations $(d\ln K)/d(1/T) = -\Delta H°/R$ and $\Delta H_3° = \Delta H_1° + \Delta H_2°$, it is found that, at the eutectoid,

$$x\frac{d\ln f_{N_2}^{\frac{1}{2}(\alpha,\gamma)}}{d(1/T)} + (1-x)\frac{d\ln f_{N_2}^{\frac{1}{2}(\gamma,\gamma')}}{d(1/T)} = \frac{d\ln f_{N_2}^{\frac{1}{2}(\alpha,\gamma')}}{d(1/T)}$$

This is identical with Eq. (15-4), it being noted that $x = 4N_N/N_{Fe}$. This equation is exact under isobaric conditions, although in the text it might appear that certain assumptions as to ideality of the gas were involved.

and

$$\frac{d \log (p_{NH_3}/p_{H_2}^{3/2})^{\alpha,\gamma'}}{d(1/T)} = 2080$$

Substituting these values in Eq. (15-5),

$$20{,}400 \frac{N_N}{N_{Fe}} + 850 \left(1 - 4 \frac{N_N}{N_{Fe}}\right) = 2080$$

The solution of this equation is $N_N/N_{Fe} = 0.072$, corresponding to 1.9 per cent N in the austenite of eutectoid composition. Figure 15-2 indicates 2.35 per cent N, and the direct measurements of Eisenhut and Kaup give a value of 2.3 per cent N for this composition. The agreement may be regarded as fair in view of the difficulty of determining the above slopes, particularly for the γ-γ' equilibrium which extends over a short temperature range.

4. **The γ'-ϵ equilibrium.** The univariant curve for this equilibrium in Fig. 15-6 illustrates well the point made previously that such a curve may exhibit marked departure from linearity in case one or more condensed phases in equilibrium are not of fixed composition. The curvature in this case appears attributable principally to the variable composition of the ϵ phase, since the γ' phase approximates the composition Fe_4N. In dealing with this equilibrium it is convenient to regard the components as Fe_4N and N. In terms of these components Eq. (14-10) becomes

$$\frac{d \ln f_{N_2}^{1/2}}{d(1/T)} = \frac{1}{R}\left[\frac{N^{\epsilon}_{Fe_4N}}{N^{\epsilon}_{N}} (\bar{H}^{\epsilon}_{Fe_4N} - H^{\gamma'}_{Fe_4N}) + (\bar{H}^{\epsilon}_{N} - H_{\frac{1}{2}N_2,g})\right] \quad (15\text{-}6)$$

As no H is infinite, it is seen that an infinite slope, $d \ln f_{N_2}^{1/2}/d(1/T)$, corresponds to a zero value of N^{ϵ}_{N}, hence to a composition of the ϵ phase equal to that of the γ' phase Fe_4N. As noted previously (page 317) the identity of composition of two phases in equilibrium necessitates a stationary value, in this case a maximum, in the temperature-composition diagram unless the isothermal derivative of the chemical potential (or activity) with composition is infinite. Since there is no indication of this special type of behavior of the ϵ phase of this composition, it probably would have been better to construct the curve representing the lower boundary of the ϵ field as passing through a maximum rather than as drawn in Fig. 15-2.

5. **The γ-ϵ equilibrium,** involving two phases of variable composition, has been investigated over only a short range. This is associated with the fact that the whole system is metastable and that the establishment of the metastable equilibrium above 600°C is difficult and above 700°C is practically impossible. The curve is so short as to appear linear in Figs.

15-4 to 15-6. Its metastable extension lies between the γ'-ϵ and the γ-γ' curves in accord with thermodynamic requirement discussed in the preceding chapter.

The Three-phase Equilibrium γ, γ', ϵ. Let us now treat the equilibria at point B, Fig. 15-6, in the same manner that we did those at A. The equations for the γ-γ' and γ'-ϵ equilibria may be written

$$4\text{Fe}(\gamma) + 4\frac{N_{\text{N}}^{\gamma}}{N_{\text{Fe}}^{\gamma}}\text{N}(\gamma) + \frac{1}{2}\left(1 - 4\frac{N_{\text{N}}^{\gamma}}{N_{\text{Fe}}^{\gamma}}\right)\text{N}_2(\text{g}) = \text{Fe}_4\text{N}(\gamma') \qquad (A)$$

and

$$\text{Fe}_4\text{N}(\gamma') + \frac{x-1}{2}\text{N}_2(\text{g}) = \text{Fe}_4\text{N}_x(\epsilon) \qquad (B)$$

where x is defined as the number of atoms of nitrogen associated with four of iron in ϵ. Upon addition these give the equation for the γ-ϵ equilibrium:

$$4\text{Fe}(\gamma) + 4\frac{N_{\text{N}}^{\gamma}}{N_{\text{Fe}}^{\gamma}}\text{N}(\gamma) + \frac{1}{2}\left(x - 4\frac{N_{\text{N}}^{\gamma}}{N_{\text{Fe}}^{\gamma}}\right)\text{N}_2(\text{g}) = \text{Fe}_4\text{N}_x(\epsilon) \qquad (C)$$

Making use of either of the methods leading to Eq. (15-5) it follows that

$$\frac{1}{2}\left(1 - 4\frac{N_{\text{N}}^{\gamma}}{N_{\text{Fe}}^{\gamma}}\right)\frac{d\log(p_{\text{NH}_3}/p_{\text{H}_2}^{\frac{3}{2}})^{\gamma,\gamma'}}{d(1/T)} + \frac{x-1}{2}\frac{d\log(p_{\text{NH}_3}/p_{\text{H}_2}^{\frac{3}{2}})^{\gamma',\epsilon}}{d(1/T)} = \frac{1}{2}\left(x - 4\frac{N_{\text{N}}^{\gamma}}{N_{\text{Fe}}^{\gamma}}\right)\frac{d\log(p_{\text{NH}_3}/p_{\text{H}_2}^{\frac{3}{2}})^{\gamma,\epsilon}}{d(1/T)} \qquad (15\text{-}7)$$

For simplicity we replace each derivative by a slope, denoted λ, appropriately superscripted.

$$\frac{1}{2}\left(1 - 4\frac{N_{\text{N}}^{\gamma}}{N_{\text{Fe}}^{\gamma}}\right)\lambda^{\gamma,\gamma'} + \frac{x-1}{2}\lambda^{\gamma',\epsilon} = \frac{1}{2}\left(x - 4\frac{N_{\text{N}}^{\gamma}}{N_{\text{Fe}}^{\gamma}}\right)\lambda^{\gamma,\epsilon}$$

Rearranging and multiplying by 2,

$$(x-1)(\lambda^{\gamma',\epsilon} - \lambda^{\gamma,\epsilon}) = \left(1 - 4\frac{N_{\text{N}}^{\gamma}}{N_{\text{Fe}}^{\gamma}}\right)(\lambda^{\gamma,\epsilon} - \lambda^{\gamma,\gamma'}) \qquad (15\text{-}8)$$

Since it is found experimentally that both factors on the right are positive and the second on the left is negative, we see that $(x - 1)$ is also negative and hence that x is less than unity. From Eq. (B), x is seen to be the number of atoms of nitrogen associated with four atoms of iron in the ϵ phase, and hence it follows that the ϵ phase at this invariant point B contains less nitrogen than the γ' phase. Using the measured values of the slopes from Fig. 15-6, it is found that $x = 0.93$, corresponding to 5.5 per cent N. Recalling that the simplifying assumption was made that the composition of γ' corresponds to the formula Fe_4N, 5.9 per cent N, the results of the above calculation may better be stated in terms of the difference in composition of γ' and ϵ; the computed difference corresponds to about $\frac{1}{2}$ per cent less nitrogen in ϵ than in the γ' in equilibrium therewith. The difference according to Fig. 15-2 is about $\frac{3}{4}$ of 1 per cent. This discrepancy can be attributed again to the difficulty of obtaining the slopes.

The Zeta Phase. Hägg[1] reported a phase of higher nitrogen content than ϵ. This phase, designated ζ, has been confirmed by Jack,[2] who prepared it by passing ammonia rapidly (to avoid cracking) over iron powder at 450°C or less. He found that alloys with a nitrogen content of 11.0 per cent were ϵ and those with 11.1 to 11.3 per cent were ζ. The crystal structure of both phases is such that the iron atoms are in a substantially hcp array. The ζ phase, corresponding closely to the composition Fe_2N, has a fully ordered structure in which half the octahedral interstices are filled by nitrogen atoms in such a manner that each has a hole above and below it. In this arrangement the nitrogen atoms are as far apart as possible even though the iron lattice is distorted from hcp. In the ϵ phase the nitrogen atoms are not fully ordered on the interstitial sites. There is a slight discontinuity in lattic parameter on passing from ϵ to ζ (even allowing for the slight difference in composition). This evidence seems to establish the existence of a ζ phase distinct from ϵ. As it decomposes rapidly above 450°C, the equilibria in which it might otherwise participate have not been investigated.

NITROGEN IN STEEL

The preceding sections of this chapter have dealt principally with the high-nitrogen portion of the iron-nitrogen system. Let us now consider the nitrogen content of iron and steel to which nitrogen has not purposely been added. The nitrogen contents of the products of the blast furnace, open hearth, and bessemer commonly fall in the following ranges:

	Nitrogen, %
Blast furnace	0.002–0.006
Open hearth	0.004–0.007
Bessemer	0.009–0.018

It will be noted that these are less than the amount of impurity commonly found in the best grade of analytical chemicals. However, the practical importance of the nitrogen in bessemer steel, for example, is considerable, and it is generally believed that the differences between bessemer and

[1] G. Hägg, *Nature,* **122,** 962 (1928); *Nova Acta Regiae Soc. Sci. Upsaliensis,* IV, **7,** 1 (1929).

[2] K. H. Jack, *Proc. Roy. Soc.* (*London*), **A195,** 34 (1948).

open-hearth steel are attributable at least in part to the higher nitrogen content of the former.

Many investigations have been performed on the effects of nitrogen in steel. The results of Tschischewski[1] on the effect of nitrogen in decreasing the elongation of wire (0.63 mm diameter) are shown in Fig. 15-7. Nitrogen produces a marked embrittling effect which is accompanied by an increase in strength. Tschischewski noted that at a nitrogen content

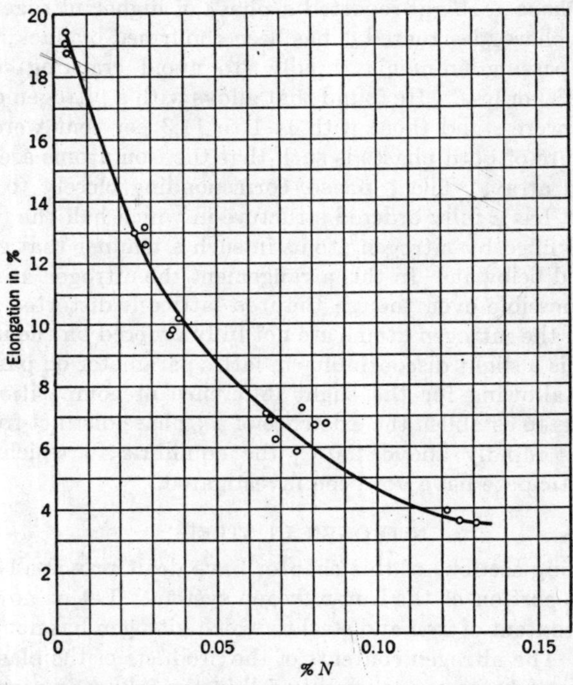

FIG. 15-7. Effect of nitrogen content on elongation of iron wire in tension. [*From N. Tschischewski, J. Iron Steel Inst.*, **92**, 47 (1915).]

of 0.3 per cent the wire broke without visible bending and at 0.4 per cent N it broke "like glass." Such observations have been frequently confirmed by other investigators. The pronounced effect of nitrogen on brittleness, particularly work-brittleness (brittleness after cold work), is shown in Fig. 15-8, taken from Work and Enzian.[2] Further work has been included in a general investigation by Enzian[3] of the effects of phosphorus and nitrogen on the properties of low-carbon steel.

[1] N. Tschischewski, *J. Iron Steel Inst.*, **92**, 47 (1915).

[2] H. K. Work and G. H. Enzian, *Trans. AIME*, **162**, 723 (1945).

[3] G. H. Enzian, *Trans. AIME*, **188**, 778 (1950).

The stiffening effect of a small amount of nitrogen is found advantageous in certain commercial applications, *e.g.*, the ends of cans. The austenitizing effect of a small amount of nitrogen (see Fig. 15-2) is of appreciable practical consequence, particularly in chromium-nickel steels. A small amount of nitrogen increases rather markedly the rate of solution of iron in acids, and the higher rate of solution of some grades of bessemer as compared with similar grades of open-hearth steel is frequently attributed to high nitrogen content. Most of the above-mentioned effects of nitrogen can be counterbalanced, if desired, by appropriate alloy additions or by lowering the carbon content.

The pronounced effects of nitrogen in steel can for the most part be interpreted in terms of the phase diagram of Fig. 15-2, the relatively high diffusivity of nitrogen, and the coherent-state theory of precipitation and hardening. The diffusivity of nitrogen in α-iron lies between that of hydrogen, which has the highest diffusivity of all elements in iron, and carbon; the diffusivity of substitutional alloying elements is many orders of magnitude smaller. From Fig. 15-3 it may be seen that iron containing 0.01 per cent nitrogen in solution will not precipitate nitride above about 300°C. At a temperature somewhat below this, nitride precipitation does occur. This is one of the few phase changes which have been observed in steel at such a low temperature and undoubtedly has significance in aging phenomena. It is certainly not surprising that a precipitate formed at this low temperature should be exceedingly fine and under appropriate circumstances should be of the coherent type believed responsible for this type of hardening.

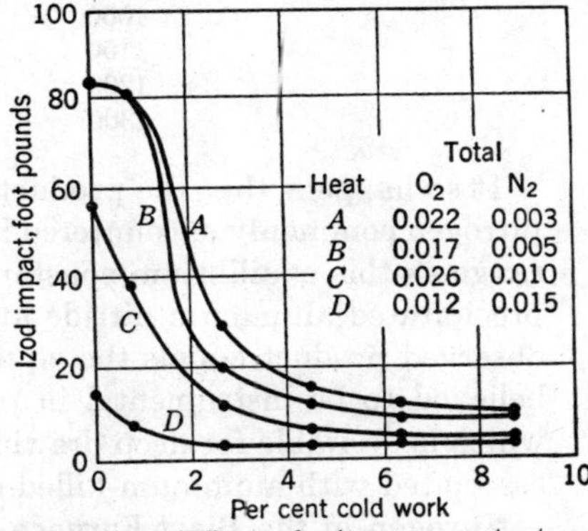

FIG. 15-8. Effect of nitrogen content on brittleness of rimmed steels. [*From Work and Enzian, Trans. AIME*, **162**, 723 (1945).]

Aluminum Nitride in Steel. The small quantity of aluminum present in aluminum-killed plain carbon steel has a pronounced effect on the behavior of nitrogen. The interaction of aluminum and nitrogen in steel is believed to be of considerable commercial importance. The equilibrium between solid aluminum nitride and aluminum and nitrogen dissolved in γ-iron has been investigated by Darken, Smith, and Filer.[1] For the equilibrium

$$AlN(s) = Al(\text{in } \gamma) + N(\text{in } \gamma)$$

[1] L. S. Darken, R. P. Smith, and E. W. Filer, *Trans. AIME*, **191**, 1174 (1951).

they find the experimentally determined equilibrium constant

$$K = [\%\text{Al}][\%\text{N}]$$

to be represented by the expression

$$\log K = -\frac{7400}{T} + 1.95$$

Values of K as calculated from this equation are given in the following table:

°C	$K \times 10^4$
900	0.35
1000	1.4
1100	3.7
1200	8.6
1300	18.

It so happens that the product of the per cent aluminum and per cent nitrogen commonly encountered in aluminum-killed steel falls in the same range as this equilibrium constant. Thus at equilibrium, steel contains precipitated aluminum nitride at temperatures below that at which the observed product equals the equilibrium product. Aluminum nitride is believed to be instrumental in producing the elongated grain structure which is desirable for deep drawing steels, as well as the grain refinement associated with aluminum-killed steel.[1]

Nitrogen in the Blast Furnace. Having seen that a small amount of nitrogen is of importance, let us now trace the course of nitrogen in the blast furnace and in the steelmaking furnace. From Fig. 15-1 the solubility of nitrogen gas at 1 atm pressure in otherwise pure liquid iron is about 0.04 per cent. Since in the bosch of a blast furnace the downcoming metal is exposed to an atmosphere in which p_{N_2} is substantially 1 atm, the question naturally arises as to why the product of this furnace contains considerably less than 0.04 per cent N. The answer to this question is to be found principally in the large effect of carbon and silicon in lowering the solubility of nitrogen.

The effect of various elements on the solubility of nitrogen gas at 1 atm pressure in iron at 1600°C is shown in Fig. 15-9.[2] It will be noted that 4 per cent carbon and 1 per cent silicon (these representing near minimum

[1] R. L. Solter and C. W. Beattie, *Trans. AIME*, **191**, 721 (1951).

[2] Data for Cr, Mn, and Ni from H. Wentrup and O. Reif, *Arch. Eisenhüttenw.*, **20**, 359 (1949); for Si from J. C. Vaughan and J. Chipman, *Trans. AIME*, **140**, 224 (1940), and M. M. Karnaukhov and A. N. Morozov, *Bull. acad. sci. U.R.S.S.*, **6**, 735 (1947); for C from L. Eklund, *Jernkontorets Ann.*, **123**, 545 (1939); for Cr and V from R. M. Brick and J. A. Creevy, *Metals Technol.*, *Tech. Paper* 1165, April, 1940.

figures for the composition of hot metal) together may be expected to lower the solubility from 0.04 to about 0.01 per cent. Since the blast-furnace product is usually considerably below 1600°C, and since as seen from Fig. 15-1 the temperature coefficient of solubility is positive (as is usual for gases in metals), the solubility of nitrogen gas in the blast-furnace product is well below 0.01 per cent and is probably in the range of the observed nitrogen content. In fact, there is some evidence that

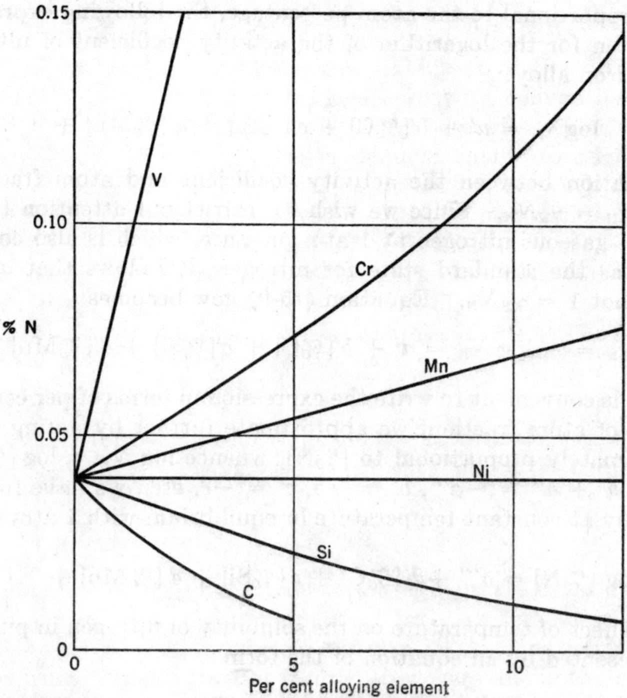

FIG. 15-9. Effect of alloying elements upon the solubility of nitrogen gas at 1 atm in iron at 1600°C.

iron direct from the blast furnace is supersaturated with respect to nitrogen gas at 1 atm. This can be explained on the basis that the iron was saturated with nitrogen in the blast furnace before it became saturated with carbon. The blast-furnace product usually loses nitrogen in its course through the mixer to the open hearth or bessemer. This is probably explained both by the approach to equilibrium of the initially supersaturated iron and by the drop in temperature, and hence in solubility, during transport.

In a ternary or multicomponent system in which one component (iron in this case) is predominant, the logarithm of the activity coefficient of any minor component at constant temperature can usually be represented with a fair degree of accuracy as a linear function of the atom percentages of each of the minor constituents. This, it will be seen, is one step better than assuming Henry's law, by which the activity coefficient would be taken as constant. If, as a further approximation for small percentages of each minor constituent, the weight percentage is assumed to be proportional to the atom percentage, the following expression may be written for the logarithm of the activity coefficient of nitrogen in a molten iron alloy:

$$\log \gamma_N = a' + b'[\%C] + c'[\%Si] + d'[\%Mn] + \cdots \qquad (15\text{-}9)$$

The relation between the activity coefficient and atom fraction is, as usual, $a_N = \gamma_N N_N$. Since we wish to restrict our attention to the solubility of gaseous nitrogen at 1 atm pressure, which is also conveniently chosen as the standard state for nitrogen, it follows that $a_N = 1$ and hence that $1 = \gamma_N N_N$. Equation (15-9) now becomes

$$\log \gamma_N = -\log N_N = a' + b'[\%C] + c'[\%Si] + d'[\%Mn] + \cdots$$

Since it is convenient to write the expression in terms of per cent nitrogen instead of atom fraction, we approximate further by noting that N_N is approximately proportional to [%N], whence $\log N_N = \log [\%N] + a''$. Setting $a' + a'' = -a'''$, $b' = -b$, $c' = -c$, etc., we have for a molten iron alloy at constant temperature in equilibrium with 1 atm of nitrogen gas:

$$\log [\%N] = a''' + b[\%C] + c[\%Si] + d[\%Mn] + \cdots \qquad (15\text{-}10)$$

The effect of temperature on the solubility of nitrogen in pure iron can be represented by an equation of the form

$$\log [\%N] = -\frac{\Delta H}{RT} + I \qquad (15\text{-}11)$$

where ΔH is the enthalpy change for the solution of 1 gram atom ($\frac{1}{2}$ mole) of nitrogen gas in the alloy. In order to consider simultaneously the effect of temperature and of alloy content, let us now make the assumption (since adequate experimental data are not available) that ΔH is independent of alloy content and that the coefficients a''', b, c, etc., of Eq. (15-10) are independent of temperature and of each other. With these approximations and the further assumption that no alloying constituents of significance other than carbon, silicon, and manganese are present, the relation at 1 atm of nitrogen becomes[1]

$$\log\,[\%\mathrm{N}] = -\frac{\Delta H}{RT} + a + b[\%\mathrm{C}] + c[\%\mathrm{Si}] + d[\%\mathrm{Mn}] \qquad (15\text{-}12)$$

Evaluating the coefficients from the data of Figs. 15-1 and 15-9, this equation becomes[2]

$$\log\,[\%\mathrm{N}] = -\frac{1000}{T} - 0.86 - 0.06[\%\mathrm{Si}] - 0.24[\%\mathrm{C}] + 0.015[\%\mathrm{Mn}] \qquad (15\text{-}13)$$

From this equation the equilibrium nitrogen content at 1250°C of a bessemer iron containing 4.1 per cent C, 2.0 per cent Si, and 0.8 per cent Mn is 0.0024 per cent N, which is comparable to the nitrogen content of the blast-furnace product given previously.

Nitrogen in the Open-hearth Process. During the part of an open-hearth heat after the melt-down, the metal is covered with slag and thus does not come in direct contact with the nitrogen of the furnace atmosphere. In the early part of the heat, scrap is exposed, but since the atmosphere is very oxidizing, the scrap is usually coated with liquid or solid oxide, which appears to be an effective barrier to nitrogen absorp-

[1] This may readily be shown from the stated assumptions and the fundamental theorem of partial differentiation. By this theorem,

$$d\log\,[\%\mathrm{N}] = \left[\frac{\partial \log\,[\%\mathrm{N}]}{\partial(1/T)}\right] d\,\frac{1}{T} + \left(\frac{\partial \log\,[\%\mathrm{N}]}{\partial[\%\mathrm{C}]}\right) d[\%\mathrm{C}] + \left(\frac{\partial \log\,[\%\mathrm{N}]}{\partial[\%\mathrm{Si}]}\right) d[\%\mathrm{Si}] + \left(\frac{\partial \log\,[\%\mathrm{N}]}{\partial[\%\mathrm{Mn}]}\right) d[\%\mathrm{Mn}]$$

Integrating this expression by evaluating the differential coefficients from Eqs. (15-10) and (15-11), regarding them as constants, Eq. (15-12) results.

[2] Actually, the coefficient for carbon in Eq. (15-13) is taken not from the single set of data incorporated in Fig. 15-9, which shows considerable scatter, but to accord with various observations indicating the solubility of nitrogen gas at 1 atm in iron at 100 to 200° above the eutectic to be about 0.003 per cent. For example, E. Eickworth [*Stahl u. Eisen*, **68,** 383 (1948)] reports that the nitrogen content fell from 0.005 per cent to 0.002 to 0.003 per cent during the first 3 min of a Thomas converter blow; whereas when the initial nitrogen content was 0.002 per cent no drop was observed. T. Kootz [*ibid.*, **68,** 378 (1948)], in accord with many observations over a period of several months, reports that iron directly from the blast furnace averaged 0.006 per cent N and from the mixer 0.004 per cent N; the 0.002 per cent drop during transfer to and from the ladle indicates supersaturation. The apparent discrepancy between these observations may be due in part to error in the temperature coefficient used in Eq. (15-13) and to the simplifying assumptions incorporated in the equation. At any rate it seems desirable to choose the numerical coefficients to fit the data which appear most reliable.

tion.[1] During the boil the dissolved nitrogen is subjected to the flushing action of the carbon monoxide bubbles which originate on the bottom of the furnace. In accord with Sievert's law and the relatively low nitrogen level, the amount of nitrogen removed by this process is not large and the nitrogen content of the bath just before tap is frequently 0.002 to 0.003 per cent N—nearly the same as that of the hot metal charged (the scrap charged, however, may be of substantially higher nitrogen content). During tap a two-point (0.002%) rise in nitrogen is not uncommon. This rapid pickup from the air, as well as the loss of nitrogen by hot metal during pouring through air, illustrates the facility of interchange of nitrogen between a gas and a ferrous medium in the absence of a protective oxide. By diminishing the access to nitrogen this pickup during tap may be decreased.

Nitrogen in the Bessemer Process. The course of the various elements during blow is shown for two cases in Figs. 15-10 and 15-11. Restricting our attention for the moment to the nitrogen content, it will be observed that this rises slowly during the beginning of the blow and more rapidly toward the end. As noted previously the solubility of nitrogen is markedly depressed by carbon and silicon and by the low temperature during the early stages of the blow. As the temperature rises and the carbon and silicon are burned out, the solubility of nitrogen increases to about 0.04 per cent toward the end of the blow. We might well inquire as to an interpretation of the observed fact that bessemer metal is commonly only one-quarter or one-third saturated with nitrogen in spite of the vigorous agitation in the presence of this gas. A not unreasonable explanation would seem to be that toward the end of the blow, when the solubility is high, the metal is separated from the gas by a thin liquid oxide or slag layer which retards nitrogen absorption. The over-all picture would then be that, at the beginning of the blow, near-equilibration with respect to nitrogen is obtained, since the oxidation produces principally gaseous oxides of carbon and solid silica; however, the solubility is markedly depressed, as mentioned previously, by the carbon, the silicon, and the low temperature. Toward the end of the blow, when the solubility is high, the departure from equilibrium is greater because of the liquid oxide film. The dashed curves representing the

[1] An analysis of nails found in the ruins of St. Pierre, Martinique, destroyed by volcanic eruption in 1902, showed exceptionally high nitrogen content. The explanation of this, subsequently verified by laboratory experiments [G. H. Enzian, H. F. Beeghly, and P. P. Tarasov, *Metal Progress*, **41**, 522 (1942)], is that during the scaling of steel nitrogen does not penetrate the scale to any significant extent. The nitrogen content of the metal remains substantially constant as the metal is consumed, and hence the percentage of nitrogen in the metal increases during scaling, whether the scaling occurs in a nitrogen atmosphere or not.

solubility of nitrogen in Figs. 15-10 and 15-11 would seem to support this contention.

Vaughan and Chipman,[1] who investigated the solubility of nitrogen in iron-silicon alloys, found that the equilibration time is much lower for iron-silicon alloys than for iron itself. This finding is in agreement with the above reasoning, but whether the rapid equilibration in the presence of silicon is to be attributed to a direct catalytic effect of this element or to

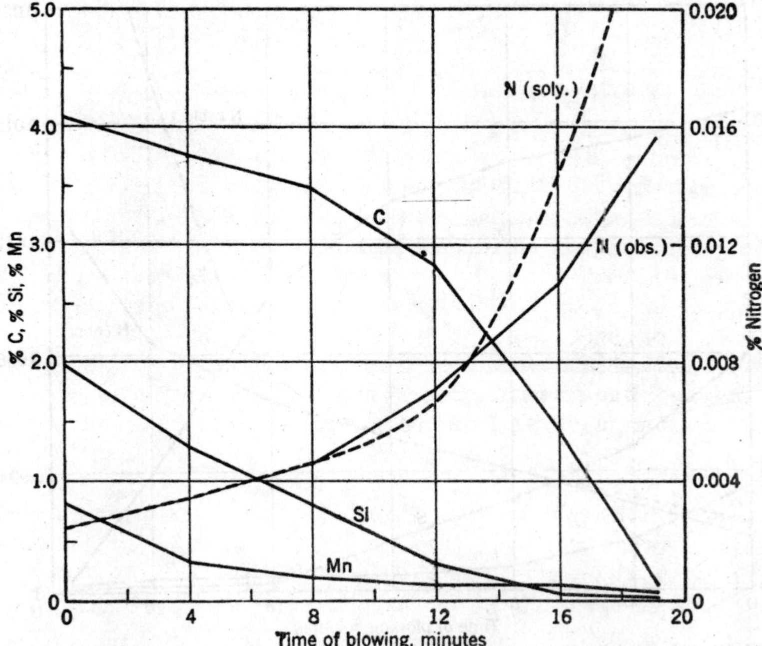

Fig. 15-10. Compositional change during bessemer blow; silicon substantially oxidized at 1.5 per cent carbon. [*From the data of T. Swinden and F. B. Cawley, Iron Steel Ind.*, **12**, 387 (1939).]

a lack of liquid oxide film remains to be settled by further experimentation. In any event, it will be noticed that, in the blow in which the silicon was burned out relatively early in comparison with the carbon (Fig. 15-11), the final nitrogen content was lower than in the blow in which silicon was not burned out until later (Fig. 15-10). This tendency appears to be quite general. Since silicon tends to burn out relatively sooner in cold than in hot blows, the above reasoning would lead to the conclusion that a low final nitrogen content is favored by a cold blow. This also is in accord with experience.

[1] J. C. Vaughan and John Chipman, *Trans. AIME*, **140, 224** (1940).

Nitrogen in the Heat-treating of Steel. Since most heat-treating operations are performed in an oxidizing atmosphere, the nitrogen content of the steel is not substantially altered, the oxide formed serving as an effective barrier. However, during bright annealing, a considerable interchange of nitrogen between the atmosphere and the steel is likely to occur. It will be seen from Fig. 15-1 that the nitrogen content of most

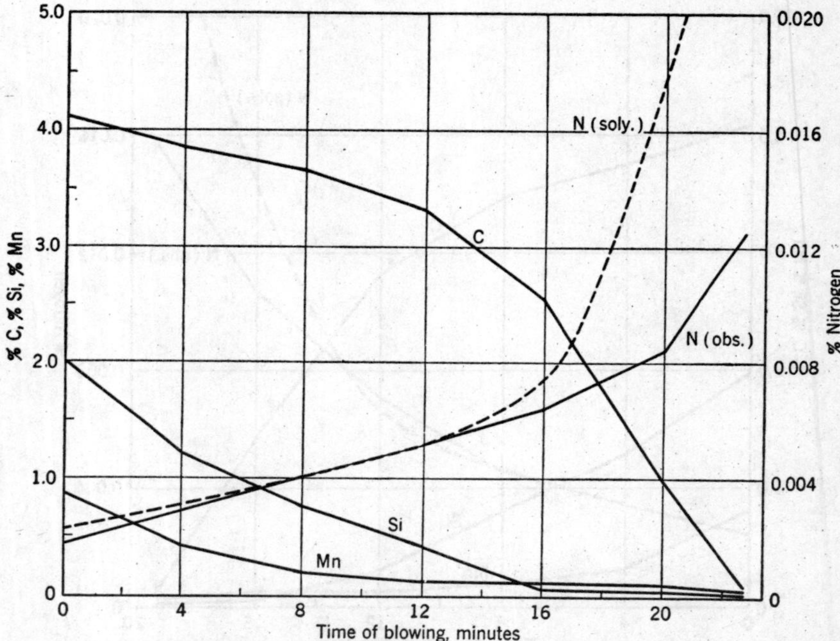

FIG. 15-11. Compositional change during bessemer blow; silicon substantially oxidized at 2.5 per cent carbon. [*From the data of T. Swinden and F. B. Cawley, Iron Steel Ind.*, **12**, 387 (1939).]

plain carbon steels lies between the solubility of nitrogen in ferrite and that in austenite. Hence such a steel, if treated in a high-nitrogen reducing atmosphere, will in general tend to pick up nitrogen in the temperature range where it is austenitic and to lose nitrogen in the temperature range where it is ferritic. For example, during the subcritical annealing of black plate in a reducing atmosphere it is not uncommon to encounter loss of nitrogen. As noted previously, the solubility of iron nitride (γ' or Fe_4N) is not usually exceeded on cooling until a relatively low temperature is reached, and it is believed that it is the coherent form of this precipitate which is responsible for the embrittling and stiffening effects of nitrogen.

Effect of Alloying Elements. The addition to steel of elements which form relatively stable nitrides usually gives rise to the precipitation of one of these nitrides rather than an iron nitride. Thus the addition of aluminum to bessemer steel is found to counteract at least partially the embrittling effect of the nitrogen. The precipitation of AlN in the temperature range 900 to 1300°C has already been discussed. Precipitation in this temperature range does not lead to an embrittling effect comparable to that caused by iron nitride, which precipitates in the range between room temperature and about 300°C.

Advantage is taken of these relatively stable nitrides in the formulation of steels intended for nitriding. Such steels usually contain appreciable proportions of elements such as chromium and aluminum, whose nitrides, being more stable than the iron nitrides, precipitate in a layer near the surface and thus harden this layer. If pure iron is nitrided at 600°C in an atmosphere containing about 10 per cent ammonia in hydrogen, the process at temperature consists of the single-phase diffusion of nitrogen in ferrite and the product contains about 0.1 per cent nitrogen at the surface. If an appreciably higher percentage of ammonia is used, an iron nitride layer will be found on the surface. Figure 15-1 shows that ferrite containing even 0.1 per cent nitrogen is highly unstable relative to nitrogen gas at 600°C, and in an extended period of time nitrogen gas tends to be liberated in imperfections, in cracks, and especially in grain boundaries. In some cases the iron swells 10 or 20 per cent by virtue of the porosity thus created and the product crumbles when bent in the hands. If a severely cold-worked mild steel is subjected to the same atmosphere, it is apt to develop severe blisters within the ferrite; the cold work apparently gives rise within the lattice to flaws which are of sufficient size to serve as nuclei for the development of the gas phase.

CHAPTER 16

THE IRON-CARBON SYSTEM

As has been mentioned previously, pure iron is capable of existing in two crystalline forms, bcc and fcc. The bcc modification, stable below 910°C, is known as α-iron or ferrite. At the temperature 910°C, the bcc and fcc forms coexist at equilibrium. For a range of several hundred degrees above this temperature the fcc form, known as γ-iron or austenite,[1] is stable. At still higher temperature a rather unusual phenomenon occurs; the bcc form again becomes stable. This second temperature at which fcc and bcc iron coexist in equilibrium is 1400°C. From this temperature up to the melting point, 1539°C, the bcc form is again stable, and in this range the bcc form is known as δ-iron or δ-ferrite.

In view of the stability of the bcc form both below 910 and above 1400°C, but not between these temperatures, it is seen that ΔF° for the reaction $\text{Fe}(\alpha,\delta) = \text{Fe}(\gamma)$ must pass through zero twice—at 910 and at 1400°C. The curve may be constructed from tabular values of $(F^\circ - H_0^\circ)/T$ for both forms of iron. This function, as well as $H^\circ - H_0^\circ$ and C_P, is given in Table 16-1. As is customary, the data of the table are carried to at least one more figure than is significant in order that the requisite precision may be obtained in taking differences. H_0° pertains to α-iron in all cases, and hence when using this table ΔH_0° is to be taken as zero for all transformations of pure iron. From this table ΔF° for the reaction $\text{Fe}(\alpha,\delta) = \text{Fe}(\gamma)$ can readily be calculated; it is shown as a function of temperature in Fig. 16-1. It will be observed that the curve crosses the axis of abscissas twice.

SOLUBILITY OF GRAPHITE AND CEMENTITE IN α-IRON

The Solubility of Cementite in α-iron. The measurement of this solubility is rather difficult, since it involves a determination of carbon content below 0.025 per cent. Reliable measurements of the solubility of cementite have, however, been reported by several investigators.[2] The equilibrium may be written

$$\text{Fe}_3\text{C(cem.)} = 3\text{Fe}(\alpha) + \text{C(in } \alpha) \qquad (16\text{-}1)$$

[1] The term austenite is more commonly used to designate the solid solution of carbon in γ-iron.

[2] L. J. Dijkstra, *Trans. AIME*, **185**, 252 (1949). J. K. Stanley, *ibid.*, **185**, 752 (1949). J. L. Ham, Private communication to J. K. Stanley, *ibid.*, **185**, 752 (1949). R. P. Smith, *J. Am. Chem. Soc.*, **68**, 1163 (1946). C. A. Wert, *Trans. AIME*, **188**, 1242 (1950).

TABLE 16-1. THERMODYNAMIC FUNCTIONS FOR IRON*

°K	$-\frac{F° - H_0°}{T}$†, cal/(gram atom)(°C)			$H° - H_0°$, cal/gram atom			C_P, cal/(gram atom) (°C)		
	α(δ)	γ	Liquid	α(δ)	γ	Liquid	α(δ)	γ	Liquid
0	0			0			0		
298.16	2.900	−1.410		1070	2932		5.98	6.39	
400	4.058	+1.347		1712	3593		6.55	6.59	
500	5.068	3.298		2388	4268		6.99	6.79	
600	5.976	4.826		3114	4952		7.54	6.99	
700	6.805	6.085		3902	5661		8.25	7.19	
800	7.575	7.167		4773	6389		9.23	7.39	
900	8.303	8.094		5765	7138		10.74	7.58	
1000	9.005	8.929		6970	7906		13.80	7.78	
1020	9.145	9.085		7262	8062		15.62	7.82	
C_P max. 1033				7481			18.00		
1040	9.285	9.238		7591	8219		13.58	7.86	
1060	9.424	9.389		7844	8377		11.92	7.90	
1080	9.563	9.537		8076	8535		11.30	7.94	
1100	9.701	9.682		8297	8695		10.83	7.98	
1120	9.836	9.824		8511	8855		10.49	8.02	
1140	9.972	9.964		8718	9015		10.23	8.06	
1160	10.106	10.102		8920	9177		10.01	8.10	
1180	10.237	10.237					9.85	8.12	
α → γ 1183.16	10.2581	10.2581		9150	9365		9.83	8.15	
1200	10.3676	10.3698		9315	9503		9.72	8.17	
1250	10.6860	10.6934		9795	9914		9.51	8.28	
1300	10.9946	11.0049	10.257	10267	10330	13431	9.41	8.38	10.28
1308	11.0431	11.0536		10342	10397	13514	9.40	8.39	
1350	11.2937	11.3052	10.647	10736	10752	13946	9.36	8.48	10.29
1400	11.5838	11.5953	11.023	11203	11178	14461	9.33	8.58	10.32
1450	11.8653	11.8759	11.385	11670	11610	14978	9.32	8.68	10.35
1500	12.1389	12.1478	11.734	12136	12046	15496	9.33	8.77	10.37
1550	12.4048	12.4116	12.074	12602	12487	16015	9.34	8.87	10.40
1600	12.6634	12.6678	12.402	13071	12933	16536	9.37	8.97	10.42
1650	12.9154	12.9171	12.720	13541	13384	17057	9.43	9.07	10.44
γ → δ 1673.16	13.0301	13.0301	12.864	13760	13595	17299	9.45	9.12	10.45
1700	13.161	13.160	13.029	14014	13841	17580	9.48	9.17	10.46
1750	13.400	13.396	13.328	14489	14302	18103	9.54	9.27	10.48
δ → l 1812	13.689	13.681	13.689	15083	14880	18753	9.60	9.39	10.50
1873	13.966	13.954	14.031	15671	15457	19394	9.68	9.51	10.53
1900	14.086	14.072	14.179	15932	15715	19678	9.72	9.57	10.54
2000	14.517	14.498	14.711	16911	16681	20734	9.86	9.77	10.58

* L. S. Darken and R. P. Smith, *Ind. Eng. Chem.*, **43**, 1815 (1951).

† $H_0°$ refers to α(δ)-iron in all cases.

for which the equilibrium constant is $K_1 = a_{Fe}^3 a_C^{\alpha(\text{cem.})}/a_{Fe_3C}$. As the solubility of carbon is low, and hence the iron nearly pure, the activity of iron may be taken as unity; also the cementite is of substantially con-

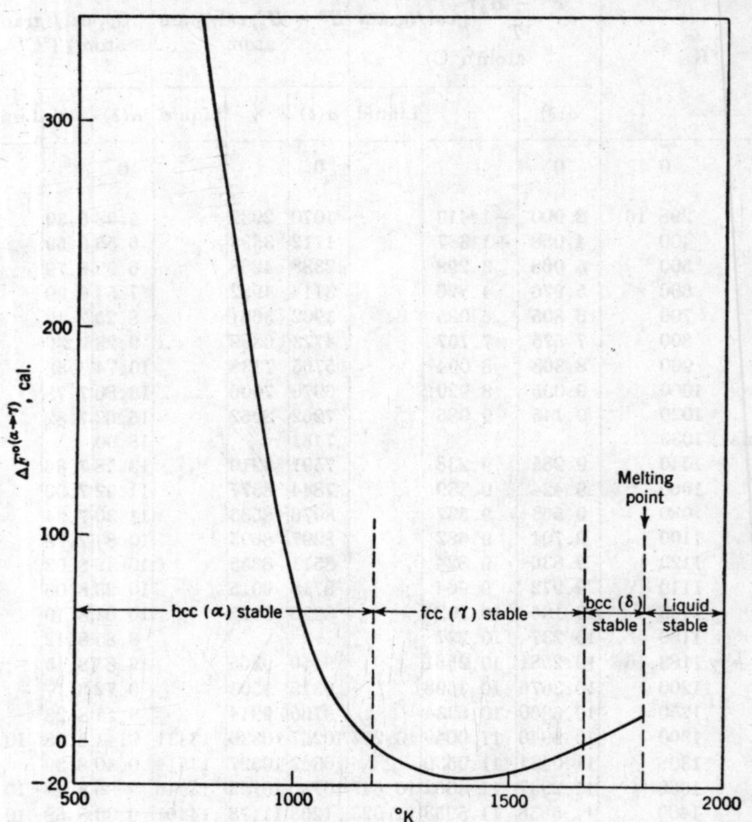

FIG. 16-1. The free-energy change accompanying the transformation of pure bcc to fcc iron.

stant composition, and its activity may be taken as unity. Hence

$$K_1 = a_C^{\alpha(\text{cem.})}$$

and

$$\Delta F_1^{\circ} = -RT \ln K_1 = -RT \ln a_C^{\alpha(\text{cem.})}$$

As the carbon concentration is so low, the activity of carbon may be safely taken proportional to the weight per cent. Choosing the standard state such that the activity equals the weight per cent at low carbon content, we may write $K_1 = [\%C]^{\alpha(\text{cem.})}$. Since ΔH is of the order of magnitude 10,000 cal and cannot reasonably be expected to vary much more than 1,000 cal in the temperature range under consideration,[1] it is to be anticipated that a plot of log $[\%C]^{\alpha(\text{cem.})}$ against $1/T$ will be substantially linear, at least within the large experimental error. Such a plot, taken from Wert's[2] review, is shown in Fig. 16-2. It is seen that a straight line is apparently adequate to express the data. This line may be represented by the equation

$$\log\,[\%C]^{\alpha(\text{cem.})} = -\frac{9700}{4.575T} + 0.41 \tag{16-1a}$$

corresponding to $\Delta H_1 = +9700$ cal.

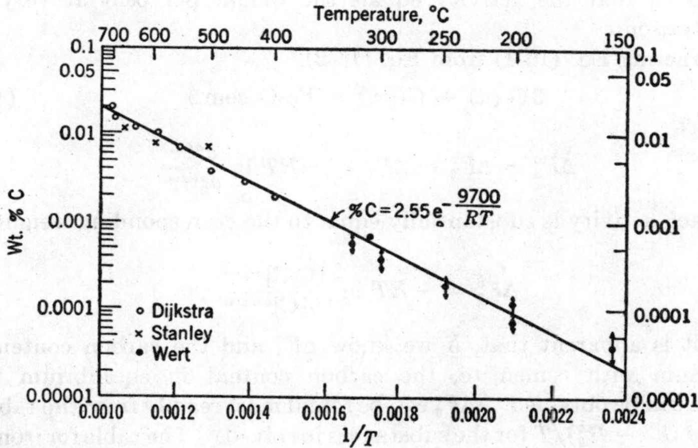

FIG. 16-2. The solubility of cementite in α-iron. [*From C. A. Wert, Trans. AIME*, **188**, 1242 (1950).]

The Solubility of Graphite in α-iron. Direct experimental data on the solubility of graphite in α-iron are not available. However, by slight extrapolation of the data on the α-γ equilibrium, discussed later, it is found that the solubility of graphite in α-iron at the α-γ-graphite eutectoid, 738°C, is 0.023 per cent. The heat of solution of graphite in α-iron

[1] This is estimated in the following way. For the above reaction,

$$\Delta C_P = 3C^{\alpha}_{P\,\text{Fe}} + \bar{C}^{\alpha}_{P\,\text{C}} - C^{\text{cem.}}_{P\,\text{Fe}_3\text{C}}$$

At, say, 800°K, $C^{\alpha}_{P\,\text{Fe}} = 9.2$ cal/deg and $C^{\text{cem.}}_{P\,\text{Fe}_3\text{C}} = 28.8$ cal/deg. The partial molal heat capacity of the dissolved carbon has not been directly measured. It would not seem unreasonable, however, to take this equal to the heat capacity of graphite or to the Dulong and Petit value. Assuming a value of 5 cal/deg, ΔC_P is then found to be 3.8 cal/deg. Over a temperature range ΔT, ΔH will change by an amount $\Delta C_P\,\Delta T$, and thus for the temperature range of about 250° considered here, ΔH_1 may be anticipated to change by 3.8 × 250 or about 1,000 cal.

[2] C. A. Wert, *Trans. AIME*, **188**, 1242 (1950).

from Smith's data is 25,700 cal/gram atom. The solubility at lower temperature could be computed from these data. This solubility can also be determined from that of cementite and other available data. The equation for the equilibrium of graphite with α-iron may be written

$$C(\text{gr.}) = C(\text{in } \alpha) \tag{16-2}$$

and

$$\Delta F_2^\circ = -RT \ln a_C^{\alpha(\text{gr.})}$$

The standard state for the carbon dissolved in α-iron is taken, as previously, such that the activity equals the weight per cent at very low concentration.

Subtracting Eq. (16-1) from Eq. (16-2),

$$3Fe(\alpha) + C(\text{gr.}) = Fe_3C(\text{cem.}) \tag{16-3}$$

for which

$$\Delta F_3^\circ = \Delta F_2^\circ - \Delta F_1^\circ = -RT \ln \frac{a_C^{\alpha(\text{gr.})}}{a_C^{\alpha(\text{cem.})}}$$

Since each activity is substantially equal to the corresponding weight per cent,

$$\Delta F_3^\circ = -RT \ln \frac{[\%C]^{\alpha(\text{gr.})}}{[\%C]^{\alpha(\text{cem.})}}$$

Hence it is apparent that, if we know ΔF_3° and the carbon content in equilibrium with cementite, the carbon content in equilibrium with graphite can be obtained. ΔF_3° can be found most readily from the tabular values of $(F^\circ - H_0^\circ)/T$ for the substances involved. The table for iron has already been given, Table 16-1; that for the other two substances is presented as Table 16-2.[1] From these tables and the value 5114 cal for $\Delta H_{0(3)}^\circ$, ΔF_3° and $[\%C]^{\alpha(\text{gr.})}/[\%C]^{\alpha(\text{cem.})}$ are found as shown in Table 16-3, and from this ratio of solubilities and the solubility of cementite from Fig. 16-2, the solubility of graphite in α-iron is calculated as shown in the last column of Table 16-3.

The enthalpy of solution of graphite in α-iron (ΔH_2) is given by the relation

$$\Delta H_2 = \Delta H_1 + \Delta H_3$$

ΔH_1 as found from Eq. (16-1*a*) is +9,700 cal and ΔH_3 as found from Table 16-2 is +5,100 cal at the eutectoid temperature. Hence ΔH_2, thus

[1] Attention should be called here to the fact that these tables were constructed not from thermal data alone; reliable equilibrium data were also considered. Thus Table 16-1 is "forced" to give $\Delta F^\circ = 0$ for the bcc-fcc transition of iron at 910 and 1400°C. The figures for graphite in Table 16-2 are from thermal data alone. The figures for cementite are forced to be in accord with the measured solubility of graphite and cementite in γ-iron near the eutectoids and with the difference in temperature of the cementite and graphite eutectics.

TABLE 16-2. ENTHALPY AND FREE-ENERGY FUNCTIONS FOR CEMENTITE AND GRAPHITE

°K	Cementite*		Graphite†	
	$H° - H°_0$, cal/gram mole	$-\frac{F° - H°_0}{T}$, cal/(gram mole) (°C)	$H° - H°_0$, cal/gram atom	$-\frac{F° - H°_0}{T}$, cal/(gram atom) (°C)
0	0	0	0	0
298.16	4,329	10.405	251.56	0.517
300	4,376	10.493	255.31	0.523
400	7,023	15.128	502.6	0.825
500	9,952	19.304	820.8	1.146
600	12,746	23.069	1198.1	1.477
700	15,452	26.410	1622.0	1.810
800	18,248	29.408	2081.7	2.138
900	21,073	32.131	2569.4	2.459
1000	23,951	34.627	3074.6	2.771
1100	26,939	36.934	3596.	3.073
1200	30,042	39.089	4130.	3.365
1300	33,165	41.113	4680.	3.647
1400	36,315	43.020	5242.	3.919
1500	39,508	44.824	5814.	4.181
1600	42,744	46.536	6400.	4.43
1700	45,970	48.165	6987.	4.68
1800	49,211	49.718	7578.	4.92
1900	52,499	51.204	8170.	5.15
2000	55,798	52.627	8780.	5.37

* From L. S. Darken and R. W. Gurry, *Trans. AIME*, **191**, 1015 (1951).
† From Rossini, Wagman, Evans, Levine, and Jaffe, *Natl. Bur. Standards, Circ.* 500 (1952).

TABLE 16-3. SOLUBILITY OF CEMENTITE IN α-IRON AND OTHER FUNCTIONS LEADING TO THE SOLUBILITY OF GRAPHITE IN α-IRON

°K	$\Delta F°_3$,* cal	Solubility of cem. in α-Fe, wt. % C	Ratio of gr. to cem. solubility in α-Fe	Solubility of gr. in α-Fe, wt. % C
298	4760	2×10^{-7}	0.00032	6×10^{-11}
400	4262	1.4×10^{-5}	0.0047	6×10^{-8}
500	3637	0.00015	0.026	4×10^{-6}
600	2915	0.00076	0.087	7×10^{-5}
700	2185	0.0024	0.207	0.0005
800	1478	0.0057	0.394	0.0022
900	827	0.0113	0.630	0.0071
1000	273	0.0195	0.870	0.0170

* For the reaction $3Fe(\alpha) + C(gr.) = Fe_3C(cem.)$.

determined, is +14,800 cal. The discrepancy with the aforementioned value (25,700 cal) given by Smith's data is large and will be discussed again later in this chapter.

THE SOLUBILITY OF GRAPHITE AND CEMENTITE IN γ-IRON

The equilibrium of γ-iron with cementite and with graphite has been discussed earlier in Figs. 13-12 and 13-13. Attention was called to the

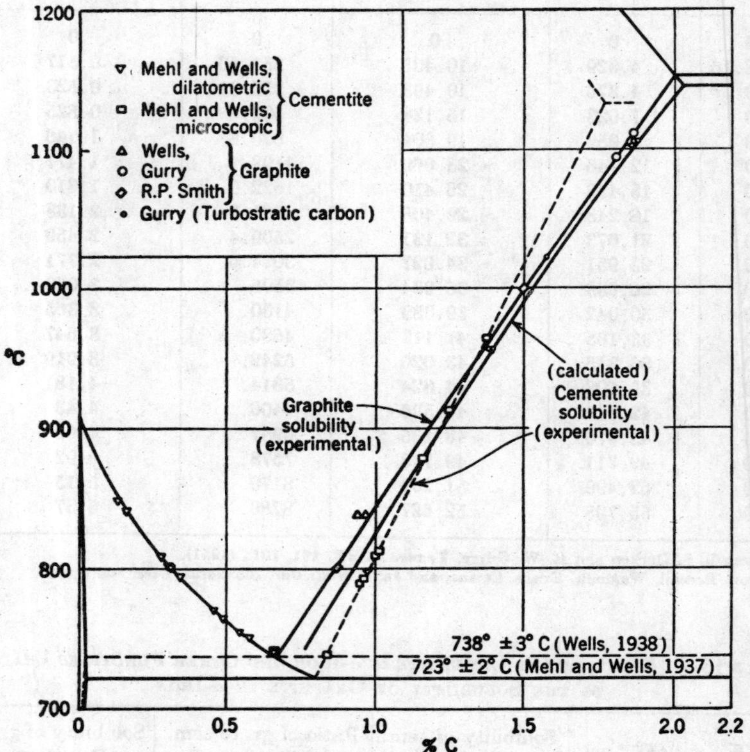

FIG. 16-3. The solubility of cementite and of graphite in γ-iron. [*From L. S. Darken and R. W. Gurry, Trans. AIME*, **191**, 1015 (1951).]

inconsistency which exists between the direct measurement of these two solubilities and the experimental finding that cementite is metastable over this whole temperature range. Pending the resolution of this apparent paradox, the present authors have chosen to accept the experimental data on graphite solubility and the observation as to the metastability of cementite. The curve for the solubility of cementite in austenite, determined on this basis, is incorporated in Fig. 16-3.

The Heat of Solution of Graphite and of Cementite in Austenite. The enthalpy change accompanying the reaction

$$C(\text{gr.}) = C(\text{in } \gamma) \qquad (a)$$

is given by the relation

$$\frac{\partial(\Delta F_a/T)}{\partial(1/T)} = \Delta H_a$$

the partial derivative indicating that *the composition of both phases is constant.* ΔF_a, the free-energy change accompanying the above nonequilibrium transfer of carbon, may be written

$$RT \ln a_C - RT \ln a_C^{(\text{gr.})} = RT \ln \frac{a_C^\gamma}{a_C^{(\text{gr.})}}$$

Since we may use any standard state for carbon, but necessarily the same for the two phases, $a_C^{(\text{gr.})}$ is equal to the activity of carbon in γ-iron in equilibrium with graphite, $a_C^{(\text{gr.})} = a_C^{\gamma(\text{gr.})}$. Denoting a_C^γ at any particular (fixed) composition N_C as $a_C^{\gamma(N_C)}$,

$$R \frac{d \ln [a_C^{\gamma(N_C)}/a_C^{\gamma(\text{gr.})}]}{d(1/T)} = \Delta H_a$$

the partial designation being omitted as the appropriate constancy requirements are indicated in the superscripts. It is convenient, as noted in the next section, to choose a nonphysical standard state such that a_C approaches N_C/N_{Fe} as both approach zero. Relative to this standard state, $a_C^{\gamma(N_C)}$ is, within the experimental error, independent of temperature; hence

$$-R \frac{d \ln a_C^{\gamma(\text{gr.})}}{d(1/T)} = \Delta H_a$$

From the data of Smith[1] on a_C^γ as a function of composition and the aforementioned data on graphite solubility, $a_C^{\gamma(\text{gr.})}$ may be evaluated as a function of temperature. A plot of log $a_C^{\gamma(\text{gr.})}$ against $1/T$ is substantially linear and has a slope which, when multiplied by 4.575, gives

$$\Delta H_a = 10{,}100 \text{ cal}$$

for the solution of graphite in austenite. It will be noted that this value is found independent of the composition of the austenite, since $a_C^{\gamma(N_C)}$ was taken independent of temperature and thus disappeared on differentiation, leaving no composition term in the above equation.

[1] R. P. Smith, *J. Am. Chem. Soc.*, **68**, 1163 (1946).

Reaction (*a*) may be combined with

$$Fe_3C(\text{cem.}) = 3Fe(\text{pure } \gamma) + C(\text{gr.}) \qquad (b)$$

to give the equation for the solution of cementite in austenite:

$$Fe_3C(\text{cem.}) = 3Fe(\text{pure } \gamma) + C(\text{in } \gamma) \qquad (c)$$

For reaction (*b*) the enthalpy change ΔH_b is -2590 cal (at 1000°C) from Tables 16-1 and 16-2; hence ΔH_c is about 7500 cal at 1000°C. ΔH_c, like ΔH_a, is found independent of the composition of austenite, and hence the designation "pure" may be omitted in reaction (*c*) as far as ΔH_c is concerned.

THE ACTIVITY OF CARBON AND OF IRON IN AUSTENITE

Before proceeding to the discussion of the equilibrium between γ and liquid iron, it is necessary to consider the activity coefficient of carbon and of iron in these phases.

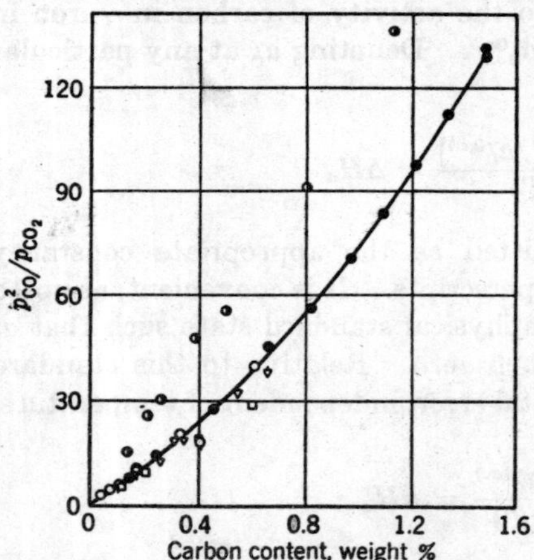

FIG. 16-4. Carbon content of austenite at 1000°C in relation to p_{CO}^2/p_{CO_2}: ● Smith, ○ Dünwald and Wagner, ◑ Bramley and Lord, ♀ Becker, □ Johansson and Von Seth, ▽ Takahashi. [*From R. P. Smith, J. Am. Chem. Soc.*, **68**, 1163 (1946).]

The Activity of Carbon in Austenite. Investigations were made by Smith[1] at 800, 1000, and 1200°C. Two methods were used, one involving equilibrium with mixtures of CO and CO_2 and the other with mixtures of CH_4 and H_2. Although there is a slight discrepancy in the results by the two methods, it is found that the results of both methods at all temperatures can be represented with adequate precision by the relation

$$\ln a_C^\gamma = \ln \frac{N_C^\gamma}{N_{Fe}^\gamma} + 6.6\,\frac{N_C^\gamma}{N_{Fe}^\gamma} \qquad (16\text{-}4)$$

the standard state being so chosen that the activity approaches the atom fraction and the atom ratio at infinite dilution.

The nature of the experimental data is illustrated in Fig. 16-4, which shows the relation between the composition of the gas and that of the iron-carbon alloy equilibrated therewith. The reaction involved here is

$$2CO = CO_2 + \underline{C}$$

[1] R. P. Smith, *J. Am. Chem. Soc.*, **68**, 1163 (1946).

for which the equilibrium constant K is $P_{CO_2}a_C^\gamma/p_{CO}^2$. It is thus apparent that, if the activity of carbon were proportional to the weight per cent, the curve of Fig. 16-4 would be a straight line. The curvature indicates a marked departure from ideality in this solution. The validity of the type of relation [Eq. (16-4)] chosen to represent the data is illustrated in Fig. 16-5; it will be noted that the ordinate is equal to

$$\log \frac{a_C^\gamma N_{Fe}^\gamma}{N_C^\gamma} - \log K$$

the first term being substantially the logarithm of the activity coefficient of carbon.

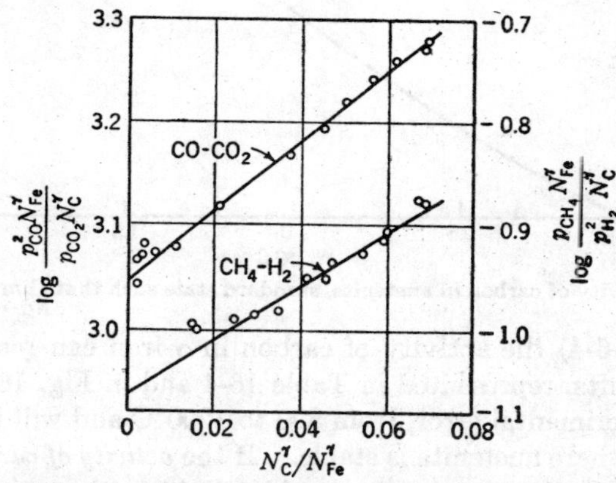

Fig. 16-5. Replot of the data of Fig. 16-4 to show the linear relation of Eq. (16-4); data on the CH_4-H_2 equilibrium are also included. *[From R. P. Smith, J. Am. Chem. Soc.*, **68, 1163 (1946).]**

Table 16-4. Activity of Carbon in Austenite

% C	a_C^*	% C	a_C^*
0.05	0.00236		
0.1	0.00479	1.1	0.0728
0.2	0.00992	1.2	0.0822
0.3	0.01537	1.3	0.0917
0.4	0.0211	1.4	0.1022
0.5	0.0273	1.5	0.1130
0.6	0.0338	1.6	0.1245
0.7	0.0407	1.7	0.1369
0.8	0.0480	1,8	0.1495
0.9	0.0559	1.9	0.1632
1.0	0.0640	2.0	0.1774

* Standard state such that $\lim_{N_C \to 0} a_C/N_C = 1$.

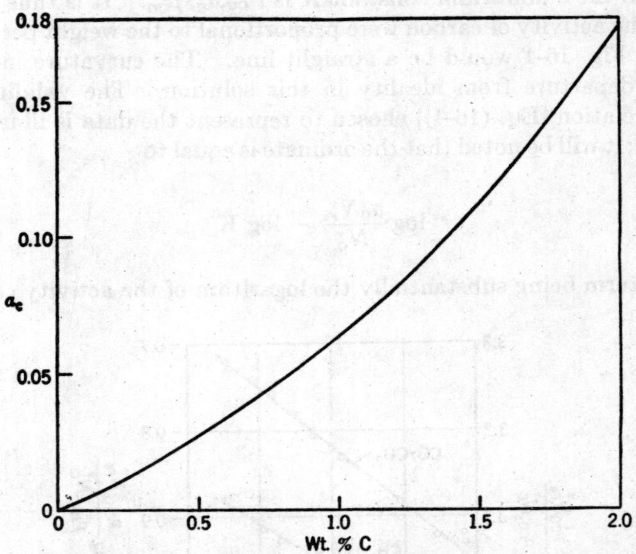

FIG. 16-6. Activity of carbon in austenite, standard state such that $\lim_{N_C \to 0} a_C/N_C = 1$.

From Eq. (16-4) the activity of carbon in γ-iron can readily be computed; the results, represented in Table 16-4 and in Fig. 16-6, are valid, within the experimental error, from 800 to 1200°C and will be used at all temperatures where austenite is stable. If the *activity of carbon relative to graphite* is desired, this may be obtained by dividing the activity of carbon at the carbon content under consideration, shown in the figure, by the activity of carbon in austenite saturated with graphite at the temperature. This latter is read from the figure at the solubility of graphite given in Fig. 16-3.

The Activity of Iron in Austenite. This may be obtained by combining Eq. (16-4) with the Gibbs-Duhem equation. The resulting relation is

$$\ln a_{Fe}^{\gamma} = -\frac{N_C^{\gamma}}{N_{Fe}^{\gamma}} - 3.3\left(\frac{N_C^{\gamma}}{N_{Fe}^{\gamma}}\right)^2 \tag{16-5}$$

It will be noted that here as well as in Eq. (16-4) the atom ratio rather than the atom fraction occurs as the independent variable. According to the derivation given by Darken and Smith[1] this is to be expected for a constituent of an interstitial solid solution.

THE ACTIVITY OF IRON IN LIQUID IRON-CARBON ALLOYS

In the absence of better knowledge as to the relationship between the activity of carbon and the composition of liquid iron-carbon alloys, it is

[1] L. S. Darken and R. P. Smith, *J. Am. Chem. Soc.*, **68**, 1172 (1946).

tentatively assumed that from zero carbon content up to saturation with graphite the activity coefficient of carbon obeys a relation of the form

$$\log \gamma_C^l = \alpha (N_{Fe}^l)^2 \qquad (16\text{-}6)$$

A similar relation, of course, will hold for iron:

$$\log \gamma_{Fe}^l = \alpha (N_C^l)^2 \qquad (16\text{-}6a)$$

where α is assumed to be independent of composition but may be a function of temperature. To determine the variation of α with temperature an evaluation will be made at the following points: the graphite eutectic, 1153°C; the point of δ-γ-liquid equilibrium; the vicinity of 1600°C; and from vapor pressure measurements at 2650°C.

The Graphite Eutectic. From Eq. (16-5) and the solubility of graphite (2.01 per cent C), the activity of iron in γ at the iron-graphite eutectic (1153°C) is 0.882. Table 16-1 supplies the free energy of fusion of γ-iron: $\Delta F°/T = 0.530$. From these the activity of iron in liquid iron saturated with graphite is 0.675. The composition of liquid iron at the graphite eutectic is given as 4.24 per cent C by Ruer and Biren.[1] Converting to the atom fraction of carbon, 0.171, it is found that the activity coefficient of iron is $0.675/(1 - 0.171) = 0.814$, whence α, the coefficient sought, is, by Eq. (16-6a), $\log 0.814/0.171^2 = -3.07$.

The δ-γ-liquid Equilibrium. From the data of Adcock[2] the composition of γ and of liquid iron in equilibrium is 0.16 and 0.52 per cent C, respectively, at 1492°C. $\Delta F°/T$ for the reaction $Fe(\gamma) = Fe(l)$ is again taken from Table 16-1. The activity of iron in γ, taken from Eq. (16-5), is found to be 0.992_3. Combination gives the activity of iron in the liquid as 0.972_5, whence the activity coefficient is 0.996_3, and it is found by Eq. (16-6a) that at this temperature $\alpha = -2.8$.

The Vicinity of 1600°C. From data on the equilibrium of CO-CO_2 mixtures with graphite it is found that at 1580°C the equilibrium constant $(p_{CO}^2/p_{CO_2})(1/a_C) = 19{,}290$, the standard state for carbon being graphite. Vacher[3] found that at this temperature liquid iron containing 0.019 per cent C is in equilibrium with a gas in which $p_{CO_2}, = 0.113$ and

$$p_{CO} = 0.887 \text{ atm.}$$

Combination of these data gives 0.409 as the activity coefficient of carbon, $\gamma_C = a_C/N_C$, where a_C is relative to graphite. At this same temperature it is found that the composition of iron saturated with

[1] R. Ruer and J. Biren, *Z. anorg. allgem. Chem.*, **113**, 98 (1920).

[2] Frank Adcock, *J. Iron Steel Inst.*, **135**, 281 (1937).

[3] H. C. Vacher, *Natl. Bur. Standards, J. Research*, **11**, 541 (1933).

graphite is 5.31 per cent C ($N_C = 0.207$). Subtraction of two equations of the type $\log \gamma_C = \alpha N_{Fe}^2$ gives the relation

$$\log \gamma_C - \log \gamma'_C = \alpha(N_{Fe}^2 - N_{Fe}'^2) \tag{16-7}$$

which applies to any choice of standard state, as the activity coefficients appear only as a ratio. The unprimed and primed quantities are now applied to the data of Vacher and to those of Ruer and Biren, respectively. By inserting the above data in this expression α is found to be -2.9. The

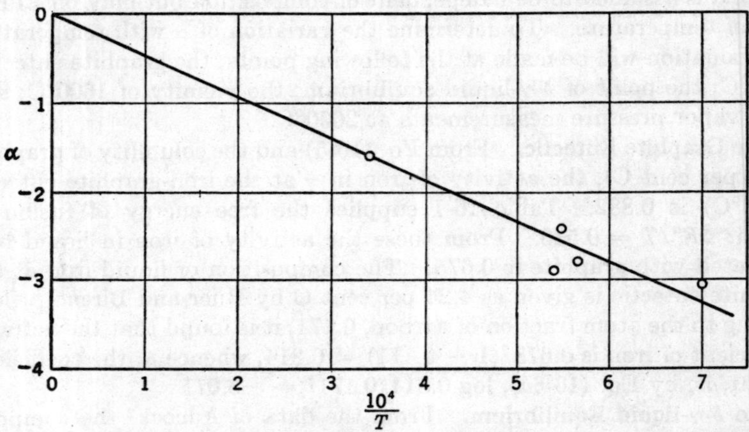

Fig. 16-7. Variation of α [Eqs. (16-6) and (16-6a)] with $1/T$ for liquid iron-carbon solution.

data of Marshall and Chipman[1] give -2.6 for α at 1600°C and -2.5 at 1700°C.

Boiling-point Measurements at 2650°C.[2] From the composition of the vapor over the iron-carbon solution at its boiling point (at reduced pressure), it is estimated that $\alpha = -1.6(\pm 0.4)$.

Variation of α with Temperature. A plot of the above-determined values of α against $10^4/T$ is shown in Fig. 16-7. The points may be represented, within the experimental error, by a straight line through the origin, indicating that the solution approaches ideality as the temperature approaches infinity or that the free energy of mixing liquid iron and liquid carbon approaches the ideal free energy at very high temperature. The line of Fig. 16-7, in conjunction with Eqs. (16-6) and (16-6a), may be used to calculate the activity coefficients for liquid iron-carbon alloys.

It will be noted that the approximation has been made throughout this section that α is independent of concentration. This assumption,

[1] S. Marshall and J. Chipman, *Trans. ASM.*, **30**, 695 (1942).

[2] O. Ruff and W. Borman, *Z. anorg. allgem. Chem.*, **88**, 397 (1914).

although crude, seems to account adequately for the available experimental data, which themselves leave something to be desired.

Since, within the experimental error, the line of Fig. 16-7 passes through the origin, the activity coefficient of iron in liquid iron-carbon alloys may be represented at all temperatures from the eutectic to the boiling point by an equation of the type

$$\log \gamma_{\mathrm{Fe}}^{l} = \frac{A}{T} (N_{\mathrm{C}}^{l})^2$$

The constant A in this equation may be evaluated from any known value of α at a particular temperature, since it is apparent by comparison with Eq. (16-6a) that $\alpha = A/T$. Of the various determinations of α given above, the one at the graphite eutectic seems most reliable. Using this value it is found that $A = \alpha T = -3.3_1(1408) = -4660$,* and the above equation thus becomes

$$\log \gamma_{\mathrm{Fe}}^{l} = -\frac{4660}{T} (N_{\mathrm{C}}^{l})^2 \tag{16-8}$$

The excess partial molal free energy of iron, $\bar{F}_{\mathrm{Fe}}^{l,xs}$ (standard-state pure liquid iron) in molten iron-carbon alloy is

$$RT \ln \gamma_{\mathrm{Fe}}^{l} = 4.575T \log \gamma_{\mathrm{Fe}}^{l} = 4.575A(N_{\mathrm{C}}^{l})^2 = -21{,}300(N_{\mathrm{C}}^{l})^2 \text{ cal/gram atom of iron}$$

If the assumptions and data leading to Eq. (16-8) are correct, the excess entropy of iron in liquid iron-carbon alloy is zero by virtue of the relation $(\partial \bar{F}_{\mathrm{Fe}}^{l,xs}/\partial T)_{P,N} = -\bar{S}_{\mathrm{Fe}}^{l,xs}$. On this basis we are led to the conclusion that the entire departure from ideality may be attributed to the enthalpy rather than the entropy component of the free energy. This is more explicitly shown by determining $\bar{H}_{\mathrm{Fe}}^{l,xs}$, which equals

$$\left(\partial \frac{\bar{F}_{\mathrm{Fe}}^{l,xs}}{T} \middle/ \partial \frac{1}{T}\right)_{P,N}$$

$\bar{H}_{\mathrm{Fe}}^{l,xs}$ is thus found to be $-21{,}300(N_{\mathrm{C}}^{l})^2$. This evaluation hardly justifies great confidence in the resulting enthalpy and entropy of iron in molten iron-carbon alloys,† and a test must await the appearance of more reliable data.

The Activity of Carbon in Liquid Iron-Carbon Alloys. In view of the reciprocal relations between the activity coefficients of iron and carbon

* The error involved in this evaluation of A is perhaps as large as 10 per cent.

† Although the preceding evaluation of the free energy may be correct within 10 per cent, the evaluation of the enthalpy and entropy is achieved by taking a derivative, thus increasing the uncertainty in the latter quantities.

[Eqs. (16-6) and (16-6*a*)], it is apparent that the foregoing evaluation of the activity coefficient of iron in liquid iron-carbon alloys implies an evaluation of the activity coefficient of carbon. At any given temperature α in Eq. (16-6) is identical with α in Eq. (16-6*a*), which was evaluated in the preceding section. The evaluation of γ^l_{Fe}, however, brings up an important point in regard to the choice of a standard state. In the above discussion of γ^l_{Fe} it was only natural to choose the standard state as pure liquid iron, so that $\gamma^l_{Fe} = 1$ when $N^l_{Fe} = 1$. Moreover, the equation used [Eq. (16-6*a*)] is empirical in nature and can hardly be expected to be valid at a carbon content much greater than saturation with graphite, even if such a supersaturated solution could be prepared.

To use Eq. (16-6) directly would imply a rather odd choice of standard state, namely, one such that γ^l_C is unity when N^l_C is unity, providing that the supersaturated solution obeys this relation over this entire range of composition. This is a nonphysical standard state. A more convenient (again nonphysical) standard state is one such that $\gamma^l_C = 1$ at infinite dilution of carbon in iron, *i.e.*, at $N^l_C = 0$.

For an indeterminate choice of the standard state Eq. (16-6) should be rewritten with an arbitrary constant, thus:

$$\log \gamma^l_C = \alpha(N^l_{Fe})^2 + I \qquad (16\text{-}9)$$

From this it follows that the above choice of standard state, such that $\gamma^l_C = 1$ when $N^l_C = 0$ $(N^l_{Fe} = 1)$, $I = -\alpha$ and

$$\log \gamma^l_C = \alpha[(N^l_{Fe})^2 - 1] \qquad (16\text{-}10)$$

Using this expression, values of α from Fig. 16-7, and the familiar relation $a^l_C = N^l_C\gamma^l_C$, the activity of carbon in liquid iron-carbon alloys can be computed at any desired temperature.

For some purposes it is convenient to choose graphite as the standard state for carbon; in this case I of Eq. (16-9) becomes a function of temperature. I is then evaluated by noting that $a^l_C = N^l_C\gamma^l_C = 1$ for the liquid alloy saturated with graphite at the particular temperature under consideration. Designating functions pertaining to such a saturated solution by the additional superscript *sat.*, and noting that $\gamma^{l,\,sat.}_C = 1/N^{l,\,sat.}_{Fe}$, I may be evaluated from Eq. (16-9); thus, $I = -\log N^{l,\,sat.}_C - \alpha(N^{l,\,sat.}_{Fe})^2$ and

$$\log \gamma^l_C = -\log N^{l,\,sat.}_C + \alpha[(N^l_{Fe})^2 - (N^{l,\,sat.}_{Fe})^2] \qquad (16\text{-}11)$$

This equation is valid when graphite is the standard state for carbon.

THE δ-LIQUID EQUILIBRIUM

The stable range of this equilibrium is so short that we shall consider only the two extremities. The upper one is the melting point of pure iron; the lower corresponds to the temperature of the γ-δ-liquid equilibrium. As a check upon the compatability of the experimentally determined diagram with the theory developed in the preceding sections let us calculate the composition of the liquid at the latter temperature, 1492°C, from the composition of δ, 0.11 per cent C, and previously discussed data. We choose to deal with the distribution of iron between the δ and liquid phases:

$$\mathrm{Fe}(\delta) = \mathrm{Fe}(\mathrm{l})$$

As usual we may write $(d \log K)/[d(1/T)] = -\Delta H°/4.575$. Since the temperature range is short, $\Delta H°$ is substantially constant[1] and has the mean value 3650 cal from Table 16-1. Inserting the value for $\Delta H°$ in the above equation and integrating,

$$\log K = -\frac{798}{T} + I$$

The integration constant is evaluated from the fact that $\Delta F°$, and hence $\log K$, is zero at the melting point, 1812°K. Thus, I is 0.440 and

$$\log K = -\frac{798}{T} + 0.440 \qquad (16\text{-}12)$$

From this relation K, which equals $a^l_{\mathrm{Fe}}/a^\delta_{\mathrm{Fe}}$, is found to be 0.973_2 at 1492°C. Since the carbon content of the δ-iron is so low, no significant error will be introduced by assuming that the activity of iron in δ is equal to its atom fraction, in accord with Raoult's law. a^δ_{Fe} is thus 0.995_0. a^l_{Fe} is obtained by using the above value for K: $a^l_{\mathrm{Fe}} = 0.973_2\, a^\delta_{\mathrm{Fe}} = 0.968_3$. The atom fraction of iron may now be computed by successive approximations using Eq. (16-8), which becomes, for this temperature,

$$\log \gamma^l_{\mathrm{Fe}} = -2.64(N^l_{\mathrm{C}})^2$$

As a first approximation N^l_{C} is taken as $(1 - 0.972_3)$, whence $\gamma^l_{\mathrm{Fe}} = 0.993$ and $N^l_{\mathrm{Fe}} = 0.968_3/0.993 = 0.975$. This gives, as a second approximation for N^l_{C}, $(1 - 0.975)$ or 0.025. The final result of the iteration process is the value 0.0270 for N^l_{C}, corresponding to a carbon content of 0.59 per cent.

The discrepancy between this and Adcock's observed value of 0.52 per cent is possibly due in part to slight impurities. Adcock found the melting point of his iron of zero carbon content to be 1533°C, whereas we

[1] To verify the fact that $\Delta H°$ is constant within the limits of error of the data, we note that ΔC_P is about 1 cal/deg, that the temperature range is 47°, and hence that $\Delta H°$ will change by about 47 cal. The experimental precision in $\Delta H°$ is hardly better than this, so the use of a mean value of $\Delta H°$ and the omission of a ΔC_P term in the equation are justified.

have accepted the Bureau of Standards melting point of 1539°C. Tentatively assuming that this discrepancy is due to impurities in Adcock's iron, it is found from Eq. (16-12) (assuming the impurities to be soluble in the liquid but insoluble in the solid) that they lower the activity of iron in the liquid by 0.0023. Making this correction the calculated carbon content of the impure liquid becomes 0.54 per cent, in reasonably good agreement with the observed value 0.52 per cent

On the other hand, the discrepancy between these two values may lie in the temperature determination. In either case the temperature of the δ-γ-liquid equilibrium needs to be raised several degrees above that (1493°C) reported by Adcock to bring this portion of the iron-carbon diagram into agreement with the tabular values (Table 16-1) for the free energy of iron and the equations given for the activity coefficients. If this temperature is taken as 1499°C (*i.e.*, if Adcock's value is raised by the same amount that the accepted value for the melting point of pure iron exceeds his determination), then the carbon content of the liquid as calculated by the preceding method is 0.53 per cent C, which agrees well with that reported by Adcock, 0.52 per cent C. The selection of 1499°C as the peritectic temperature, which fits also the data on the composition of γ, seems to be the best that can be made at the present time.

THE EQUILIBRIUM BETWEEN AUSTENITE AND LIQUID

Of the several possible ways in which this heterogeneous equilibrium may be treated, perhaps the easiest is to consider the distribution of iron and carbon between the two phases. The equations are

$$\mathrm{Fe}(\gamma) = \mathrm{Fe}(\mathrm{l})$$
$$\mathrm{C}(\gamma) = \mathrm{C}(\mathrm{l})$$

for which we may write the equilibrium constants

$$K_{\mathrm{Fe}} = \frac{a^l_{\mathrm{Fe}}}{a^\gamma_{\mathrm{Fe}}} \tag{16-13}$$

$$K_{\mathrm{C}} = \frac{a^l_{\mathrm{C}}}{a^\gamma_{\mathrm{C}}} \tag{16-14}$$

It is convenient to choose the standard states for iron as pure γ and pure liquid iron and the standard state for carbon such that a_{C} approaches N_{C} (or $N_{\mathrm{C}}/N_{\mathrm{Fe}}$) as N_{C} approaches zero in both phases. K_{Fe} may be evaluated from the tabular functions for iron (Table 16-1) by use of the relation $4.575T \log K_{\mathrm{Fe}} = F^{\circ,\gamma} - F^{\circ,1}$ or its equivalent,

$$\log K_{\mathrm{Fe}} = \frac{1}{4.575}\left[\left(\frac{F^{\circ,\gamma} - H^\circ_0}{T}\right) - \left(\frac{F^{\circ,1} - H^\circ_0}{T}\right)\right].$$

K_C may be evaluated from the compositions and from the previously considered activity coefficients for carbon in γ and in liquid at any pair of isothermal points on the solidus and liquidus under consideration. K_C may conveniently be evaluated at the two most widely separated temperatures, the peritectic,[1] 1499°C, and the graphite eutectic, 1153°C. Assuming a linear relation, $\log K_C = A + B/T$, K_C may be evaluated therefrom at intermediate temperatures.

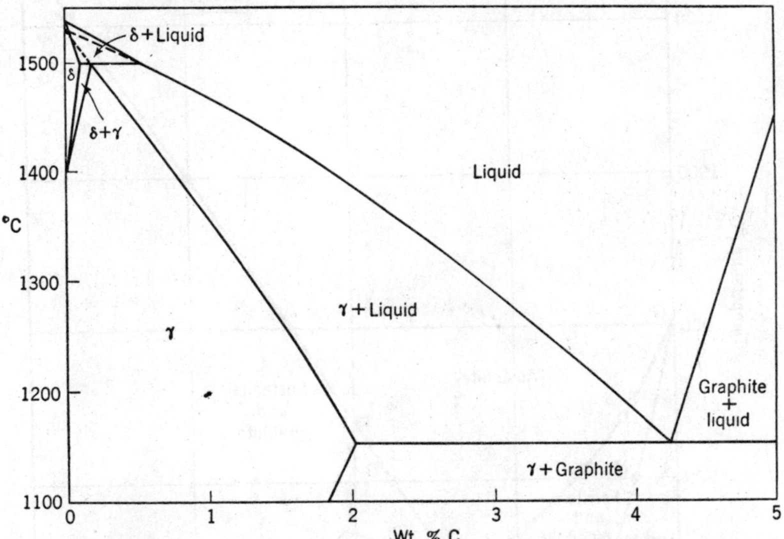

Fig. 16-8. Temperature-composition diagram for the iron-carbon system; γ-liquid equilibrium and δ region.

The general procedure is then to find compositions of γ and liquid which simultaneously satisfy Eqs. (16-13) and (16-14) at a given temperature, using the expressions previously developed between composition and activity for each component in each phase. The procedure is somewhat tedious but may be performed graphically by plotting (1) the locus of points (%C^l vs. %C^γ) which satisfy Eq. (16-13) and (2) the locus of points which satisfy Eq. (16-14). The intersection of these two curves gives the composition of the solidus and of the liquidus at the chosen

[1] The composition of γ-iron at the peritectic may be computed in the following manner. By interpolation in Table 16-1 it is found that $\Delta F^{(\delta-\gamma)}/T = 0.0058$, whence $\log a^\gamma_{Fe}/a^\delta_{Fe} = -0.0012_7$. Taking $a^\delta_{Fe} = N^\delta_{Fe} = 0.995_0$ (from Adcock's composition of δ considered in the preceding section) it is found that $a^\gamma_{Fe} = 0.992_1$. Using Eq. (16-5), it is found that $N^\gamma_C/N^\gamma_{Fe} = 0.0078_6$, corresponding to 0.16_9 per cent C in γ-iron at the peritectic. This agrees well with Adcock's experimental determination of 0.16 per cent C.

temperature. The results are subject to the uncertainty of the above-assumed relation between K_C and T and particularly to the uncertainty of the activities for the liquid [Eqs. (16-8) and (16-9)].[1] Hence the computation can hardly be regarded as more than a refined interpolation

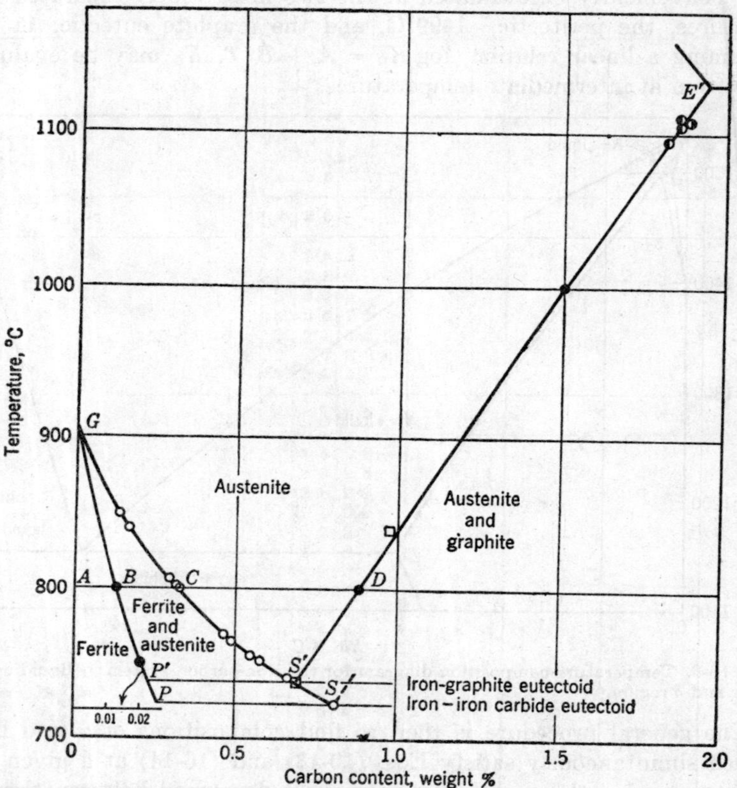

FIG. 16-9. Temperature-composition plot of the α-γ equilibrium of the iron-carbon system and the solubility of graphite in austenite: ● Smith; ○ Mehl and Wells; □ Wells; ◐ Gurry, equilibrium with graphite; ◑ Gurry, equilibrium with carbon from toluene. Note the expanded scale for the line *GP*. [*From R. P. Smith, J. Am. Chem. Soc.*, **68,** 1163 (1946).]

between the two terminal temperatures. Reproduction of the details does not seem warranted. The results are incorporated in Fig. 16-8. In this figure the austenite solidus and liquidus have been extrapolated (dashed curves) to the melting point of pure γ iron, 1532°C.[2] The equi-

[1] In fact, since the numerical coefficient of these equations does not exactly satisfy the data for the eutectic (see Fig. 16-7), an adjustment must be made in their use.

[2] This temperature is obtained from Table 16-1. It is the temperature at which $\Delta F^{\circ(\gamma-l)}$ is equal to zero.

librium between pure γ and pure liquid iron at this temperature is, of course, metastable; accordingly, this melting point lies below that of pure δ iron which coexists in stable equilibrium with the liquid at 1539°C.[1]

THE α-γ OR γ-δ EQUILIBRIUM

Of the many experimental investigations of the composition of austenite in equilibrium with ferrite in the range of temperature from the eutectoid to the α-γ transition of pure iron, that of Mehl and Wells[2] appears the most reliable. The corresponding equilibrium composition of α-iron has been investigated by Smith,[3] by Dijkstra,[4] and by Stanley;[5] all three are in relatively good agreement. This portion of the iron-carbon diagram is shown in Fig. 16-9.

The Heat of Transfer of Iron from $\alpha(\delta)$ to γ. Darken and Smith,[6] using the data of this diagram in conjunction with the previously mentioned activity data of Smith, determined the equilibrium constant for the reaction

$$Fe(\alpha,\delta) = Fe(\gamma)$$

From this constant the heat of transformation of pure iron is found as a function of temperature by use of the familiar relation

$$\Delta H^\circ = -4.575 \left[\frac{d \log K}{d(1/T)}\right]$$

The results are shown in Table 16-5. The value at 910°C is in excellent agreement with the calorimetric determination of Awberry and Griffiths,[7] 215 cal/gram atom, this being the value incorporated in Table 16-1.

TABLE 16-5. HEAT OF TRANSFORMATION OF PURE IRON FROM α TO γ

°C	ΔH°, cal/gram atom	
	From Smith	From Table 16-1
910	215	215
850	330	335
800	495	485
750	825	770
723	1000	950

[1] W. F. Roeser and H. T. Wensel, *Natl. Bur. Standards, J. Research*, **26**, 273 (1941).
[2] R. F. Mehl and Cyril Wells, *Trans. AIME*, **125**, 429 (1937).
[3] R. P. Smith, *J. Am. Chem. Soc.*, **68**, 1163 (1946).
[4] L. J. Dijkstra, *Trans. AIME*, **185**, 252 (1949).
[5] J. K. Stanley, *Trans. AIME*, **185**, 752 (1949).
[6] L. S. Darken and R. P. Smith, *J. Am. Chem. Soc.*, **68**, 1172 (1946).
[7] J. H. Awberry and E. Griffiths, *Proc. Roy. Soc. (London)*, **174**, 1 (1940).

The reader will observe that the heat of this reaction is appreciably greater in the vicinity of the eutectoid than at 910°C. It is this rather high heat which gives rise to the recalescence phenomenon. This phenomenon may be illustrated by an electrically heated steel wire: When the current is shut off, the incandescence first fades out and then reappears for a brief period. The effect is occasioned by the relatively high heat of the transformation at low temperature; the low temperature of trans-

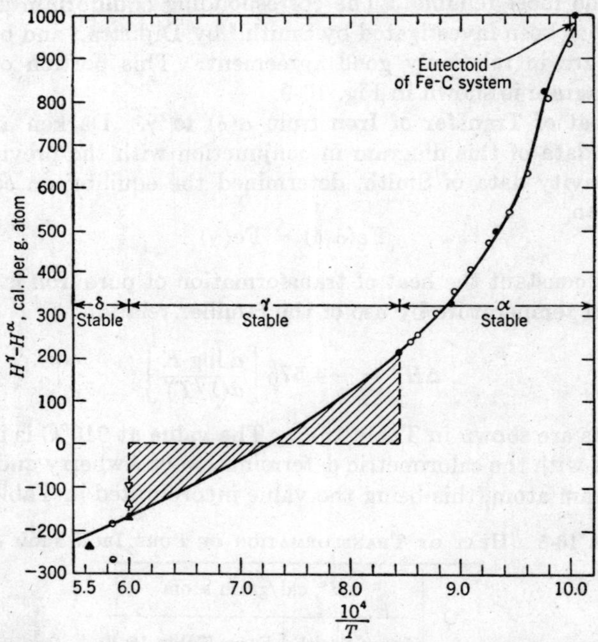

FIG. 16-10. Enthalpy change for the transformation of iron from bcc to fcc; shaded areas constructed equal, since the two forms may be in equilibrium at either of the terminal temperatures. ○ Smith; ● Smith, recalculated; ▽ Austin; △ Oberhoffer and Grosse; ▲ from enthalpy data of Umino and Zuithoff; ◓ or ◒ estimated from equilibrium diagram of Adcock or Epstein. [*From L. S. Darken and R. P. Smith, Ind. Eng. Chem.*, **43,** 1815 (1951).]

formation, in turn, is associated with high carbon content and rapid initial cooling rate. The wire should be of small diameter, perhaps B. & S. gage 20.

It is of interest to note the method used to determine ΔH° of Table 16-1 at temperatures above 910°C. The fundamental principle involved is that ΔF° is zero both at 910 and at 1400°C. By integration of the relation $[d(\Delta F^\circ/T)]/[d(1/T)] = \Delta H^\circ$ between these two temperatures it is found that $\int_{1183}^{1673} \Delta H^\circ \, d(1/T) = \Delta F^\circ/T \Big]_{1183}^{1673} = 0$. Thus the curve of

Fig. 16-10 must be constructed in such a manner that the two shaded areas are equal, so that they cancel in evaluating the integral. Under the assumption that the curve is smooth, this condition alone determines fairly well its course over the entire region. The construction is aided somewhat by the experimental data on the heat of the γ-δ transition, although these are rather poor, and by the available heat capacities in the vicinity of the transformation temperatures which give the slope of the curve at these temperatures by means of the relation

$$\frac{d\,\Delta H^\circ}{d(1/T)} = -T^2\,\Delta C_P$$

Heat of Transfer of Carbon from $\alpha(\delta)$ to γ. For the distribution of carbon between α or δ and γ phase we may write the chemical equation

$$C(\alpha,\delta) = C(\gamma)$$

The corresponding equilibrium constant is

$$K = \frac{a_C^\gamma}{a_C^\alpha}$$

where the standard state for carbon is taken such that the activity approaches the atom fraction or atom ratio at infinite dilution. It will be noted that, since K involves only a ratio of activities, it would have the same numerical value if the standard state were chosen so that the activity approaches the weight per cent at infinite dilution. K may be evaluated from the data determining the temperature-composition diagram and from Smith's data on the activity coefficient of carbon in austenite, both previously discussed. In view of the low concentration of carbon in α- and δ-iron and the lack of experimental evidence of the departure of the activity coefficient from unity, a_C is taken equal to N_C/N_{Fe} in both α- and δ-iron. The logarithm of K, thus evaluated, is plotted against $10^4/T$ in Fig. 16-11. It will be observed that the single point (lower left corner) for δ-iron, which is calculated from the data of Adcock at the temperature of the γ-δ-liquid equilibrium, lies on the same straight line with the points for the α-γ equilibrium (upper right corner).

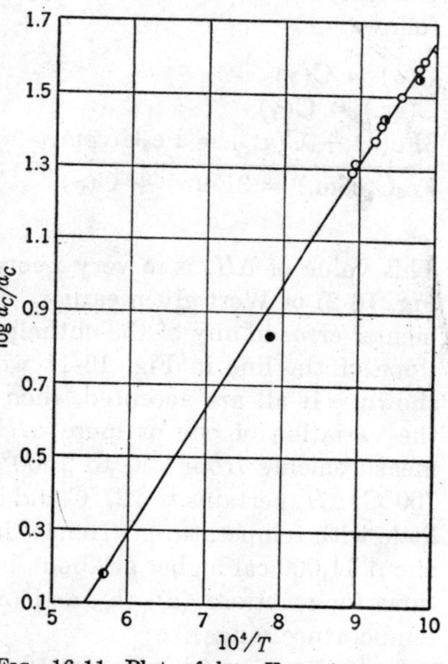

FIG. 16-11. Plot of log K against $10^4/T$ for the equilibrium of carbon between α- and γ-iron: ◑ Smith; ○ Mehl and Wells, per Smith; ◐ Adcock; ● Smith, extrapolation of data for silicon alloys. [*From R. P. Smith, J. Am. Chem. Soc.*, **70**, 2724 (1948).]

The lone central point was calculated by extrapolation of data on iron-carbon-silicon alloys at the indicated temperature. The enthalpy change accompanying the transfer of 1 gram atom of carbon from α or δ to γ-iron at low concentration may be evaluated as -4.575 times the slope of the line of Fig. 16-11 and is thus found to be $-15{,}600$ cal/gram atom.

This datum may be combined with previously obtained enthalpy changes to obtain the enthalpy of solution of cementite in α-iron as follows:

$$C(\alpha) = C(\gamma) \qquad \Delta H_a = -15{,}600^{*} \qquad (a)$$

$$C(\text{gr.}) = C(\gamma) \qquad \Delta H_b = +10{,}100^{\dagger} \qquad (b)$$

$$3Fe(\alpha) + C(\text{gr.}) = Fe_3C(\text{cem.}) \qquad \Delta H_c = 5100^{\ddagger} \qquad (c)$$

$$Fe_3C(\text{cem.}) = 3Fe(\alpha) + C(\alpha) \qquad \Delta H_d = \Delta H_b - \Delta H_a - \Delta H_c = +20{,}600 \qquad (d)$$

This value of ΔH_d is in very poor agreement with that (+9700 cal from Fig. 16-2) of Wert given earlier. It hardly seems likely that the experimental error in any of the enthalpy determinations is this large; *e.g.*, the slope of the line in Fig. 16-11 would have to be one-third as great as shown. If all are accepted, then the discrepancy would be blamed on the variation of one or more ΔH's with temperature; ΔH_a is based on measurements from 750 to 1500°C; ΔH_b, 740 to 1100°C; ΔH_d, 150 to 700°C; ΔH_c pertains to 727°C and is known (from Table 16-2) to vary but little with temperature. Hence there remains the possibility that ΔH_d is about 11,000 cal higher at about 1000°C [the rough mean of the temperatures for reactions (*a*), (*b*), and (*c*)] than at 400°C (the mean of Wert's temperature range).

In conjunction with Tables 16-1 and 16-2 it is found that this implies that the mean partial molal heat capacity of carbon dissolved in α-iron is about 16 cal/deg. Although this is very high as compared with the Dulong and Petit value of 6 cal/deg, it cannot be discounted as impossible. The acceptance of this possibility would mean that the curve of Fig. 16-2 should not be linear but concave upward—a tendency which the data themselves indicate slightly. A satisfactory solution must await further investigation.

THE SOLUBILITY OF GRAPHITE IN LIQUID IRON

The most reliable measurements of the solubility of graphite in liquid iron appear to be those of Ruer and Biren,§ given in Table 16-6. There

* From Fig. 16-11.

† From solubility data.

‡ From Table 16-2.

§ R. Ruer and J. Biren, *Z. anorg. allgem. Chem.*, **113**, 98 (1920).

being no calorimetric data available on the heat of solution of graphite in liquid iron, an attempt will now be made to determine this from solubility data.

TABLE 16-6. SOLUBILITY OF GRAPHITE IN LIQUID IRON

°C	Wt. % C	°C	Wt. % C
1135	4.20	2000	6.76
1150	4.24	2100	7.31
1200	4.36	2200	7.77
1300	4.62	2300	8.30
1400	4.88	2400	8.89
1500	5.15	2500	9.52
1600	5.41	2600	11.51
1700	5.66		
1800	5.96		
1900	6.33		

Let us consider the nonequilibrium dissolution of graphite in a liquid iron-carbon alloy of some fixed carbon content.

$$\text{C(graph.)} = \underline{\text{C}}$$

From the definitional expression for activity, the free-energy change of this reaction is seen to be $RT \ln a_2/a_{2(sat.)}$, where a_2 is the activity of carbon at the fixed carbon content and $a_{2(sat.)}$ is the activity at saturation with graphite. The standard state, necessarily the same in both cases, is chosen such that the activity approaches the atom fraction at infinite dilution. In terms of activity coefficients the above expression for the free-energy change may be written

$$\Delta F = RT \ln \frac{\gamma_2 N_2}{\gamma_{2(sat.)} N_{2(sat.)}}$$

Expressing the activity coefficient as before, $\log \gamma_2 = \alpha N_1^2 + I$,

$$\Delta F = RT \left\{ \ln \frac{N_2}{N_{2(sat.)}} + 2.303\alpha[N_1^2 - N_{1(sat.)}^2] \right\}$$

By virtue of the relation $d(\Delta F/T)/d(1/T) = \Delta H$, it is found that[1]

$$\Delta H = 4.575 \frac{d}{d(1/T)} \{ -\log N_{2(sat.)} + \alpha[N_1^2 - N_{1(sat.)}^2] \}$$

[1] Note that N_2 is, in this case, an arbitrary fixed composition and hence a constant which thus disappears on differentiation. $N_{2(sat.)}$ varies with temperature.

If the particular solvent in which we choose to determine the heat of solution of graphite is pure liquid iron, this equation becomes [noting that $N_1 = 1$ and $N_{1(sat.)} = 1 - N_{2(sat.)}$]

$$\Delta H = 4.575 \frac{d}{d(1/T)} \{-\log N_{2(sat.)} + \alpha N_{2(sat.)}[2 - N_{2(sat.)}]\}$$

The expression in braces, evaluated from Table 16-6 and the previously developed relation $\alpha = -4660/T$ [see Eq. (16-8)], is plotted against $1/T$ in Fig. 16-12. The curve appears to be substantially straight only in the

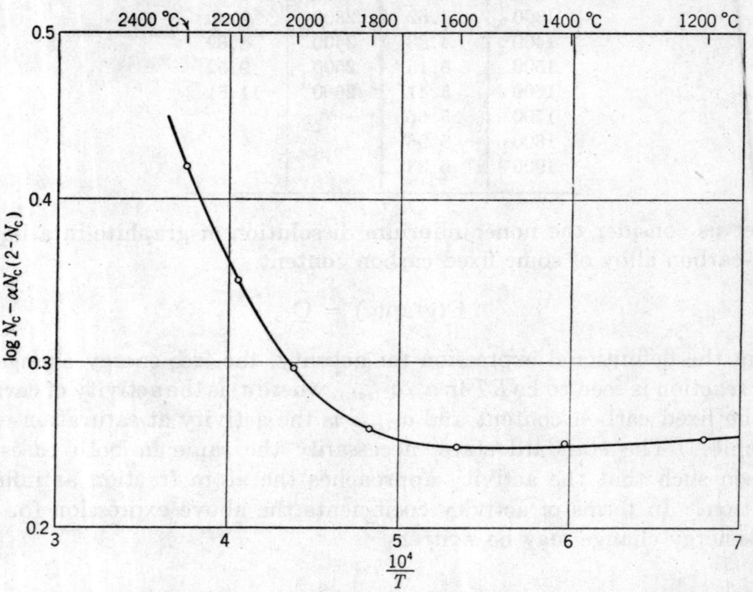

FIG. 16-12. Plot for evaluating the heat of solution of graphite in liquid iron.

region of lower temperature; here its slope is negligibly small, corresponding to $\Delta H = 0$ for the enthalpy change on dissolving graphite in a large quantity of pure liquid iron. Similar treatment gives +6500 cal for the enthalpy of solution of graphite in the liquid of eutectic composition, and in general $\Delta H = 21{,}000(1 - N_1^2)$.

OTHER CARBIDES OF IRON

In addition to the familiar orthorhombic cementite (Fe_3C), other carbides of iron have been reported in the literature. Of these the best known is the Hägg carbide,[1] approximately Fe_2C. This is encountered in

[1] G. Hägg, *Z. Krist.*, **89**, 94 (1934).

iron catalysts for the Fisher-Tropsch synthesis and may be prepared[1] by treating reduced iron oxide, preferably with a promoter, with a carburizing gas such as butane at 250 to 300°C. The Hägg carbide decomposes[2] to cementite and free carbon above 450°C; it has been reported[3] to react with iron to produce cementite at even lower temperature (300°C).

The equilibrium

$$2Fe(\alpha) + CH_4 = Fe_2C(\text{Hägg}) + 2H_2$$

has been investigated by Browning, DeWitt, and Emmett.[4] The equilibrium constant $p_{H_2}^2/p_{CH_4}$ is lower than that for the corresponding cementite equilibrium by a factor of about 1.5 in the range 300 to 360°C. Hence, in this range, Hägg carbide is metastable relative to cementite and α-iron and the solubility of Hägg carbide in α-iron is about 1.5 times that of cementite. The temperature coefficient of the free-energy change for the Hägg carbide-cementite reaction is uncertain, but since it is known that cementite is more stable at higher temperature, as noted above, it is not unreasonable to suppose that Hägg carbide becomes stable relative to cementite (but not relative to graphite) at some temperature under 300 and probably under 200°C.

A hexagonal carbide of composition approximately Fe_2C has also been reported.[5] This was prepared by carburizing reduced iron oxide (containing 10 per cent Cu and $\frac{1}{3}$ per cent K_2O) with CO for several weeks at 170°C. It is reported to transform spontaneously to Hägg carbide at 420°C. Eckstrom and Adcock[6] have reported a carbide composition FeC.

On the basis of X-ray observations Jack[7] has reported the occurrence of the above-mentioned hexagonal carbide, designated by him as ϵ-iron carbide, during the first stage of tempering of martensite. Roberts, Averbach, and Cohen[8] confirm this and conclude from dilatometric measurements that the composition of the ϵ carbide corresponds approximately to $Fe_{2.4}C$. They further report that the composition of the martensite drops to 0.25 per cent C upon precipitation of the ϵ carbide.

[1] L. C. Browning, T. W. DeWitt, and P. H. Emmett, *J. Am. Chem. Soc.*, **72**, 4211 (1950).

[2] E. C. Cohn and L. J. E. Hofer, *J. Am. Chem. Soc.*, **72**, 4662 (1950).

[3] L. J. E. Hofer and E. M. Cohn, *J. Chem. Phys.*, **18**, 766 (1950).

[4] L. C. Browning, T. W. DeWitt, and P. H. Emmett, *J. Am. Chem. Soc.*, **72**, 4211 (1950).

[5] E. C. Cohn and L. J. E. Hofer, *J. Am. Chem. Soc.*, **72**, 4662 (1950). L. J. E. Hofer, E. M. Cohn, and W. C. Peebles, *ibid.*, **71**, 189 (1949).

[6] H. C. Eckstrom and W. A. Adcock, *J. Am. Chem. Soc.*, **72**, 1042 (1950).

[7] K. H. Jack, *J. Iron Steel Inst.*, **169**, 26 (1951).

[8] C. S. Roberts, B. L. Averbach, and M. Cohen, ASM preprint No. 11 (1952).

CHAPTER 17

ELECTROCHEMISTRY; REVERSIBLE CELLS

A reaction is usually classified as electrochemical if it is associated with a flow of electric current over a finite distance (greater than interatomic). The flow of electric current involves the motion of charged particles—ions, electrons, or both. Thus the electroplating of copper is regarded as an electrochemical phenomenon: Measurable electric current flows through a finite circuit and a chemical reaction occurs. On the other hand, the neutralization of an alkaline solution by an acid is not regarded as electrochemical, even though the reaction may be expressed in terms of ions, since the reaction between H^+ and OH^- ions does not involve the flow of current over a distance greater than that between these two particles.

The corrosion of most metals is electrochemical, since it involves a flow of current over a distance which is definitely finite, although short (perhaps of the order of magnitude of the grain size); this current is measurable under appropriate circumstances. The fact that substantially all corrosion phenomena are regarded as electrochemical leads one to infer correctly that a large proportion of all chemical reactions is electrochemical in nature. Even the passage of current through a metal at high temperature has an electrochemical aspect. Although in this case the major part of the current is carried electronically, a small fraction, perhaps a few parts per million, is carried by metallic ions.

This chapter will be limited to a small portion of the electrochemical field, namely, that concerned with reversible cells. These serve as an important means of determining the free energy of electrochemical reactions and the activities of components in solution.

REVERSIBLE CELLS

It will be recalled from Chap. 7 that one of the combined statements of the first and second laws, for a reversible process, is

$$dF = V\,dP - S\,dT - \delta w'$$

At constant pressure and temperature this reduces to

$$dF = -\delta w'$$

where $\delta w'$ refers to all work done by the system other than that against pressure. The work we now wish to consider is the electrical work which may be accomplished during the transport of an electrical charge from one potential to another. The above equation integrates to give, at constant temperature and pressure,

$$\Delta F = -w' \tag{17-1}$$

From elementary electrical theory, the work done in transporting an electrical charge from one potential to another is equal to the product of the charge and the potential difference.

A chemical system which gives rise to, or which is capable of producing, an electric current accompanying a chemical reaction is called a galvanic cell or simply a *cell*. For any cell the over-all chemical reaction may be represented by an equation called the *cell reaction*. The amount of charge transferred during an electrochemical reaction is equal to the product of the faraday $\mathfrak{F}$ (the charge of 1 gram electron, 96,494 abcoulombs) and the number $\mathfrak{N}$ of gram electrons transferred through the external circuit when the reaction occurs to the extent of the number of gram moles represented by the equation. The external work done by any cell operating at potential $\mathcal{E}$ for a period of time such that $\mathfrak{N}$ gram electrons of current flow is thus $\mathfrak{N}\mathfrak{F}\mathcal{E}$, the product of the charge and the potential. If the cell is reversible, this work is related to the ΔF of the cell reaction by Eq. (17-1) and we may write

$$\Delta F = -\mathfrak{N}\mathfrak{F}\mathcal{E} \tag{17-2}$$

Criterion of a Reversible Cell. The employment of the electrochemical method to determine the free-energy change of a reaction requires the experimental realization of a cell in which this reaction occurs electrochemically. Although, as pointed out previously, natural and artificial electrochemical reactions are common, reversible reactions suitable for cell measurement are indeed rare. It is found by experience that, in general, one can by no means simply construct a cell of the phases involved in a certain reaction and obtain the free-energy change of this reaction by measuring the potential and applying Eq. 17-2.

The fundamental reason for this ever-present difficulty in working with reversible cells is that a cell with a finite potential, considered as a system, is not in a state of complete equilibrium. If it were, the potential would be zero. In attempting to construct a reversible cell the aim is to prevent completely all nonelectrochemical reactions and all electrochemical reactions other than the one under consideration. In addition, the reaction under consideration must proceed only when current flows in the external circuit. When reversibility is achieved, it is found that there is a certain value of the externally applied potential, supplied, for example, by a potentiometer connected to the cell, such that, whenever the applied potential exceeds this value, current flows in one direction and, when it is less current flows in the reverse direction. If this condition is fulfilled over a period of time and in spite of a small flow of current in either or both directions, the reversible potential is known as the *electromotive force* of the cell (commonly abbreviated emf). The symbol $\mathcal{E}$ is used for this emf. Although in the previous section this symbol was used rather loosely, it will henceforth be applied only to a reversible emf. It is apparent that a cell coupled with an external potential which exactly balances it constitutes a system which is in partial equilibrium.

Conventions. Let us consider a simple cell composed of two solutions of copper sulfate, one more concentrated than the other (kept separated by gravity or by a porous cup), and two electrodes of pure copper, one inserted in each of the two solutions. This cell is conventionally represented in the following manner:

$$\mathrm{Cu(fcc)} \mid \mathrm{CuSO_4(aq.\ soln.,\ conc.\ 1)} \mid \mathrm{CuSO_4(aq.\ soln.,\ conc.\ 2)} \mid \mathrm{Cu(fcc)}$$

Sometimes commas or semicolons are used instead of the bars to designate the phase boundaries within the cell. Such a cell which has identical electrodes inserted in solutions differing only in concentration is called a *concentration cell.* When the cell reaction is allowed to proceed in the spontaneous direction, *i.e.*, the direction delivering external power, the cell reaction consists principally of the dissolution of copper from the electrode in the weaker solution and the plating of copper on the electrode in the stronger solution. Thus the over-all cell reaction is equivalent to the transfer of copper sulfate from the stronger to the weaker solution.

The following convention regarding the *sign of the potential* of a cell is now almost universally accepted: If the electrode on the left of the cell as written is negative, the emf is assigned a positive value. Thus the emf is taken positive if positive current tends to flow spontaneously from left to right through the cell as written. The *cell reaction* is written so that it corresponds to the reaction which occurs when positive current flows from left to right through the cell as written. These two conventions together are seen to be in accord with the relation $\Delta F = -\mathfrak{N}\mathfrak{F}\mathcal{E}$ [Eq. (17-2)].

If, in reference to the cell described above, the copper sulfate concentration 1 is less than concentration 2, then the left electrode tends more strongly to go into solution. The left electrode thereby tends to become negative, and positive current tends to flow spontaneously from left to right within the cell. Accordingly the emf will be positive in this case. If concentration 1 is greater than concentration 2, the emf will be negative. For the cell as written above, the cell reaction, regardless of the relative strengths of the two solutions, will be

$$\mathrm{CuSO_4(aq.\ soln.,\ conc.\ 2)} = \mathrm{CuSO_4(aq.\ soln.,\ conc.\ 1)}$$

so that, if concentration 1 is less than concentration 2, ΔF, according to the expression $\Delta F = -\mathfrak{N}\mathfrak{F}\mathcal{E}$, will be negative, and if concentration 1 is greater than concentration 2, ΔF will be positive. These signs correspond to a free-energy decrease accompanying the dilution of copper sulfate solution—a process which obviously tends to occur spontaneously. For the cell reaction above the number of gram electrons $\mathfrak{N}$ involved at each electrode is 2. If the cell reaction is written with a number of gram moles of copper sulfate other than unity, both $\mathfrak{N}$ and ΔF will be changed proportionately.

LIQUID JUNCTION

Let us extend the discussion of the concentration cell of the previous section for which the electrode reactions consist of the reversible solution and plating out of copper. In addition to the reaction at the electrodes there is a direct diffusion from the stronger to the weaker solution in the vicinity of the boundary between the two. This cannot be considered as a reversible reaction; diffusion is by nature an irreversible process. The ions moving under the influence of the concentration gradient in this vicinity give rise to a potential known as a *liquid-junction potential.* Since electrodes of some sort are indispensable in a cell, it is apparent that the liquid-junction potential cannot be measured independent of the electrode potentials. It is thus seen that the free energy of transfer of copper sulfate from one solution to the other cannot be calculated directly from Eq. (17-2) without knowledge of the liquid-junction potential, since the measured potential includes the liquid-junction potential. A great deal of work has been done on the measurement and calculation of liquid-junction potentials. A detailed discussion is beyond the scope of this book.

Frequently in experimental work the liquid-junction potential is a nuisance, and our interest in it is focused on how to make a correction therefor in order to calculate the potential corresponding to the electrode reactions. Although the liquid-junction potential is never zero (except in special cases), it is often small compared with the measured potential of the cell as a whole. At times, for lack of knowledge as to its correct value, it is neglected.

In a cell involving aqueous solutions the liquid-junction potential is frequently minimized by inserting a *salt bridge* between the two solutions. This consists of a tube filled with strong potassium chloride solution, this electrolyte being chosen because the mobilities of its ions are practically identical.

Daniell-type Cells. The Daniell cell consists of a copper electrode in an aqueous copper sulfate solution and above this, separated by gravity, a less dense zinc sulfate solution in which is suspended a zinc electrode.

The cell reaction involves the dissolution of metallic zinc and the formation of metallic copper.

$$Zn(hcp) + CuSO_4(aq.\ soln.) = ZnSO_4(aq.\ soln.) + Cu(fcc)$$

The liquid junction potential in this case is only a small part of the measured potential.

In a similar manner, at elevated temperature, the free energy of the reaction

$$K(l) + NaCl(l) = KCl(l) + Na(l)$$

may be determined approximately by measuring the potential of the cell

$$Na(l) \mid NaCl(l) \mid KCl(l) \mid K(l)$$

adequate precaution being taken to prevent mixing of the electrolytes. Ignoring the potential difference at the junction of the two electrolytes, the measured potential corresponds to the free energy of the above reaction. However, this type of cell, sometimes known as the Daniell type, leaves much to be desired as a means of accurately determining ΔF by potential measurement because of the uncertainty of the junction potential. It is used less frequently now than in the past.

CELLS WITHOUT LIQUID JUNCTION

Let us next consider a different type of cell in which liquid junction is not purposely introduced. Such a cell, used by Salstrom and Hildebrand,[1] employs a molten lithium bromide–silver bromide solution as the electrolyte, pure silver as one electrode, and bromine gas bubbling over graphite (which is inert) as the other.

$$Ag(fcc) \mid AgBr(\text{in } LiBr) \mid Br_2(g,\ 1\ atm,\ gr.)$$

The over-all cell reaction is

$$Ag(fcc) + \tfrac{1}{2}Br_2(g, 1\ atm) = AgBr(\text{fused salt soln.})$$

With suitable cell construction they found the cell to be reversible and Eq. (17-2) to be directly applicable.

Although no liquid junction was mentioned in the above description, a moment's reflection discloses that the solution in the immediate vicinity of the bromine electrode is saturated with bromine and the solution in the immediate vicinity of the silver electrode is saturated with silver. Thus the electrolyte departs slightly from the stoichiometric ratio of silver to bromine, and there is in principle a liquid junction even here. However, the liquid junction potential in this case is small and introduces no appreciable experimental error. Another difficulty arising from the same source is that, if the electrodes are not adequately separated, bromine

[1] E. J. Salstrom and J. H. Hildebrand, *J. Am. Chem. Soc.*, **52**, 4650 (1930).

will be carried by diffusion and convection to the silver electrode, thus permitting direct and nonelectrochemical reaction. This difficulty was overcome by suitable cell design involving the usual expedient of a capillary between the cathode and anode[1] portions of the cell.

It is thus seen that, although the type of cell just described is commonly termed "without liquid junction," a liquid junction actually exists,[2] and the cell must be designed to minimize the associated irreversible effects. In a sufficiently long time diffusion will ultimately lead to some irreversibility in any event. Judicious choice of electrodes and electrolyte and good cell design can extend the period of substantial reversibility to a very long time—even to many years in the case of the Weston standard cell. On the other hand, an unfortunate choice of electrodes or electrolyte or poor cell design may result in a cell whose voltage either changes so rapidly as to be virtually unmeasurable or levels off at a value which is not the true emf corresponding to any definite cell reaction.

The foregoing reaction involving the formation of silver bromide might have been investigated by use of a cell with liquid junction, of the type

$$\text{Ag} \mid \text{AgBr} \mid \text{AgBr(in LiBr)} \mid \text{Ag}$$

The liquid junction involved, however, would considerably affect the accuracy of the computed results. A comparison of the emf of such a cell with the corresponding emf determined by Salstrom and Hildebrand would furnish an accurate determination of the liquid-junction potential involved.

Nearly all the difficulties which have been mentioned in connection with the construction of a reversible cell are accentuated at high temperature. The reason for this becomes evident when we recall that a cell involves a partial equilibrium and depends upon the suppression of one or more reactions. Obviously such a suppression becomes more difficult at high temperature, and greater ingenuity is required in the construction of a cell whose emf is related to the free energy of a definite reaction by Eq. (17-2).

ELECTRODE REACTIONS

As mentioned previously, for every reversible cell there is a definite cell reaction. In the above case this was written

$$\text{Ag(fcc)} + \tfrac{1}{2}\text{Br}_2\text{(g,1 atm)} = \text{AgBr(soln.)}$$

[1] The *cathode* is that electrode toward which the positive ions (cations) migrate, and the *anode* is that electrode toward which the negative ions (anions) migrate. It will be noticed that the cathode is negative in an electrolytic cell (one which is consuming power) but positive in a galvanic or voltaic cell (one delivering power).

[2] Indeed, as may be inferred from this discussion, there is no such thing as a cell without liquid junction, or more generally, to include solid solutions also, there is no cell without concentration gradients.

From the fundamental definition of an electrochemical reaction it follows that this reaction, or any other, cannot occur electrochemically in a single location. In the ideal reversible cell the reaction occurs in two parts, one at each electrode. Thus it is common to write two *single-electrode reactions,* the sum of which is the over-all cell reaction. The two electrode reactions for this case are

At the silver electrode, $Ag(fcc) = Ag^+(soln.) + e$
At the bromine electrode, $\frac{1}{2}Br_2(g, 1\text{ atm}) + e = Br^-(soln.)$

where e designates an electron. The sum of these is the over-all cell reaction given above.[1] The symbol $\mathfrak{N}$ in Eq. (17-2) refers to the number of gram electrons involved in each electrode reaction. Obviously, the same number of gram electrons is involved in both electrode reactions, since they must cancel on addition. The same amount of electric charge must enter and leave the solution by the two electrodes.

Single-electrode Potentials. Although it is patently impossible to construct a cell with a single electrode, the term "single-electrode potential" is common, particularly in reference to aqueous solutions. To give meaning to this concept it is necessary to choose arbitrarily a particular standard electrode which is assigned a zero potential and against which all others are compared. When dealing with aqueous solutions the electrode almost universally chosen as the standard is the standard hydrogen electrode, a hydrogen electrode with the gas at 1 atm in contact with a solution of unit hydrogen ion activity. The metallic electrode whose standard potential is to be evaluated by comparison with this consists of the pure metal surrounded by a solution of the corresponding metal ion at unit activity.

Thus, for example, the standard electrode potential of cadmium may be measured with an aqueous cell of the type

$$Cd \mid CdCl_2(a = 1) \mid HCl(a = 1) \mid H_2(1\text{ atm}, Pt)$$

the symbol Pt meaning that the hydrogen is bubbled over platinum which is inert. The *standard electrode potential* of cadmium is equal to the potential of this cell when both activities are unity. It will be observed that a liquid-junction potential is involved here. Due correction is made for this, or more commonly, the potential of the corresponding cell with-

[1] Actually, addition of these electrode reactions gives silver and bromine ions; these constitute an alternative representation of the silver bromide in the cell reaction, since the silver bromide is substantially completely ionized. Even if it were not completely ionized but existed partly in molecular form, no difficulty would be introduced as long as the ionized and unionized portions were in equilibrium. Ionic reactions are well known to be rapid, and this equilibrium is commonly established.

out liquid junction is determined by combining the measurements on two or more separate cells. The potential of the above cell, for example, may be obtained by addition of the potentials of the cells[1]

$$\mathrm{Cd} \mid \mathrm{CdCl_2}(a = 1) \mid \mathrm{AgCl,Ag}$$

and

$$\mathrm{Ag,AgCl} \mid \mathrm{HCl}(a = 1) \mid \mathrm{H_2}(1\ \mathrm{atm,Pt})$$

Actually, in any case, measurement is made on a series of cells to permit extrapolation to the hypothetical state of unit activity.

The Electrochemical Series. The standard electrode potentials corresponding to several electrode reactions for aqueous solutions at 25°C are collated in Table 17-1.[2] This constitutes the well-known electromotive or electrochemical series. Although it is commonly said that a metal in this series is capable of reducing from solution any metal below it in the series, this statement is strictly true only under the stated conditions, and caution is to be observed in the application of the series. The potentials given apply only to the standard states of the metals and to aqueous solutions of their ions at unit activity and at 25°C. The formation of a complex ion in the electrolyte may alter considerably the potential of a metal; *e.g.*, tin is indicated as electropositive with respect to iron (its use in tin cans hinges on this fact), but under certain conditions involving complex ion formation it may even be electronegative. The general rough similarity between the electrochemical series and Pauling's electronegativity scale was pointed out in Chap. 4.

Even greater caution must be exercised in the application of this series

[1] It is sometimes convenient to use an electrode consisting of two or more phases each of which enters into the electrode reaction. Such electrodes are designated as those of the second or third, etc., kind, depending upon the number of phases entering the electrode reaction. In this nomenclature the simpler electrodes already considered would be classed as those of the first kind. The silver–silver chloride electrode frequently used in electrochemical investigations of aqueous chloride solutions is composed of metallic silver and silver chloride which is deposited thereon by electrolytic means. The electrode reaction is

$$\mathrm{Ag} + \mathrm{Cl^-} = \mathrm{AgCl} + e$$

and the electrode is seen to be of the second kind. The mercury-calomel and mercury–mercuric oxide electrodes are others of this kind. These are commonly designated simply in terms of the slightly soluble salt involved, *e.g.*, the silver chloride or calomel electrode. Such electrodes have not been used in high-temperature work on account of the obvious difficulty of finding a salt or oxide of low solubility in a fused electrolyte. Electrodes of two metallic phases, however, may prove useful in the investigation of heterogeneous equilibria in metallic systems.

[2] From Eastman and Rollefson, "Physical Chemistry," McGraw-Hill Book Company, Inc., New York, 1947.

TABLE 17-1. STANDARD POTENTIALS FOR VARIOUS ELECTRODE REACTIONS AT 25°C*

Reaction	Potential
$Cs = Cs^+ + e$	3.02
$Li = Li^+ + e$	3.02
$Rb = Rb^+ + e$	2.99
$K = K^+ + e$	2.922
$Na = Na^+ + e$	2.712
$\frac{1}{3}Al + \frac{4}{3}OH^- = \frac{1}{3}H_2AlO_3 + \frac{1}{3}H_2O + e$	2.35
$\frac{1}{2}Mg = \frac{1}{2}Mg^{++} + e$	2.34
$\frac{1}{3}Al = \frac{1}{3}Al^{3+} + e$	1.67
$\frac{1}{2}H_2 + OH^- = H_2O + e$	0.828
$\frac{1}{2}Zn = \frac{1}{2}Zn^{++} + e$	0.762
$\frac{1}{2}Fe = \frac{1}{2}Fe^{++} + e$	0.440
$\frac{1}{2}Pb + \frac{1}{2}SO_4^{--} = \frac{1}{2}PbSO_4 + e$	0.355
$\frac{1}{2}Pb + Cl^- = \frac{1}{2}PbCl_2 + e$	0.268
$\frac{1}{2}Sn = \frac{1}{2}Sn^{++} + e$	0.136
$\frac{1}{2}Pb = \frac{1}{2}Pb^{++} + e$	0.126
$\frac{1}{2}HO_2^- + \frac{1}{2}OH^- = \frac{1}{2}O_2 + \frac{1}{2}H_2O + e$	0.076
$\frac{1}{2}H_2 = H^+ + e$	0.000
$\frac{1}{2}Sn^{++} = \frac{1}{2}Sn^{4+} + e$	−0.15
$\frac{1}{2}H_2SO_3 + \frac{1}{2}H_2O = \frac{1}{2}SO_4^{--} + 2H^+ + e$	−0.20
$Ag + Cl^- = AgCl + e$	−0.2222
$Hg + Cl^- = \frac{1}{2}Hg_2Cl_2 + e$	−0.2676
$\frac{1}{2}Cu = \frac{1}{2}Cu^{++} + e$	−0.3448
$OH^- = \frac{1}{4}O_2 + \frac{1}{2}H_2O + e$	−0.401
$\frac{1}{2}I^- + OH^- = \frac{1}{2}IO^- + \frac{1}{2}H_2O + e$	−0.49
$\frac{3}{2}I^- = \frac{1}{2}I_3^- + e$	−0.5355
$Hg + \frac{1}{2}SO_4^{--} = \frac{1}{2}Hg_2SO_4 + e$	−0.6151
$\frac{1}{2}H_2O_2 = \frac{1}{2}O_2 + H^+ + e$	−0.682
$Fe^{++} = Fe^{3+} + e$	−0.771
$Hg = \frac{1}{2}Hg_2^{++} + e$	−0.7986
$Ag = Ag^+ + e$	−0.7995
$\frac{3}{2}OH^- = \frac{1}{2}HO_2^- + \frac{1}{2}H_2O + e$	−0.87
$\frac{1}{2}Hg_2^{++} = Hg^{++} + e$	−0.910
$\frac{1}{3}NO + \frac{2}{3}H_2O = \frac{1}{3}NO_3^- + H^+ + e$	−0.96
$NO + H_2O = HNO_2 + H^+ + e$	−0.99
$\frac{1}{2}H_2O = \frac{1}{4}O_2 + H^+ + e$	−1.229
$\frac{1}{2}Mn^{++} + H_2O = \frac{1}{2}MnO_2 + 2H^+ + e$	−1.28
$Cl^- = \frac{1}{2}Cl_2 + e$	−1.3583
$\frac{1}{3}Cr^{3+} + H_2O = \frac{1}{6}Cr_2O_7^{--} + \frac{7}{3}H^+ + e$	−1.36
$\frac{1}{2}Cl^- + \frac{1}{2}H_2O = \frac{1}{2}HClO + \frac{1}{2}H^+ + e$	−1.49
$\frac{1}{5}Mn^{++} + \frac{4}{5}H_2O = \frac{1}{5}MnO_4^- + \frac{8}{5}H^+ + e$	−1.52
$H_2O = \frac{1}{2}H_2O_2 + H^+ + e$	−1.77
$Co^{++} = Co^{3+} + e$	−1.84

* Latimer, "Oxidation Potentials," 1938.

to nonaqueous solutions such as fused salts or slags. Although one might expect that potassium would substantially displace sodium from solution, it is found that the equilibrium between sodium-potassium alloys and fused salts is markedly affected by the nature of the anion. The equilibrium

$\underline{K}$(fused alloy) + NaX(fused salt) = $\underline{Na}$(fused alloy) + KX(fused salt)

where X represents a halogen anion, may be shifted to the right or left by appropriate choice of anion.

ELECTROMOTIVE FORCE AND CELL REACTION

For a reversible cell it was shown [Eq. (17-2)] that the free-energy change of the cell reaction is related to the emf of the cell by the expression $\Delta F = -\mathfrak{N}\mathfrak{F}\mathcal{E}$. It should be stressed, perhaps, that the utilization of the ΔF so obtained demands an exact knowledge of the cell reaction; from lack of such knowledge many errors and false conclusions have resulted from emf measurements.

General Expression for the Potential of a Cell. In order to transform the above relation into one relating activities to the emf, we must first write the cell reaction. In its most general form the cell reaction taking place when $\mathfrak{N}$ faradays are passed may be represented

$$l\text{L} + m\text{M} + \cdots = q\text{Q} + r\text{R} + \cdots$$

as in Eq. (9-14), and by transposition of Eq. (9-15) it is found for the nonequilibrium reaction that

$$\Delta F = \Delta F^\circ + RT \ln \frac{a_\text{Q}^q a_\text{R}^r \cdots}{a_\text{L}^l a_\text{M}^m \cdots}$$

It will be observed that the activity quotient has the same form as an equilibrium constant but that in the present case it is not an equilibrium constant, as the activities involved are not those pertinent to equilibrium (except in the special case that the emf of the cell and hence ΔF are equal to zero). Combining this expression with Eq. (17-2),

$$\mathcal{E} = -\frac{\Delta F^\circ}{\mathfrak{N}\mathfrak{F}} - \frac{RT}{\mathfrak{N}\mathfrak{F}} \ln \frac{a_\text{Q}^q a_\text{R}^r \cdots}{a_\text{L}^l a_\text{M}^m \cdots}$$

It is also customary to define $\mathcal{E}^\circ$ as the emf of the corresponding cell in which all the reactants and products are in their standard states. Thus $\Delta F^\circ = -\mathfrak{N}\mathfrak{F}\mathcal{E}^\circ$, and the above equation becomes

$$\mathcal{E} = \mathcal{E}^\circ - \frac{RT}{\mathfrak{N}\mathfrak{F}} \ln \frac{a_\text{Q}^q a_\text{R}^r \cdots}{a_\text{L}^l a_\text{M}^m \cdots} \tag{17-3}$$

This equation forms the basis of all computations of activities from emf measurements. For $\mathcal{E}$ in absolute volts and T in degrees Kelvin the evaluation of the constants and conversion to common logarithms gives

$$\mathcal{E} = \mathcal{E}^\circ - \frac{T}{5040\mathfrak{N}} \log \frac{a_\text{Q}^q a_\text{R}^r \cdots}{a_\text{L}^l a_\text{M}^m \cdots} \tag{17-4}$$

At 25°C, the equation becomes

$$\mathcal{E} = \mathcal{E}^\circ - \frac{0.05915}{\mathfrak{N}} \log \frac{a_Q^q a_R^r \cdots}{a_L^l a_M^m \cdots} \tag{17-5}$$

From Eq. (17-4) it is seen in general that the number of different activity products obtainable by potential measurements is limited only by the number of chemical reactions which by the skill of the experimenter can be made to take place electrochemically and reversibly. An electrochemical reaction is one in which at least one of the reactants or products is separated from the others by a finite distance (a distance large compared with interatomic distances) as the reaction proceeds. In a reversible cell designed for potential measurement this distance is commonly that between the electrodes.

From any of the foregoing equations it is seen that, as the concentration, and hence the activity, of any product or reactant approaches zero, the potential $\mathcal{E}$ of the cell increases without limit. An infinite potential is, of course, never actually observed. A potential observed at sufficiently low concentration is not a true emf but reflects irreversible processes, *i.e.*, polarization. If the measured potential of a cell is to correspond to the written cell reaction, each reactant and product must be present in finite quantity. Also, at low concentration the drawing of a small current, with the accompanying composition change, produces a much larger fluctuation in the potential than it would at higher concentration.

Concentration Cells. Any cell whose cell reaction consists simply of a concentration change is called a concentration cell. One of the simplest types may be illustrated by that used in determining the activities in binary metallic systems as mentioned in Chap. 10. In such a cell, *e.g.*, that used in the investigation of the molten cadmium-lead system, one electrode is pure cadmium, the other a cadmium-lead alloy, and the electrolyte a fused alkali chloride solution containing a small percentage of cadmium chloride. The cell may be written

$$\text{Cd(pure l)} \mid CdCl_2\text{(in alkali chlorides)} \mid \text{Cd(in l Cd-Pb)}$$

The chloride mixture provides a low-melting electrolyte that does not react appreciably with either electrode; *i.e.*, there is no measurable side reaction of the type $2NaCl + Cd = CdCl_2 + 2Na$. Another reason for using a low cadmium-chloride concentration is to prevent side reactions of the type $CdCl_2 + Pb = PbCl_2 + Cd$.

The cell reaction of this cell is

$$\text{Cd(pure l)} = \text{Cd(in l Cd-Pb)}$$

At each electrode a conversion of metallic cadmium to cadmium ion Cd^{++} or the reverse takes place and hence $\mathfrak{N}$ is 2 for this case. If the activity of pure liquid cadmium is taken as unity, Eq. (17-4) becomes

$$\mathcal{E} = \mathcal{E}^\circ - \frac{T}{5040 \cdot 2} \log a_{\text{Cd}}$$

where a_{Cd} is the activity of cadmium in the alloy. In the special case in which the alloy is also pure cadmium, it is evident that the last term above is zero, since $a_{\text{Cd}} = 1$; $\mathcal{E}$ is zero, since the electrodes are identical; and hence $\mathcal{E}^\circ$ is evaluated as zero. The relation between the measured potential and the activity of cadmium thus becomes

$$\mathcal{E} = -\frac{T}{10{,}080} \log a_{\text{Cd}}$$

The activities in the cadmium-lead system discussed in Chap. 10 were obtained by use of this relation.

Concentration Cells with Liquid Junction. Another type of concentration cell involves a liquid junction between two electrolytes, usually called simply a cell with liquid junction. An example of this is the cell[1]

$$O_2(\text{Pt}) \mid Na_2O,B_2O_3(\text{l II}) \mid Na_2O,B_2O_3(\text{l I}) \mid O_2(\text{Pt})$$

The reaction at the right electrode is $\frac{1}{2}O_2 + 2e = O^{--}(\text{I})$ and at the left electrode $O^{--}(\text{II}) = \frac{1}{2}O_2 + 2e$. The over-all cell reaction consists not only of these two electrode reactions but also of the electrolytic transfer across the junction between the two liquids. If the oxygen ion is much more mobile than any other ion in these solutions, so that O^{--} may be regarded as the only constituent that moves as current flows, then the *transport number* of O^{--}, which is defined as the fraction of the total current carried thereby, is unity and the reaction occurring at the liquid junction is $O^{--}(\text{I}) = O^{--}(\text{II})$. The sum of these three reactions, which is the over-all cell reaction, is in this case zero; O_2 gas is merely consumed at one electrode and produced at the other. Hence ΔF and $\mathcal{E}$ would also be zero if the reactions are properly thus represented.

On the other hand if the transport number of sodium ion is unity, the reaction at the liquid junction may be written $2Na^+ (\text{II}) = 2Na^+ (\text{I})$. Adding this to the two electrode reactions the over-all cell reaction is now $2Na^+ (\text{II}) + O^{--}(\text{II}) = 2Na^+ (\text{I}) + O^{--}(\text{I})$ or $Na_2O\ (\text{II}) = Na_2O\ (\text{I})$. In this case the potential would be written

$$\mathcal{E} = -\frac{RT}{2\mathfrak{F}} \ln \frac{a^{\text{I}}_{Na_2O}}{a^{\text{II}}_{Na_2O}}$$

[1] W. Stegmaier and A. Dietzel, *Glastech. Ber.*, **18**, 353 (1940).

Other possibilities such as transfer by borate ion also exist. It is evident that the actual cell reaction is none of these limiting cases but is some intermediate one and that the potential is a function of the transport numbers of the various ions. In multicomponent systems the liquid-junction potential depends on the nature of the boundary. It is apparent that the interpretation of the potential of cells with liquid junction is difficult. This subject will not be pursued further here; the reader, if interested, should consult a text on electrochemistry.[1]

Concentration Cells by Combination. Any cell may be paired with another similar cell so that the over-all reaction of the combination consists only of a concentration change. The pairing may be accomplished experimentally or may consist of the subtraction of the two individual potentials. Thus the cell considered previously,

$$\mathrm{Ag(fcc)} \mid \mathrm{AgBr(conc.\ I\ in\ LiBr)} \mid \mathrm{Br_2(1\ atm,\ gr.)} \qquad (A)$$

may be paired with another in which the fused salt composition is different, namely,

$$\mathrm{Ag(fcc)} \mid \mathrm{AgBr(conc.\ II\ in\ LiBr)} \mid \mathrm{Br_2(1\ atm,\ gr.)} \qquad (B)$$

The combination may be represented

$$\mathrm{Ag(fcc)} \mid \mathrm{AgBr(conc.\ I\ in\ LiBr)} \mid \mathrm{Br_2(1\ atm,\ gr.)} \mid \mathrm{AgBr(conc.\ II\ in\ LiBr)} \mid \mathrm{Ag(fcc)} \qquad (C)$$

It will be seen that by this means it is possible to avoid the difficulties of the liquid junction involved in the cell

$$\mathrm{Ag(fcc)} \mid \mathrm{AgBr(conc.\ I\ in\ LiBr)} \mid \mathrm{AgBr(conc.\ II\ in\ LiBr)} \mid \mathrm{Ag(fcc)}$$

The reactions and the potentials of the three cells may be represented as follows:

$$\mathrm{Ag(fcc)} + \tfrac{1}{2}\mathrm{Br_2(1\ atm.)} = \mathrm{AgBr(conc.\ I)} \qquad (a)$$

$$\mathcal{E}_a = \mathcal{E}^\circ - \frac{RT}{\mathfrak{N}\mathfrak{F}} \ln a^{\mathrm{I}}_{\mathrm{AgBr}}$$

$$\mathrm{Ag(fcc)} + \tfrac{1}{2}\mathrm{Br_2(1\ atm.)} = \mathrm{AgBr(conc.\ II)} \qquad (b)$$

$$\mathcal{E}_b = \mathcal{E}^\circ - \frac{RT}{\mathfrak{N}\mathfrak{F}} \ln a^{\mathrm{II}}_{\mathrm{AgBr}}$$

$$\mathrm{AgBr(conc.\ II)} = \mathrm{AgBr(conc.\ I)} \qquad (c)$$

$$\mathcal{E}_c = -\frac{RT}{\mathfrak{N}\mathfrak{F}} \ln \frac{a^{\mathrm{I}}_{\mathrm{AgBr}}}{a^{\mathrm{II}}_{\mathrm{AgBr}}}$$

Reaction (c) is the difference of reactions (a) and (b), and similarly $\mathcal{E}_c = \mathcal{E}_a - \mathcal{E}_b$, since $\mathcal{E}^\circ$ has the same value for cells A and B, the same

[1] For example, S. Glasstone, "An Introduction to Electrochemistry," D. Van Nostrand Company, Inc., New York, 1942.

standard states being chosen for both. It will be noted that the cell reaction and the expression for the potential of the combined cell are identical in form with those given previously for a concentration cell.

If a solution II of cell B is chosen as pure liquid AgBr, then it is found from the expression for the potential of reaction (c) that

$$\mathcal{E}_c = -\left(\frac{RT}{\mathfrak{N}\mathfrak{F}}\right) \ln a^{\mathrm{I}}_{\mathrm{AgBr}},$$

thus permitting a direct evaluation of the activity of AgBr as a function of concentration. Experimentally the cell C would not be used; rather its potential would be evaluated from those of A and B. In fact, the experimental measurement would consist of measuring the potential of a series of cells of the type of A; $\mathcal{E}°$ would be regarded as given by the end member (pure AgBr) of this series. Hence it is not necessary, although it is common, to think in terms of the combined cell C.

Temperature Coefficient of a Cell. The usual procedure when working with cells is to measure the emf of each individual cell or set of cells at several temperatures without otherwise altering the experimental conditions. This procedure is easier than covering the concentration range at one temperature and then preparing new cells for the next temperature, and so on. Thus $d\mathcal{E}/dT$ at fixed concentration is usually obtainable directly from the experimental measurements.

By differentiation of Eq. (17-2) it is found that $d\mathcal{E}/dT$ is simply related to the free-energy change of the cell reaction by the equation

$$\frac{d\,\Delta F}{dT} = -\mathfrak{N}\mathfrak{F}\frac{d\mathcal{E}}{dT} \tag{17-6}$$

The combined statement of the first and second laws, $dF = V\,dP - S\,dT$ [Eq. (7-29)], which for a reaction at constant composition of all reactants and products becomes $d\,\Delta F = \Delta V\,dP - \Delta S\,dT$, reduces at constant pressure to

$$\frac{d\,\Delta F}{dT} = -\Delta S \tag{17-7}$$

Under the usual condition that the potential measurements all pertain to the same pressure these two relations may be combined to give

$$\Delta S = \mathfrak{N}\mathfrak{F}\frac{d\mathcal{E}}{dT} \tag{17-8}$$

an explicit expression for the entropy change of the cell reaction in terms of the temperature coefficient of a given cell.

In cells involving fused alloys and salts the plot of $\mathcal{E}$ against T is usually linear within the experimental error and the slope $d\mathcal{E}/dT$ is commonly assigned to the mean temperature of the range covered. Any departure from linearity in the $\mathcal{E}$-T plot is represented as $d^2\mathcal{E}/dT^2$ which by differentiation of Eq. (17-8) is $(1/\mathfrak{N}\mathfrak{F})(d\,\Delta S/dT)$, or from Eq. (7-12) is $\Delta C_P/\mathfrak{N}\mathfrak{F}T$. If the reaction involves solutions, as is usually the case, ΔC_P involves partial molal heat capacities which are rarely known from calorimetric or other source. Sufficiently precise cell measurements could be used to evaluate these.

From the definitional expression $\Delta F = \Delta H - T\,\Delta S$ the enthalpy change of the cell reaction is also readily found with the aid of Eqs. (17-2) and (17-8).

$$\Delta H = \Delta F + T\,\Delta S = -\mathfrak{N}\mathfrak{F}\mathcal{E} + \mathfrak{N}\mathfrak{F}T\,\frac{d\mathcal{E}}{dT} \tag{17-9}$$

For a concentration cell such as

Cd(pure l) | $CdCl_2$(in alkali chlorides) | Cd(in l Cd-Pb)

discussed previously, in which the cell reaction involves the transfer of cadmium from pure liquid to a solution, $\Delta F = \bar{F}_{Cd} - F^\circ_{Cd} = RT \ln a_{Cd}$. Similarly $\Delta S = \bar{S}_{Cd} - S^\circ_{Cd}$ and $\Delta H = \bar{H}_{Cd} - H^\circ_{Cd} = \bar{L}_{Cd}$. Thus, for such a cell, partial molal quantities are readily evaluated.

CHAPTER 18

FORMAL BASIS OF DIFFUSION THEORY[1]

Over a century ago, Parrot (1815) observed qualitatively that gases tend to intermingle no matter how carefully mechanical agitation and convection be avoided. Further observations of diffusion phenomena in gases and in liquids were reported by Graham in the decade 1860–1870; in 1870 Loschmidt first performed experiments on gaseous diffusion in such a way as to permit mathematical treatment of the data.

The theoretical work of Adolf Fick[2] published in 1855 precedes any quantitative experimental work and is still regarded as the phenomenological or descriptive basis of diffusion theory. This basis may best be understood by considering first Fick's original formulation and later its many limitations and modifications.

FICK'S LAWS

Let us first restrict our attention to an isothermal, isobaric, binary system of well-defined state, consisting of a single-phase region in which diffusion occurs in only one direction. The nature of these restrictions is of considerable importance, as exceptions are frequent if not common; some are considered later. One type of diffusion process which usually conforms to these restrictions is that which occurs in a vertical column of uniform cross section, such as is generally used in the experimental investigation of diffusion in a gas or liquid or which presumably occurs in the type of welded specimen commonly used in the investigation of diffusion in metals. For such a system, as illustrated in Fig. 18-1, Fick's first law states that the quantity of diffusing substance which passes per unit time through unit area of a plane at right angles to the direction of diffusion, known as the flux J, is proportional to the concentration gradient of the diffusing substance. Representing the concentration, or amount per unit volume, as C and taking the x direction to coincide with the direction in which diffusion occurs, *Fick's first law* may be written:

$$J = -D \frac{\partial C}{\partial x} \tag{18-1}$$

[1] This chapter originally appeared in substantially its present form as a paper by L. S. Darken in the monograph, "Atom Movements," American Society for Metals, Cleveland, 1951. It is here reproduced by kind permission of the American Society for Metals.

[2] Adolf Fick, *Pogg. Ann.*, **94**, 59 (1855).

The factor D, known as the *diffusivity* or *coefficient of diffusion*, is introduced as a proportionality factor. It will be noted that it has the dimensions of (length)^2/(time); it is usually given in square centimeters per second. This law is similar in form to Fourier's law stating that the flow of heat is proportional to the temperature gradient and to Ohm's law according to which the electric current is proportional to the gradient of the electrical potential.

Fick's law, like many natural laws, was formulated well before adequate experimental confirmation existed. For some time thereafter

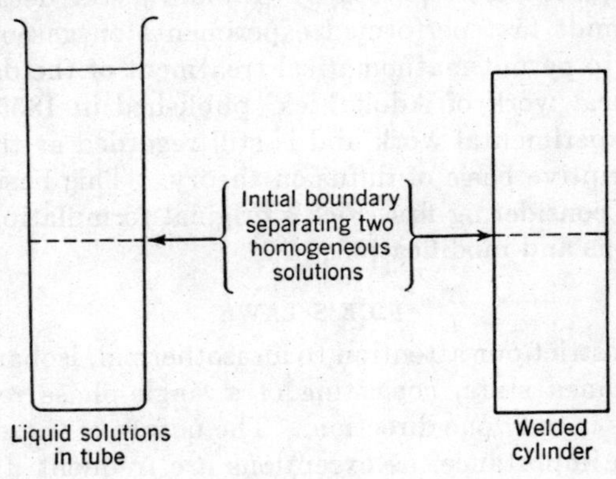

Fig. 18-1. Schematic representation of specimen for non-steady-state linear diffusion.

experimental investigation centered on diffusion in gases and in dilute liquid solutions. This work was long regarded as confirming the validity of Fick's laws. Early work on the diffusivity of gases (summarized by A. Lonius in 1909—his tabulation is also given by Loeb[1]) showed that the variation of the diffusivity D with composition was small, in no case over 8 per cent; in fact the variation was considerably less than expected, the expectation being based on the early approximate kinetic theory of gases. The fact that the diffusivity in liquid and solid solutions may vary widely with composition was not fully appreciated until considerably later.

Fick's second law is derived from the first by means of the apparently simple consideration that the rate of accumulation of diffusible substance in a given volume element is the difference between the inward and outward flux. In other words, what goes in and does not go out, stays there. Although Fick considered D as constant, the derivation is no more com-

[1] L. B. Loeb, "Kinetic Theory of Gases," McGraw-Hill Book Company, Inc., New York, 1934.

plicated if this restriction is removed. The reasoning for diffusion in a single direction is as follows: We consider two parallel unit planes a

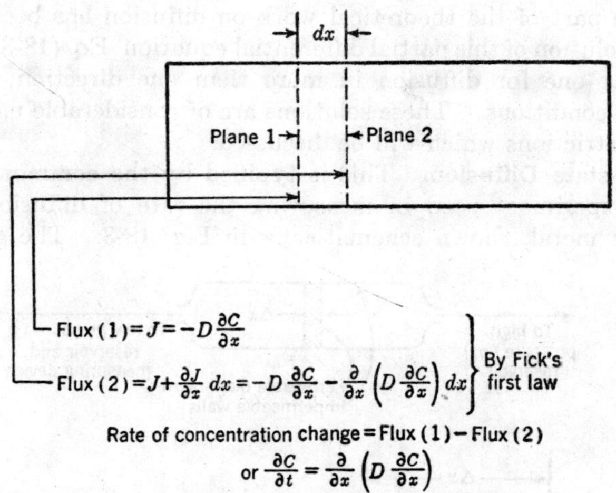

FIG. 18-2. Derivation of Fick's second law.

distance dx apart, as illustrated in Fig. 18-2. By Fick's first law the flux through the first plane is

$$J = -D \frac{\partial C}{\partial x}$$

and the flux through the second plane is

$$J + \frac{\partial J}{\partial x} dx = -D \frac{\partial C}{\partial x} - \frac{\partial}{\partial x}\left(D \frac{\partial C}{\partial x}\right) dx$$

By subtraction

$$\frac{\partial J}{\partial x} = -\frac{\partial}{\partial x}\left(D \frac{\partial C}{\partial x}\right)$$

But $\partial J/\partial x$, the difference in flux, equals the negative rate of concentration change $-\partial C/\partial t$. Thus we arrive at *Fick's second law:*

$$\frac{\partial C}{\partial t} = \frac{\partial}{\partial x}\left(D \frac{\partial C}{\partial x}\right) \tag{18-2}$$

Originally, D was considered constant (at constant temperature) and thus was taken outside the differential,

$$\frac{\partial C}{\partial t} = D \frac{\partial^2 C}{\partial x^2} \tag{18-3}$$

SOLUTIONS OF FICK'S LAW FOR A FEW BOUNDARY CONDITIONS (D CONSTANT)

A large part of the theoretical work on diffusion has been concerned with the solution of this partial differential equation, Eq. (18-3), or the corresponding one for diffusion in more than one direction, for various boundary conditions. These solutions are of considerable use in spite of certain restrictions which will be discussed.

Steady-state Diffusion. This is typified by the common diaphragm type of experiment used in measuring the rate of diffusion of a gas[1] through a metal, shown schematically in Fig. 18-3. The gas is main-

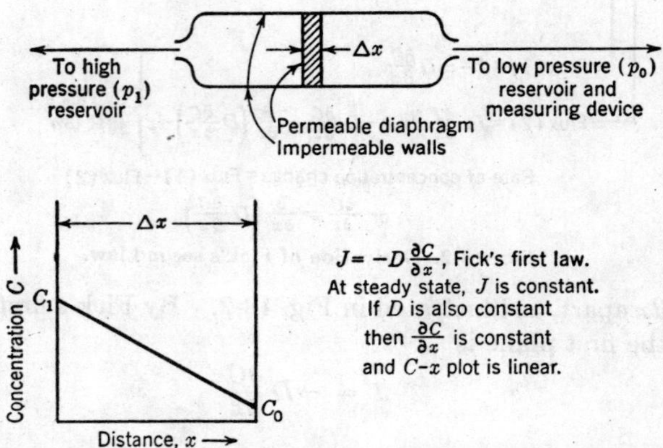

FIG. 18-3. Schematic representation of steady-state diffusion of gas and relation between concentration and distance for constant diffusivity.

tained at a higher pressure p_1 on one side of the diaphragm and at a lower pressure p_0, which may be substantially zero, on the other side. After sufficient lapse of time, it is found that a steady state is achieved; *i.e.*, the rate at which gas enters the metal on the high-pressure side and the rate at which it leaves the low-pressure side become equal, and this common rate does not change with time; similarly the concentration of dissolved gas at each point within the diaphragm does not change with time. As C is constant, the left side of Eq. (18-3) is 0, and this equation thus reduces to Eq. (18-1) with the flux J a constant independent of time and of position within the diaphragm. Under the assumption that D is also constant, it follows that the concentration gradient dC/dx is constant and hence may be written $\Delta C/\Delta x$ or $(C_0 - C_1)/\Delta x$, where C_1 and C_0 represent the concentration on each side of the diaphragm and Δx the

[1] Incidentally, this is a rather loose and perhaps misleading usage of terms, since after a gas has dissolved in a metal, it is no longer a gas.

thickness of the diaphragm. To describe this situation it is customary to invent a new term, the permeability designated P.

The *permeability* is the diffusion flux J under certain defined or standard conditions. There is no universal or even general agreement as to these standard conditions. It might seem simplest to choose this standard condition as corresponding to unit negative concentration gradient, in which case $P = D$. Sometimes the standard dimensions of the diaphragm are chosen as 1 cm² cross section and 1 mm thickness; in this case for unit concentration change from one side of the diaphragm to the other, $P = 10D$. Frequently the concentration of the diffusing substance in the diaphragm is unknown. If each side of the diaphragm is substantially in equilibrium with the gas in contact with it, then the concentration at each side of the diaphragm is related to the pressure. For a diatomic gas such as hydrogen or nitrogen this relation is given by Sievert's law: $C = s\sqrt{p}$, where s is a proportionality constant, equal to the solubility at unit pressure. The flux may then be expressed in terms of the pressure:

$$J = -Ds\,\frac{\sqrt{p_0} - \sqrt{p_1}}{\Delta x} \tag{18-4}$$

The permeability is frequently taken as the flux per unit area for unit value of $(\sqrt{p_1} - \sqrt{p_0})/\Delta x$. If p_0 is taken as zero, p_1 as 1 atm, and Δx as 1 mm, then $P = 10Ds$ where s, the Sievert's law constant, is the solubility of the gas in the metal at 1 atm pressure. It is thus seen that, in comparing the behavior of various gases and metals under these or similar fixed conditions, the flux or permeability is proportional to both the diffusivity and the solubility; *i.e.*, the steady-state rate of diffusion through the diaphragm cannot be determined from either the diffusivity or solubility alone. Care must be taken in employing Eq. (18-4) that proper units are used; to obtain P in grams per second per square centimeter, D must be in square centimeters per second and s in grams per cubic centimeter.

Since the permeability is merely a flux under certain defined conditions, it is clear that, if the conditions are adequately defined, then the flux or permeability has a definite unambiguous meaning. As much cannot always be said of the diffusivity; the reason for this will soon become apparent to the reader.

Non-steady-state Diffusion in the Semi-infinite Medium. This condition is exemplified by the so-called free diffusion which occurs in the liquid column or welded pair depicted in Fig. 18-1. Initially each portion is uniform and the boundary is sharp. The treatment is somewhat simplified by considering only one-half of such a couple, which may be considered as a semi-infinite solid or fluid, *i.e.*, one whose dimension in the direction of diffusion is large compared with the distance over which measurable composition change occurs during the time of the experiment. The composition is initially uniform, and the surface is considered as being brought instantaneously at the beginning of the experiment to a specific surface concentration C_s, which is maintained constant during the whole time of the experiment. These conditions imposed upon the specimen are known as boundary conditions; in this case they may be written

$$C = C_0 \quad \text{at t} = 0 \text{ and } 0 < \text{x} < \infty$$
$$C = C_s \quad \text{at x} = 0 \text{ and } 0 < \text{t} < \infty$$

The problem is to determine the distribution of the diffusing substance at any time after the start of the experiment. The mathematical problem involved is the solution of the partial differential equation (18-3), Fick's second law, corresponding to these boundary conditions for the case of constant D.

Let us try as solution $C = f(x/\sqrt{t})$. Setting $y = x/\sqrt{t}$ and

$$\frac{dC}{dy} = p$$

and noting that

$$\left(\frac{\partial C}{\partial t}\right)_x = \frac{-y}{2t}\left(\frac{\partial C}{\partial y}\right)_x \quad \text{and} \quad \left(\frac{\partial^2 C}{\partial x^2}\right)_t = \frac{1}{t}\left(\frac{\partial^2 C}{\partial y^2}\right)_t$$

it is found that

$$-\frac{py}{2} = \frac{D\,dp}{dy} \tag{18-5}$$

which proves that the trial solution is a correct one, since x and t have disappeared. The original partial differential equation has thus been reduced to an ordinary differential equation. It remains to show that this solution satisfies the boundary conditions. Integration gives

$$-\frac{y^2}{4} = D \ln p + I \tag{18-6}$$

or setting $I = -D \ln A$,

$$p = \frac{dC}{dy} = Ae^{-y^2/4D}$$

Integrating again

$$C - C_s = 2A\sqrt{D}\int_0^{x/2\sqrt{Dt}} e^{-\lambda^2}\,d\lambda \tag{18-7}$$

where C_s is the value of C at the surface ($x = 0$) and λ is introduced as an integration variable. The constant A is evaluated by considering a very great distance x at which $C = C_0$, the uniform initial concentration of the diffusing substance:

$$C_0 - C_s = 2A\sqrt{D}\int_0^\infty e^{-\lambda^2}\,d\lambda = 2A\sqrt{D}\,\frac{\sqrt{\pi}}{2} \tag{18-8}$$

whence,

$$A = \frac{C_0 - C_s}{\sqrt{\pi D}}. \tag{18-9}$$

Substituting this value of A into Eq. (18-7),

$$\frac{C - C_s}{C_0 - C_s} = \frac{C_s - C}{C_s - C_0} = \frac{2}{\sqrt{\pi}}\int_0^{x/2\sqrt{Dt}} e^{-\lambda^2}\,d\lambda \tag{18-10}$$

Inspection shows that the boundary conditions are satisfied.

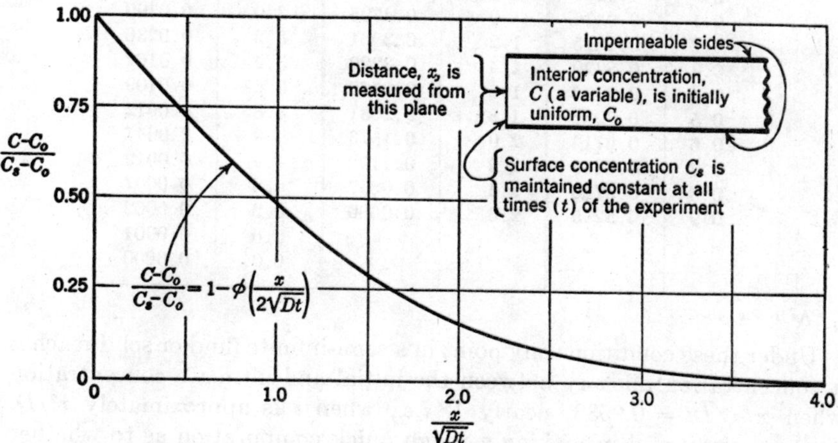

FIG. 18-4. Penetration curve for unidimensional diffusion in semi-infinite medium. Diffusivity constant; boundary conditions indicated.

The entire expression on the right is known as the normalized probability integral, designated φ. Values of $1 - \varphi$, which is equal to $(C - C_0)/(C_s - C_0)$ or $(C_0 - C)/(C_0 - C_s)$, are given in Table 18-1 and are plotted in Fig. 18-4 as a function of $x/\sqrt{Dt}$. The concentration function $(C - C_0)/(C_s - C_0)$ is easily visualized by noting that each point of the specimen starts with an initial concentration C_0 and after a very long time approaches a final concentration C_s. The fraction of this total concentration range which has been traversed in time t is equal to the concentration function $(C - C_0)/(C_s - C_0)$. It is particularly noteworthy that this fraction is a function only of the single dimensionless variable $x/\sqrt{Dt}$. Thus the penetration curve, or plot of concentration against distance, is for this case of such nature that it is of precisely the same form at all times; *i.e.*, any one curve may be imagined as obtained from any other by uniformly stretching or contracting the abscissa. This particular feature is maintained even if the diffusivity is not constant. For constant diffusivity, the same equation and curve are valid whether diffusion is inward or outward.

TABLE 18-1. CONCENTRATION DURING DIFFUSION IN A SEMI-INFINITE MEDIUM*

C = local concentration at distance x from surface after time t
C_0 = initial uniform concentration
C_s = concentration maintained at surface

$\frac{x}{\sqrt{Dt}}$	$\frac{C - C_0}{C_s - C_0}$	$\frac{x}{\sqrt{Dt}}$	$\frac{C - C_0}{C_s - C_0}$	$\frac{x}{\sqrt{Dt}}$	$\frac{C - C_0}{C_s - C_0}$
0	1.0000	0.9538	0.5000	2.8	0.0477
0.1	0.9436	1.0	0.4795	3.0	0.0399
0.2	0.8875	1.2	0.3961	3.2	0.0236
0.3	0.8320	1.4	0.3222	3.4	0.0162
0.4	0.7773	1.6	0.2579	3.6	0.0109
0.5	0.7237	1.8	0.2031	3.8	0.0072
0.6	0.6714	2.0	0.1573	4.0	0.0047
0.7	0.6206	2.2	0.1198	4.4	0.0019
0.8	0.5716	2.4	0.0897	4.8	0.0007
0.9	0.5245	2.6	0.0660	5.2	0.0002
				5.6	0.0001
				6.0	0.0000

* As in Fig. 18-4.

Under these conditions any point in a semi-infinite fluid or solid reaches a concentration midway between the initial and ultimate concentration when $x/\sqrt{Dt} = 0.9538$, nearly 1, *i.e.*, when t is approximately x^2/D. This is very useful in making a rough quick computation as to whether appreciable compositional change has occurred by diffusion under a given set of circumstances. For example, if a mild steel surface is to be exposed to strongly decarburizing conditions at 1000°C, we may wish to inquire as to how much time it will take for appreciable decarburization to take place $\frac{1}{8}$ in. in from the surface. Taking D for carbon in iron at this temperature as 3×10^{-7} cm²/sec and expressing the distance in centimeters, it is found from the expression x^2/D that nearly 4 days will be required.

The total amount of diffusing substance per unit area which has entered the semi-infinite fluid or solid from $t = 0$ to t is

$$\int_0^\infty (C - C_0)\,dx \quad \text{or} \quad 2\sqrt{Dt}\int_0^\infty (C - C_0)\,d\left(\frac{x}{2\sqrt{Dt}}\right)$$

Substituting $(1 - \varphi)(C_s - C_0)$ for $(C - C_0)$, by virtue of Eq. (18-10), this total amount is found to be

$$2\sqrt{Dt}\,(C_s - C_0)\int_0^\infty (1 - \varphi)d\left(\frac{x}{2\sqrt{Dt}}\right)$$

The definite integral has the value $1/\sqrt{\pi}$. Thus the total amount of diffusing substance which has crossed unit area of the bounding surface up to time t is $1.1284\sqrt{Dt}\,(C_s - C_0)$.

Non-steady-state Diffusion in Objects of Certain Geometric Shapes. A great deal of the work on diffusion has been concerned with the solution of Fick's second law for various boundary conditions, the diffusivity being considered constant. As the mathematics is the same as for heat-flow problems, many of these solutions appear in texts and papers on heat flow, and for this reason some of this work is not widely known among some workers in the field of diffusion. It will merely be mentioned in passing that for nonlinear problems Fick's second law is written in vector notation

$$\frac{\partial C}{\partial t} = \nabla(D\,\nabla C) \tag{18-2a}$$

or, if D is independent of concentration,

$$\frac{\partial C}{\partial t} = D\,\nabla^2 c \tag{18-3a}$$

Solutions of the latter are available in tabular form for diffusion into or out of variously shaped geometric objects initially of uniform composition C_0 and whose surface is brought instantaneously to and maintained at a constant surface concentration C_s. Among such geometric objects for which tables are available may be mentioned the sphere, the finite and infinite cylinder, the infinite slab, the infinite rectangular rod, and rectangular solids of bricklike shape. In all these cases the fractional saturation, *i.e.*, the ratio of the amount of diffusing substance that has entered the object at a given time to the amount that will have entered after a very long (infinite) time, is found to be a function of $L/\sqrt{Dt}$ or more conveniently of Dt/L^2, where L is a characteristic length of the object. L may be the thickness or half thickness of a slab or the radius of a cylinder or sphere; in more complicated cases with several characteristic lengths L_1, L_2, L_3—as for the brick shape—the fractional saturation is a function of Dt/L_1^2, Dt/L_2^2, and Dt/L_3^2. The fractional saturation, as well as the mean or average concentration of the slab, cylinder, or sphere, C_m, under these conditions is represented graphically in Fig. 18-5 and tabularly in Table 18-2. It will be noted that these curves or tables may be used, providing D is constant, either to determine D experimentally or to evaluate the mean concentration if D is known.

For the slab or sheet it may be noted that the diffusion process is nearly (93 per cent) complete when $Dt/L^2 = 1$. This is easy to keep in mind and permits a rapid estimate as to whether a contemplated diffusion process occurs slightly or appreciably or perhaps goes substantially to equilibrium. For example, if a bright mild steel sheet of 1 mm thickness is exposed at 600°C to a hydrogen atmosphere for 1 min, we may inquire as to what extent the diffusion of hydrogen from the atmosphere into the steel (or conversely) has occurred. Taking the diffusivity D of hydrogen as 10^{-4} cm²/sec, Dt/L^2 is found to be 2.4. Reference to Table 18-2 shows

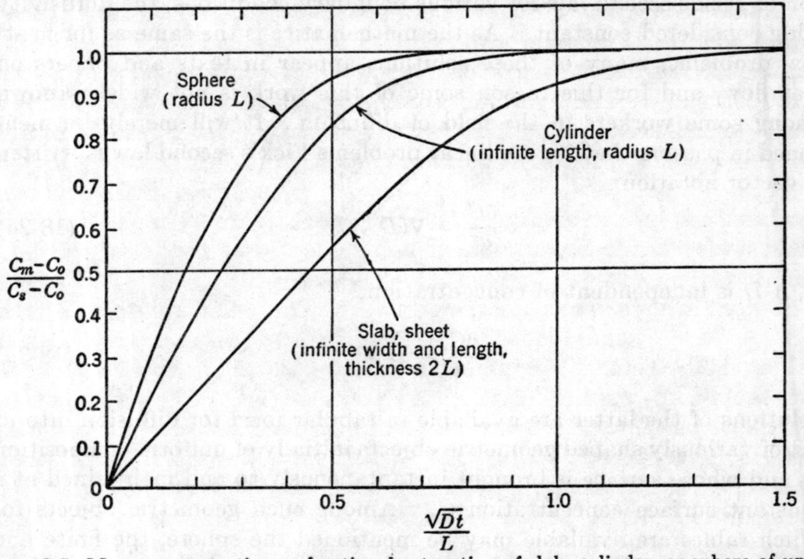

FIG. 18-5. Mean concentration or fractional saturation of slab, cylinder, or sphere of uniform initial concentration C_0 and constant surface concentration C_s. C_m is the mean concentration at time t.

that for a value of 2.4 the concentration fraction is nearly unity. Hence, providing the boundary condition is actually fulfilled and providing there is no serious violation of the other assumptions involved, it may be concluded that the sheet, if initially devoid of hydrogen, becomes substantially saturated during a 1-min exposure at 600°C.

To compute the concentration C_c at the center of objects of certain simple geometric shapes, for similar boundary conditions, Table 18-3 is convenient; in this table L represents the thickness rather than the half thickness of the slab. For these same boundary conditions, local concentrations at points other than the center may be found for the cylinder and slab from Table 18-4; in this table, x represents the distance from the center.

TABLE 18-2. FRACTIONAL SATURATION (AVERAGE COMPOSITION) OF SLAB, CYLINDER OR SPHERE DURING DIFFUSION*

$$\text{Fractional saturation } F = \frac{C_m - C_0}{C_s - C_0}$$

where C_m = mean concentration at time t
C_0 = uniform initial concentration
C_s = constant surface concentration

L is the radius of the cylinder or sphere, or the half-thickness of the slab. Table and equations valid only for constant D.

Dt/L^2	$(C_m - C_0)/(C_s - C_0)$		
	Slab	Cylinder	Sphere
0.005	0.078	0.157	0.226
0.01	0.110	0.216	0.310
0.02	0.161	0.302	0.421
0.03	0.195	0.360	0.500
0.04	0.227	0.412	0.560
0.05	0.251	0.452	0.604
0.06	0.275	0.488	0.648
0.08	0.320	0.550	0.720
0.10	0.357	0.606	0.774
0.15	0.438	0.708	0.861
0.20	0.503	0.781	0.916
0.25	0.560	0.832	0.948
0.30	0.612	0.878	0.969
0.40	0.702	0.9316	0.988
0.50	0.764	0.9616	0.9957
0.60	0.816	0.9785	0.9984
0.70	0.856	0.9879	0.9994
0.80	0.887	0.9932	0.9998
0.90	0.912	0.9960	0.9999
1.00	0.931	0.9979	
1.50	0.980	0.9999	
2.00	0.9942		
3.00	0.9995		

At short times:

For the slab $F = 1.128\,\sqrt{Dt}/L$
For the cylinder $F = 2.256\,\sqrt{Dt}/L$
For the sphere $F = 3.385\,\sqrt{Dt}/L$

At long times:

For the slab $Dt/L^2 = -0.0851 - 0.933 \log (1 - F)$
For the cylinder $Dt/L^2 = -0.06375 - 0.3988 \log (1 - F)$
For the sphere $Dt/L^2 = -0.05043 - 0.2333 \log (1 - F)$

* A. B. Newman, *Trans. Am. Inst. Chem. Engrs.*, **27**, 310 (1931). B. Serin and R. T. Ellickson, *J. Chem. Phys.*, **9**, 742 (1941). D. H. Andrews and J. Johnston, *J. Am. Chem. Soc.*, **46**, 640 (1924).

TABLE 18-3. CONCENTRATION AT CENTER OF CERTAIN GEOMETRIC FORMS DURING DIFFUSION*

To find $(C_s - C_c)/(C_s - C_0)$ (where C_c = local concentration at geometric center of object, C_s = surface concentration, C_0 = uniform initial concentration), consult Table A for function(s) required; these functions are tabulated in Table B. Table and equations valid only if D is constant.

Table A

Geometric form	$\frac{C_s - C_c}{C_s - C_0} = 1 - \frac{C_c - C_0}{C_s - C_0}$
Infinite cylinder	$C(Dt/r^2)$
Finite cylinder	$C(Dt/r^2) \times S(Dt/a^2)$
Infinite slab	$S(Dt/a^2)$
Infinite rectangular rod	$S(Dt/a^2) \times S(Dt/b^2)$
Brick	$S(Dt/a^2) \times S(Dt/b^2) \times S(Dt/c^2)$

r = radius of cylinder; a = length of cylinder, thickness of infinite slab, rectangular rod, or brick; b = width of rectangular rod or brick; c = length of brick. $L = a, b, c$, or r.

Table B

Dt/L^2	$S[Dt/(a,b, \text{ or } c)^2]$	$C(Dt/r^2)$
0	1.00000	1.00000
0.01	0.99919	1.00000
0.02	0.97516	0.99999
0.03	0.91755	0.99953
0.04	0.84580	0.99627
0.05	0.77231	0.98710
0.06	0.70220	0.97054
0.07	0.63722	0.94697
0.08	0.57776	0.91772
0.09	0.52363	0.88436
0.10	0.47449	0.84836
0.12	0.38953	0.77293
0.14	0.31976	0.69798
0.16	0.26248	0.62692
0.18	0.21546	0.56126
0.20	0.17687	0.50145
0.22	....	0.44755
0.25	....	0.37684
0.30	....	0.28249
0.35	....	0.21161
0.40	....	0.15849

Beyond the range of the table, the following equations may be used:

$$\log S[Dt/(a,b, \text{ or } c)^2] = 0.10491 - 4.28631 Dt/L^2$$

$$\log C(Dt/r^2) = 0.204654 - 2.51161 Dt/L^2$$

* F. C. W. Olson and O. T. Schultz, *Ind. Eng. Chem.*, **34**, 874 (1942).

TABLE 18-4. LOCAL CONCENTRATION DURING DIFFUSION IN INFINITE CYLINDER OR SLAB*

C = local concentration at distance x from center of slab or cylinder of mean relative saturation $(C_m - C_0)/(C_s - C_0)$
C_0 = initial uniform concentration
C_s = constant surface concentration
C_m = mean concentration

x/L	0.95	0.9	0.8	0.7	0.6	0.5	0.3	0.0
$\frac{C_m - C_0}{C_s - C_0}$	Values of $\frac{C - C_0}{C_s - C_0}$ for Cylinder of Radius L							
0.1	0.435	0.097	0.006	0.001	0.000	0.000	0.000	0.000
0.2	0.726	0.475	0.112	0.025	0.010	0.002	0.000	0.000
0.3	0.823	0.653	0.363	0.160	0.059	0.016	0.001	0.000
0.4	0.875	0.755	0.525	0.330	0.185	0.095	0.023	0.003
0.5	0.908	0.821	0.647	0.483	0.342	0.230	0.095	0.035
0.6	0.931	0.864	0.732	0.605	0.483	0.378	0.228	0.140
0.7	0.950	0.901	0.805	0.712	0.619	0.535	0.405	0.320
0.8	0.968	0.937	0.873	0.811	0.750	0.695	0.605	0.541
0.9	0.984	0.968	0.937	0.904	0.874	0.847	0.800	0.767
	For Slab of Half-thickness L							
0.1	0.703	0.435	0.100	0.020	0.004	0.000	0.000	0.000
0.2	0.842	0.688	0.420	0.223	0.105	0.045	0.007	0.000
0.3	0.890	0.787	0.590	0.420	0.282	0.180	0.063	0.015
0.4	0.918	0.842	0.690	0.550	0.422	0.320	0.177	0.090
0.5	0.935	0.874	0.754	0.640	0.536	0.443	0.305	0.223
0.6	0.950	0.902	0.807	0.718	0.633	0.558	0.447	0.376
0.7	0.963	0.928	0.856	0.788	0.724	0.667	0.582	0.530
0.8	0.976	0.951	0.905	0.860	0.817	0.780	0.720	0.687
0.9	0.987	0.975	0.951	0.929	0.908	0.890	0.861	0.843

* R. W. Gurry, *Trans. AIME*, **150**, 172 (1942).

Another type of boundary condition for which solution of Eq. (18-3) is available corresponds to an initial concentration of diffusing substance at a point, line, or plane source.[1] This solution is applicable, for example, if a metallic surface is plated with a very thin layer of another metal and subsequently held at elevated temperature for sufficient time that the penetration distance is large compared with the original thickness of the plate.

An even more complicated set of boundary conditions is encountered in the edgewise growth of pearlite. A mathematical solution of Eq. (18-3)

[1] R. M. Barrer, "Diffusion in and through Solids," The Macmillan Company, New York, 1931.

for this case was published by Brandt.[1] He assumed that diffusion flux is principally through austenite; subsequent evidence casts considerable doubt on this assumption.

VARIATION OF DIFFUSIVITY WITH COMPOSITION

As pointed out previously a change of D with composition in gaseous systems was called for by the early kinetic theory of gases. For gases this change was found to be less than was originally anticipated and was small enough to be of only limited interest. Attention was brought to the variation of the diffusivity with concentration of electrolytes in dilute aqueous solution by the experimental work of Clack[2] in 1924. Theoretical and experimental interest in diffusion of electrolytes and macromolecules (high polymers) in aqueous solutions has fruitfully continued as evidenced by the excellent conference on this subject held by the New York Academy of Sciences[3] in 1944. Although the variation of diffusivity with composition in metallic systems has been brought forcibly to our attention by the outstanding work of Mehl and coworkers,[4] it is still true that much of the work on diffusion in nonmetallic fields is overlooked or discarded by metallurgists—even by those whose interest is focused on metallic diffusion.

The variation of diffusivity with composition is most readily illustrated and understood by means of a steady-state phenomenon. According to Fick's first law for unidimensional diffusion, the flux per unit area is $J = -D(dC/dx)$. For steady-state diffusion through a plane diaphragm the flux through every plane parallel to the faces must be the same. Thus J is a constant, and it follows that the diffusivity is inversely proportional to the concentration gradient,

$$D = -\frac{J}{dC/dx}$$

D can thus be determined by experimental determination of the flux and of the concentration. To determine whether D varies, it is necessary only to find whether dC/dx varies, *i.e.*, whether a plot of C against x is a straight line.

The steady-state method has been employed in this manner to investigate the diffusivity of carbon in iron-carbon alloys.[5] A hollow cylinder was employed rather than a plane diaphragm. A carburizing gas was passed through the inside, and a decarburizing gas over the outside. The

[1] W. H. Brandt, *J. Applied Phys.*, **16**, 139 (1945).
[2] B. W. Clack, *Proc. Phys. Soc.* (*London*), **36**, 313 (1924).
[3] *Ann. N.Y. Acad. Sci.*, **46**, 209 (1945).
[4] R. F. Mehl, *J. Applied Phys.*, **8**, 174 (1937).
[5] R. P. Smith, to be published.

flux per unit length of cylinder through a cylindrical surface of radius r is $2\pi rD(dC/dr)$. Thus at steady state $D(dC/d \log r)$ is a constant and D is inversely proportional to the slope of a plot of C against $\log r$, a plot which would be linear if D were constant. Such a plot is shown in Fig. 18-6. It will be noted that the departure from linearity is pronounced, indicating that the diffusivity changes many fold in this range of composition (0 to 1.5 per cent C).

As mentioned previously the extensive work of Mehl and coworkers first brought this phenomenon to the prominent attention of metallurgists. In some cases the variation of the diffusivity with composition is even more pronounced. A tenfold variation in D is exhibited by the continuous series of solid solutions of the nickel-copper system[1] as shown in Fig. 18-7. Matano's calculation[2] of the diffusivity for the continuous series of solid solutions of gold with nickel, platinum, and palladium is shown in Fig. 18-8. An even more pronounced example is found in the copper-aluminum system. In this system at 500°C, the diffusivity for alloys of nearly pure aluminum is about 2000 times that for alloys of nearly pure copper; here, although both elements crystallize in the fcc system, a number of intermediate phases exist.

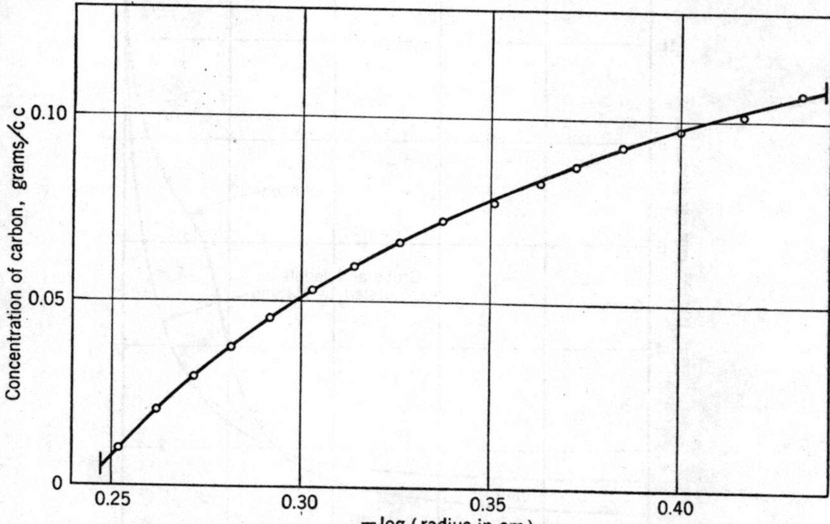

FIG. 18-6. Steady-state diffusion of carbon at 1000°C through wall of hollow iron cylinder. The diffusivity is inversely proportional to the slope of the curve.

[1] W. A. Johnson, *Trans. AIME*, **166**, 114 (1946).

[2] C. Matano, *Japan J. Phys.* (*Trans.*), **8**, 109 (1933).

At this point it might well be asked as to how Fick's laws survived such experimental findings at all. These laws were originally premised on a proportionality between flux and concentration gradient. Such proportionality, as illustrated above, is in general not found experimentally. In part Fick's laws [Eqs. (18-1) and (18-2)] have degenerated to a definition of the diffusivity D, now recognized to be a function of concentration. In other words, Fick's formulation provides a convenient way of describ-

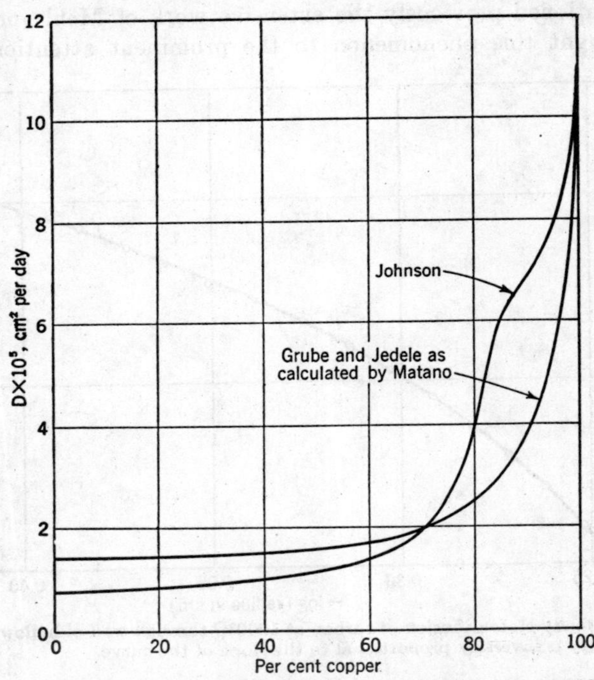

FIG. 18-7. Variation of D with composition for the nickel-copper system at 1052°C. [*From W. A. Johnson, Trans. AIME*, **166**, 114 (1946).]

ing, at least partially, diffusion phenomena in a binary system. However, his laws still retain some aspect of physical law quite distinct from definition. Existing data support Fick's laws as limiting laws, valid at sufficiently low concentration or for small concentration differences. As such, their validity is of the same nature as that of the ideal-gas law or of Henry's law stating that the vapor pressure or activity of a solute is proportional to its concentration providing the concentration is sufficiently low. From this aspect, Fick's law might be revised to state that the diffusivity as defined by Eq. (18-1) is a variable which approaches a finite constant at zero concentration. Furthermore, the implication is definitely carried that the function D is a function only of state variables such as temperature (which throughout this whole discussion is considered constant), stress (which has been considered to consist only of a uniform pressure), and composition. It is usually taken for granted that the diffusivity is not a function of the concentration gradient, *e.g.*, that the diffusivity as found under uniform boundary conditions by the steady-state diaphragm method is independent of the diaphragm thickness. This implication of Fick's law is based mainly on tradition and elementary considerations of atomistics; faith therein seems so well established that no one has conducted a rigorous experimental test of it for diffusion in a solid metal.

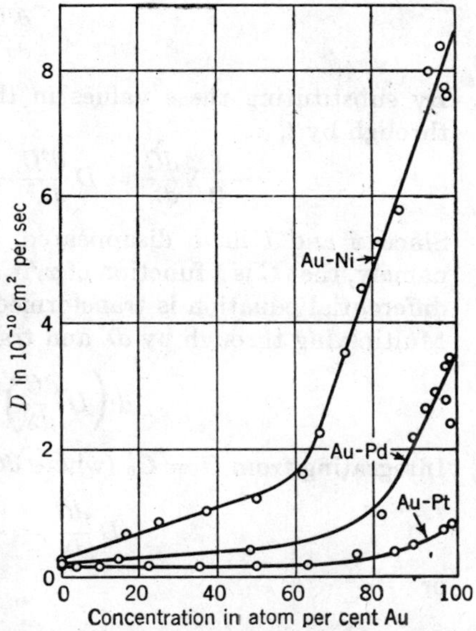

FIG. 18-8. Variation of D with composition for the Au-Ni, Au-Pd and Au-Pt systems at 900°C. [*From C. Matano, Japan J. Phys.* (*Trans.*), **8**, 109 (1933).]

Fick's first and second laws might better be called two forms of one law. The second is sometimes said to be derived by inserting the first in the equation of continuity. This latter, however, is merely a statement of the law of conservation of atomic or molecular species, *e.g.*, that hydrogen atoms are not created or destroyed in the process. It is also assumed that, at constant temperature and pressure, the diffusivity is a function of concentration only and not of the concentration gradient.

For a variable diffusivity, the modified form of Fick's law [Eq. (18-2)] cannot be directly integrated, even for the case of unidimensional diffusion into a semi-infinite solid. However, Boltzmann,[1] in 1894, showed that in this case the two variables time t and distance x may be replaced by a single variable $\lambda = x/\sqrt{t}$. The general form [Eq. (18-2)] may also be written

$$\frac{\partial C}{\partial t} = \frac{\partial}{\partial x}\left(D\,\frac{\partial C}{\partial x}\right) = D\,\frac{\partial^2 C}{\partial x^2} + \frac{\partial D}{\partial x}\,\frac{\partial C}{\partial x}$$

[1] L. Boltzmann, *Ann. Physik*, **53**, 960 (1894).

Let us try as solution, as for the case of constant D, $C = f(x/t^{\frac{1}{2}})$ and define the variable λ as $x/t^{\frac{1}{2}}$. If this trial solution is correct, then

$$\frac{\partial C}{\partial t} = \frac{dC}{d\lambda}\frac{\partial \lambda}{dt} = -\frac{\lambda}{2t}\frac{dC}{d\lambda}$$

$$\frac{\partial C}{\partial x} = \frac{dC}{d\lambda}\frac{\partial \lambda}{\partial x} = \frac{1}{t^{\frac{1}{2}}}\frac{dC}{d\lambda}$$

$$\frac{\partial D}{\partial x} = \frac{1}{t^{\frac{1}{2}}}\frac{dD}{d\lambda}$$

$$\frac{\partial^2 C}{\partial x^2} = \frac{1}{t}\frac{d^2C}{d\lambda^2}$$

By substituting these values in the original equation and multiplying through by t,

$$-\frac{1}{2}\lambda\frac{dC}{d\lambda} = D\frac{d^2C}{d\lambda^2} + \frac{dD}{d\lambda}\frac{dC}{d\lambda} = \frac{d}{d\lambda}\left(D\frac{dC}{d\lambda}\right)$$

Since x and t have disappeared, it is proved that the trial solution, namely, that C is a function of $x/t^{\frac{1}{2}} = \lambda$, is correct, and the original partial differential equation is transformed to an ordinary differential equation. Multiplying through by $d\lambda$ and transposing,

$$d\left(D\frac{dC}{d\lambda}\right) = -\frac{1}{2}\lambda\, dC$$

Integrating from $C = C_0$ (where $dC/d\lambda = 0$) to C,

$$D\frac{dC}{d\lambda} = -\frac{1}{2}\int_{C_0}^{C}\lambda\, dC$$

or

$$D = \frac{-\frac{1}{2}\int_{C_0}^{C}\lambda\, dC}{dC/d\lambda} \qquad (18\text{-}11)$$

This is the relation generally used in evaluating D from a diffusion penetration curve, such as shown in Fig. 18-9, obtained from a unidimensional diffusion experiment such as shown schematically in Fig. 18-1. Boltzmann's original derivation of this relation received little publicity, and this relation was some time later derived empirically by Matano.[1] It is now widely used for the determination of the diffusivity for binary metallic solutions by the non-steady-state method.

CHOICE OF REFERENCE FRAME FOR DIFFUSION

In the earlier diffusion experiments on the linear diffusion of gases or liquids in a tube, the coordinate axis was naturally taken as fixed relative to the containing tube. For dilute solutions no occasion arose to suggest

[1] C. Matano, *Japan J. Phys. (Trans.)*, **8**, 109 (1933).

any different choice. Many texts treat the matter of choice of coördinate axis as self-evident.

For the welded pairs, so commonly used in investigation of metallic diffusion, it seems obvious to choose an x axis rigidly fixed in the pair, although volume change could introduce some uncertainty. In this case of the welded pair it was found convenient to choose the origin, or point from which the distance x is measured, in such a manner that the two areas of Fig. 18-9 are equal. This figure is constructed from the data of Rhines and Mehl[1] on the copper-aluminum system. This choice of

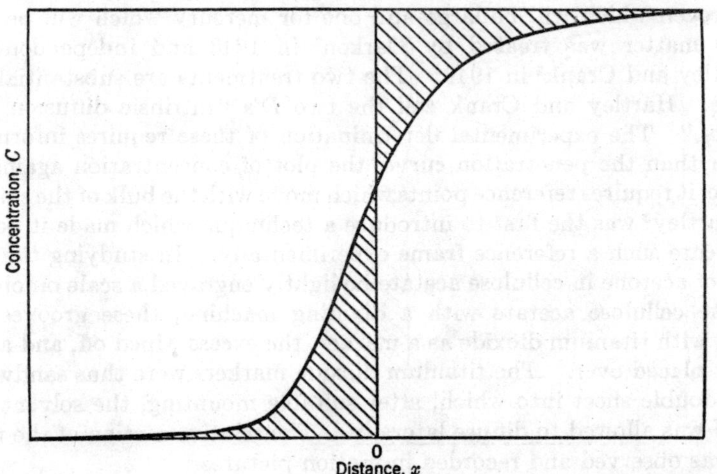

FIG. 18-9. Penetration curve. The Matano interface, the origin for the distance axis as shown, is so chosen that the two shaded areas are equal.

origin was suggested on empirical grounds by Matano[2] and is known as the *Matano interface*. It has intuitive appeal, since it corresponds to the choice one would make on the assumption that the amount of component 1 which diffuses in one direction is equal to the amount of component 2 which diffuses in the other. This choice of origin was later[3] placed on a firmer basis for the case where there is no volume change. The problem involved in the change of total volume during diffusion will not be considered here, as this problem may be regarded as one imposed upon but not inherent in the problem of choice of reference frame.

The seat of the problem lies in the subjective matter of what type of motion we choose to call diffusion and how we wish to describe the observed phenomena. As an illustration we may consider the inter-

[1] F. N. Rhines and R. F. Mehl, *Trans. AIME*, **128**, 185 (1938).
[2] C. Matano, *Japan J. Phys.* (*Trans.*), **8**, 109 (1933).
[3] L. S. Darken, *Trans. AIME*, **175**, 184 (1948).

diffusion of a light gas such as hydrogen and a heavy gas such as mercury vapor. From kinetic theory it is well known that the thermal agitational velocity of the hydrogen molecule is much greater than that of mercury. Hence we may picture the fast hydrogen molecules as agilely penetrating the space between those of mercury, whereas the mercury is so slow that its role in the process is limited principally to a passive submission to a general expansion accompanying the penetration of hydrogen. This viewpoint, which applies equally well to liquids or solids, leads to the definition of *two* coefficients of diffusion, in the above case one for hydrogen which will be large and one for mercury which will be small. This matter was treated by Darken[1] in 1948 and independently by Hartley and Crank[2] in 1949. The two treatments are substantially the same. Hartley and Crank call the two D's "intrinsic diffusion coefficients." The experimental determination of these requires information other than the penetration curve, the plot of concentration against distance; it requires reference points which move with the bulk of the solution.

Hartley[3] was the first to introduce a technique which made it possible to locate such a reference frame experimentally. In studying the diffusion of acetone in cellulose acetate he lightly engraved a scale on one side of the cellulose acetate with a dividing machine; these grooves were filled with titanium dioxide as a marker, the excess wiped off, and a plain sheet placed over. The titanium dioxide markers were thus sandwiched in a double sheet into which, after suitable mounting, the solvent (acetone) was allowed to diffuse laterally. The relative motion of the markers was observed and recorded in motion pictures.

Shortly after Hartley's work, Smigelskas and Kirkendall[4] published their work on the motion of molybdenum markers during diffusion in the copper-zinc system. Metallurgists were generally unfamiliar with Hartley's work. These observations came as a surprise, and the phenomenon is known as the *Kirkendall effect.* It will be noted that Smigelskas' and Kirkendall's markers were present only in one plane, that of the original interface. Hartley's markers were distributed at many distances from the original interface; more information can, of course, be gained in this case.

As indicated in the preceding, a formal understanding of the phenomenon of diffusion in a lattice that is marked is not so difficult as is sometimes believed. Bardeen[5] has gone further and presented a detailed

[1] L. S. Darken, *Trans. AIME,* **175,** 184 (1948).

[2] G. S. Hartley and J. Crank, *Trans. Faraday Soc.,* **45,** 801 (1949).

[3] G. S. Hartley, *Trans. Faraday Soc.,* **42,** 6 (1946).

[4] A. D. Smigelskas and E. O. Kirkendall, *Trans. AIME,* **171,** 130 (1947).

[5] J. Bardeen, *Phys. Rev..* **76,** 1403 (1949).

treatment of a possible set of atomistics of the process. His conclusions are in accord with those of the phenomenological treatment.

If we wish to recognize or define as diffusion only motion relative to the markers and thus employ for each volume element coordinate axes rigidly attached to a marker, then we are led to two intrinsic diffusion coefficients. If, on the other hand, we choose to take the attitude that diffusion and bulk motion are all part of one and the same process and that our interest is limited to the relation between composition distance and time, then a single diffusivity serves, just as it did prior to the work of Hartley and of Smigelskas and Kirkendall. In terms of the intrinsic diffusivities D_1 and D_2, the diffusivity as commonly used is

$$D = N_2D_1 + N_1D_2$$

the N's being mole or atom fractions. Although we may get along in a formally correct manner with a single diffusivity, part of the observable phenomena (such as the motion of markers) cannot be described thereby. An extended description may be given in terms of one diffusivity and marker velocities or alternatively in terms of two intrinsic diffusivities, D_1 and D_2; in general all are functions of concentration.

LIMITATIONS OF FICK'S LAWS IN BINARY SYSTEMS

If, in a given case, diffusion is predominantly along grain boundaries, it is obvious that Fick's laws cannot apply directly, since the diffusion flux will then depend upon the nature and extent of the grain boundary regions, as well as upon temperature, composition, and composition gradient. In several cases the effect of grain boundaries has been demonstrated to be absent, within the experimental error.

In other cases the reason for anticipation of failure of the laws is less obvious. An illustration of such arises in the investigation of the diffusion of hydrogen at room temperature in mild steel. If cylinders of mild steel are immersed in dilute acid at room temperature, it is well known that their hydrogen content increases. It might be thought that such experiments could be employed to determine the diffusivity of hydrogen in iron or steel by use of Fig. 18-5. However, it was found[1] that the diffusivity so determined differed widely (by a factor of perhaps 100) from one to another of nominally similar steels. It was subsequently found that the prior history of cold work and heat-treatment had a great deal to do with the variation of the apparent diffusivity. In fact, if we insist on retaining the formal method of representing diffusion phenomena as presented previously in this chapter, then we are forced to the conclusion that the diffusivity D varies widely with concentration, even

[1] L. S. Darken and R. P. Smith, *Corrosion*, **5**, 1 (1949).

at the low concentrations involved, and shows no indication of approaching constancy at vanishingly small concentration. D also varies widely with prior history of the steel specimen. Thus it may be said that Fick's law fails in this case. A more fruitful point of view is that part of the hydrogen is trapped in lattice imperfections and is not to be included in the concentration whose gradient appears in Fick's law. That is, part of the hydrogen is regarded as semi-inert; Fick's law is applied to the remainder. This treatment[1] seems to give satisfactory results. The approximate treatment is as though a separate phase were precipitated; this is presumably inobservable by direct means, since it occurs on an atomic scale. The method of treatment is similar to that of diffusion processes accompanied by phase change, developed by Darken[2] and by Rhines, Johnson, and Anderson.[3]

Although the foregoing is an example of the failure of D to approach constancy at low concentration, a more severe failure of Fick's law has recently been observed by Hartley.[4] In an investigation of the diffusion of solvents in high polymers he found what is probably the first reported case of the failure of Fick's law in its most general form for a binary system. Here, not only was D found to vary but it was found to be not even a single-valued function of composition; it varied with time. The polymer slowly became anisotropic as diffusion proceeded; this was interpreted in terms of the rotation of the large polymer molecules. Although the small size and approximate spherical symmetry of the atoms in metals might predispose us to think that we are unlikely to encounter related phenomena here, perhaps we should not feel too certain of this.

As mentioned previously, Fick's law cannot be expected to apply if grain boundary diffusion occurs. This can hardly be called a failure, since homogeneity is, or should be, stated as prerequisite to the application of Fick's law; grain boundaries which serve as a transport medium are certainly inhomogeneities in this sense. However, if the diffusion process itself creates lattice imperfections (holes, dislocations, or boundaries), then we may anticipate a failure of Fick's law analogous to that found by Hartley in high polymers; that is, D will no longer be a single-valued function of composition.

DIFFUSION IN MULTICOMPONENT SYSTEMS

Although diffusion phenomena are involved in many small- and large-scale processes involving multicomponent systems, very little attention has been given to either theoretical or experimental investigation of

[1] L. S. Darken and R. P. Smith, *Corrosion*, **5**, 1 (1949).

[2] L. S. Darken, *Trans. AIME*, **150**, 157 (1942).

[3] F. N. Rhines, W. A. Johnson, and W. A. Anderson, *Trans. AIME*, **147**, 205 (1942).

[4] G. S. Hartley, *Trans. Faraday Soc.*, **45**, 820 (1949).

diffusion in systems of more than two components. The experimental investigation by Mehl and Rhines[1] of the simultaneous diffusion of nickel and silicon in solid copper deserves mention. However, as Onsager[2] said, "It is a striking symptom of the common ignorance in this field that not one of the phenomenological schemes which are fit to describe the general case of diffusion is widely known."

It might well be asked why we do not treat our experimental data for multicomponent systems just like those for binary systems, namely, by dividing the flux by the concentration gradient and recording this ratio as the diffusivity. The answer is that a diffusivity so defined is in general a function of so many variables that its utility is negligibly small. Even for the simple case of an ideal solution this ratio is a function not only of composition but also of the derivatives of the compositions with respect to each other. Thus an entirely new method of systematically reporting the results of experiment is required for multicomponent systems. Only in special pseudobinary cases does this ratio have the fruitful meaning for ternary or multicomponent diffusion that it does for binary.

As a generalization[3] of Fick's law it may be assumed that the flux of each component is a linear function of the concentration gradients:

$$J_i = -\sum_k D_{ik} \nabla C_k$$

or for linear diffusion, to which this discussion is limited,

$$J_i = -\sum_k D_{ik} \frac{\partial C}{\partial x}$$

in which it will be noted that S^2 coefficients, D_{ik}, appear, S being the number of components.

For two components this type of representation becomes

$$J_1 = -D_{11} \frac{\partial C_1}{\partial x} - D_{12} \frac{\partial C_2}{\partial x}$$

$$J_2 = -D_{21} \frac{\partial C_1}{\partial x} - D_{22} \frac{\partial C_2}{\partial x}$$

[1] R. F. Mehl and F. N. Rhines, *Trans. AIME*, **137**, 301 (1940).

[2] L. Onsager, *Ann. N.Y. Acad. Sci.*, **46**, 241 (1945).

[3] The discussion here is based principally on Onsager's treatment [*Ann. N.Y. Acad. Sci.*, **46**, 241 (1945)]. Treatment of diffusion in multicomponent systems has also been developed by Kirkwood [*J. Chem. Phys.*, **14**, 180 (1946)] and by Leaf [*Phys. Rev.*, **70**, 748 (1946)].

Four coefficients D_{ik} which in general are functions of composition appear here; however, in this case two can be eliminated by virtue of the relation that the sum $(C_1 + C_2)$ of the concentrations is the reciprocal of the density, leaving only two which may be regarded as the intrinsic coefficients mentioned earlier. Of these two, one can be eliminated by appropriate choice of coordinate axes, thus, however, losing the ability to treat the motion of markers.

An easy way of viewing this elimination is by noting that one of the concentrations and one of the fluxes can always be eliminated, thus reducing the number of coefficients to $(S - 1)^2$. For a three-component system it will thus be noted that four coefficients are left, as compared to one for a binary system.

Application of the Principle of Microscopic Reversibility. Onsager[1] has developed a very important general principle known as the *principle of microscopic reversibility* which has been verified for thermoelectric phenomena, electrolytic cells with liquid junction, and other cases. This principle states that for a system in thermodynamic equilibrium every type of micromotion occurs just as often as its reverse; the gap between equilibrium and an irreversible process is bridged by considering the micro-fluctuations in a system at equilibrium. This is believed to be a very important principle of general applicability except for nonconservative force fields, such as the motion of a charged particle in a magnetic field.

As mentioned previously, scientific problems may in general be treated by one or both of two distinct general methods. The first and more customary involves a detailed analysis of all steps from the beginning to the end of the process under consideration. The second method involves consideration only of the initial and final state of the system under consideration and ignores the nature of the path or intermediate states. This second method is very powerful; the outstanding example is thermodynamics. Scientific laws may be classified, though more loosely, in the same way. Some require for their application a detailed knowledge of the atomistics of the particular situation; others do not. The outstanding examples of the latter class are the laws of classical thermodynamics. The principle of microscopic reversibility is also of this class. It involves no assumption as to detailed atomistic behavior. However, like the laws of thermodynamics its fruitfulness can be enhanced by a knowledge of atomistics. An example of one of the simplest deductions from this principle is that a catalyst for any reaction must be an equally good catalyst for the reverse reaction under similar conditions. The principle of microscopic reversibility bears another resemblance to the classical

[1] L. Onsager, *Phys. Rev.*, **37**, 405 (1931); **38**, 2265 (1932).

thermodynamic laws. Just as the original statements of these latter are closely connected with the invention of the then new functions energy and entropy, so the principle of microscopic reversibility is coupled to a function known as the dissipation function.

Utilization of the principle in the present case of diffusion leads to $\frac{1}{2}(S-1)(S-2)$ relations between the coefficients D_{ik} and thermodynamic quantities. These relations are

$$\sum_j \frac{\partial \mu_i}{\partial C_j} D_{jk} = \sum_j \frac{\partial \mu_k}{\partial C_*} D_{ji}$$

where the μ's are the chemical potentials (partial molal free energies). Thus, if the thermodynamic properties are known, the number of independent coefficients D_{ik} for a ternary system may be reduced to three as compared with one for a binary system. The corresponding number of independent coefficients for a multicomponent system is $\frac{1}{2}S(S-1)$. It may be noted incidentally that information of thermodynamic nature may conversely be obtained from diffusion data.

Since the foregoing method utilizes the thermodynamic functions anyway, it seems simpler to start with these, viewing the virtual force acting as derived from the chemical potentials. For a ternary system, then,

$$J_1 = R_{11}\frac{\partial \mu_1}{\partial x} + R_{12}\frac{\partial \mu_2}{\partial x} + R_{13}\frac{\partial \mu_3}{\partial x}$$

$$J_2 = R_{21}\frac{\partial \mu_1}{\partial x} + R_{22}\frac{\partial \mu_2}{\partial x} + R_{23}\frac{\partial \mu_3}{\partial x}$$

$$J_3 = R_{31}\frac{\partial \mu_1}{\partial x} + R_{32}\frac{\partial \mu_2}{\partial x} + R_{33}\frac{\partial \mu_3}{\partial x}$$

These nine coefficients could be reduced to four in a manner involving reasoning similar to that of the preceding paragraphs. Let us, however, consider J as the flux through a plane which is stationary relative to embedded markers. Then we might be tempted to consider that only R_{11}, R_{22} and R_{33} have nonzero values, since it would seem reasonable to suppose that the flux of the ith component would depend on the gradient of no chemical potential other than its own. This guess would reduce the number of coefficients to three or for an S-component system to S. In any event $R_{12} = R_{21}$, $R_{13} = R_{31}$, and $R_{23} = R_{32}$ by the principle of microscopic reversibility. Whether each of these three pairs is negligibly small remains to be determined by experiment, since there is at present no theory beyond the principle of microscopic reversibility to predict the extent of the coupling of these irreversible processes. The experimental investigation of this coupling would seem to be a matter of considerable theoretical and practical interest; to the best of the authors' knowledge no one has yet conducted such an experiment.

THERMODYNAMIC FUNCTIONS AS AN AID IN INTERPRETING DIFFUSION PHENOMENA

It was first suggested by Einstein[1] for ideal solutions, and subsequently by many others including Hartley[2] for nonideal solutions, that the virtual force acting on a diffusing atom or ion in a binary solution may be regarded as the negative gradient of the chemical potential or partial molal free energy.[3] This virtual force gives rise to a drift velocity which it seems reasonable to take as relative to fixed markers such as those of Hartley

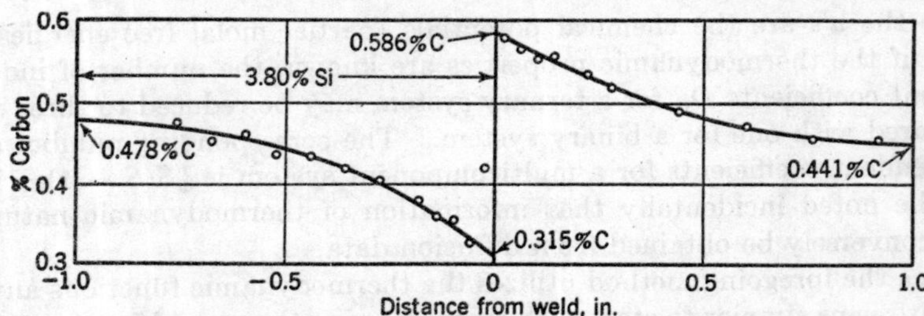

FIG. 18-10. Nonuniform distribution of carbon produced by diffusion from an initially uniform distribution. Carbon migrated from Fe-Si-C alloy to Fe-C alloy. Treatment: 13 days at 1050°C. [*From L. S. Darken, Trans. AIME*, **180**, 430 (1949).]

or of Smigelskas and Kirkendall. This leads immediately to the diffusion flux of the ith species as

$$J_i = -\frac{1}{N}\frac{\partial \bar{F}_i}{\partial x} B_i C_i$$

$\bar{F}_i$ being the partial molal free energy and B_i the mobility, a function only of composition at constant temperature and pressure.

This expression suggests for multicomponent solutions that the diffusion flux of an atom of the ith kind depends directly not upon the concentration gradient but upon the free-energy gradient. This is experimentally illustrated by the fact that diffusion may occur when the concentration gradient is zero if the free-energy gradient is not zero. If an iron-silicon-carbon bar be welded to an iron-carbon bar of the same carbon content, it has been shown[4] that the initially uniform carbon content is disturbed to give a decidedly nonuniform carbon distribution.

[1] A. Einstein, *Ann. phys.*, **17** (4), 549 (1905).

[2] G. S. Hartley, *Trans. Faraday Soc.*, **27**, 10 (1931).

[3] The validity of this assumption has been confirmed with a very high degree of experimental precision by Harned and Nuttall [*J. Am. Chem. Soc.*, **69**, 736 (1947)] for aqueous solutions of potassium chloride.

[4] L. S. Darken, *Trans. AIME*, **180**, 430 (1949).

as illustrated in Fig. 18-10. In simple language this may be explained by saying that carbon prefers iron to silicon; carbon prefers manganese even more.

Comparison of the foregoing expression with the phenomenological expression embodied in Fick's law gives the intrinsic diffusion coefficient as

$$D_i = kTB_i\left(1 + N_i \frac{\partial \ln \gamma_i}{\partial N_i}\right)$$

where k is Boltzman's constant and γ_i the activity coefficient, the gram atomic volume being considered constant. This type of treatment evaluates the intrinsic diffusivity D for a binary system in terms of thermodynamic properties and other known coefficients except for B_i. If we adhere to a pure phenomenology, all that can be said for B_i is that it is in general a function of composition. It has been proposed empirically[1] for certain cases that B_i is proportional to γ_i; however, this seems doubtful in general and certainly has not been adequately demonstrated experimentally. However, independent of the nature of B_i, this type of treatment involves the intrinsic diffusivities (*i.e.*, diffusivities relative to inert markers) and leads to an indirect check of the Kirkendall effect by other methods, in particular by radioactive tracers. On this basis it has been shown[2] for a binary system of constant molal volume that the diffusivity D may be represented as

$$D = N_1D_2 + N_2D_1 = (N_1D_2^* + N_2D_1^*)\left(1 + N_2 \frac{\partial \ln \gamma_2}{\partial N_2}\right)$$

where D_1^* and D_2^* are the diffusivities of the radioactive tracers. It is assumed in the derivation that local equilibrium prevails and that the state of a small-volume element is adequately described by its temperature, pressure, and composition. Complications, as discussed in the preceding section, arising from the fact that the introduction of a radioactive tracer is in fact the introduction of a third component, have been ignored. The omitted terms, involving D_{ik} or R_{ik}, are so small that the relation has been verified within reasonable experimental error.

Now it might seem reasonable to assume for a ternary or multicomponent system that we might apply the same reasoning. In other words, we might assume for such a system that the diffusion flux of the ith component may be represented as

$$J_i = -\frac{1}{N}\frac{\partial \bar{F}_i}{\partial x}B_iC_i$$

[1] J. C. Fisher, J. H. Hollomon, and D. Turnbull, *Trans. AIME*, **175**, 202 (1948).

[2] L. S. Darken, *Trans. AIME*, **175**, 184 (1948).

and is independent of the gradients of the other partial molal free energies. Such an assumption, that there is no cross effect, would be analogous to the assumption that there is no thermoelectric effect or Hall effect or thermal diffusion effect. In other words this assumption is in need of experimental justification. Such a critical experiment might be of the following type. Let it be supposed that two experiments be conducted involving the diffusion of both carbon and nitrogen through an iron diaphragm. In one experiment they flow in the same direction; in the other, in opposite directions. There will be one composition which is common to the two experiments. If B_i is a function of composition only, as tentatively assumed, then the ratio of the flux J_i to the gradient $\partial \bar{F}_i/\partial x$ should be the same in the two cases. Such an experiment has never been reported. If the two ratios should turn out to be the same in general, then a certain element of simplification is possible for diffusion in multicomponent systems; otherwise each flux must in general be represented as a function of all the partial molal free-energy gradients and the only simplification possible is that arising from Onsager's principle. In such case it would be impossible to represent diffusion in a multicomponent system with a simplicity even remotely approaching that for a binary system.

It is seen that the limited form of Fick's law with a constant diffusivity is qualitatively or semiquantitatively fruitful for many isothermal binary or pseudobinary systems. The extended form of Fick's law, with a diffusivity which is dependent on composition, is quantitatively applicable to many more binary systems; even in a binary system, a single diffusivity does not give a complete description of all possible diffusion phenomena. The complete description requires besides the diffusivity another parameter such as the velocity of inert markers; alternatively the phenomena may be described in terms of intrinsic diffusivities. Additional complications are introduced in the mathematical treatment if there is appreciable total volume change accompanying diffusion.

In the general case of diffusion in a multicomponent system, although general phenomenological relations are available, these are by no means simple and it is not known to what extent simplification is permissible. Some simplification is achieved by the introduction of the gradient of the partial molal free energy as the virtual driving force in diffusion. In any event it is clear that diffusion in a multicomponent system cannot be described adequately by a simple extension of the method currently used for binary systems.

CHAPTER 19

KINETICS OF METALLURGICAL PROCESSES[1]

As discussed in the first chapter of this book, there are two general types of scientific approach to a problem. One approach, which is the more widely known, involves an analysis of the situation at the beginning of the experiment; this is followed by an analysis of the situation at each instant of time during the experiment and leads to a detailed description of the prevailing conditions at the end of the experiment. Suppose, for example, that we wish to predict the final rest position on a billiard table of a ball moving initially in a certain known direction with a known speed. The obvious approach is to estimate or measure off its course by consideration of angles of rebound, taking account of the "English," or spin, if necessary, and finally estimating from experience (or calculating from coefficients of friction) the slowing down occasioned by friction, thus establishing the point of the course at which the ball may be expected to stop. Any interference with the calculated course because of collision with an unanticipated object would obviously cause us to revise drastically our estimate of the rest position. This general method of approach characterized by detailed description of how one situation develops from another seems intuitively appealing and is the most widely known method. There is no doubt that it gives very satisfying results when it can be used.

There is a second general method of approach. In this method, the final state of a system is predicted from knowledge of general broad principles and their application, considering only the restrictions imposed upon the system. It is unnecessary to follow the course in detail; intermediate steps are not even considered. For example, if the billiard table mentioned in the first paragraph were tilted to one corner, one could predict that the final resting place of the ball would be in that corner; this prediction can be made independent of any particular course the ball may follow. This second general method of attacking problems is the one used in thermodynamics. There its use is so outstanding and fruitful that the method is sometimes called *the* thermodynamic method.

[1] This chapter originally appeared in substantially its present form as Chap. 15 of "Basic Open Hearth Steelmaking" by L. S. Darken, American Institute of Mining and Metallurgical Engineers, New York, 1951. It is here reproduced by kind permission of the American Institute of Mining and Metallurgical Engineers and of its Committee for the Physical Chemistry of Steelmaking.

A typical application of the thermodynamic method is involved in the prediction of the equilibrium state of a system of specified components at a particular temperature and pressure. In order to make this prediction, no knowledge whatsoever of the detailed nature or course of the reaction is required; in fact, by the aid of catalysts, the reaction may proceed in different cases by different mechanisms and yet reach the same final equilibrium state. This may be likened to a man of his word who has an appointment; we may feel reasonably certain that he will keep it and that, if one road is blocked, he will take another. This analogy may be carried further: If sufficiently serious difficulties arise, he may be late or even fail completely to keep the appointment. Similarly, the equilibrium state may sometimes be achieved very slowly or even not at all during the allotted time.

The general problem of determining the equilibrium state can be handled adequately by the thermodynamic method, as set forth in Chaps. 6 through 9. However, the problems involved in the manner and rate of approach to the equilibrium state clearly necessitate a different method. On the basis of the general principles outlined in the preceding paragraphs, we should expect that this different method would call for either (1) a new broad general principle or (2) a detailed knowledge of the atomistics of each situation, *i.e.*, a detailed knowledge of the behavior of individual atoms during the course of a reaction. Until the last decade or two, neither of these alternative requisites for an understanding of chemical-rate phenomena was available, and rate phenomena, except for gaseous reactions, were treated on an essentially phenomenological or descriptive basis.

RATE THEORY

The foundation of present rate theory goes back historically to Arrhenius, who found that for many processes the specific reaction-rate constant k may be written as a function of temperature in the following way:

$$k = Ae^{-Q/RT}$$

Alternatively, log k is a linear function of $1/T$. The quantity Q, although usually determined only from the slope of such a plot, is regarded as the energy or heat of activation of the reaction. The formal identity of this relation to the corresponding one involving the equilibrium constant is apparent [see Eq. (9-31)].

For gaseous reactions, it has been shown by comparison of experiment with kinetic theory that the constant A in the Arrhenius equation includes a probability factor; the probability of a collision resulting in reaction is relatively high when size and orientation factors are favorable and relatively low when these factors are unfavorable. Extension of this type of treatment is usually known as the classic or collision theory. In spite of its partial success over a period of about fifty years in dealing with gaseous reactions, its inherent difficulties in dealing with rate phenomena in condensed phases have long been apparent. Although we are still a long way from a complete understanding of rate phenomena and although the theory of this whole field is still in a somewhat fluid state, far from perfection, it is the feeling of the authors and of many others that the theory of absolute reaction rates is intellectually more satisfactory than the classical theory. Hence this theory will be briefly described in the following pages. Those wishing to apply the classical theory can find it in many texts.[1]

The Theory of Absolute Reaction Rates. This theory is also known as the *semiempirical theory of rates*, the *activated complex theory*, and the *transition-state method*. Many have contributed to the development of this theory; among the many may be mentioned London,[2] Marcelin, Lindemann, Polanyi, Rice, Rodebush, Dushman, and Eyring.[3] An outstanding number of successful applications have been made by Eyring and coworkers.

This theory involves both of the requisites mentioned in a preceding paragraph: A new broad principle is introduced, and a detailed knowledge of atomic behavior is required for a complete solution. However, in spite of the major advances in quantum mechanics, the knowledge of atomic behavior during reaction is still inadequate to enable us to calculate rates from atomic properties alone. Even in the simplest case, that of gases, a complete theoretical solution has not been achieved. For solids and liquids, the theory has provided a model or picture and the general form of equation, just as thermodynamics provides the general form of equation for the variation of the equilibrium constant with temperature. In fact, the theory is in part thermodynamic, and the concept of free energy plays a very important role.

The modern rate theory assumes that any observable process (usually chemical but sometimes in a broader sense) may be described adequately in terms of the energy of the atomic configurations involved. For any conceivable way the reaction might occur, we may imagine a plot of energy (for the unit atomic process) vs. a distance coordinate; this latter in the case of single atom may be visualized as the distance traveled by

[1] For example, J. W. Mellor, "Chemical Statistics and Dynamics," Longmans, Green & Co., Inc., New York, 1904, or C. N. Hinshelwood, "Kinetics of Chemical Change," Oxford University Press, New York and London, 1942.

[2] F. London, "Probleme der Modernen Physik," p. 104, Sommerfeld Festschrift, S. Hirzel, Leipzig, 1928.

[3] H. Eyring, *J. Chem. Phys.*, 3, 107 (1934).

the atom along whatever curved or zigzag path it may be conceived as taking. Any such curve will exhibit a maximum energy at some point. Of all possible paths, there will be one path for which the maximum energy is lower than for all other paths. This path is considered to be the predominant actual path taken during the course of reaction and is known in general as the reaction coordinate. A plot of energy against the reaction coordinate is shown schematically in Fig. 19-1. This sort of generalization may readily be understood in terms of a mechanical analogue. The system may be regarded as a ball or marble, free to roll under the influence of gravity on a track whose height (gravitational

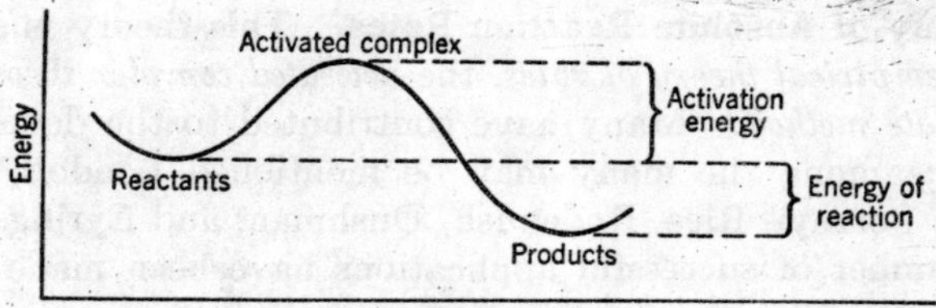

FIG. 19-1. Schematic potential-energy curve for a reaction.

energy) is given by the curve. Any point where the marble may rest is obviously a minimum of the curve; such a point corresponds to a state of the atoms, which persists long enough to be observed as represented by reactants or products. The marble, if initially at rest at the minimum designated *Reactants*, would, of course, never of its own accord move to the minimum designated *Products*, even though this direction is "downhill" and energy is lower there, since it must first acquire the *activation energy* necessary to surmount the peak designated *Activated complex*. A chemical system at the absolute zero of temperature would be subject to the same restriction; hence, since no energy can be acquired, it is readily inferred that chemical reactions do not occur at this temperature. At any finite temperature, however, any small assemblage of atoms has a finite probability of acquiring the necessary activation energy from the energy of thermal agitation of the entire system; it is obvious that the higher the activation energy, the lower is this probability.

The two fundamental principles of the modern reaction-rate theory are both concerned with this activated complex. They are as follows:

1. The activated complex, though of exceedingly short life, may be treated as any other chemical species (*e.g.*, as having a definite set of thermodynamic functions) and *is in equilibrium with the reactants*. The chemical equation for its formation from the reactants may be set up, and the corresponding thermodynamic equilibrium constant may be written. The superscript dagger † is commonly used to designate a quantity per-

taining to an activated complex; the equilibrium constant for the formation of the activated complex from the reactants is written $K^\dagger$. If $K^\dagger$ and the concentrations of the reactants are known, the concentration of the activated complex is readily calculable therefrom in the same way as for any other equilibrium (strictly, *activity* should be written in place of *concentration*). The free energy of formation of the activated complex in the standard state is written $\Delta F^\dagger$. (The superscript zero to designate standard state is commonly omitted here.) The two are related in the usual way, $\Delta F^\dagger = -RT \ln K^\dagger$.

2. The specific rate of decomposition of the activated complex into products is a universal rate independent of the nature of the particular reaction or particular activated complex. The rate is RT/Nh, where R is the gas constant, h is Planck's constant, and N is Avogadro's number.[1] The average life of the activated complex is the reciprocal of this, namely, Nh/RT, which is $(4.77/T) \times 10^{-11}$ sec, or at 1873°K is 2.55×10^{-14} sec. Thus the life period of an activated complex is indeed very short compared with the usual time of observation of reactions, and it is obvious that we can never hope to isolate an activated complex from the reaction mixture. The high magnitude of this universal rate also makes it clear that reactions which require measurable time to proceed do so not on account of slowness of an individual group of atoms in passing through the activated complex stage, but rather because there are so few activated complexes; that is, $K^\dagger$ is small.

If the reactants be designated R_1, R_2, etc., and the activated complex A, the chemical reaction for formation of the activated complex becomes

$$R_1 + R_2 + \cdots = A$$

and the equilibrium constant is

$$K^\dagger = \frac{a^\dagger}{a_{R_1}a_{R_2} \cdots} = \frac{C^\dagger}{C_{R_1}C_{R_2} \cdots} \frac{\gamma^\dagger}{\gamma_{R_1}\gamma_{R_2} \cdots} \qquad (19\text{-}1)$$

where a represents activity, C concentration (usually in moles per unit volume), and γ the activity coefficient. The reaction rate, or number of moles of activated complex per unit volume[2] that decompose per unit time, is $(RT/Nh)C^\dagger$. Substituting for $C^\dagger$ from the expression for $K^\dagger$, it is then found that

$$\text{Reaction rate} = \left(\frac{RT}{Nh} K^\dagger \frac{\gamma_{R_1}\gamma_{R_2} \cdots}{\gamma^\dagger}\right) C_{R_1}C_{R_2} \cdots \qquad (19\text{-}2)$$

[1] This rate is frequently written as kT/h, where $k = R/N$, Boltzmann's constant.

[2] Providing the concentration is expressed as moles per unit volume. Otherwise the unit is that of the concentration per unit time.

The coefficient $[(RT/Nh)K^\dagger(\gamma_{R_1}\gamma_{R_2}/\gamma^\dagger)]$ of the concentration terms is known as the *specific reaction rate constant*, which is usually designated k. The expression relating $\Delta F^\dagger$ and $K^\dagger$ may be rewritten in exponential form as

$$K^\dagger = e^{-\Delta F^\dagger/RT} = e^{-\Delta H^\dagger/RT}\, e^{\Delta S^\dagger/R}$$

the second equality following from the definitional equation

$$\Delta F^\dagger = \Delta H^\dagger - T\,\Delta S^\dagger$$

Inserting this expression for $K^\dagger$ in Eq. (19-2), the reaction rate may now be written:

$$\text{Reaction rate} = \left(\frac{RT}{Nh}\frac{\gamma_{R_1}\gamma_{R_2}\cdots}{\gamma^\dagger}\, e^{-\Delta H^\dagger/RT}\, e^{\Delta S^\dagger/R}\right) C_{R_1}C_{R_2}\cdots \qquad (19\text{-}3)$$

and the specific reaction-rate constant (the coefficient of the concentration terms) is

$$k = \frac{RT}{Nh}\frac{\gamma_{R_1}\gamma_{R_2}\cdots}{\gamma^\dagger}\, e^{-\Delta H^\dagger/RT}\, e^{\Delta S^\dagger/R} \qquad (19\text{-}4)$$

If the activity coefficients are unity, this simplifies to

$$k = \frac{RT}{Nh}\, e^{-\Delta H^\dagger/RT}\, e^{\Delta S^\dagger/R}$$

If, further, $\Delta H^\dagger$ and $\Delta S^\dagger$ are essentially independent of temperature, it will be noted that the expression is of the form $k = aTe^{-\Delta H^\dagger/RT}$ where the constant $a = (R/Nh)e^{\Delta S^\dagger/R}$. This expression gives the expected temperature dependence of the specific reaction-rate constant. Relatively little error, over a small temperature range, is involved in the approximation[1]

$$k = aeT_m e^{-(\Delta H^\dagger + RT_m)/RT}$$

where T_m is the mean temperature of the range under consideration. This form is identical with the old formulation of Arrhenius, who wrote $k = Ae^{-Q/RT}$.

Returning now to the more general formulation but considering $\Delta H^\dagger$ and $\Delta S^\dagger$ constant, the expression for the reaction rate is conveniently written in five terms:[2]

[1] This approximation results from noting that T may be written:

$$T = T_m\left(1 + \frac{T - T_m}{T_m}\right) \cong T_m\, e^{(T-T_m)/T} = T_m\, e \cdot e^{-T_m/T}$$

[2] For the sake of simplicity, a sixth term, the transmission coefficient, is omitted from the discussion. The transmission coefficient is the probability that the activated complex will decompose to products rather than to the reactants from which it was formed.

Reaction rate =

$$\left(\frac{RT}{Nh}\right)(e^{-\Delta H^\dagger/RT})(e^{\Delta S^\dagger/R})\left(\frac{\gamma_{R_1}\gamma_{R_2}\cdots}{\gamma^\dagger}\right)(C_{R_1}C_{R_2}\cdots) \quad (19\text{-}5)$$

The following facts are to be noted about these terms (the first four together constitute the specific reaction-rate constant):

1. The first term contains only universal constants and the absolute temperature and hence is independent of the particular reaction under consideration.
2. The second term is the principal temperature-dependent term, involving the enthalpy (heat) of activation.
3. The third term depends on the entropy of formation of the activated complex. In a general way, it is to be expected that, if the activated complex is relatively simple, $\Delta S^\dagger$ is small and, if it is complicated (as compared with the reactants) in atomic configuration, $\Delta S^\dagger$ is large and negative.
4. The fourth term involves the activity coefficients, which may or may not remain essentially constant (or unity) during the course of reaction.
5. The fifth term, which is the only one of the five not included in the specific reaction-rate constant, is the product of the concentrations of the reactants.

It is to be emphasized that the foregoing discussion of reaction rate may be applied at any one time to only one particular step in a chemical reaction. The usual over-all chemical reaction consists in general of several chains of reactions all proceeding simultaneously. The complications introduced thereby may be so great as to make the calculation prohibitively difficult. However, in especially favorable cases, there is only one important sequence of reactions leading to the final products, and this sequence has one step which is so much slower than all others that it may be considered as responsible for the observed rate.

Let us now compare the accepted mechanisms of two relatively simple gas reactions:

$$H_2 + I_2 \rightarrow H_2I_2 \rightarrow 2HI \quad (a)$$

$$Cl_2 \rightarrow 2Cl;\ Cl + H_2 \rightarrow H_2Cl \rightarrow HCl + H;$$
$$H + Cl_2 \rightarrow HCl_2 \rightarrow HCl + Cl;\ Cl + H_2 \rightarrow H_2Cl \rightarrow HCl + H, \text{etc.} \quad (b)$$

The first reaction proceeds in the two steps characteristic of the unit process as considered above; the activated complex H_2I_2 forms directly from the reactants H_2 and I_2 and decomposes directly to the product HI. The second reaction involves the primary photochemical dissociation of Cl_2 to atoms; this is followed by a reaction sequence each step of which yields HCl and an atom of hydrogen or chlorine; these atoms then act as reactants in the next step of the sequence. The rate of the reaction involving Cl and H_2 is considerably slower than that involving H and Cl_2; hence the former is considered as the rate-determining step.

The elementary steps in chemical reactions are usually simple, as in the foregoing examples. Difficulty in treatment arises principally in the determination of the elementary steps, particularly in cases where a large number of different steps occur in a sequence or where the over-all reaction proceeds by more than one sequence.

Classification of Reactions. A reaction may be classified according to its *molecularity*, *i.e.*, the number of molecules or molecules + atoms, taking part as reactants in the formation of the activated complex. On this basis, the elementary steps of any over-all reaction are designated monomolecular, bimolecular, trimolecular, and so forth. As mentioned previously, most known elementary reactions are simple and hence are usually monomolecular or bimolecular (counting atoms as molecules). Clearly, an over-all reaction involving many steps cannot properly be thus classified; only the individual elementary steps of an over-all reaction may be classified in this manner.

Over-all reactions are commonly classified empirically according to the number of concentrations to the product of which the rate is proportional. The *reaction order* may be defined as the sum of the powers to which the concentrations of the reactants must be raised to give a satisfactory solution to the rate equation; reactions are designated as of first order, second order, and so on, according as the sum of the powers is equal to one, two, etc. Zero and fractional orders are sometimes found. For a simple over-all reaction such as $H_2 + I_2 \rightarrow H_2I_2 \rightarrow 2HI$, in which only one elementary step is involved, the order and molecularity are identical; this particular reaction is bimolecular and of second order. This identity of molecularity and order in such a simple case is readily understood by reference to the fifth term of the reaction-rate equation derived previously [Eq. (19-5)]. It will be noted that a concentration term appears here for every molecule of reactant appearing in the chemical equation for the formation of the activated complex.

In a purely formal way, the over-all reaction rate (in contrast to the rate of a step) is sometimes written as proportional to powers of the concentrations of the reactants even for complicated reactions. This is a gross oversimplification and in general leads to powers that are not even constant; for short concentration ranges, however, the powers may be substantially constant, though frequently fractional.

Integrated Form of Equation for First-order Reaction. In the most ortunate cases, as mentioned previously, the observed over-all reaction proceeds principally by one chain or sequence of steps, and one of these steps is so much slower than the others that it is responsible for the observed rate. This slow step involves, as do all steps, the formation of an activated complex and its decomposition. If this step is unimolecular, the rate of the over-all reaction will be of the first order. The reaction rate is the rate of disappearance of reacting species, which is then proportional at constant temperature to the concentration thereof, or

$$-\frac{dC}{dt} = kC$$

Upon integration, this gives

$$-\ln\frac{C}{C_0} = k(t - t_0)$$

where C_0 is the initial concentration at time t_0.

It sometimes happens in bimolecular or more complicated reactions that one reactant is present in much smaller concentration than the others, so that the concentration of these others changes relatively little during the reaction. Thus, although the expression for the rate

$$-\frac{dC}{dt} = kC_1C_2C_3 \cdots$$

contains several concentration terms, all except C_1 in such a case may be regarded as constant, so that integration yields the same logarithmic form as for unimolecular reactions. Throughout this discussion, it has been assumed that the reverse reaction does not occur to any appreciable extent, *i.e.*, that the system is so far from equilibrium that the reaction goes in one direction only and that activated complexes are formed from reactants only, not from products.

The Decomposition of ϵ Iron Nitride—a Second-order Reaction. Goodeve and Jack[1] measured the rate of decomposition of the ϵ phase (the high-nitrogen limit of this phase approximates the composition Fe_2N) of the iron-nitrogen system in the temperature range 350 to 500°C. Using a powder composed of nearly spherical particles (diameter 4×10^{-4} cm) they found that the reaction was of second order, *i.e.*, that the rate of loss of nitrogen was proportional to the square of the nitrogen content. Representing the nitrogen content, expressed as atoms of nitrogen per 100 atoms of iron, as (N),

$$\frac{d(\mathrm{N})}{dt} = -k'(\mathrm{N})^2$$

[1] C. Goodeve and K. H. Jack, *Discussions of the Faraday Society*, **4**, 82 (1948).

Integrating from zero time when (N) = $(N)_0$ to any arbitrary time t,

$$\frac{1}{(N)} - \frac{1}{(N)_0} = k't$$

or

$$\frac{(N)_0 - (N)}{(N)} = N_0 k't$$

That the observations fit this expression for a second-order reaction is verified by plotting $[(N)_0 - (N)]/(N)$ against t. A typical plot is shown in Fig. 19-2. The slope of this line divided by $(N)_0$ gives k', which in this case is 1.08×10^{-8} (nitrogen atoms per 100 iron atoms)$^{-1}$ sec^{-1}. The variation of log k' with $1/T$ as found by a series of such experiments *in vacuo* at temperatures from 350 to 500°C is shown in Fig. 19-3. From this plot it is found that

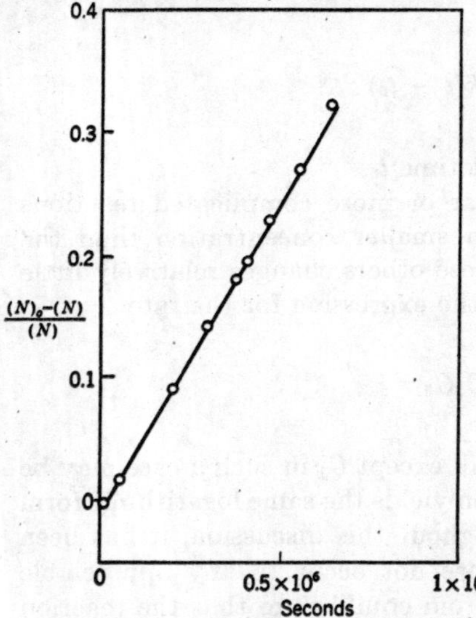

FIG. 19-2. Decomposition of ε iron nitride at 350°C.

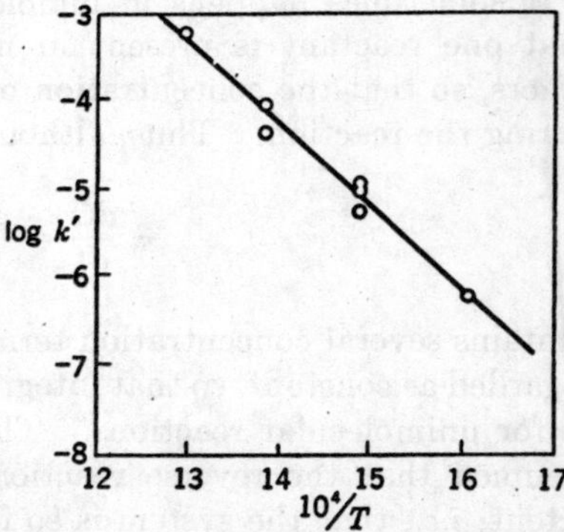

FIG. 19-3. Variation with temperature of the reaction rate constant k' for the evolution of nitrogen from ε iron nitride.

$$k' = 6.3_5 \times 10^6 e^{-42,100/RT}$$

Let us now attempt to develop an approximate expression for k' by means of the previously developed theory and a model. It will be assumed that the reaction proceeds at the surface by the combination of two "dissolved" nitrogen atoms uniting to form ultimately an N_2 molecule (the distinction in properties between nitrogen atoms in the interior and on the surface is tentatively ignored). Since there is a strong tendency for nitrogen atoms in iron and iron nitrides to keep as far apart as possible, as evidenced by the ordered structure of the nitrides, we know there is a repulsive force between them which becomes an attractive force only at distances sufficiently short that the triple bond of the nitrogen molecule may be formed. At some distance between the normal spacing (5 A) of nitrogen atoms in the nitride and the separation (1.5 A) in the nitrogen molecule, the energy required passes through a maximum. This position of maximum energy requirement may be considered as corresponding to the activated complex.

The reaction-rate constant, ignoring activity coefficients, is

$$k = \frac{RT}{Nh} K^{\dagger}$$

where $K^{\dagger} = C_{N_2}{}^{\dagger}/C_N^2$ is the equilibrium constant for the surface reaction

$$2N = N_2{}^{\dagger}$$

and concentrations are conveniently expressed in moles, or gram atoms, per mole of total available interstitial spaces on the surface. With this choice of standard state and with this relatively simple reaction, it seems as though no gross error is involved in setting $\Delta F^{\dagger} \cong \Delta H^{\dagger}$, whence

$$k = \frac{RT}{Nh} e^{-\Delta H^{\dagger}/RT} = \frac{RT_m e}{Nh} e^{-(\Delta H^{\dagger}+RT_m)/RT}$$

The rate constant k is seen to have the dimension (moles of N_2 per mole of surface interstitial positions)$^{-1}$ per second. To convert to the units in which k' and concentration are expressed by Goodeve and Jack, k must be multiplied by the ratio of the total number of hundred interstitial positions on the surface of the particle to four times the total number of iron atoms in the particle.[1] This ratio, calculated from data on the crystallography given by Goodeve and Jack, is 6.1×10^{-6}. Thus the calculated value of k' is

$$6.1 \times 10^{-6} \frac{RT_m e}{Nh} e^{-(\Delta H^{\dagger}+RT_m)/RT} = 12.2 \times 10^7 e^{-(\Delta H^{\dagger}+RT_m)/RT}$$

By comparison with the empirical equation it is seen that the coefficient of the exponential is too large by a factor of about 20. It seems quite possible that the principal error in the theoretical treatment lies in the assumption that the distribution of nitrogen atoms among available positions is the same on the surface as in the interior of a particle. In view of the large attractive force between iron and nitrogen atoms it seems only

[1] This relation is found by expressing the rate in the theoretical equation in the same units as in the empirical and then equating the two rates. The factor 4 arises as the square of the ratio of the number of iron atoms to the number of interstitial positions.

reasonable to expect the relative concentration of nitrogen to be greater in the interior where each nitrogen atom is surrounded by iron atoms than on the surface where each nitrogen atom can at best be only partially surrounded by iron atoms. A correction of one-twentieth for this factor does not seem at all unreasonable.[1]

The energy of activation may also be estimated by Eyring's empirical rule that this energy is about one-quarter of the energy of the bonds broken in forming the activated complex. The bonds broken in this case are the bonds from two nitrogen atoms to the surrounding iron atoms, and this energy is about 169,000 cal.[2] One-fourth of this is about 42,000 cal, in good agreement with the observed value of 42,100 ± 1400 cal.

Thus if we include our estimate for the relative depletion of nitrogen at the surface, the estimated theoretical expression for k' is

$$k'_{\text{theor.}} = 6.1 \times 10^6 e^{-42,100/RT}$$

whereas the relation based on direct observation is

$$k'_{\text{obs.}} = 6.3_5 \times 10^6 e^{-42,100/RT}$$

The agreement is rather fortuitous (perhaps somewhat forced by prior knowledge of the observed values); order-of-magnitude agreement is all that can usually be expected of such calculations.

Goodeve and Jack calculated the coefficient of the exponential by classical kinetic theory and also obtained good agreement; they made no correction for the depletion of nitrogen at the surface. A discussion of their results is incomplete without mention of their further experiments on nitrogen evolution from the ϵ phase in an atmosphere of hydrogen. In this case the rate is 10,000 times greater. The conclusion is reached that hydrogen removes the nitrogen as ammonia (which was verified experimentally) and that in this case diffusion of nitrogen to the surface is the slow step. The rate was found to be substantially the same in an atmosphere of nitrogen or carbon monoxide as *in vacuo*.

Reaction Sequence with Two Slow Steps. Let us now consider an over-all reaction with two slow steps, neither of which is sufficiently slow

[1] The three nitrogen bonds may resonate among eight positions in the interior and among only about half as many at the surface. The ratio of the permutations indicative of the relative distribution is $(4 \cdot 3 \cdot 2)/(8 \cdot 7 \cdot 6) = \frac{1}{14}$. Again, the distribution of nitrogen between ferrite (few positions for bonds) and austenite (many positions for bonds) is about $\frac{1}{20}$.

[2] This value is found as the heat of solution of monatomic nitrogen in austenite:

$$\begin{array}{lll} 2N(g) = N_2(g) & \Delta H = 169,000 \text{ cal} \\ N_2(g) = 2N(\text{in } \gamma\text{-Fe}) & \Delta H \cong 0 \\ \hline 2N(g) = 2N(\text{in } \gamma\text{-Fe}) & \Delta H \cong 169,000 \text{ cal} \end{array}$$

that the other may be considered very fast in comparison. Let it be supposed, as a simple example, that the concentrations of reactants A, B, and C are held constant by continuous supply, that A and B react to give an intermediate product X, which may be isolated, and that this intermediate product reacts with another reactant C to give a final product D. The observable reactions are

$$A + B \rightarrow X$$
$$X + C \rightarrow D$$

It is to be understood that there may be other products besides X and D. The simplest mechanism would perhaps be

$$+ B \rightarrow AB \rightarrow X \qquad (a)$$
$$+ C \rightarrow XC \rightarrow D \qquad (b)$$

The rate r_1 of the first reaction is $k_1C_AC_B$, which is constant under the supposed condition that C_A and C_B are held constant. The rate of the second reaction r_2 is $k_2C_XC_C$. The intermediate product X is thus produced by the first reaction at the rate $k_1C_AC_B$ and consumed by the second at the rate $k_2C_XC_C$, and hence its net rate of accumulation is

$$\frac{dC_X}{dt} = k_1C_AC_B - k_2C_XC_C$$

As we have arbitrarily taken all these concentrations except C_X as constant, the expression is readily integrated (taking $C_X = 0$ at $t = 0$) to give

$$1 - \frac{k_2}{k_1}\frac{C_XC_C}{C_AC_B} = e^{-k_2C_Ct}$$

From this it is seen that, as time progresses, the term on the right becomes smaller and a steady state is approached as this term approaches zero. At steady state,

$$1 - \frac{k_2}{k_1}\frac{C_XC_C}{C_AC_B} = 0$$

and C_X is a constant, $(k_1/k_2)(C_AC_B/C_C)$, dependent on the two reaction-rate constants and on the arbitrary choice of C_A, C_B, and C_C. The rate of formation of the final product D at the steady state is

$$r_2 = k_2C_XC_C = k_1C_AC_B$$

The carbon oxidation in the open hearth bears a strong superficial resemblance to the pair of consecutive reactions discussed above. The first reaction consists of the transfer of oxygen from the slag to the metal; the second consists of the reaction of dissolved carbon and oxygen to give gaseous oxides of carbon. However, this resemblance is somewhat misleading, as the carbon-oxygen reaction in the open hearth is not very far from equilibrium; similarly, the reaction involving the transfer of oxygen from slag to metal is probably not so far from equilibrium that the reverse reaction can be completely ignored. The discussion so far has not treated such reversible reactions.

Reactions Near Equilibrium. If a system is sufficiently near equilibrium, the reverse reaction must also be considered. Considering only first order reactions, we may write, for the chemical reactions (1) $A \rightarrow B$ and (2) $B \rightarrow A$, the corresponding rate equations $r_1 = k_1C_A$ and $r_2 = k_2C_B$. The over-all rate corresponding to the gross progress according to reaction (1) is then

$$r = r_1 - r_2 = k_1C_A - k_2C_B$$

At equilibrium $r_1 = r_2$ and

$$\frac{k_1}{k_2} = \frac{C_B}{C_A}$$

which, except for the absence of activity coefficients, is the equilibrium constant K. In general for a reaction not at equilibrium there exists for any concentration of A a corresponding equilibrium concentration of B designated C_B^E, which is equal to KC_A, or $C_A = C_B^E(k_2/k_1)$. The over-all reaction rate may thus be written

$$r = k_2(C_B^E - C_B)$$

or, denoting as C_A^E the concentration of A that would be in equilibrium with a given concentration B, we may similarly write

$$r = k_1(C_A - C_A^E)$$

Let us now consider again a sequence of reactions with two slow steps but this time with both of the slow steps near equilibrium. Under the same restrictions as previously (constant concentrations of all reactants but one), we have

$$A \rightarrow X \qquad X \rightarrow A \qquad (a)$$
$$X \rightarrow B \qquad B \rightarrow X \qquad (b)$$

Substances at constant concentration have been omitted from the equation. From the preceding development, the over-all rates for these two reactions in the forward direction are

$$r_a = k_a(C_X^E - C_X)$$
$$r_b = k_b(C_X - C_X^{E'})$$

where C_X^E and $C_X^{E'}$ represent the concentrations of X that would be in equilibrium with the existing concentration of A and B, respectively.

Again we note that, in sufficient time under the stipulated condition of constant concentrations of reactants, a steady state will be reached at which $r_a = r_b$ and

$$\frac{C_X^E - C_X}{C_X - C_X^{E'}} = \frac{k_b}{k_a}$$

Application to the Oxidation of Carbon in the Open Hearth. It would not seem unreasonable to suppose that this method might be applied to the oxidation of carbon in the open hearth. The reactions corresponding to reactions (*a*) and (*b*) are

$$\text{O(slag)} \rightarrow \underline{\text{O}} \qquad \underline{\text{O}} \rightarrow \text{O(slag)} \qquad (a)$$
$$\underline{\text{O}} + \underline{\text{C}} \rightarrow \text{CO(g)} \qquad \text{CO(g)} \rightarrow \underline{\text{O}} + \underline{\text{C}} \qquad (b)$$

If the reasonable assumption is made that these are both surface reactions, the total rates are proportional to slag-metal and gas-metal surfaces. Designating the actual surface per unit cross section as S_{sm} and S_{gm}, respectively, the rate equations become

$$r_a = k_a S_{sm}([\text{O}]_{se} - [\text{O}])$$
$$r_b = k_b S_{gm}([\text{O}] - [\text{O}]_{ce})$$

where the subscripts *se* and *ce* refer to slag equilibrium and carbon equilibrium, respectively; that is, $[\text{O}]_{se}$ and $[\text{O}]_{ce}$ are the concentrations of oxygen that would exist if the metal were in equilibrium with the existing slag or carbon, respectively. The rate of reaction (*b*) is regarded as limited by oxygen rather than carbon, as the oxygen concentration is normally much the smaller.

In view of the low oxygen content of the metal at any instant as compared with the total amount of oxygen required for the oxidation of the carbon, it would not seem unreasonable to suppose that a substantially steady state prevails under favorable conditions of steady boil. If this is so, the two rates may be equated:

$$\frac{[\text{O}] - [\text{O}]_{ce}}{[\text{O}]_{se} - [\text{O}]} = \frac{k_a S_{sm}}{k_b S_{gm}}$$

The numerator of the left side has been designated $\Delta[\text{O}]$ by Brower and Larsen,[1] who have found this a very useful quantity in interpreting the kinetics of this reaction. Setting $[\text{O}] - [\text{O}]_{ce} \equiv \Delta[\text{O}]$, the foregoing equation may be rewritten:

$$\Delta[\text{O}] = \frac{k_a S_{sm}}{k_b S_{gm}} ([\text{O}]_{se} - [\text{O}]) \qquad \textbf{(19-6)}$$

[1] T. E. Brower and B. M. Larsen, *Trans. AIME*, **172**, 137, 164 (1947).

At a constant temperature, k_a and k_b are substantially constant, and for a given furnace the slag-metal interface area probably does not vary much. Furthermore, as the actual oxygen content [O] of the metal is usually rather small compared with the oxygen content $[O]_{se}$, which would be in equilibrium with the slag, $[O]_{se}$ is the predominant term of the two in parentheses. Hence the equation leads us to expect that for a particular furnace at constant temperature $\Delta[O]$ will depend principally upon

1. The slag, or, more precisely, the oxide layer immediately over the metal; *e.g.*, $\Delta[O]$ should be high after an ore addition.

2. The area of gas-metal interface. This interface is generated principally by the reaction series itself and is somewhat difficult to predict; however, this area is undoubtedly increased by a lime boil; the equation predicts that such increase of S_{gm} leads to a decrease of $\Delta[O]$.

Both of these expectations, as well as the implied independence of $\Delta[O]$ of carbon content, were verified by the findings of Brower and Larsen[1] in their investigations of many open-hearth heats.

This application to oxidation of carbon in the open hearth leaves much to be desired but does take into consideration the all-important fact that the observed oxygen content of the bath is not in equilibrium with either the slag or the carbon but is between the two. Neither the transfer of oxygen from the slag nor its liberation as CO and CO_2 is a step sufficiently slower than the other that either may be considered alone as rate controlling. For example, it is well known that the over-all reaction rate, *i.e.*, rate of carbon drop, may be increased by oreing, which speeds up reaction (*a*), or by poling, which speeds up reaction (*b*). At least two steps must be considered in order to obtain even a superficial understanding of carbon oxidation.

Aside from the several simplifying assumptions already mentioned in the treatment of carbon oxidation, attention should be called particularly to the fact that the atomistic nature of the process has been ignored. The nature of the activated complexes has not been considered. Hence we have been forced to the expedient of assuming that the concentrations of $\underline{O}$ and $\underline{C}$ are equal or proportional to the respective concentrations of the actual reactants in the formation of the activated complex. Attention has already been called to the high universal specific rate RT/Nh of decomposition of the activated complex. The actual surface reactions involved in the oxidation of carbon in the open hearth would not seem to be such as to offer any particular hindrance to the formation of such a complex once the reactants are present. Thus, if we visualize the process of CO evolution into any already existing bubble, the steps we might consider would be the migration of the dissolved $\underline{C}$ and $\underline{O}$ to adjacent

[1] T. E. Brower and B. M. Larsen, *Trans. AIME*, **172**, 137, 164 (1947).

positions on the surface, the formation of a molecule of CO on the surface, and finally its departure into the gas phase.

If we postulate, in accord with the findings of Marshall and Chipman,[1] that at high carbon content at least one carbon atom is nearly always in the vicinity of each oxygen atom in liquid iron, the reaction rate to form CO at the surface is nearly unimolecular, involving the transfer by an oxygen atom of its bonds from the normal positions in the liquid to a carbon atom. According to the previously developed theory [see Eq. (19-5)], the rate of this reaction is approximately

$$\frac{RT_m e}{Nh} e^{-(\Delta H^{\dagger}+RT_m)/RT} C_{ox}$$

where C_{ox} is taken as the atom fraction of oxygen on the surface at which CO is evolved, and hence the rate given by the above expression is in atoms of oxygen or carbon per atom of iron at the surface per unit time. To convert to a rate based on the total number of iron atoms, this expression must be divided by the ratio of the number of iron atoms on the surface to the total number. Estimating this as the ratio of the thickness of the surface layer (2×10^{-8} cm) to the bath depth (70 cm) and converting from atom fraction per second to percentage of carbon per hour, the rate is found to be approximately

$$10^8 e^{-\Delta H^{\dagger}/RT}[\mathrm{O}]_s$$

where $[\mathrm{O}]_s$ is the oxygen content of the surface layer in weight per cent. Estimating $\Delta H^{\dagger}$ as about one-half the corresponding value found by Goodeve and Jack[2] for nitrogen evolution, namely, as 20,000 cal, and tentatively considering $[\mathrm{O}]_s$ to be identical with the corresponding value ($[\mathrm{O}] \cong 0.04$ per cent) for the mass of the metal, the rate is found to be roughly 10^4 per cent carbon per hour. As this value is absurdly high compared with observed rates of less than 1 per cent per hour, we are forced to the conclusion that this reaction depletes the surface of oxygen to such an extent that $[\mathrm{O}]_s$ is very much smaller than [O]. Hence this reasoning suggests that diffusion is probably the rate-controlling step in the over-all reaction. This subject will be treated again later.

DIFFUSION

The formal development of the fundamental diffusion equations was discussed in Chap. 18. The diffusivity (or coefficient of diffusion) D of one substance in another is defined by the following relation:

$$\text{Flux by diffusion through unit area} = J = -D\frac{\partial C}{\partial x}$$

[1] S. Marshall and John Chipman, *Trans. ASM*, **30**, 695 (1942).

[2] C. Goodeve and K. H. Jack, *Discussions of the Faraday Society*, **4**, 82 (1948).

where C is the concentration and x the distance in the direction in which diffusion occurs. This is known as Fick's first law. By consideration of the conservation of mass, *i.e.*, the fact that the rate of increase of any element in a given volume is equal to the difference between the inward and outward diffusion rates, it is readily shown that

$$\frac{\partial C}{\partial t} = \frac{\partial}{\partial x}\left(D\,\frac{\partial C}{\partial x}\right)$$

which is known as Fick's second law.

Representation of the Diffusivity, D, in Terms of Atomic Quantities by Means of the Reaction-rate Theory.[1] Considering, for simplicity, only interstitial diffusion, the fundamental step is the passage of a solute atom from one interstitial position to an adjacent interstitial position. A solute atom midway between two such positions is in the activated state. From the general reaction-rate equation, the rate of passage of atoms from one interstitial position to another, considering activity coefficients constant, is $(RT/Nh)K^{\dagger}C$. If $1/m$ represents the fraction of adjacent interstitial positions corresponding to motion in the forward direction (the direction of diffusion) and λ the distance between adjacent interstitial positions, the flux J^{+} of atoms in the forward direction is

$$J^{+} = \frac{RT}{Nh}\,K^{\dagger}C\,\frac{\lambda}{m}$$

and the *net* flux J in the forward direction, which is the difference between the flux in the forward direction and that in the reverse direction, is[2]

$$J = -\,\frac{RT}{Nh}\,K^{\dagger}\,\frac{\partial C}{\partial x}\,\frac{\lambda^2}{m}$$

By comparison with Fick's first law,

$$D = \frac{\lambda^2}{m}\,\frac{RT}{Nh}\,K^{\dagger}$$

If activity coefficients be retained, it is found that

$$D = \frac{\lambda^2}{m}\,\frac{RT}{Nh}\,\frac{K^{\dagger}}{\gamma^{\dagger}}\left(\gamma + C\,\frac{d\gamma}{dC}\right)$$

It is found in the diffusion of carbon in austenite that the terms involving activity coefficients vary enough to give rise to a doubling of the diffusivity as carbon content rises from 0 to about 1 per cent. The varia-

[1] The treatment here follows that of J. C. Fisher, J. H. Hollomon, and D. Turnbull, *Trans. AIME*, **175**, 202 (1948).

[2] The steps omitted in this derivation may be found in the paper by J. C. Fisher, J. H. Hollomon, and D. Turnbull, *Trans. AIME*, **175**, 202 (1948).

tion of the diffusivity with composition usually becomes pronounced when a wide composition range is considered. Although γ can be determined by independent experiments, $\gamma^{\dagger}$ cannot; hence the variation of D with composition cannot fully be predicted and can be determined only by experimental determination of D.

Diffusivity of Elements in Liquid Iron. Holbrook, Furnas, and Joseph[1] and Paschke and Hautmann[2] have measured the diffusivity of several elements in liquid iron. Holbrook, Furnas, and Joseph carried out their experiments by placing relatively pure iron in a small tube, 2- to 4-mm bore, of graphite or porcelain. The iron was melted, and at the zero time for diffusion the tube was immersed in a bath of iron, usually saturated with graphite and containing also manganese, phosphorus, silicon, or sulfur. After about an hour, the tube was removed from the bath and from the furnace. The chilled sample in the small hole was broken into segments, which were analyzed, thus giving the penetration curves of Fig. 19-4. The similarity to the ideal curve of Fig. 18-4 is not pronounced. The departures from the ideal curve are probably due to experimental errors (it is difficult to avoid convection completely) and to variation of D with composition. The diffusivity of each element may be determined by comparison of its penetration curve with the ideal curve of Fig. 18-4. On account of the difference in form, however, the value of D so determined would depend on the point used in the comparison. The selection of the mid-point, *i.e.*, the point at which

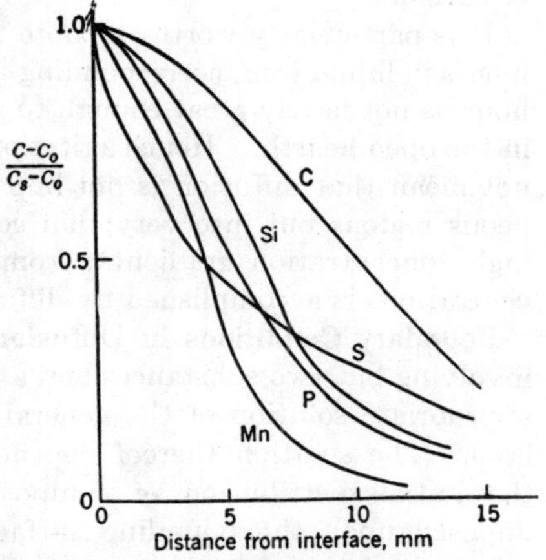

FIG. 19-4. Average diffusion penetration into liquid iron at 1200 to 1600°C; ½ hr for *C* and *S*, about 1 hr for the others. [*From Holbrook, Furnas, and Joseph, Ind. Eng. Chem.*, **24**, 993 (1932).]

$$\frac{C - C_0}{C_s - C_0} = \frac{1}{2},$$

tends to minimize experimental errors. The nature of the data would not seem to justify a more refined method.[3] From Fig. 18-4, it is found

[1] W. F. Holbrook, C. C. Furnas, and T. L. Joseph, *Ind. Eng. Chem.*, **24**, 993 (1932).

[2] M. Paschke and A. Hautmann, *Arch. Eisenhüttenw.*, **9**, 305 (1935).

[3] However, it is of interest to note that those elements (C, Si, P) whose binary solutions with iron exhibit negative departures from Raoult's law give diffusion penetration curves that exhibit a rather marked inflection. On the other hand, sulfur,

that $x/\sqrt{Dt} = 0.954$ at this point, whence $D = 1.099x_1^2/t$ where x_1 is the penetration distance at which $(C - C_0)/(C_s - C_0) = \frac{1}{2}$. By this method, it is found that for carbon $D = 6 \times 10^{-4}$ cm²/sec; for Si, P, S, and Mn, D is about 1×10^{-4} cm²/sec. Paschke and Hautmann's results ($D_{Mn} = 1.2 \times 10^{-4}$; $D_{Si} = 0.4 \times 10^{-4}$) are in fair agreement. The variation of D with temperature is small in contrast to the case with solids and is less than the experimental error. These values are averages over the temperature range 1200 to 1700°C; they apply to iron saturated with graphite, except in the experiments for the determination of the diffusivity of carbon.

It is particularly worthy of note that the rate of diffusion of these elements in liquid iron, corresponding to a penetration of roughly 1 cm in an hour, is not nearly great enough to permit equalization of concentrations in the open hearth. Hence agitation is extremely important. This does not mean that diffusion is not important. Stirring stretches inhomogeneous regions out into very thin convoluted strata with large area and high concentration gradients; from this point the equalization of concentrations is accomplished by diffusion.

Boundary Conditions in Diffusion Problems. All diffusion problems involving but two substances and a constant diffusivity can be solved by appropriate solution of the general differential equation (Fick's second law). The solution thereof depends on the boundary conditions, *i.e.*, the initial distribution of composition and the arbitrary conditions imposed upon the bounding surfaces of the system. Many different solutions of this equation appropriate to different conditions are available in the literature; a number were discussed in the preceding chapter.

At this point it is desired to direct attention to a set of boundary conditions that may be exemplified by those commonly encountered in the tarnishing of silver by traces of sulfur in the atmosphere. The rate of this tarnishing reaction is controlled by the diffusion of sulfur through the atmosphere to the exposed silver surface. All sulfur that actually reaches the surface rapidly and firmly combines as Ag_2S. It is observed that in a given geographic locality the rate of this tarnishing reaction is constant with time; the weight gain (sulfur pickup) of a silver specimen when plotted against time of exposure gives a substantially straight line. This is a markedly different behavior from what one might expect if he pictured the atmosphere around the specimen as static and hence treated the problem as one of diffusion through a semi-infinite medium; by such treatment the rate would drop off rapidly with time just as the rate of

whose binary solution with iron exhibits positive departure from Raoult's law, gives a penetration curve that exhibits a type of departure from the ideal which is opposite to that of C, Si, and P. Manganese, which is believed to form an essentially ideal solution with iron, yields a nearly ideal penetration curve.

penetration of carbon during carburizing does. It is apparent, then, that the tarnishing rate of silver exposed to the atmosphere depends upon convection and air currents to bring sulfur to the vicinity of the specimen.

Let us then change our picture of the conditions of the atmosphere in the vicinity of the silver specimen to another idealization involving (1) a uniform, well-mixed atmosphere of constant composition at a distance from the specimen and (2) a "dead" film of air immediately adjacent to the specimen. Let us assume that the sulfur diffuses only through the dead film. This model is very much like that used in considering heat transfer from a gas stream to a wall. If the film is of constant thickness Δl and the boundary condition constant, the rate of diffusion is seen immediately from Fick's first law to be constant, thus,

$$J = -D \frac{C}{\Delta l}$$

where C is the concentration of sulfur in the bulk of the atmosphere. The thickness Δl of the film is in general rather difficult to estimate. For purposes of calculating heat conduction from a vertical wall to "still air," the apparent film thickness is about $\frac{1}{10}$ in. In treating the problem of heat flow from a solid to a boiling liquid, the film thickness in the liquid is commonly about $\frac{1}{1000}$ in. for natural convection, *i.e.*, with no stirring other than that occasioned by the boiling itself.

Application of Preceding Boundary Condition to Carbon Boil in Open Hearth. It would not seem unreasonable that a similar situation prevails in the open hearth—that oxygen from the slag diffuses through a thin layer or film of the metal and is mixed by convection beyond this film. In order to check this hypothesis, let us postulate tentatively that we should anticipate a film thickness of about 0.003 cm,[1] as noted above for many boiling liquids, and determine whether the rate of carbon drop calculated on this hypothesis from the diffusivity agrees in order of magnitude with the observed rate.

From Fick's first law, it may be shown[2] that the rate of carbon drop on the basis of this model is

[1] On the basis of laboratory experiments on the oxidation of elements dissolved in molten iron, Samarin and Shvartzman [*J. Phys. Chem. U.S.S.R.*, **22**, 565 (1948)] arrived at the conclusion that the film thickness in their experiments was about 0.01 cm. These authors also concluded that the rate of carbon oxidation is determined by diffusion, as also did Shvartzman, Samarin, and Temkin [*ibid.*, **21**, 1027 (1947)].

[2] The derivation of this relation is as follows: From Fick's first law, the rate of transfer of oxygen across unit area of "dead" layer in terms of mass per unit time is $D[([O]_{se} - [O])/100\Delta l]\,\rho$, ρ being the density. Assuming that one atom of carbon reacts with each atom of oxygen, the rate of carbon consumption is $\frac{12}{16}$ of this. Each unit mass of carbon reacted comes from a volume of unit cross section and of length equal to the bath depth and hence decreases the percentage of carbon by $100/l\rho$. The product of these three factors gives the right side of the equation.

$$-\frac{d[\mathrm{C}]}{dt} = D\,\frac{[\mathrm{O}]_{se} - [\mathrm{O}]}{l\,\Delta l}\,\frac{12}{16}$$

where l is the bath depth and Δl the film thickness. For conditions during oreing, the following average or approximate values are used:

$[\mathrm{O}]_{se} = 0.23$ per cent (solubility limit at 1600°C)
$[\mathrm{O}] = 0.04$ per cent
$l = 30$ in. $= 76$ cm (average bath depth)
$D = 3 \times 10^{-4}\ \mathrm{cm^2/sec}$ (this diffusivity of oxygen is selected as intermediate between that of C and that of S, Mn, P, and Si)
$\Delta l = 0.003$ cm

On this basis,

$$-\frac{d[\mathrm{C}]}{dt} = 3 \times 10^{-4}\left(\frac{0.19}{76 \times 0.003}\right)\frac{12}{16} = 1.87 \times 10^{-4}\%\ \text{per sec}$$
$$= 0.67\%\ \text{per hr}$$

This fortuitously good agreement with the observed carbon drop during heavy oreing lends some credence to the mechanism postulated. A similar calculation for the rate of carbon drop during normal boil conditions, setting[1] $([\mathrm{O}]_{se} - [\mathrm{O}]) = 0.04$, gives 0.14 per cent/hr—a figure very close to the observed average rate.

Adopting, at least tentatively, this mechanism for the oxidation of carbon in the open hearth, we now see that the reaction-rate constants previously designated k_a and k_b may be evaluated in terms of the diffusivity D of oxygen, the film thickness and the bath depth l; Δl as above represents the film thickness of the dead layer of metal adjacent to the slag, and $\Delta l'$ the film thickness of the dead layer of metal adjacent to the CO bubbles.[2]

$$k_a = \frac{D}{l\,\Delta l}$$

$$k_b = \frac{D}{l\,\Delta l'}$$

The expression previously given [Eq. (19-6)] for $\Delta\mathrm{O} \equiv [\mathrm{O}] - [\mathrm{O}]_{ce}$ becomes

$$\Delta[\mathrm{O}] = \frac{\Delta l'}{\Delta l}\,\frac{S_{sm}}{S_{gm}}\,([\mathrm{O}]_{se} - [\mathrm{O}])$$

[1] T. E. Brower and B. M. Larsen, *Trans. AIME*, **172,** 137 (1947).

[2] These expressions are found by comparing the rates per unit area as given by the two expressions. For example, r_a was previously given as $k_a([\mathrm{O}]_{se} - [\mathrm{O}])$; as just derived from diffusion considerations, $r_a = (D/l\,\Delta l)([\mathrm{O}]_{se} - [\mathrm{O}])$. Equating these two gives the expression for k_a.

It would not seem unreasonable to assume that the two film thicknesses Δl and $\Delta l'$ are nearly identical, in which case, at steady state,

$$\Delta[\mathrm{O}] \cong \frac{S_{sm}}{S_{gm}} ([\mathrm{O}]_{se} - [\mathrm{O}]) \tag{19-7}$$

Under average conditions during oreing, $\Delta[\mathrm{O}]$ is about 0.04 per cent,[1] and again taking the rough average value of $([\mathrm{O}]_{se} - [\mathrm{O}])$ as 0.19 per cent, Eq. (19-7) predicts that $S_{gm}/S_{sm} = 5$; in other words, that at any given instant during steady-state ore boil, the total surface area of all the gas bubbles in the metal is about five times the area of the slag-metal interface. Similarly, it may be predicted for conditions during a normal boil that the total bubble surface is about twice the area of the slag-metal interface. These conclusions do not seem unreasonable; however, they should be interpreted merely as provisional confirmation of the validity of the assumed mechanism.

Application to Manganese Transfer in the Open Hearth. The film theory may also be applied to the rate of approach to equilibrium of the manganese distribution between slag and metal. It is now assumed that the slow step is the diffusion of manganese through a thin film of metal adjacent to the slag, that the top side of this film is in equilibrium with the slag by the reaction $\underline{\mathrm{Mn}} + \mathrm{FeO} = \mathrm{MnO} + \underline{\mathrm{Fe}}$, and that the bottom side has the composition characteristic of the bath as a whole. Proceeding by the same method as for carbon, the rate of transfer of manganese from slag to metal may be represented

$$-\frac{d[\mathrm{Mn}]}{dt} = \frac{D}{l\,\Delta l} ([\mathrm{Mn}]_{se} - [\mathrm{Mn}])$$

where $[\mathrm{Mn}]_{se}$ represents the per cent manganese at the top of the film in equilibrium with the slag and [Mn] that characteristic of the entire bath. Assuming $[\mathrm{Mn}]_{se}$ as constant, integration from an arbitrary zero time gives

$$\log ([\mathrm{Mn}]_{se} - [\mathrm{Mn}]^{t=0}) - \log ([\mathrm{Mn}]_{se} - [\mathrm{Mn}]) = \frac{Dt}{2.3l\,\Delta l}$$

The departure of this reaction from equilibrium is measured by $([\mathrm{Mn}]_{se} - [\mathrm{Mn}])$ which may conveniently be designated $\Delta[\mathrm{Mn}]$, whence

$$\log \frac{\Delta[\mathrm{Mn}]^{t=0}}{\Delta[\mathrm{Mn}]} = \frac{Dt}{2.3l\,\Delta l}$$

Let us now suppose that a manganese ore addition was made to the furnace, thus producing a departure from slag-metal equilibrium equal to

[1] T. E. Brower and B. M. Larsen, *Trans. AIME*, **172**, 164 (1947).

$\Delta[\text{Mn}]^{t=0}$, and let us inquire as to the time required to reduce the departure from equilibrium, $\Delta[\text{Mn}]$, to $\frac{1}{10}\Delta[\text{Mn}]^{t=0}$. For this case

$$\log \frac{\Delta[\text{Mn}]^{t=0}}{\Delta[\text{Mn}]} = 1$$

and the time required is, by the above equation,

$$t = \frac{2.3l\,\Delta l}{D}$$

Setting $D = 10^{-4}$ cm^2/sec and using the same numerical values of l and Δl as in the preceding section, t is found to be 40 min. This is in very good accord with experience which indicates that about a half hour is required to reestablish manganese equilibrium after it has been disturbed.

Substantially the same reasoning and conclusion apply to the approach to equilibrium of the sulfur distribution between metal and slag.

THERMAL CONDUCTIVITY AND DIFFUSION

The subject of heat transfer embraces so many details that no attempt will be made here to include a comprehensive treatment; the interested reader should consult a standard text[1] or the short monograph by Austin.[2] Reference has already been made to the similarity between diffusion and heat transfer. In fact, the fundamental equations for thermal conductivity are identical in form with Fick's laws, and solutions of the differential equations that have been developed for one case may be applied to the other. The thermal diffusivity of solids and liquids is usually very much greater than the mass diffusivity of any constituent. For ideal gases, however, the two are numerically identical when expressed in the same units. This identity allows us to draw some rather interesting conclusions.

Suppose, for example, that we wish to dry some air by passing it over a cold plate. The identity of the diffusivities of water vapor and heat allows us immediately to conclude that, under such conditions when half the water is removed from the air, the mean temperature of the air will be halfway between the initial temperature and the plate temperature, provided that condensation occurs only on the plate and that the plate

[1] W. H. McAdams, "Heat Transmission," 2d ed., McGraw-Hill Book Company, Inc., New York, 1942. A. Schack, "Heat Transfer and Transmission," translated by Goldschmidt and Partridge, John Wiley & Sons, Inc., New York, 1933. L. R. Ingersoll and O. J. Zobel, "An Introduction to the Mathematical Theory of Heat Conduction," Ginn & Company, Boston, 1913.

[2] J. B. Austin, "Heat Flow in Metals," American Society for Metals, Cleveland, 1942.

temperature is sufficiently low that the vapor pressure of the condensed water may be neglected.

As another example, let us consider the blowing of air through a molten metal, as in the bessemer. Oxygen diffuses through the air bubbles and oxidizes constituents of the metal at the boundary; heat diffuses from the metal into the initially cold air in the bubble. It might be thought that, if air be blown in such a way (*i.e.*, in large rather than small bubbles) that some of the oxygen does not have time to diffuse to the metal, some heat might be lost to this unreacted air. However, from the equalities of thermal and material diffusivities, it is readily seen that this is not so, and if oxygen does not have time to diffuse out of the bubble, heat does not have time to diffuse into that portion from which oxygen does not diffuse.

The effect of the size of a specimen on the time required for heating is of particular importance. Let us consider the heating or cooling of a homogeneous solid object initially of uniform temperature and subject to fixed boundary conditions—such as might occur when a metal or other specimen is immersed in boiling water which brings the surface to 100°C immediately—or which are approximated during a good quench. As shown in Chap. 18, the solution of the differential diffusion equation for any given geometrical form under these conditions is such that the temperature at any specified point at any time is a function only of Dt/x^2, where x is a linear dimension of the particular geometrical form. This important fact tells us that, for a series of objects of the same material and of geometrically similar shape, the time required to heat (under similar boundary conditions) to a similar temperature distribution is proportional to the square of any linear dimension thereof. If, for example, during quench, 1 min is required for the center of a 1-in. rod to cool from 1000 to 200°C, 4 min will be required for the center of a rod twice as large in all linear dimensions to cool through the same temperature range.

The above-mentioned nature of the solution of the diffusion equation under constant boundary conditions has many applications. It leads to the so-called parabolic law for rate of scale formation,

$$X = a\sqrt{t}$$

where X is the scale thickness and a is a constant depending on the particular metal, boundary conditions, and temperature. The depth of penetration during carburizing is given by an equation of the same form, X then being interpreted as the distance from the interface to a point having some arbitrary carbon content. Care must be taken in the practical application of the diffusion equation that the boundary conditions used in the mathematical solution do in fact correspond to the actual conditions in experiment or practice; many examples of failure to observe such precaution may be found in the literature.

The rate of many endothermic metallurgical reactions is limited by the rate of supply of the required heat. For example, during the burning of limestone each lump consists essentially of an unburned core of $CaCO_3$ and a burned shell of porous CaO. This shell is a fairly good insulator, and hence the time required to burn the lump is determined principally by the rate of heat transfer from the outer surface to the interface where reaction occurs. Proposed processes for the direct reduction of iron oxides with hydrogen have never attained large-scale operation because of, among other reasons, the difficulty of supplying the required heat through a muffle. The time required for an open-hearth heat depends mainly on the amount of heat required by the metallurgical reactions and the rate at which this heat can be transferred to the bath. Although a large part of this chapter has dealt with the rate of reactions at constant temperature, it should be borne in mind that, in a large number of practical metallurgical rate problems, the principal difficulty is in attaining or maintaining the desired temperature where the reaction actually occurs.

NUCLEATION AND GROWTH PROCESSES

Many heterogeneous reactions proceed by a process involving the nucleation of a new phase in the parent phase and subsequent growth of the nuclei. Thus CO bubbles in the open hearth are first nucleated, a process to be discussed directly, and subsequently grow by diffusion of carbon and oxygen thereto in the manner previously discussed. The decomposition of austenite at subequilibrium temperatures presents examples of the nucleation and growth processes, discussed by Mehl.[1] It is believed by Mehl and coworkers that substantially all ageing and overageing processes proceed in this manner. The nucleation and growth processes in crystalline metals have been investigated in several cases, and Geisler[2] has proposed a general sequence of events for the formation of a crystalline precipitate in a crystalline metal.

Nucleation of Gas Bubbles. Bubbles do not form readily within the body of a liquid even when it is considerably supersaturated with a gas. The explanation of this phenomenon is in the surface tension of the liquid. The liquid exerts a pressure on a spherical gas bubble as though its surface layer were an elastic membrane. This pressure, which is in addition to any other pressure to which the system is subject, is given by the relation

$$P = \frac{2\sigma}{r}$$

[1] R. F. Mehl, *J. Iron Steel Inst.*, **159**, 113 (1948).
[2] A. H. Geisler, *Trans. AIME*, **180**, 230 (1949).

where σ is the surface tension of the liquid and r the radius of the bubble. Taking the surface tension[1] of liquid steel as 1500 dynes/cm, it follows, for example, that a bubble 1 mm in diameter is subject to a pressure of 0.06 atm more than the liquid around it. A bubble of $\frac{1}{16}$ mm diameter would similarly be subject to an extra pressure of 1 atm. Unless the liquid is supersaturated with gas to at least a corresponding amount, the bubble will not be stable but will dissolve in the liquid.

The question arises at this point as to how bubbles are ever nucleated if such high degrees of supersaturation are required even to stabilize them. There are two known ways[2] in which bubbles start in liquids. One way is by means of vortices resulting from agitation; vortex motion gives rise to a very high negative pressure at the center of the vortex, thus enabling a bubble to start. More commonly, bubbles start at an interface where there is already a gas film; the bubble grows thereon, breaks loose, and leaves a starting nucleus of gas for the next bubble. Such a stream of bubbles issuing from the same point may be observed in a glass of carbonated beverage. It is believed that the bubbles in a normal open-hearth boil originate in this way on the furnace bottom.[3] Sometimes bubbles form more readily on impurities or dirt in the liquid; this is probably principally because of sorbed gas on the surface or trapped gas in capillary spaces. Körber and Oelsen[4] found that, under laboratory conditions, general bubble formation throughout liquid steel did not occur until the degree of supersaturation was over 10 atm. This extent of supersaturation does not commonly occur in the open hearth, though it may under some conditions; "frothing" of a heat may be accounted for in this manner.

[1] "Metals Handbook," p. 200, American Society for Metals, Cleveland, 1948.

[2] R. B. Dean, *J. Applied Phys.*, **15**, 446 (1944).

[3] T. E. Brower and B. M. Larsen, *Trans. AIME*, **172**, 164 (1947).

[4] F. Körber and W. Oelsen, *Mitt. Kaiser-Wilhelm-Inst. Eisenforsch. Düsseldorf*, **17**, 39 (1935).

PROBLEMS

Chapter 2

2-1. (*a*) Evaluate $(\partial V/\partial T)_P$ for an ideal gas from the two equations following Eq. (2-2). (*Hint:* If $x = F(y,z)$, then $(\partial z/\partial y)_x = -(\partial x/\partial y)_z/(\partial x/\partial z)_y$ as demonstrated on page 180.) (*b*) Verify your answer to (*a*) by a method involving partial differentiation of the mathematical formulation of the ideal-gas law [Eq. (2-3)].

2-2. Dissociated ammonia is used as a protective atmosphere in certain heat-treating operations. Estimate in cubic feet, measured at 70 and 1500°F and at atmospheric pressure, the volume of completely dissociated gas produced from 1 lb of liquid ammonia.

2-3. Experimental accuracy may require the correction of weight readings to vacuum by allowing for the buoyancy of air. (*a*) Calculate the order of magnitude of this correction for a sample of 5 g of beryllium weighed at 21°C and 760 mm Hg against brass weights (density = 8.4 g/cm^3). (*b*) How would the correction differ, if 5 g of gold were weighed instead of beryllium? (*c*) What is the percentage correction for weighing water in air against brass weights?

2-4. If a 100-g sample of silver is weighed in air to 0.1 mg with brass weights of density 8.4 (g/cm^3), what is the correction for buoyancy?

2-5. The solubility of nitrogen gas in liquid iron at 1 atm pressure and 1540°C is 0.039 per cent by weight. Calculate the volume of nitrogen (in cubic centimeters) dissolved by 100 g of iron under these conditions, measured at (*a*) standard temperature and pressure (STP) and (*b*) at 1540°C and 1 atm pressure.

2-6. Copper dissolves approximately 0.0001 per cent carbon at 1100°C. [*Trans. AIME*, **166**, 128 (1946).] In the presence of oxygen, the carbon is evolved as carbon monoxide and dioxide. Calculate the maximum volume of gas evolved in this manner from 100 g of copper saturated with carbon at 1100°C. Assume that the gas is measured at 1 atm pressure and (*a*) 0°C and (*b*) 1100°C.

2-7. It is sometimes stated that all isotopes of a given element possess essentially identical physical and chemical properties, except with respect to nuclear phenomena. Test this generalization for the process of effusion of (*a*) hydrogen and deuterium and (*b*) the natural isotopic mixture of silver (atomic weight 107.88) and the radioactive isotope Ag^{110}.

2-8. Verify the units and dimensions of the gas constant R by analyzing the units and dimensions of the equation $u = \sqrt{3RT/M}$. (See text for meaning of notation.)

2-9. (*a*) What are the rms velocity u_{rms}, the arithmetic mean velocity $\bar{u}$, and the most probable velocity u_p of the molecules of monatomic mercury vapor at 357°C? What are these quantities at 672°C? (*b*) What are the corresponding values for lithium vapor?

2-10. (*a*) Using only the information given in the text, derive the general equation for the velocity of sound in ideal gases: $u_s = \sqrt{\gamma PV/M}$, where V is the volume of 1 gram mole of gas, M is its molecular weight, and the other symbols have the meanings explained in the text. [*Hint:* $R = 8.314 \times 10^7$ ergs/(gram mole)(°K) and 1 erg = 1 g-cm^2/sec^2.] (*b*) From the equation given in (*a*) calculate the velocity of sound in air at 20°C and 1 atm pressure and in mercury vapor at 357°C and 1 atm pressure.

2-11. In the thermal-transpiration experiment described in the text (page 11), a pressure difference of 1 cm of mercury was observed. Calculate the theoretical pressure difference which would have been possible with the hot compartment at normal barometric pressure and at 1000°C.

2-12. In experimental equipment nitrogen and mercury vapor initially at identical pressure and temperature are separated by a porous wall, the pores of which are small compared to the mean free paths of the gases. The temperature of the two compartments is maintained constant. In which compartment will the pressure tend to increase? In what respect, if at all, would the case differ, if the mercury vapor were replaced by helium?

2-13. What is the heat capacity C_V at constant volume of 1 mole of monatomic mercury vapor at 400°C and 1 atm pressure? At 600°C and 1 atm pressure?

2-14. Why would you expect the portion of the van der Waals' curve which has a positive slope in Fig. 2-2 not to have a physical counterpart?

2-15. Calculate the pressure at which 1 mole of SO_2 occupies a volume of (*a*) 10 liters and (*b*) 1 liter at a temperature of 500°K. Use the ideal-gas and the van der Waals' equation and calculate the percentage differences between the results. The following values of the van der Waals' constants may be used: $a = 6.71$ atm-liter² and $b = 0.0564$ liter.

2-16. In an installation for the electric smelting of zinc a gas mixture consisting of 45 per cent by volume zinc vapor and 55 per cent carbon monoxide is passed into a condenser. The gases enter the condenser at 850°C; the pressure may be assumed to be approximately atmospheric. Estimate the volume of gases in cubic feet entering the condenser per ton of zinc produced.

Chapter 3

3-1. (*a*) Prepare a table for lead similar to that given for tin in the text, and show the various possible arrangements of nonbonding electrons, bonding electrons, and metallic orbitals. Include in this table a few pairs of 5d electrons and the four outer electrons. Which arrangement corresponds to a valence of 2.0? (*b*) Prepare a similar table for magnesium. What arrangement agrees with a valence of 2.0?

3-2. Calculate the theoretical ratio c/a for a hexagonal close-packed (hcp) structure from the fundamental geometrical relations illustrated in Fig. 3-6.

3-3. From simple geometric relations in the hcp structure derive a general equation in terms of the parameters a and c for the distance between two neighboring lattice sites lying in adjacent planes parallel to the basal plane. Show that this distance is a in the ideal hcp structure.

3-4. (*a*) Find in terms of the lattice parameter a (the edge of the unit cell shown in Fig. 3-6) the distances between nearest neighbors and next-nearest neighbors in the face-centered cubic (fcc) structure. (*b*) Apply the results of (*a*) to aluminum ($a = 4.041$ kx units) and locate the nearest and next-nearest neighbor distances in Fig. 5-9.

3-5. (*a*) Find in terms of the lattice parameter a (the edge of the unit cell shown in Fig. 3-6) the distances between nearest neighbors and next-nearest neighbors in the body-centered cubic (bcc) structure. (*b*) Apply the results of (*a*) to potassium ($a = 5.31$ kx units) and locate the nearest and next-nearest neighbor distances in Fig. 5-10.

3-6. As a general criterion it has been suggested that the electrical resistivity of solid metals increases with temperature while that of nonmetals decreases. Apply this criterion to some of the elements in Hume-Rothery's class II and to such elements as silicon, germanium, antimony, and bismuth in class III.

3-7. Body-centered titanium ($a = 3.32$ kx units) is stable above approximately 880°C; hcp titanium ($a = 2.95$ kx units, $c = 4.73$ kx units) is stable below that temperature. Compare the distances between equatorial neighbors and between neighbors in adjacent planes (parallel to the basal plane) in the hcp structure. Calculate also the interatomic distance in the bcc structure.

3-8. Using Pauling's equation, $r(1)$ from Table 3-4, and crystallographic data in Table 3-3, calculate the bond number n for nearest and next-nearest neighbors of potassium. Add the bond numbers of the 14 bonds to compare the resulting valence v with the accepted value.

3-9. (*a*) The interatomic distance between nearest neighbors in gray tin is 2.797 kx units. Calculate the single-bond radius $r(1)$ by Pauling's equation, assuming that the four electron pairs are shared between the four nearest neighbors. (*b*) In white tin each atom has four neighbors at 3.016 and two at 3.175 kx units. Calculate from $r(1)$ as obtained in (*a*) and from Pauling's equation the bond number n for each type of neighbor. Add the bond number for each of these neighbors to obtain the valence v for white tin. (*Note:* The result differs slightly from the value 2.44 given in Table 3-4.)

3-10. (*a*) From Pauling's equation, the valence $v = 3$, and crystallographic data, calculate the single bond radius $r(1)$ for aluminum, assuming that only nearest neighbors form bonds. (*b*) Using the value of $r(1)$ found in (*a*), calculate the bond number of the next-nearest neighbors. Note that the resulting value is not large enough to justify a correction of the bond number of the nearest neighbor bonds used in (*a*).

3-11. (*a*) Calculate the axial ratio c/a for zinc on the basis of Pauling's equation and the assumption that the bond number is $\frac{1}{2}$ for the six equatorial neighbors and $\frac{1}{4}$ for the other six neighbors. Take $r(1)$ from Table 3-4. Find the ratio c/a using the relation that the ratio of the distance of the farther neighbors to that of the equatorial neighbors is $\sqrt{a^2/3 + c^2/4}$ to a. Determine the percentage difference between the calculated and the observed axial ratio. (*b*) Repeat this calculation using $\frac{1}{6}$ as the bond number for the six neighbors in adjacent planes.

3-12. Perform similar calculations for cadmium.

3-13. (*a*) From the lattice parameter of α-iron find the radius of the iron atom in the bcc structure on the basis that the atoms behave as hard spheres tangent to their nearest neighbors. Compute the lattice parameter of γ-iron, assuming again that the atoms behave as hard spheres of the radius computed for α-iron. What is the percentage discrepancy between the computed and observed values? (*b*) Using Pauling's equation, the valence v, and the lattice parameter of α-iron, calculate the number of bonds associated with the eight nearest neighbors and the number associated with the six next nearest neighbors in the bcc structure. From this result compute $r(1)$, $r(\text{CN}12)$ [equal to $r(\frac{1}{2})$], and the lattice parameter of γ-iron. What is the percentage discrepancy between the computed and observed values of the latter?

3-14. (*a*) Using Pauling's equation calculate $r(1)$ for iron from the nearest neighbor distance (as obtained from the lattice parameter) of the atoms in γ-iron. (Assume the bond number n is equal to $\frac{1}{2}$.) (*b*) From Pauling's equation, the lattice parameter and the value of $r(1)$ as calculated in (*a*), find for γ-iron the bond number for an atom and its next-nearest neighbor at distance a. What fraction of the valence $v = 5.78$ thus appears assignable to next-nearest neighbor bonds?

Chapter 4

4-1. Silver forms a continuous series of solid solutions with gold and a eutectic with limited terminal solid solubilities with copper. Is this consistent with the rule for the size factor?

4-2. Plot the radius of the metallic elements in the third, fourth, and fifth period against atomic number. Enter the upper and lower limits for favorable size factors for copper. Verify for various binary copper systems the statements in the text concerning the role of the size factor.

4-3. Plot on a single graph of temperature versus composition the limits of the primary solid solutions of silver with the three elements immediately following it in the Periodic Table. Discuss any significant relations observed.

4-4. Near room temperature the solubility of cadmium in silver is over 40 weight per cent, while the solubility of cadmium in copper is less than 3 per cent. Discuss this difference in solubilities in terms of current theory.

4-5. Apply the criteria for extensive solid solubility to tantalum and molybdenum, which are known to form a complete series of solid solutions. (Include a calculation of the electronegativities from Gordy's equation and data in Table 3-4.)

4-6. Predict on which side of each of the following binary systems the more extensive terminal solid solution is likely to occur: (*a*) Al-Cu; (*b*) Al-Ni; (*c*) Al-Zn; (*d*) Ag-Mg; (*e*) Ag-Zn; (*f*) Cu-In. What, if any, discrepancies exist between published phase diagrams and your predictions?

4-7. Predict the relative corrosion resistance, electrical resistivity, hardness, and elastic limit of ordered and disordered gold-nickel alloys, assuming a behavior similar to that of the gold-copper system.

4-8. Construct a chart similar to Figs. 4-8*ff*. for elements whose solubility in zinc is known.

4-9. The valence used by Gordy (Table 4-1) differs from that given by Pauling (Table 3-4) for the following metals for which values of electronegativity have been determined: Ag, Au, Cd, Cu, Ga, Hg, In, Pb, Sn, Tl, Zn. Using Gordy's equation and Pauling's value of v in place of n', calculate the electronegativity of these metals and tabulate the results together with the values given in the fourth and fifth columns of Table 4-1.

4-10. The following densities have been determined for tantalum-molybdenum solid solutions [*J. Inst. Metals*, **80**, 528 (1952)]:

Composition, atomic % Ta	Density, g/cm³
0	10.16
10.10	10.83
20.08	11.65
30.02	12.25
39.96	12.91
49.47	13.43
61.21	14.47
71.47	14.92
82.77	15.44
91.42	15.96
100.00	16.55

What values of the lattice parameters correspond to these densities? Does the system comply with Vegard's law?

4-11. For an interstitial solid solution based on an fcc structure calculate the ratio of the radius of the solute atom to that of the solvent atoms on the assumption that the atoms behave as hard spheres tangent to each other. Compare this value with Hägg's "critical ratio."

Chapter 5

5-1. The literature contains data for several properties of liquid copper-tin alloys. The following table gives information on density [*Z. Metallkunde*, **14**, 145, (1922)] and electrical resistivity [*Ferrum*, **11**, 289, (1914)] of these alloys at 1200°C:

Weight % tin	Density, g/cm³	Electrical resistivity, microhm-cm
0.0	7.81	22.41
2.0	....	26.53
4.76	....	33.63
8.0	7.79	
18.0	7.71	
20.0	....	59.11
29.0	7.68	
38.0	7.64	
38.43	....	72.08
42.5	....	73.40
60.0	....	73.10
65.0	7.09	
80.0	....	72.10
85.0	6.63	
100.0	6.4	69.45

Plot these data and discuss the relation of this graph to Fig. 5-14.

5-2. For metallic elements, Richard's rule as stated in the text postulates the relation $\Delta H_f = kT_f$. Test this relation for Ag, Al, Au, Cd, Cu, Hg, In, Sn, and Zn. Does k become more nearly constant if the comparison is restricted to (*a*) the same group or period in the periodic system or to (*b*) the same crystal system of the solid?

5-3. The solubility of various metals in liquid iron at 1600°C is as follows:
Completely soluble: Ni Co Mn Cu Au Si Sb Ti Al
Partly soluble: Sn
Relatively insoluble: Ag Bi Pb

Draw a chart similar to Figs. 4-8*ff*. Is any sort of grouping apparent in this chart? If Ag is assigned a hypothetical electronegativity of 1.4, suggested by Fig. 4-10, is the grouping more definite? What range of electronegativity is suggested by this plot as compatible with liquid miscibility at this elevated temperature?

Chapter 6

6-1. A gas enclosed in a cylinder fitted with a frictionless and weightless piston is initially at 100°C and 10 atm pressure and undergoes each of the following processes:

(*a*) The gas is maintained at constant volume, and its temperature is reduced to 50°C.

(b) The gas is expanded at constant pressure, and its temperature is raised to 200°C.

(c) The gas is compressed adiabatically until its temperature has risen to 200°C.

(d) The gas is compressed isothermally until the pressure has risen to 20 atm.

(e) The gas undergoes the compression described in (c) and is then returned to its initial state.

Write the first law of thermodynamics for each of these processes. Without reference to an equation of state, reduce the expressions to a simplified form, for example by eliminating a term which is equal to zero; also indicate for each term whether it is positive or negative.

Do the answers depend upon whether these processes are carried out reversibly?

6-2. Write the first law for the following processes and simplify the resulting expressions as far as possible. (a) A gas is initially enclosed in a container connected through a valve to a second evacuated container of the same volume as the first. The valve is opened and the temperature and pressure are allowed to equalize adiabatically. (b) A similar expansion is carried out under isothermal conditions. (c) Does the assumption of the ideal-gas law lead to a simplification of the answers to (a) and (b)?

6-3. Derive from the first law the equation $(\partial E/\partial T)_P = C_P - PV\alpha$, where

$$\alpha \equiv \frac{1}{V}\left(\frac{\partial V}{\partial T}\right)_P$$

6-4. Derive the relation $C_P - C_V = R$ for 1 mole of an ideal gas by applying the first law in the form of Eq. (6-1) to the following alternative processes: the gas is heated through a temperature interval dT (1) in a single step at constant pressure and (2) at constant volume followed by an isothermal expansion, in which no work is done. Equate the first-law expressions for the two processes.

6-5. Show that for common gases the first term on the right of Eq. (6-11) is more important than the second. Is this also true for solids and liquids? How would the absolute level of the pressure affect the answers?

6-6. The hydrostatic pressure on 100 g of silver is increased isothermally and reversibly from 0 to 500 atm. Calculate the work of compression. (*Hint:* $V\beta\, dP$ may be substituted for dV, where β is the isothermal compressibility; changes in β and in V with pressure may be neglected.) Assume for silver a density of 10.5 g/cm^3 and a compressibility β of 1.0×10^{-6} per atmosphere.

6-7. How does the work of reversible isothermal compression of a condensed phase between two pressures vary with (1) the mass, (2) the density, (3) the compressibility of the material, (4) the pressure difference and (5) the temperature? (*Hint:* See Prob. 6-6.)

6-8. Calculate the work done when the hydrostatic pressure on a cube of copper measuring 2 cm along each edge is increased reversibly and isothermally at 0°C from 1 to 100 atm. The isothermal compressibility β is 7.6×10^{-6} per atmosphere. Does the shape of the block affect the work done?

6-9. A strong capillary tube filled with mercury and closed at 0°C is heated to 10°C. Assuming the volume of the capillary remains constant, calculate the resulting pressure. For mercury near room temperature the expansion coefficient α is 18×10^{-5} per degree and the compressibility β is 3.7×10^{-6} per atmosphere.

6-10. (a) Derive an equation analogous to Eq. (6-3) for the application of a unidimensional stress σ to a metal bar of length 1. How do the signs account for the two cases of extension and compression? (b) Evaluate the work term in the equation derived in (a) for the process in which the stress on a wire of unit cross section A is increased from σ_1 to σ_2 at constant temperature. (*Hint:* Substitute $d\sigma$ for dl by use of Young's modulus $Y = \frac{1}{l} \cdot \frac{dl}{d\sigma}$; assume that changes in Y, l, and A are negligible.) (*c*) How would the work change with the cross section? With the temperature?

6-11. In a hydrometallurgical plant 75 gal of a leaching solution must be pumped each minute through a vertical distance of 22 ft. (*a*) Assuming that the specific gravity of the solution is 1.15, calculate the minimum power required in horsepower and kilowatts. (*b*) What additional factors would an engineer have to consider in specifying the power required?

6-12. It has been stated that, if 1 g of platinum at 100°C is immersed into 10 g of water at 10°C, the rise in temperature is approximately one-ninth as large as when 10 g of platinum at 100°C is immersed into 10 g of water at 10°C; temperature equilibrium is assumed in both cases. Without making specific calculations, show whether or not this statement seems probable.

6-13. The following low-temperature specific heats were measured for γ-manganese [*U.S. Bur. Mines Tech. Paper* 686 (1946)]:

T, °K	C_P, cal/(gram atom)(°K)
53.7	1.413
61.7	1.829
71.8	2.346
80.3	2.745
94.7	3.340
155.5	4.979

(*a*) Find the mean value of the Debye characteristic temperature θ from the first five of these values. How does this value of θ compare with the value based on 155.5°K? (*b*) Find the heat capacity of γ-manganese at 20 and at 40°K.

6-14. The following low-temperature specific heats were measured for α-manganese [*U.S. Bur. Mines Tech. Paper* 686 (1946)]:

T, °K	C_P, cal/(gram atom)(°K)
54.2	1.380
62.1	1.806
82.2	2.928
95.9	3.585
101.2	3.538
104.3	3.624
111.5	3.840

Plot these values of the heat capacity against temperature and notice any irregularities. Calculate the Debye characteristic temperature from each of these heat capacities.

6-15. In an investigation of the thermodynamic properties of manganese, the specific heat of α-manganese at 298.16°K was found to be 6.29 cal/(gram atom)(°K); the following heat contents were also determined [*U.S. Bur. Mines, Tech. Paper* 686 (1946)]:

$$H^\circ_{700} - H^\circ_{298.16} = 2895 \text{ cal/gram atom}$$
$$H^\circ_{1000} - H^\circ_{298.16} = 5450 \text{ cal/gram atom}$$

Find a suitable equation for $H^\circ_T - H^\circ_{298.16}$ and also for C_P as a function of temperature.

6-16. Find the heat capacity of γ-manganese from the following values of the heat contents:

T, °K	$H_T^\circ - H_{298.16}^\circ$, cal/gram atom
1374	9,940
1410	10,330

6-17. The following table gives representative values of the heat contents of titanium carbide, determined experimentally by the "drop method" [*J. Am. Chem. Soc.*, **68**, 370 (1946)]:

T, °K	$H_T^\circ - H_{298.16}^\circ$, cal/gram mole
360.5	576
587.8	2,943
790.5	5,270
1001.7	7,860
1205.1	10,400
1454	13,570
1735	17,120

The heat capacity C_P of titanium carbide at 298.16°K determined by another method [*Ind. Eng. Chem.*, **36**, 865 (1944)] is 8.04 [cal/(gram mole)(°K)]. Find an equation of the form $a + bT - cT^{-2}$ for the heat capacity. [For a method of calculation, see K. K. Kelley, Contributions to the Data on Theoretical Metallurgy X, *U.S. Bur. Mines Bull.* 476 (1949), pp. 7 and 8.]

6-18. (*a*) Using the equation for the low-temperature heat capacity of zinc given in the text, show how in the temperature range of Fig. 6-5 the ratio of the contributions of the Debye or lattice term to the contribution of the electron term varies. (*b*) Perform the same calculation for tungsten.

6-19. Draw schematic sketches of (*a*) the temperature dependence of the molar heat capacity C_P of solid copper from $T = 0$ to $T = 1000$°K and (*b*) the quantity $H_T^\circ - H_{298.16}^\circ$ for copper at $P = 1$ atm from $T = 0$ to $T = 1500$°K.

6-20. The following heat capacities have been reported for copper-zinc alloys [*Z. Metallkunde*, **32**, 191 (1940)]:

°C	Heat capacities, cal/(g)(°K)			
	0% Zn	10% Zn	20% Zn	30% Zn
100	0.095	0.095	0.095	0.095
700	0.110	0.110	0.110	0.110

Compare these data with values calculated from published heat-capacity equations of copper and zinc and assume the Kopp-Neumann rule according to which the heat capacity of a solid solution equals the weighted sum of the heat capacities of the components.

6-21. (*a*) Calculate the heat required in calories to heat 1 gram mole of Al_2O_3(s) from 300 to 1300°K. (*b*) Perform a corresponding calculation for heating 1 gram mole of N_2(g) from 300 to 1300°K.

6-22. What is the heat in kilocalories required to raise the temperature of 1000 kg of MgO from 0 to 1773°C?

6-23. From heats of formation given in Table 1, page 523, find the heats of reaction $\Delta H°$ at 25°C for the following reactions:

(*a*) $FeO(s) + CO(g) = Fe(s) + CO_2(g)$

(*b*) $ZnO(s) + C(s) = Zn(s) + CO(g)$

(*c*) $ZnO(s) + \frac{1}{2}C(s) = Zn(s) + \frac{1}{2}CO_2(g)$

6-24. Assume that in order to determine the heat of formation of carbon dioxide, CO_2, the following data have been obtained:

1. Upon the combustion of 0.6 g of graphite to CO_2 with excess oxygen in a bomb calorimeter, a temperature increase $\Delta T' = 2.35°K$ was found.

2. Upon heating the same calorimeter electrically, an adiabatic temperature increase of 0.80°K was found. The resistance of the heating coil was 30 ohms, the current 1.67 amp, and the time 80 sec.

Calculate $\Delta E°$ and $\Delta H°$ for the reaction $C(\text{graphite}) + O_2(g) = CO_2(g)$.

6-25. Calculate $\Delta H°$ for the reaction

$$ZnO(s) + \{2HCl + 200H_2O\}(l) = \{ZnCl_2 + 201H_2O\}(l)$$

assuming that the following observations have been made:

(*a*) A temperature increase $\Delta T' = 3.51°K$ occurs in a calorimeter if 0.1 mole ZnO reacts adiabatically.

(*b*) A temperature increase $\Delta T'' = 0.80°K$ is measured if the same calorimeter, filled with the reaction products, is heated for 60 sec by a current of 2 amp through a heating coil having a resistance of 8 ohms and placed inside the calorimeter.

6-26. The heat of formation of tungsten carbide, WC, at 298.16°K can be found from its heat of combustion (−285.80 Kcal/gram mole) and the heats of formation of WO_3 and CO_2 at the same temperature. The literature contains the following two values for the heat of formation of WO_3 at 298.16°K: −195,700 ± 900 and −200,160 ± 100 cal/gram mole. Calculate the heat of formation of WC from each of these values. (See the discussion of these data in O. Kubaschewski and E. Ll. Evans, "Metallurgical Thermochemistry," pp. 79–80, Academic Press, Inc., New York, and Butterworth-Springer, Ltd., London, 1951.)

6-27. Express the enthalpy increment $H_T^\circ - H_{298.16}^\circ$ of gold (solid) as a function of the temperature from the equation C_P [cal/(gram atom)(°K)] $= 5.66 + 1.24 \times 10^{-3}T$.

6-28. (*a*) Express the heat of fusion of gold as a function of temperature by means of Eq. (6-20), which applies to a phase change as a special type of reaction. (*b*) Perform a corresponding calculation for copper.

6-29. (*a*) Express the heat of evaporation of liquid zinc as a function of temperature. What, if any, assumptions are contained in this expression? (*b*) Perform the corresponding calculation for liquid mercury.

6-30. Express the heat of reaction $\Delta H°$ for the following reaction as a function of temperature: $C(\text{graphite}) + O_2(g) = CO_2(g)$.

6-31. Express the heat of reaction $\Delta H°$ for the following reaction as a function of temperature: $2Cu(l) + \frac{1}{2}O_2(g) = Cu_2O(s)$. (*Hint:* Hess' law discussed in the text may be applied to phase changes; use the result of Prob. 6-28*b*.)

6-32. In an installation for the electric smelting of zinc, a gas mixture consisting of 45 per cent by volume zinc vapor and 55 per cent carbon monoxide is passed into a condenser. The gases enter the condenser at 850°C, the zinc may be assumed to condense at 600°C, and the carbon monoxide leaves the condenser at this same temperature. How much heat must be dissipated per ton of zinc condensed to keep the temperature of the condenser constant?

6-33. Express the heat of formation of aluminum oxide from aluminum (solid and liquid) and gaseous oxygen as a function of temperature.

6-34. (*a*) Find the heat of reaction ΔH° as a function of temperature for the reaction $ZnO(s) + C(graphite) = Zn(g) + CO(g)$. (*b*) If the reduction is carried out in a retort, estimate how many Btu per pound of zinc produced must be transmitted through the walls, assuming that pure ZnO and the stoichiometric amount of carbon are charged at 27°C and that the zinc is distilled at 907°C?

6-35. In an advertisement of an electric melting furnace an "over-all average power consumption of 275 kwhr per ton" of bronze melted is claimed. Estimate the thermal efficiency of this furnace, assuming that "90–10" bronze scrap is to be melted and heated to a casting temperature of 1950°F.

6-36. In an experiment a uniform heat input balances the heat losses from a bath of 100 g of copper. It is desired to add 10 g of copper, which is at room temperature. The final temperature of the bath should be 1127°C. Neglecting the heat stored in the crucible and furnace, calculate the temperature to which the bath must be superheated before the addition is made. (Preliminary question: Would the answer differ if twice as large an addition were to be made to twice as large a bath?)

6-37. In the second stage of copper converting, "white metal" is blown to copper according to the equation $Cu_2S(l) + O_2(g) = 2Cu(l) + SO_2(g)$. (*a*) Find the heat effect ΔH° for this reaction as a function of temperature. (*b*) Assume the converter charge to be at 1200°C. If air at 27°C is blown into this charge, will the temperature of the charge rise or fall? (*c*) Calculate how the use of oxygen in place of air would affect the bath temperature.

6-38. The thermite process is carried out with aluminum and the equivalent amount of hematite, Fe_2O_3. (*a*) If in a given case the reaction attains a temperature of 1700°C, what fraction of the total heat of reaction has been lost? (*b*) What weight of iron per gram of aluminum must be added to the reacting mixture to give a maximum temperature of 1600°C, if the total heat loss is assumed to be the same as in (*a*)?

6-39. (*a*) Calculate the maximum flame temperature for the combustion of a blast-furnace gas analyzing 25 per cent CO, 12.5 per cent CO_2, 62.5 per cent N_2. Assume that the theoretical amount of air is used and that the combustion reaction goes to completion. (*b*) What is the calorific value of this gas based on the heat of combustion of the carbon monoxide expressed in Btu per cubic foot (STP) of the gas.

6-40. A natural gas containing 95 per cent CH_4 and 5 per cent N_2 is burned with the stoichiometric amount of air to produce CO_2 and $H_2O(g)$; all available heat is used to maintain a muffle operating at a temperature of 1700°F (1200°K). Assuming complete combustion, calculate the amount of heat given to the muffle per cubic foot (STP) of natural gas burned.

6-41. (*a*) Calculate the difference $(\Delta H - \Delta E)$ for the reaction $O_3(g) = \frac{3}{2}O_2(g)$ occurring at 50°C and 1 atm pressure. How does this difference change with temperature? (*b*) Perform a similar calculation for the reaction

$$NH_3(g) = \tfrac{1}{2}N_2(g) + \tfrac{3}{2}H_2(g)$$

at 500°C and 1 atm pressure.

Chapter 7

7-1. One mole of an ideal diatomic gas is compressed reversibly and adiabatically from 1 atm and 27°C to 5 atm pressure. Calculate (*a*) the temperature rise, (*b*) the final volume of the gas, (*c*) the work done on the gas.

7-2. (*a*) Calculate the work necessary to compress 1 lb of air reversibly and isothermally at 81°F from 1 to 5 atm pressure. (*b*) If the compression were carried out adiabatically, what would the temperature rise be? The minimum work required? Assume for air $C_P = 7$ cal/(gram mole)(°K) and an equivalent molecular weight of 29.0.

7-3. One mole of an ideal diatomic gas is compressed reversibly and isothermally at 27°C from 1 to 10 atm pressure. (*a*) Calculate the work required. (*b*) Is the corresponding adiabatic work larger or smaller than the isothermal work? What is the reason for the difference?

7-4. (*a*) If a gas is compressed adiabatically from V_1 to V_2, its temperature rises from T_1 to T_2. The adiabatic work $w_{q=0}$ required is larger than the isothermal work of compression at T_1, but smaller than the isothermal work at T_2: $|w_{T_1}| < |w_{q=0}| < |w_{T_2}|$. Check this statement by a schematic *P-V* diagram and discuss its meaning. (*b*) If a gas is compressed adiabatically from a pressure P_3 to a pressure P_4, the temperature rises from T_3 to T_4. Determine whether the following equation is true:

$$|w_{T_3}| > |w_{q=0}| > |w_{T_4}|$$

7.5 One mole of a perfect monatomic gas is initially at $T_1 = 273$°K and $P_1 = 3$ atm pressure and occupies a volume V_1. The final state of the gas is given by a pressure $P_2 = 2$ atm, a temperature T_2, and a volume V_2. Disregard any irreversible aspects and assume that the expansion from V_1 to V_2 is carried out in the following ways:

(*a*) Isothermally, against an external pressure, $P_{ex} = 2$ atm
(*b*) Isothermally, against an external pressure, $P_{ex} = 1$ atm
(*c*) Isothermally, against an external pressure, $P_{ex} = 0$ atm
(*d*) Isothermally and reversibly
(*e*) Adiabatically and reversibly
(*f*) Adiabatically, against an external pressure, $P_{ex} = 0$ atm

Calculate the following quantities for each of these processes and arrange the results in tabular form: V_2, T_2, q, w, ΔE, ΔH. (See also Probs. 7-12 and 9-3.)

7-6. Sketch schematic diagrams of the Carnot cycle in (*a*) *P-V*; (*b*) *T-S*; and (*c*) *T-V* coordinates. What important quantity, if any, does the area represent in each case?

7-7. Express the entropy increment $\Delta S°$ as a function of T upon heating a substance from $T_1 = 298$°K to T_2 assuming (*a*) no phase changes in the temperature interval T_1 to T_2; (*b*) a phase change having a heat effect $\Delta H°$ at T_{tr} where $T_1 < T_{tr} < T_2$. The heat capacity of each phase is of the form $a + bT - cT^{-2}$.

7-8. Show that, if the standard enthalpy change $\Delta H°$ of a reaction is constant over a range of temperature, the standard entropy change $\Delta S°$ is also constant.

7-9. Calculate the entropy of (*a*) a gram atom of liquid aluminum at 1032°K; (*b*) 100 g of nickel at 1370°C, taking $S°_{298.16}$ from Kelley's tables [*U.S. Bur. Mines Bull.* 477 (1950)] or other published source.

7-10. Find the entropy of 1 mole of hydrogen at 1000°K and 0.1 atm pressure.

7.11. Find the entropy of 1 gram atom of zinc (gas) at a pressure of 0.5 atm and a temperature of 1200°K.

7-12. Calculate the changes in the entropy of the gas for the processes (*a*) to (*f*) described in Prob. 7-5.

7-13. Calculate the entropy change $\Delta S°$ which takes place when 1 g of liquid water at 0°C and 1 g of liquid water at 100° are mixed. A constant heat capacity of 1 cal/(g)(°C) may be assumed for water from 0 to 100°C.

7-14. Find the entropy of fusion for the metals in Group I*B* and for cobalt, iron, and nickel. Compare these entropies of fusion with those of gallium, bismuth, antimony, and germanium. What bearing, if any, do these results have on Richard's rule? (See Prob. 5-2.)

7-15. Small droplets of gold have been observed to supercool by a maximum amount of approximately 230°C. What is the entropy change associated with the isothermal solidification of 1 gram atom of such supercooled gold? What is the entropy change of the surroundings if they are assumed to be (and to remain) at the same temperature as the supercooled gold?

7-16. Calculate the entropy change of (*a*) the system and (*b*) the surroundings for the isothermal solidification of 1 gram atom of aluminum supercooled (in the form of small droplets) by 130°C below its normal freezing point.

7-17. (*a*) Calculate the entropy change associated with the following reactions for (1) the system and (2) the surroundings:

$$\text{(i) } 2Al(s) + Fe_2O_3(s) = 2Fe(s) + Al_2O_3(s) \text{ at } 298°K$$
$$\text{(ii) } 2Al(l) + Fe_2O_3(s) = 2Fe(s) + Al_2O_3(s) \text{ at } 1000°K$$

(*b*) Is it thermodynamically possible for these reactions to proceed spontaneously in the direction written?

7-18. Calculate the coefficient of thermal expansion α for an ideal gas at 273°K from the ideal-gas law.

7-19. Evaluate (*a*) the coefficient of thermal expansion α and (*b*) the coefficient $(dP/dT)_V$ for liquid water at the temperature of its greatest density (3.9°C).

7-20. Find the difference $(C_P - C_V)$ between the heat capacity at constant pressure and that at constant volume for copper at room temperature and at 100°C.

7-21. (*a*) What can be said about the relative magnitudes of the heat capacities at constant pressure C_P and at constant volume C_V of an alloy such as Invar, which has a vanishingly small expansion coefficient? (*b*) Can the difference between the heat capacities at constant pressure and at constant volume $(C_P - C_V)$ assume a negative value?

7-22. (*a*) Calculate the work, the heat effect, and the change in internal energy for the isothermal compression at room temperature of 1 gram atom of iron from 1 to 1000 atm pressure. (*b*) Assuming now adiabatic conditions and a heat capacity C_P of 6 cal/(gram atom)(°K), calculate the temperature change ΔT.

7-23. What is the ratio of the heat effect to the work done in an isothermal elastic hydrostatic compression of copper at room temperature and atmospheric pressure? What is the ratio of the energy change to the elastic work? How would these ratios change with an increase in pressure? An increase in temperature?

7-24. Calculate the adiabatic thermoelastic temperature change which a tensile stress of 17,700 psi causes in a bar of nickel at room temperature and at 527°C. (The observed temperature change at 527°C was −0.4°C.)

7-25. When a bar of iron is subjected to a tensile stress under adiabatic conditions, the temperature of the bar falls. When Armco iron was stressed at different temperatures by a stress of 16,950 psi, the following observations were made:

t, °C	Δt, °C
105	−0.15
320	−0.24
590	−0.30
700	−0.27*
760	−0.15*

* By extrapolation from lower stresses.

(*a*) How do the first two values compare with calculated values? (*b*) Interpret the decrease in the magnitude of Δt above 590°C. What value would you expect for the effect at 800°C?

7-26. (*a*) Assume that a block of copper undergoes the following alternative processes entirely in the elastic range and under adiabatic conditions: a hydrostatic compression, a unidimensional compression, a unidimensional extension. In what direction does the temperature of the block tend to change in each of these processes? (*b*) Answer this question also for a block of idealized Invar ($\alpha \approx 0$) and for a material with a negative coefficient of thermal expansion.

7-27. Derive Eq. (7-16) from Eq. (7-15).

7-28. Derive the relation $C_P - C_V = R$ for an ideal gas from Eq. (7-20).

7-29. Consider a metal bar as a thermodynamic system with length l and unidimensional tension σ as the relevant state variables, corresponding to V and $-P$. (Note sign.) (*a*) Derive equations corresponding to Eqs. (7-14) and (7-16) for this system. (*b*) Derive equations corresponding to Eqs. (7-30) to (7-33).

7-30. Starting with the defining equations of the thermal expansion coefficient α and the isothermal compressibility β, prove that $(\partial S/\partial V)_T = \alpha/\beta$.

7-31. Using the appropriate Maxwell relation, evaluate numerically the isothermal change of entropy with pressure for mercury vapor at 357°C and 1 atm pressure.

7-32. Show that the change in the molar free energy of graphite accompanying changes in the total pressure is negligibly small for most purposes.

Chapter 8

8-1. Zirconium dioxide has the following heat capacities at low temperature [*Ind. Eng. Chem.*, **36**, 377 (1944)]:

T, °K	54.3	71.4	103.6	144.9	195.4	245.5	295.0
C_P, cal/(gram mole)(°K)	1.473	2.573	4.770	7.399	10.03	11.93	13.25

(*a*) Find the entropy increment $\Delta S°$ between 50.1 and 298.16°K. From Debye and Einstein functions the entropy of zirconium dioxide at 50.1°K is calculated to be 0.445 EU. What is the entropy of zirconium dioxide at 298°K? (*b*) Find the free energy of formation of zirconium dioxide from the elements at 298°K.

8-2. Find the entropy at 298.16°K of vanadium monoxide from the following data [*J. Am. Chem. Soc.*, **73**, 3894 (1951)]:

T, °K	C_P, cal/(gram mole)(°K)	T, °K	C_P, cal/(gram mole)(°K)
54.97	1.061	160.46	6.553
68.07	1.671	206.39	8.373
80.49	2.298	245.79	9.636
94.98	3.055	275.58	10.39
114.72	4.122	296.36	10.84
135.96	5.262	298.16	(10.86)

The Debye characteristic temperature is 398°K, leading to a value of the entropy at 52°K of 0.33 EU. (Verify this calculation.)

8-3. The heat capacity of titanium carbide at low temperatures is as follows [*Ind. Eng. Chem.*, **36**, 865 (1944)]:

T, °K	C_P, cal/(gram atom)(°K)	T, °K	C_P, cal/(gram atom)(°K)
55.1	0.312	156.2	4.010
65.9	0.571	176.2	4.750
74.2	0.821	196.8	5.474
86.5	1.242	226.3	6.380
95.7	1.587	256.0	7.168
115.6	2.389	285.7	7.830
135.6	3.190	294.9	7.987

Since the heat capacity is very low at 55.1°K, the T^3 relation may be used for extrapolation. Find the entropy of titanium carbide at 298.16°K and compare the value with an estimate of 5.5 EU, made by analogy with silicon carbide. [*U.S. Bur. Mines Bull.* 407 (1937).]

8-4. At temperatures of less than 2.19°K and pressures of at least 25 atm, helium II freezes without any latent heat of fusion. What is the approximate value of the entropy of helium II? [See, L. Tisza, *Physics Today*, **1**, No. 4 (August, 1948).]

8-5. What is the difference in heat capacities $C_P - C_V$ of a solid crystalline material in thermal equilibrium at absolute zero temperature?

Chapter 9

9-1. Derive Eq. (9-1) for an ideal gas from the equation $F = H - TS$.

9-2. Find the change in free-energy associated with the isothermal expansion of 1 mole of ideal gas from 2 to 1 atm pressure at 298°K.

9-3. Calculate the changes in free energy of the gas for the processes described in Prob. 7-5. (See also Prob. 7-12.)

9-4. Calculate the free-energy change associated with the expansion of 1 mole of hydrogen from 100 to 50 atm at 273°K and compare it with the value for the corresponding expansion of an ideal gas. (Use data in Table 9-1.)

9-5. Calculate values of f/P from the data in Table 9-1 for hydrogen and methane at 0 and 200°C, and plot these values against pressure including zero pressure. Discuss these curves.

9-6. Derive Eq. (9-17) for the reaction $MgCO_3(s) = MgO(s) + CO_2(g)$ by substituting this equation for Eq. (9-14) and continuing the derivation in terms of $MgCO_3$, MgO, and CO_2.

9-7. In some foreign publications the equilibrium constant for a reaction

$$A = 3B + 2C$$

is defined as $K' = a_A/a_B^3 \cdot a_C^2$. Examine various thermodynamic equations in which the equilibrium constant appears, and verify that the difference in definitions of the constants K and K' can be canceled out by defining the change in such state functions as the free energy as (initial value − final value).

9-8. Using thermochemical data and the fact that the dissociation pressure equals 1 atm at 884°C, express the free-energy change $\Delta F°$ as a function of the temperature T for the dissociation of calcium carbonate.

9-9. For the reaction Fe(δ) = Fe(l)

$$\Delta F° = 3{,}130 - 0.3T \ln T + 0.518T$$
$$\Delta C_P = 0.3 \text{ cal/(gram atom)(°K)}$$

Why are two temperature-dependent terms present in the $\Delta F°$ equation although ΔC_P itself is independent of temperature? Do both of these terms drop out if $\Delta C_P = 0$ for a reaction?

9-10. Assume that M is a typical metal and melts at 600°K; MO_2 is a typical metal oxide and melts at 1000°K. Sketch schematic diagrams for $\Delta H°$, $\Delta F°$, and $\Delta S°$ as functions of the temperature for the reaction $M + O_2(g) = MO_2$ in the range 300 to 1100°K. Consider only the stable phases.

9-11. For the reaction $Co(s) + \frac{1}{2}O_2(g) = CoO(s)$

$$\Delta F° = -59{,}850 + 19.6T$$

(*a*) Calculate the oxygen equilibrium pressure in atmospheres over Co and CoO at 1000°C. (*b*) Find the uncertainty of the value calculated in (*a*) if the error in the $\Delta H°$ term is estimated to be ±500 cal.

9-12. The following equilibrium data have been determined for the reaction

$$NiO(s) + CO(g) = Ni(s) + CO_2(g)$$

t, °C	663	716	754	793	852
$K \times 10^{-3}$	4.535	3.323	2.554	2.037	1.577

(*a*) Find $\Delta H°$, K, and $\Delta F°$ at 1000°K by use of a plot. (*b*) Would an atmosphere of 15 per cent CO_2, 5 per cent CO, and 80 per cent N_2 oxidize nickel at 1000°K?

9-13. (*a*) Convert the ratios of the gases in equilibrium with iron and wüstite (FeO) given in Table 9-2 to volume per cent CO and per cent H_2 at a total pressure of 1 atm and plot each against temperature. How would these curves change with an increase of the pressure to 2 atm? (*b*) Why is it ambiguous to say that "a gas consisting of 60 volume per cent CO and 40 per cent CO_2 is reducing to iron"? (*c*) Would an atmosphere consisting of 8 per cent hydrogen, 2 per cent water vapor, and 90 per cent nitrogen be oxidizing to iron at 900°F? At 1400°F?

9-14. Calculate the equilibrium constants for the following reaction at $T = 1000$°K from values of $\Delta H°$ at 298°K, standard molar entropies at 298°K, and entropy and enthalpy increments given by Kelley [*U.S. Bur. Mines Bull.* 476 (1949)]:

(*a*) $NiO(s) + CO(g) = Ni(s) + CO_2(g)$
(*b*) $CoO(s) + CO(g) = Co(s) + CO_2(g)$
(*c*) $Co(s) + \frac{1}{2}O_2(g) = CoO(s)$

9-15. Derive Eq. (9-28) from the equation $\Delta F° = \Delta H° - T\Delta S°$ and equations for $\Delta H°$ and $\Delta S°$ as functions of the temperature. What does this derivation show concerning the constant I in Eq. (9-28)?

9-16. According to calculations from thermochemical data [*U.S. Bur. Mines Tech.*

Paper 676 (1945)] the reaction $MgCl_2 \cdot 6H_2O(s) = MgCl_2 \cdot 4H_2O(s) + 2H_2O(g)$ is accompanied by the free-energy change:

$$\Delta F^\circ = 27{,}820 - 3.85T \log T + 5.56 \times 10^{-3}T^2 - 57.98T$$

What is the pressure of H_2O in equilibrium with $MgCl_2 \cdot 6H_2O$ and $MgCl_2 \cdot 4H_2O$ at 100°C?

9-17. In the dehydration of $MgCl_2 \cdot H_2O$ by heating, the following hydrolysis reaction may occur:

$$MgCl_2 \cdot H_2O(s) = Mg(OH)Cl(s) + HCl(g)$$

The following pressures of HCl result from this reaction [*U.S. Bur. Mines Tech. Paper* 676 (1945)]:

T, °K	483.2	503.2	523.2	543.2
P_{HCl}, mm Hg	50	100	200	360

A value of $\Delta C_P = -1.65 - 4.14 \times 10^{-3}T$ can be estimated for the reaction, but heat of formation and entropy data for Mg(OH)Cl are lacking. (*a*) Using the definition of Σ in the text and Eq. (9-30), make a plot to find ΔH_0° and apply the value thus found to the individual Σ-values to obtain individual values of I for averaging. Write an equation for ΔF°. (*b*) What is the heat effect ΔH° for the hydrolysis reaction at 298°K and at 600°K? (*c*) Could the hydrolysis reaction be suppressed by the presence of 0.2 atm of HCl at 503.2°K? At 523.2°K? What pressure of HCl would be required at 600°K?

9-18. The heat of formation of silver oxide is −7300 cal/gram mole. Calculate the approximate temperature at which silver oxide begins to decompose on heating (*a*) in pure oxygen and (*b*) in air. Use the following data:

	Entropy S_{298}°, EU	Approx. heat capacity C_P, cal/(gram mole)(°K)
Ag_2O	29.1	15.7
O_2	49.0	7.0
Ag	10.2	6.1

9-19. In a certain region in an iron blast furnace at which the temperature is 1700°F, the reaction $FeO + CO = Fe + CO_2$ occurs. The value of the equilibrium constant $K = P_{CO_2}/P_{CO}$ is 0.44. The gas reaching this region contains 11 per cent CO_2 and 30 per cent CO. Assuming that equilibrium between gas and ore is attained, how much carbon must be burned to CO at lower levels in the furnace for each 55.85 lb of iron reduced in this region?

9-20. In a laboratory test lead monoxide (litharge, PbO) is reduced with solid carbon by placing the mixture in a silica crucible and heating in a muffle furnace to about 627°C. (*a*) Is it likely that this charge is heated for other reasons than the necessity of obtaining an adequate rate of reaction? (*b*) If the muffle contains an inert atmosphere, is there any advantage in placing a salt cover over the charge? (*Hint:* The salt melts to form a liquid seal across the top of the charge.)

9-21. Magnesium carbonate is to be heated in air at 100°C. The carbon dioxide content of atmospheric air is about 0.03 per cent. Is this concentration of carbon dioxide sufficient to repress the decomposition of magnesium carbonate?

	Entropy S°_{298}, EU	Heat of formation ΔH°_{298}, Kcal/gram mole	Approx. heat capacity C_P, 273–373°K, cal/(gram mole)(°K)
$MgCO_3$	15.7	−266.0	18.1
MgO	6.6	−143.8	9.0
CO_2	51.1	− 94.1	9.1

9-22. Magnesium can be produced by the following reaction:

$$MgO(s) + C(s) = Mg(g) + CO(g)$$

(*a*) Calculate ΔF° for this reaction at 1300, 1500, 1700, 1900, and 2100°C from published thermochemical data. (*b*) What are the equilibrium pressures of magnesium and carbon monoxide at each of these temperatures?

9-23. (*a*) Show by calculations from thermochemical data under what conditions magnesium oxide can be reduced to the metal at high temperatures with aluminum or silicon as the reducing agent. (*b*) How would you modify your appraisal of the process if the silicon to be used is in the form of ferrosilicon analyzing 20 per cent Fe?

9-24. One step in the manufacture of specially purified nitrogen is the removal of small amounts of residual oxygen by passing the gas over copper gauze at approximately 500°C. The following reaction takes place: $2Cu(s) + \frac{1}{2}O_2(g) = Cu_2O(s)$. (*a*) Assuming that equilibrium is reached in this process, calculate the amount of oxygen present in the purified nitrogen. The following free-energy equation may be used: $\Delta F^\circ = -39{,}850 + 15.06T$. (*b*) What would be the effect of raising the temperature to 800°C? Of lowering it to 300°C? What is the probable reason for using 500°C? (*c*) What would be the effect of increasing the pressure?

9-25. In a gas-analysis train which is part of a vacuum fusion apparatus, carbon monoxide is oxidized to carbon dioxide by cupric oxide at about 300°C. (*a*) Calculate the concentration of residual carbon monoxide in equilibrium with carbon dioxide, assuming the latter to be at 0.05 atm pressure. (*b*) In the operation of this equipment, the gases are passed through an efficient absorber of carbon dioxide and recirculated over the cupric oxide. How does this affect the residual carbon monoxide concentration?

9-26. Molten magnesium is protected from atmospheric oxidation by burning sulfur above the bath. It has been suggested that a protective film of magnesium sulfate is formed. Is the formation of magnesium sulfate under these conditions thermodynamically possible?

9-27. (*a*) Calculate the partial pressure of zinc over liquid zinc at 600°C from the appropriate equation given in the text. (*b*) What is the heat effect ΔH° of condensation of 1 gram atom of zinc at 600°C?

9-28. The partial pressure of oxygen in equilibrium with Cu_2O and CuO has been found to be 0.0208 atm at 900°C and 0.1303 atm at 1000°C. Calculate ΔH° for the reaction $2Cu_2O(s) + O_2(g) = 4CuO(s)$.

9-29. The following equilibrium data have been determined for the reaction

$$CoO(s) + CO(g) = Co(s) + CO_2(g)$$

T, °K.	836	888	936	1028	1079
log K.	2.241	2.036	1.879	1.618	1.499

(*a*) Find $\Delta H°$, $\Delta S°$, and $\Delta F°$ at 1000°K for this reaction. (*b*) If the standard free-energy change for the dissociation of carbon dioxide into carbon monoxide and oxygen is 46,720 cal/mole of CO_2 at 1000°K, calculate the standard free-energy change of formation of CoO at this temperature.

9-30. The standard free-energy change for the reduction of chromium oxide by hydrogen is given approximately as follows:

$$Cr_2O_3(s) + 3H_2(g) = 2Cr(s) + 3H_2O(g)$$
$$\Delta F° = 97{,}650 - 28.6T$$

(*a*) Find the maximum partial pressure of water vapor in otherwise pure hydrogen at 1 atm in which chromium can be heated without oxidation at 1500°K. (*b*) Is the oxidation of chromium by water vapor exothermic or endothermic? What, if anything, can be said from the above data about the heat effect at 1500°K of the oxidation of chromium by pure oxygen? (*c*) Would the equilibrium in the above reaction be affected by a change in the pressure of the hydrogen–water-vapor mixture from 1 to 2 atm? To 200 atm?

9-31. Using data in Table 9-2 calculate the heat of reaction $\Delta H°$ for the reduction of magnetite, Fe_3O_4 to iron at 1000°F by (*a*) CO and (*b*) H_2.

Perform corresponding calculations for wüstite, Fe_xO, at 1300°F.

9-32. The adiabatic reduction of iron oxide (wüstite) by carbon monoxide has been reported to result in a temperature rise of about 125°C, while adiabatic reduction by hydrogen causes a temperature drop of about 350°C. Suggest a process for carrying out the reduction without a temperature change.

9-33. The mean specific heat of manganese (*g*) is 4.98 cal/(gram atom)(°K) and that of manganese (*l*) is 11.0 cal/(gram atom)(°K). The entropy of evaporation is 25.35 EU at the melting point (1517°K). The vapor pressures of liquid manganese are as follows:

T, °K.	1517	1600	1900	2200	2500
P, atm.	1.17×10^{-3}	3.20×10^{-3}	5.56×10^{-2}	4.15×10^{-1}	1.81

(*a*) Find the heat and free energy of vaporization of liquid manganese as functions of temperature. (*b*) Calculate the boiling point from the answer to (*a*). What is the heat of evaporation at this temperature? What is the entropy of evaporation at this temperature and by how much does it differ from Trouton's constant?

9-34. Answer the following questions by means of the principle of Le Chatelier. (*a*) The removal of phosphorus from liquid steel by oxidation evolves heat. How does the equilibrium change with an increase in temperature? (*b*) The so-called carbon-deposition reaction in the blast furnace is $2CO(g) = C(s) + CO_2(g)$. How does an increase in pressure at constant temperature change the equilibrium? Is the equilibrium constant changed?

9-35. If a solute lowers the surface tension, do you expect it to be present at the surface in larger or in smaller concentration than in the interior?

9-36. If a soap film is extended adiabatically, do you expect its temperature to rise or fall? (*Hint:* The surface tension of the film may be assumed to decrease with a rise in temperature.)

9-37. It has been stated that the adiabatic thermoelastic temperature change is always in the direction which stiffens the body being stressed. Verify this statement for the extension and compression of a metal bar. Discuss the relation of this statement to the principle of Le Chatelier.

9-38. The solubility of hydrogen in most solid metals increases as the temperature is increased, but a few metals, for example titanium and palladium, dissolve decreasing amounts of hydrogen with increasing temperature. How does the heat effect of the solution of hydrogen in these two classes of metals differ?

9-39. From the following values for the heat of formation $\Delta H°$ at 298.16°K (cal gram mole) and data in Table 2, page 523, calculate the heats of formation at 0°K. Compare the values calculated in this manner with the values in the column headed $\Delta H_0°$ in Table 2.

Substance	$\Delta H°_{298.16}$
$H_2O(g)$	−57,798
$CO(g)$	−26,416
$CO_2(g)$	−94,052
$H(g)$	+52,089

9-40. Find the free energy of formation at 298°K of H_2, H_2O, CO, and CO_2 from the data in Tables 2 and 3, pages 523 and 524.

9-41. For the reaction $CO_2(g) + C(\text{graphite}) = 2CO(g)$, calculate $\Delta F°$ and K for 1000 and 1500°K using data in Tables 2 and 3, pages 523 and 524.

9-42. Using Tables 2 and 3, pages 523 and 524, calculate the equilibrium constant K at 1000 and 1500°K for the reaction $CO(g) + H_2O(g) = CO_2(g) + H_2(g)$.

9-43. (*a*) Calculate and plot $\Delta F°$ vs. T and $\log K$ vs. $1/T$ for the reaction

$$CO(g) + H_2O(g) = CO_2(g) + H_2(g)$$

at several temperatures above and below 1500°K. Obtain data from the Bureau of Standards Tables. (*b*) Find $\Delta H°$ and $\Delta S°$ at 1500°K from these curves.

9-44. Evaluate the equilibrium constant K for the reaction

$$CO(g) + H_2O(g) = CO_2(g) + H_2(g)$$

at each of the several temperatures from the following data:

$$FeO(s) + CO(g) = Fe(s) + CO_2(g);\ K_1 = p_{CO_2}/p_{CO}$$
$$FeO(s) + H_2(g) = Fe(s) + H_2O(g);\ K_2 = p_{H_2O}/p_{H_2}$$

t, °C	$\log K_1$	$\log K_2$
600	−0.046	−0.479
700	−0.172	−0.375
800	−0.272	−0.302
900	−0.344	−0.226
1000	−0.402	−0.175

Plot the results as log K vs. $1/T$ and as $\Delta F°$ vs. T and compare them with the results of Prob. 9-42.

9-45. Carry through a derivation [similar to that leading to Eq. (9-35)] to evaluate $(\Delta F° - \Delta H^0_{298})/T$ for use in preparing a table of this function. In the preparation of such a table what type of data are needed in addition to those required in preparing a table of $(\Delta F° - \Delta H^\circ_0)/T$? Also what data are not required? What are the advantages and disadvantages of the two types of tables?

Chapter 10

10-1. (*a*) Calculate the atom fractions of chromium, nickel, and iron in an 18-8 stainless steel, disregarding such minor constituents as manganese or silicon. (*b*) Perform a corresponding calculation for a "90-10" copper-tin bronze.

10-2. Calculate the atom fractions N_C of carbon and iron in a solid solution of the two elements containing 0.01, 0.1, and 1 per cent C. Determine for these compositions the ratio $N_C/\%C$ and also the limiting value of this ratio as $\%C \rightarrow 0$.

10-3. If the molal heat of mixing is represented by $\Delta H = aN_1N_2$, find the partial molal heats.

10-4. Discuss the meaning of the partial molal heat capacity at constant pressure $\bar{C}_{P,i}$. Write equations which relate this quantity to (*a*) the heat capacity C_P and (*b*) the enthalpy H of the solution.

10-5. Derive Eq. (10-16) from Eq. (10-12).

10-6. Assume that 1 part of liquid nickel and 2 parts of liquid copper mix without an appreciable heat of solution and without a volume change. Is there a change in internal energy E? In enthalpy H? In entropy S? In free energy F? Is any work done in this process if the mixing is carried out in vacuum—under an atmosphere of argon?

10-7. (*a*) Calculate the heat effect ΔH for dissolving 1 gram atom of solid nickel in 9 gram atoms of liquid copper at 1200°C, assuming an ideal solution is formed. (*b*) Calculate the entropy change of the system and of the surroundings for this process.

10-8. The following pressures of zinc have been determined for copper-zinc alloys at 1060°C:

Atom fraction of Zn	1.0	0.45	0.30	0.20	0.15	0.10	0.05
p_{Zn}, mm Hg	3040	970	456	180	90	45	22

(*a*) Does this system obey Raoult's law? Henry's law? Over what concentration ranges? (*b*) What is the free-energy change when 1 gram atom of liquid zinc dissolves in a very large amount of copper-zinc alloy in which the atom fraction of zinc is 0.30? (*c*) Perform a corresponding calculation for a copper-zinc alloy in which the atom fraction of zinc is 0.60. Find the difference in free energy of 1 gram atom of zinc in liquid copper-zinc alloys at atom fractions of 0.45 and 0.15.

10-9. The activity coefficient of zinc in liquid brass has been expressed for the temperature range 1000 to 1500°K by the equation:

$$RT \ln \gamma_{Zn} = -4600\, N^2_{Cu}$$

Calculate the partial pressure of zinc, P_{Zn}, over a solution of 60 per cent copper and 40 per cent zinc at 1500°K.

10-10. (*a*) A stream of nitrogen gas is passed in a closed system over a boat containing mercury at 100°C. The flow rate of the nitrogen is slow enough to allow this gas to become saturated with mercury vapor. The total volume of nitrogen used is 22 liters measured at 20°C and 1 atm. The nitrogen was found to contain 0.0674 g of mercury. Calculate the vapor pressure of mercury at 100°C. (*b*) When the same experiment is carried out with a sodium amalgam (Na + Hg) in which the atom fraction of sodium is 0.122, assume that 22 liters of nitrogen gas saturated with mercury is found to contain 0.0471 g of mercury. Take pure mercury as the standard state and calculate the activity a, the activity coefficient γ, and the difference $(\mu - \mu^\circ)$ equal to $(\bar{F} - \bar{F}^\circ)$ of mercury in this amalgam.

10-11. Estimate the partial pressure of magnesium at 700°C over a liquid alloy consisting of 90 per cent aluminum and 10 per cent magnesium. State any assumptions made.

10-12. (*a*) Will a gas mixture containing 97 per cent H_2O and 3 per cent H_2 oxidize nickel at 1000°K? Base your calculations on the following data:

$$Ni + \tfrac{1}{2}O_2 = NiO \qquad \Delta F^\circ_{1000} = -35{,}400$$
$$H_2 + \tfrac{1}{2}O_2 = H_2O \qquad \Delta F^\circ_{1000} = -45{,}600$$

(*b*) An alloy containing 10 atom per cent Ni and 90 atom per cent Au is a solid solution at 1000°K. It is found that this solution reacts with water vapor to form NiO. Assume that approximate measurements indicate that the reaction reaches equilibrium when the water vapor–hydrogen mixture contains 0.35 per cent hydrogen by volume. Find the corresponding value of the activity coefficient γ of Ni in the alloy.

10-13. The standard free-energy change for the reaction

$$Cr_2O_3(s) + 3H_2(g) = 2Cr(s) + 3H_2O(g)$$

is given approximately by the equation

$$\Delta F^\circ = 97{,}650 - 28.6T$$

Assume that chromium is more readily oxidized than metal M. Assume also that chromium forms solid solutions with metal M and that Cr_2O_3 is insoluble in these solid solutions. Estimate the maximum partial pressure of water vapor in otherwise pure hydrogen at 1 atm pressure in which a solid solution of 10 atom per cent Cr and 90 atom per cent M can be heated without oxidation at 1500°K. State any assumptions.

10-14. In lead refining, antimony is removed from liquid lead by selective oxidation with air at about 900°C. Assuming Raoult's law, estimate how low the removal of antimony can be carried.

10-15. The following solubilities of oxygen in 100 g of liquid silver at 1075 C have been measured [*Z. physik. Chem.*, **68**, 115 (1909)]:

Pressure of O_2, mm Hg	Oxygen dissolved, cm^3/100 g Ag
128	81.5
488	156.9
760	193.6
1203	247.8

(*a*) Show by a plot or by suitable calculations whether these observations agree with Sieverts' law for the solubility of gases in metals. (*b*) How much oxygen does 100 g of silver absorb at 1075°C from atmospheric air? (*c*) What pressure of air corresponds to 1 atm of oxygen with respect to the solubility of oxygen in silver at 1075°C? Does this ratio of air to oxygen pressure apply only to this temperature? Is this ratio restricted to liquid silver as a solvent of oxygen?

10-16. At 1 atm pressure and 1750°C, 100 g of iron dissolves 35 cm^3 (STP) of nitrogen. Under the same conditions, 100 g of iron dissolves 35 cm^3 of hydrogen. Thus at 1750°C the solubility of hydrogen in iron equals that of nitrogen. Argon is insoluble in molten iron. How much gas will 100 g of iron dissolve at 1750°C and 760 mm pressure under an atmosphere which consists of (*a*) 50 per cent nitrogen and 50 per cent hydrogen? (*b*) 50 per cent argon and 50 per cent hydrogen? (*c*) 33 per cent nitrogen, 33 per cent hydrogen, and 34 per cent argon?

10-17. At 800°C, 100 g of silver dissolves 3.3 cm^3 (STP) of oxygen at 1 atm pressure. How much oxygen does silver dissolve at 800°C under (*a*) 0.1 and (*b*) 10 atm pressure?

10-18. Find the partial pressures of nitrogen and hydrogen in equilibrium at 1540°C with molten iron containing 0.010 per cent nitrogen and 0.0005 per cent hydrogen. At that temperature molten iron contains 0.039 per cent nitrogen in equilibrium with nitrogen at 1 atm and 0.0025 per cent hydrogen in equilibrium with hydrogen at 1 atm pressure.

10-19. The following solubility values in cubic centimeters of H_2 dissolved by 100 g of metal at 1 atm pressure have been found [*Trans. AIME*, **156**, 149 (1944)]:

	1000°C	1100°C	1200°C	1300°C
100% Cu	....	5.73	7.34	9.37
88.5% Cu, 11.5% Sn	3.09	4.11	5.35	6.85
78.3% Cu, 21.7% Sn	2.11	2.97	3.94	5.10

Find K, $\Delta H°$, and $\Delta F°$ for the solution of hydrogen in copper and the copper-tin alloys.

10-20. List those of the following quantities which can be larger than unity for a constituent of a binary solution: the activity a_1, the activity coefficient γ_1, the fugacity f_1. Specify the conditions to which each answer applies.

10-21. From the equation for the activity coefficient of zinc in copper-zinc alloys $RT \ln \gamma_{Zn} = -4600N_{Cu}^2$ find the activity coefficient of copper at 1500°K by means of the Gibbs-Duhem equation.

10-22. If $a_2 = kN_2(1 + bN_2)$, find a_1 by the Gibbs-Duhem equation.

10-23. From electromotive cell measurements at 435°C the following equation has been found for the activity coefficient of zinc in cadmium-zinc alloys:

$$\ln \gamma_{Zn} = 0.87(1 - N_{Zn})^2 - 0.30(1 - N_{Zn})^3$$

Find an equation for the activity coefficient of cadmium in these alloys by means of the Gibbs-Duhem equation.

10-24. Four grams of an alloy of 75 weight per cent gold and 25 weight per cent silver at 0°C is dropped into a calorimeter which contains liquid tin at approximately 240°C. After dissolution the temperature of the calorimeter decreases by 0.10°C. When a mechanical mixture of 3 g gold and 1 g silver at 0°C is added to a fresh batch of tin equal in amount to that used for the addition of the alloy, the temperature of the system increases by 0.63°C. The heat capacity (or energy equivalent) of the calorimeter is 33.6 cal/°C. Assume that the final temperature of the calorimeter is exactly 240°C in both cases. Find the heat of formation in calories per gram atom of the alloy. At what temperature (0°C or 240°C) does this value of the heat of formation of the alloy apply?

10-25. (*a*) Express the free-energy change for the transfer of 1 mole of cobalt from pure liquid to a 1 per cent solution in liquid iron as a function of temperature. (*b*) Find a corresponding equation for the solution of solid chromium in liquid iron. For the fusion of chromium, $\Delta F^\circ = 4350 - 2.10T$.

Chapter 11

11-1. The relations of the chemical potential μ_i to the thermodynamic functions E, H, A, and F are given by Eq. (11-6). Is there a corresponding useful relation between μ_i and the entropy S of a phase?

11-2. A system consists of silver-rich solid solution α, copper-rich solid solution β in equilibrium with vapor and eutectic liquid at the eutectic temperature. How many degrees of freedom does the system have?

11-3. Apply the phase rule to a system consisting of molten iron in equilibrium with δ-Fe and iron vapor. Does the state of this system represent a triple point?

11-4. Criticize the question "How many phases are present in a hypoeutectoid plain-carbon steel?"

11-5. What is the number of components in a system consisting of ferrite and cementite (Fe_3C)?

11-6. In a ternary solution, the concentration of two components can be changed independently. Does this fact introduce one or two variables in the sense of the phase rule?

11-7. The concentration of solute in a solid solution of fixed over-all composition under equilibrium conditions may differ at the surface from that in the interior. (See Prob. 9-35.) A range of composition, therefore, appears to coexist at equilibrium, whereas the phase rule as expressed by Eq. (11-13) indicates that only one phase (composition) is stable. How can this apparent difficulty be resolved?

Chapter 12

12-1. How does the change in volume associated with the decomposition of austenite enter into the equation for the free-energy change involved in the nucleation of this process?

12-2. The vapor pressure of zinc is 0.1 atm at 990°K and 1.0 atm at 1180°K. Calculate (*a*) the heat of vaporization of zinc and (*b*) the coefficient dP/dT at the boiling point.

12-3. Indicate how the triple point of gallium can be calculated. Is the result of this calculation affected by the fact that this metal expands on freezing?

12-4. A solid undergoes an allotropic transformation in which a low-temperature form α changes to a high-temperature form β. Show by simple sketches that two possibilities exist for the location of this triple point, depending on whether the equilibrium line $\alpha - \beta$ cuts the melting or the sublimation curve. What is the effect of the relative densities of α, β and the liquid?

12-5. (*a*) The following statement has been made: "In heterogeneous fields of the binary phase diagram the partial pressure of a component is controlled by the phase or compound containing the larger amount of that component; for example in the heterogeneous ($\alpha + \beta$) field in the copper-zinc system the pressure of zinc is equal to that of the limiting β-phase." Discuss this statement critically. Formulate a corresponding statement for the partial pressure of a component of a binary system under invariant conditions, as at the temperature of a eutectic reaction. Give a specific example. (*b*) When the vapor pressure of mercury over the system cerium-mercury is determined at 340°C, it is found that a uniform pressure of about 550 mm Hg exists in the composition range 0 to 14.9 per cent Ce. At this composition the pressure drops sharply to about 110 mm Hg and continues at this value up to at least 55 weight per cent of Ce [*Z. anorg. allgem. Chem.*, **176,** 23 (1928)]. What conclusion can be drawn from these observations regarding the phase diagram of the cerium-mercury system at 340°C?

12-6. The density of solid lead at the normal melting point of 327.3°C has been reported as 11.005 (g/cm³) and that of liquid lead at 327.3°C as 10.645 (g/cm³). Assuming a heat of fusion ΔH_f of 1190 cal/gram atom, estimate the melting point of lead at 100 atm pressure.

12-7. If a metal undergoes an allotropic transformation at room temperature and high pressure and the same transformation occurs at low temperature at 1 atm pressure, is the volume change associated with the transformation positive or negative?

12-8. Bismuth has a density of 9.80 (g/cm³) at room temperature. Its coefficient of linear expansion α_{lin} is 14.6×10^{-6} per degree. The density of the liquid metal at the melting point T_m(271°C) has been reported as 10.07 (g/cm³). The heat of fusion is 2600 cal/gram atom. (*a*) What is the sign of dT_m/dP? (*b*) At what temperature will bismuth melt under 100 atm of pressure?

12-9. From the data and discussion in the text, evaluate the industrial possibilities of using hydrostatic pressure to suppress the pearlite reaction in plain carbon steel.

12-10. (*a*) Assuming that the solid phase crystallizing from silver-lead solutions is pure silver, find the heat of fusion ΔH_f of silver from the following data:

Atom fraction Ag.........	1.000	0.973	0.916	0.794
Liquidus T_1, °K.........	1234.6	1213	1148	1010

(*b*) If ΔH_f of silver is known from other measurements to be 2690 cal/gram atom, find the activity of silver a_{Ag} in each solution.

12-11. The crystals which solidify on cooling silver-rich alloys in the silver-zinc system are solid solutions of zinc in silver. The following data are given by K. K. Kelley, Contributions to the Data on Theoretical Metallurgy V, *U.S. Bur. Mines Bull.* 393 (1936) p. 5, where N_{Ag} is the atom fraction of silver in the liquid solutions, T_1 is, the temperature at which the first crystals appear, and T_2 is the temperature at which solidification is complete.

N_{Ag}	T_1, °K	T_2, °K
1.000	1,234.6	1,234.6
0.845	1,145	1,073
0.774	1,105	1,015
0.708	1,039	993
0.686	1,023	983

Find the heat of fusion ΔH_f of silver from these data. (*Hint:* Plot first N_{Ag} against T_2 and read values of composition corresponding to T_1 from this curve.)

12-12. What is the melting point lowering caused by the solution of 0.160 g of oxygen in 100 g of silver? The solubility of oxygen in solid silver may be neglected. The melting point of pure silver is 960.8°C, and its heat of fusion is 2690 cal/gram atom.

12-13. The Fe-C peritectic occurs at 1492°C. The peritectic liquid contains 0.50 per cent carbon. Calculate the composition of δ-Fe in equilibrium with this liquid, and compare the result with the corresponding value taken from a published phase diagram.

12-14. (*a*) One per cent silicon lowers the liquidus of iron by 12°C. What is the composition of the first solid to solidify from the melt? (*b*) One per cent chromium lowers the liquidus of iron by 1°C. What is the composition of the first solid? (*c*) Using only the information given in this problem and the results of (*a*) and (*b*), draw simple sketches of the iron-rich side of the iron-silicon and iron-chromium systems.

Chapter 13

13-1. For the binary system shown in the schematic temperature-composition diagram of Fig. 12-7*a*, draw schematic free-energy-composition diagrams at several representative temperatures.

13-2. Draw schematic free-energy-composition diagrams for the system shown in Fig. 12-7*b*. Include the temperature of the minimum among the several temperatures to be represented.

13-3. Draw schematic free-energy-composition diagrams for the system shown in Fig. 12-7*c*. Include the eutectic temperature, several temperatures above and one temperature below this eutectic temperature among the temperatures to be represented.

13-4. In the binary system A-B an A-rich primary solid solution α and the intermediate phases β and β' occur. β and β' are of the same composition, but β' is metastable with respect to β. Draw schematically (*a*) a portion of the temperature-composition diagram showing the phase boundaries of α in equilibrium with β and β', respectively, and (*b*) a free-energy-composition diagram showing the relative free energies of α, β, and β' at some temperature T. Explain the relation of diagram (*b*) to diagram (*a*). Does it matter whether at the temperature T the free energy of β (or β') is larger or smaller than that of α?

Chapter 14

14-1. Where would the curve for the formation of silver oxide, Ag_2O, fall in a plot of the type of Fig. 14-4? What would the slope of this curve be?

14-2. Calculate the data required for adding to Fig. 14-4 a curve for the reaction $Sn + O_2(g) = SnO_2$.

14-3. Using Fig. 14-4, find solutions for the following problems: (*a*) Prob. 9-11, (*b*) Prob. 9-22, (*c*) Prob. 9-24, (*d*) Prob. 9-30. Compare the results with previously calculated solutions to these problems.

14-4. Using Fig. 14-4, find approximate values of the standard entropy change $\Delta S°$ at 298°K for the following reactions:

(*a*) $Ni(s) + \frac{1}{2}O_2(g) = NiO(s)$

(*b*) $2C(\text{graphite}) + O_2(g) = 2CO(g)$

Compare these values with exact values calculated from data given in tables of standard entropies at 298°K.

14-5. Using Fig. 14-4, find the equilibrium constant (*a*) for the reaction of Fe_2O_3 with aluminum at 1600°C according to the equation

$$Fe_2O_3(s) + 2Al(l) = Al_2O_3(s) + 2Fe(l)$$

How does this constant change with an increase in temperature?

14-6. From Fig. 14-4, find the approximate value of the equilibrium constant at 950°C for the reaction

$$H_2(g) + CO_2(g) = H_2O(g) + CO(g)$$

14-7. (*a*) From Fig. 14-9, find the equilibrium constant for the reaction of silver with hydrogen sulfide at room temperature. (*b*) Can hydrogen be used efficiently to reduce silver sulfide at room temperature? How does the equilibrium shift as the temperature is raised?

14-8. Determine from Fig. 14-9 the relative stabilities at room temperature of (*a*) CuS and FeS_2, (*b*) MnS and FeS, (*c*) H_2S and SO_2.

14-9. Using Fig. 14-12, find $\Delta F°$ and K at 800°C for the reduction of solid chromium chloride, $CrCl_2$, by hydrogen.

14-10. From Figs. 14-4 and 14-12, find the standard free-energy change $\Delta F°$ at 500 and 1000°C for the following reactions:

(*a*) $SnO + Cl_2(g) = SnCl_2 + \frac{1}{2}O_2(g)$
(*b*) $\frac{1}{2}TiO_2 + Cl_2(g) = \frac{1}{2}TiCl_4 + \frac{1}{2}O_2(g)$
(*c*) $FeO + 2HCl(g) = FeCl_2 + H_2O(g)$

State also whether the metal compounds are solid, liquid, or gaseous at 500 and 1000°C.

14-11. Taking data from relevant graphs, find the standard free-energy change $\Delta F°$ at 500 and 1000°C for the reaction,

$$\tfrac{1}{2}TiO_2(s) + \tfrac{1}{2}C(s) + Cl_2(g) = \tfrac{1}{2}TiCl_4(g) + \tfrac{1}{2}CO_2(g)$$

Compare the results with the results of Prob. 14-10*b*.

14-12. Check the following statement by a specific example, taking data from the relevant graphs: "Chlorination of metal sulfides is far less energetic with HCl than with Cl_2 because the free energy of formation of H_2S is small and the free energy of formation of 2HCl is large."

14-13. Appraise the thermodynamic aspects of a process for separating cadmium and zinc by selective chlorination of their oxides. How does temperature affect this process?

Chapter 15

15-1. Carry out in detail the derivation of Eq. (15-4) by the method outlined in the footnote on page 381.

15-2. From the data in the text, estimate the maximum amount of nitrogen which can remain unprecipitated in a plain-carbon steel containing 0.05 per cent of free aluminum at 900°C.

15-3. From the data for the solubility of nitrogen gas at 1 atm pressure in iron (Fig. 15-1), find the heat of solution in γ-iron for 1 gram mole of gaseous nitrogen.

15-4. At a pressure of nitrogen of 1 atm liquid iron dissolves approximately 0.039 per cent of nitrogen at 1540°C and 0.043 per cent at 1760°C. (*a*) Estimate the heat of solution. (*b*) What are the limits of uncertainty in the heat of solution, if both solubility values are uncertain to ±0.0005 per cent?

Chapter 16

16-1. Derive Eq. (16-5) from Eq. (16-4) and the Gibbs-Duhem equation.

16-2. Show how Eq. (16-6*a*) follows from Eq. (16-6).

16-3. (*a*) Using the method suggested in the text, derive an equation for the equilibrium constant K_C, defined in Eq. (16-14), as a function of temperature. What is the value of the heat of reaction? (*b*) From the value of the heat of reaction found in (*a*) and an assumed value of $\Delta C_P = 1$ cal/(gram atom)(°K), derive an equation for $\ln K_C$ as a function of temperature. (*c*) Evaluate K_C at 1300°C from each of the above equations and calculate the per cent difference.

16-4. The following partial pressures of CO and CO_2 are in equilibrium with austenite of the stated carbon contents at 1000°C according to the equation:

$$2CO(g) = C(\text{in austenite}) + CO_2(g)$$

$\%C^\gamma$	p_{CO}	p_{CO_2}
0.13	0.891	0.109
0.45	0.9660	0.0340
0.96	0.9862	0.0138
1.50 (saturated with graphite)	0.9928	0.0072

(*a*) Find the activity of carbon, in each case taking graphite as the standard state. (*b*) Plot the activities against carbon concentrations. Analyze the relation of this graph to Fig. 16-6. Take several values of the activity from Fig. 16-6 (*e.g.*, at carbon = 0.20, 0.50, and 1.0 per cent), convert them to activities relative to graphite, and compare the results with corresponding values read from your plot.

16-5. From measurements of the equilibrium $2CO(g) = C(\text{in austenite}) + CO_2(g)$, the following values of the activity of carbon have been found as a function of atom fraction N_C of carbon in austenite at 1000°C with graphite as standard state:

N_C	0.01	0.02	0.03	0.04	0.05	0.06
a_C	0.09	0.20	0.33	0.47	0.66	0.85

The equilibrium constant K of the reaction $2H_2(g) + C(\text{graphite}) = CH_4(g)$ is $9.6 \cdot 10^{-3}$ at 1000°C. (*a*) Find the atom fraction and weight per cent of carbon in austenite, if a sample is brought into equilibrium with a mixture containing 99.5 per cent H_2 and 0.5 per cent CH_4 at a total pressure of 1 atm. (*b*) Will steel containing 0.4 weight per cent carbon be carburized or decarburized by a gas mixture containing 99 per cent H_2 and 1.0 per cent CH_4 at a total pressure of 2 atm and a temperature of 1000°C?

16-6. Calculate the solubility of diamond (*a*) in γ-iron at 1200°K and (*b*) in α-iron at 900°K. (Preliminary question: Is the solubility of diamond larger or smaller than that of graphite?)

16-7. Determine the metastable melting point of pure γ-iron using data in Table 16-1.

16-8. By the method used in the text, find the heat of solution of graphite in iron-carbon alloys of eutectic composition.

Chapter 17

17-1. A laboratory installation for the investigation of copper refining consists of four electrolytic cells in series. After a run of 7.33 hr with an average current of 34.2 amp, the increase in cathode weight is found to be 1089 g. What is the efficiency of the process?

17-2. How many kilowatt-hours are required for the electrolytic refining of 1 ton of copper, if the voltage is 0.4 volt and 100 per cent efficiency is assumed? How would the energy requirements change if the voltage were lowered to 0.2 volt? How much heat is liberated in each case?

17-3. In a laboratory installation, a current of 2.0 amp flows for 48,250 sec; 0.45 gram atom of a bivalent metal is deposited. What is the current efficiency of the process? Can you tell how much heat (in joules or calories) is liberated?

17-4. If the cell discussed in the text is written Br_2 | AgBr, LiBr | Ag, what is the sign of the emf? Write the cell reaction and verify the sign in Eq. (17-2).

17-5. The emf of the galvanic cell

$$\text{Cd(pure l)} \mid \text{KCl-NaCl-LiCl-CdCl}_2\text{(l)} \mid \text{Cd(in l. Cd-Sn, } N_{Cd} = 0.258)$$

has been found to be +0.0324 volt at 483°C. (*a*) Calculate $\bar{F}_{Cd} - F^\circ_{Cd}$ (in calories) at 483°C. (*b*) What is the activity of cadmium in the alloy, relative to pure cadmium as standard state? (*c*) Calculate the vapor pressure of cadmium over the alloy, if the vapor pressure of pure cadmium at 483°C is calculated as 9.2 mm Hg. (*d*) At the concentration of cadmium considered, does the system Cd-Sn show positive or negative deviation from Raoult's law?

17-6. The emf of the cell Ag(s) | AgCl(s) | Pt, Cl_2 (1 atm) as a function of temperature has been found to be $\mathcal{E}(\text{volt}) = 0.977 + 5.7 \cdot 10^{-4}(350 - t) - 4.8 \cdot 10^{-7}(350 - t)^2$ valid between $t = 120$ and 450°C. Solid AgCl is an ionic conductor. Find the value of ΔH° at 350° for the reaction $Ag(s) + \frac{1}{2}Cl_2(g) = AgCl(s)$.

17-7. The emf of the cell Ag(s) | AgCl(s) | Pt, Cl_2 (1 atm) has been found to be +1.003 volts at 306°C. AgCl is an ionic conductor. (*a*) Calculate ΔF° (in calories) for the reaction $Ag(s) + \frac{1}{2}Cl_2(g) = AgCl(s)$ at 306°C. (*b*) Calculate the equilibrium partial pressure of chlorine over Ag(s) and AgCl(s) at 306°C.

17-8. For a cell in which the reaction is

$$\tfrac{1}{2}H_2(g, 1\ \text{atm}) + AgCl(s) = HCl(aq, a = 1) + Ag(s)$$

the standard emf at 25°C is 0.2225 volt. What is the value of the standard free-energy change ΔF°? The value of ΔA°?

17-9. An electrolytic cell is set up in which one electrode is liquid thallium, the other liquid thallium-lead alloy. The electrolyte is a fused LiCl-KCl-TlCl mixture. When the mole fraction of thallium in the alloy is 0.20, the emf of the cell is 115.2 mv at 438°C. (*a*) Calculate the activity and the activity coefficient of thallium in the alloy. (*b*) Calculate the vapor pressure of Tl in the alloy. For the reaction

$$\text{Tl(l)} = \text{Tl(g)}$$
$$\Delta F^\circ = 42{,}530 + 4.95T \log T - 40.61T$$

17-10. In the cell Ag(pure, s) | AgCl | Ag(in sol. Ag-Au) the following emf values were measured at 400°C:

N_{Ag}	1.0	0.9	0.8	0.7	0.6	0.5	0.4	0.3	0.2
$\mathcal{E}$, (volts) . .	0	0.0110	0.0231	0.0367	0.0527	0.0725	0.0963	0.1292	0.1782

(*a*) Calculate the activity of silver at each composition. (*b*) Calculate the activity of gold at each composition, using the Gibbs-Duhem equation.

17-11. The following data were found for the cell Pb(l) | $PbCl_2$ in LiCl-KCl(l) | Pb(in l Pb-Bi):

N_{Pb}	$\mathcal{E}°$ at 700°K, mv	$10^6\ d\mathcal{E}°/dT$, volt/°C
0.848	5.32	7.4
0.720	11.48	14.4
0.600	19.29	20.8
0.496	27.82	27.8
0.414	35.94	37.6
0.328	45.40	46.4
0.230	59.76	64.4
0.111	86.15	102.0

(*a*) Write the cell reaction and calculate $\Delta S°$, $\Delta F°$, $\Delta H°$. (*b*) Calculate the activities of lead at the concentrations given. (*c*) Calculate the activities of bismuth for N_{Bi} equal to 0, 0.2, 0.4, 0.6, 0.8, and 1.0.

17-12. The following data apply to the cell Cd(l) | $CdCl_2$ in LiCl | Cd(in l Cd-Sb) operated at 753°K:

N_{Cd}	0.8960	0.8180	0.7497	0.6760	0.5880	0.5590	0.4340	0.4006	0.3745	0.3444
$\mathcal{E}$, mv	3.31	7.20	11.56	18.96	29.69	33.85	52.00	56.51	59.90	63.95

(*a*) Calculate the activity of cadmium in each solution. (*b*) Find the vapor pressure of pure cadmium at 753°K from the following data:

$$\mathrm{Cd(l)} = \mathrm{Cd(g)}$$
$$\Delta F° = 26{,}110 + 4.97T \log T - 40.15T$$

What is the vapor pressure of cadmium at 753°K over a cadmium-antimony alloy in which the mole fraction of cadmium is 0.75? 0.40?

17-13. An electrolytic cell is set up in which one electrode is liquid cadmium and the other, liquid cadmium-lead alloy. The electrolyte is a fused LiCl-KCl-$CdCl_2$ mixture. When the mole fraction of cadmium in the alloy is 0.50, the emf of the cell is 10.80 mv at 500°C, and the temperature coefficient $d\mathcal{E}/dT$ is 36×10^{-3} mv/°C. (*a*) Calculate the activity and the activity coefficient of cadmium in the alloy. (*b*) Calculate $\bar{S}_{Cd} - S°_{Cd}$ for the above cell reaction.

17-14. (*a*) From Fig. 14-12, find the reversible decomposition voltages at 540°C of $ZnCl_2$ and $PbCl_2$. (*b*) What reaction or reactions may be expected to take place in a cell containing a mixed electrolyte of $PbCl_2 - ZnCl_2 - NaCl$, operated at a voltage not exceeding 1.6 volts at 540°C?

Chapter 18

18-1. Find the mean concentration (fractional saturation) $(c_m - c_o)/(c_s - c_o)$ for the diffusion of hydrogen into a sheet of bright mild steel, 5 mm thick. Samples of this sheet are exposed to hydrogen at 1 atm pressure at 500, 600, and 700°C and for periods of 1 min, 1 hr, and 1 day. For hydrogen in iron, $D = 1.65 \times 10^{-2}e^{-9200/RT}$ (cm²/sec).

18-2. In certain laboratory equipment operating at 725°C, hydrogen at 10 atm is separated by a disk of nickel from a continuously evacuated space. The area of the disk is 10 cm^2, and its thickness is 3 mm. Calculate the loss of hydrogen in cubic centimeters per hour. The solubility of hydrogen in nickel at 1 atm pressure and 725°C is approximately 7.0 cm^3 (STP) per 100 g of nickel. The density of nickel at 725°C is approximately 8.9 (g/cm^3), and the diffusivity D is 6×10^{-5} cm^2/sec for hydrogen in nickel at 725°C. (*Hint:* To obtain the answer in volume units, the concentration of hydrogen in nickel must be expressed in cubic centimeters of H_2 per cubic centimeter of Ni.)

18-3. Using the approximate method given in the text, estimate the depth below the surface of a plate of mild steel at which the concentration of carbon may be expected to decrease to one-half of its initial value as a result of exposure to strongly decarburizing conditions at 1000°C for (*a*) 1 hr and (*b*) 1 day. Take

$$D = 3 \times 10^{-7} \text{ cm}^2/\text{sec}.$$

18-4. A steel having an original concentration of carbon $c_o = 0.20$ per cent is exposed at 1700°F for 1 hr to a gas which maintains a concentration c_s at the surface of the steel of 0.50 per cent carbon. Using the master plot for $\frac{c - c_o}{c_s - c_o}$ vs. $x/\sqrt{Dt}$ (Fig. 18-4), find the carbon concentration at $x = 0.01$ cm and 0.04 cm.

$$D = 1.0 \times 10^{-7} \text{ cm}^2/\text{sec at } 1700°\text{F}$$

18-5. Using the plot for the concentration fraction as a function of x, D, and t (Fig. 18-4), plot a curve showing carbon concentration as a function of distance (x) after 8 hr at 927°C for carburizing a steel having an original carbon content c_o of 0.10 per cent C with a gas which maintains a concentration c_s of 1.20 per cent C at the surface. $D = 0.49 \cdot e^{-36,600/RT}$. Discuss any simplifying assumptions made.

18-6. A diffusion experiment was conducted with two butt-welded steels having the following compositions [*Trans. AIME*, **180**, 430 (1949)]:

	C	Si	Mn	S	P
A	0.49	3.80	0.25	0.006	0.011
B	0.45	0.05	0.88	0.008	0.020

(*a*) After 13 days at 1050°C, the carbon contents at the interface were 0.32 per cent on the *A* side and 0.59 per cent on the *B* side. Explain the apparent "uphill" diffusion of carbon. (*b*) Assume that after an otherwise identical diffusion experiment carried out for many months the carbon content was found to be equal at both sides of the interface. Explain the apparent reversal in the diffusion direction.

Chapter 19

19-1. Show that the equations for the diffusivities as functions of temperature in Prob. 18-1 and 18-5 are special cases of the Arrhenius equation.

19-2. (*a*) Starting with the observation that hydrogen accelerates the evolution of nitrogen from the ϵ phase in the iron-nitrogen system, would you expect this effect to increase or to decrease as the temperature is raised? (*b*) Analyze the assumption made in the footnote on page 476, according to which the heat of solution is negligibly small for the reaction $N_2(g) = 2N$ (in γ-iron).

19-3. A bubble of inert gas is present 10 cm below the surface of a bath of liquid silver. The diameter of the bubble is 0.1 mm. The surface tension of liquid silver is 910 dynes/cm. Calculate the total pressure exerted on the bubble.

TABLE 1. SELECTED VALUES OF HEATS OF FORMATION AT 298.16°K*
(Kcal/gram mole)

Ag_2O	− 7.306
Al_2O_3	−399.09
$CO(g)$	− 26.416
$CO_2(g)$	− 94.052
CoO	− 57.2
CuO	− 37.1
Cu_2O	− 39.84
Cu_2S	− 19.0
$Fe_{0.95}O$	− 63.7
Fe_3O_4	−267.
Fe_2O_3	−196.5
$H_2O(g)$	− 57.798
MgO	−143.84
$MgCO_3$	−266.
$MgSO_3$	−241.
$MgSO_4$	−305.5
NiO	− 59.3
PbO	− 52.07
PbO_2	− 66.12
$SO_2(g)$	− 70.96
$SO_3(g)$	− 94.45
Sb_2O_3	−166.5
Sb_2O_5	−234.4
SiO_2(qu)	−205.4
SnO_2	−138.8
ZnO	− 83.17
ZrO_2	−258.2

* Selected Values of Chemical Thermodynamic Properties, *Nat. Bur. Standards.*

TABLE 2. SELECTED VALUES OF ΔH_0° AND $H_T^\circ - H_0^\circ$*
(cal/gram mole)

Substance	ΔH_0°	$H_T^\circ - H_0^\circ$		
		298.16°K	1000°K	1500°K
$H_2O(g)$	−57,104	2367.7	9608.0	13848
$CO(g)$	−27,201.9	2072.6	7257.0	11358.0
$CO_2(g)$	−93,968.6	2238.1	10222.0	17004
$O_2(g)$	0	2069.8	7497.0	11776.5
$H_2(g)$	0	2023.8	6965.8	10694.2
H(g)	51,620	1481.2	4968.0	7451.9
C(graphite)	0	251.56	3075.0	5814

* Selected Values of Chemical Thermodynamic Properties, *Nat. Bur. Standards.*

TABLE 3. SELECTED VALUES OF $(F^\circ_T - H^\circ_0)/T$*
[cal/(gram mole)(°K)]

Substance	298.16°K	1000°K	1500°K
$H_2O(g)$	−37.165	−47.010	−50.598
$CO(g)$	−40.350	−48.860	− 51.864
$CO_2(g)$	−43.555	−54.109	−58.481
$O_2(g)$	−42.061	−50.697	−53.808
$H_2(g)$	−24.423	−32.738	−35.590
$H(g)$	−22.4247	−28.4365	−30.4508
C(graphite)	− 0.5172	− 2.771	− 4.181

* Selected Values of Chemical Thermodynamic Properties. *Nat. Bur. Standards.*

INDEX

K

L

M

N

O

P

Q

R

S